Mineral Deposits and Earth Evolution

Special Publication reviewing procedures

The Society makes every effort to ensure that the scientific and production quality of its books matches that of its journals. Since 1997, all book proposals have been refereed by specialist reviewers as well as by the Society's Books Editorial Committee. If the referees identify weaknesses in the proposal, these must be addressed before the proposal is accepted.

Once the book is accepted, the Society has a team of Book Editors (listed above) who ensure that the volume editors follow strict guidelines on refereeing and quality control. We insist that individual papers can only be accepted after satisfactory review by two independent referees. The questions on the review forms are similar to those for *Journal of the Geological Society*. The referees' forms and comments must be available to the Society's Book Editors on request.

Although many of the books result from meetings, the editors are expected to commission papers that were not presented at the meeting to ensure that the book provides a balanced coverage of the subject. Being accepted for presentation at the meeting does not guarantee inclusion in the book.

Geological Society Special Publications are included in the ISI Index of Scientific Book Contents, but they do not have an impact factor, the latter being applicable only to journals.

More information about submitting a proposal and producing a Special Publication can be found on the Society's web site: www.geolsoc.org.uk.

GEOLOGICAL SOCIETY SPECIAL PUBLICATION NO. 248

Mineral Deposits and Earth Evolution

EDITED BY

I. McDONALD
Cardiff University, UK

A. J. BOYCE
Scottish Universities Environmental Research Centre, Glasgow, UK

I. B. BUTLER
Cardiff University, UK

R. J. HERRINGTON
Natural History Museum, London, UK

and

D. A. POLYA
University of Manchester, UK

2005
Published by
The Geological Society
London

THE GEOLOGICAL SOCIETY

The Geological Society of London (GSL) was founded in 1807. It is the oldest national geological society in the world and the largest in Europe. It was incorporated under Royal Charter in 1825 and is Registered Charity 210161.

The Society is the UK national learned and professional society for geology with a worldwide Fellowship (FGS) of 9000. The Society has the power to confer Chartered status on suitably qualified Fellows, and about 2000 of the Fellowship carry the title (CGeol). Chartered Geologists may also obtain the equivalent European title, European Geologist (EurGeol). One fifth of the Society's fellowship resides outside the UK. To find out more about the Society, log on to www.geolsoc.org.uk.

The Geological Society Publishing House (Bath, UK) produces the Society's international journals and books, and acts as European distributor for selected publications of the American Association of Petroleum Geologists (AAPG), the American Geological Institute (AGI), the Indonesian Petroleum Association (IPA), the Geological Society of America (GSA), the Society for Sedimentary Geology (SEPM) and the Geologists' Association (GA). Joint marketing agreements ensure that GSL Fellows may purchase these societies' publications at a discount. The Society's online bookshop (accessible from www.geolsoc.org.uk) offers secure book purchasing with your credit or debit card.

To find out about joining the Society and benefiting from substantial discounts on publications of GSL and other societies worldwide, consult www.geolsoc.org.uk, or contact the Fellowship Department at: The Geological Society, Burlington House, Piccadilly, London W1J 0BG: Tel. +44 (0)20 7434 9944; Fax +44 (0)20 7439 8975; E-mail: enquiries@geolsoc.org.uk.

For information about the Society's meetings, consult *Events* on www.geolsoc.org.uk. To find out more about the Society's Corporate Affiliates Scheme, write to enquiries@geolsoc.org.uk.

Published by The Geological Society from:
The Geological Society Publishing House
Unit 7, Brassmill Enterprise Centre
Brassmill Lane
Bath BA1 3JN,
UK
(*Orders*: Tel. +44 (0)1225 445046
        Fax +44 (0)1225 442836)
Online bookshop: http://bookshop.geolsoc.org.uk

British Library Cataloguing in Publication Data
A catalogue record for this book is available from the British Library.

ISBN 1-86239-182-3

Distributors

USA
  AAPG Bookstore
  PO Box 979
  Tulsa
  OK 74101-0979
  USA
Orders: Tel. +1 918 584-2555
        Fax +1 918 560-2652
        E-mail bookstore@aapg.org

India
  Affiliated East–West Press PVT Ltd
  G-1/16 Ansari Road, Darya Ganj,
  New Delhi 110 002
  India
Orders: Tel. +91 11 2327-9113/2326-4180
        Fax +91 11 2326-0538
        E-mail affiliat@nda.vsnl.com

Japan
  Kanda Book Trading Company
  Cityhouse Tama 204
  Tsurumaki 1-3-10
  Tama-shi
  Tokyo 206-0034
  Japan
Orders: Tel. +81 (0)423 57-7650
        Fax +81 (0)423 57-7651
        E-mail geokanda@ma.kcom.ne.jp

Typeset by Type Study, Scarborough, UK
Printed by Cromwell Press, Trowbridge, UK

Contents

Preface

Mineral deposits are the source of all the metals, industrial and bulk minerals that feed the global economy. In addition to being key primary sources of wealth generation, mineral deposits are also valuable windows through which to view the evolution and interrelationships of the Earth system. Unlike hydrocarbon deposits that are largely restricted to more recent phases of geological time, mineral deposits have formed throughout the last 3.8 billion years of the Earth's history. As such they preserve key evidence for early magmatic and tectonic processes, the state of the atmosphere and hydrosphere, and the evolution of life. Furthermore, the very activities of exploration, evaluation and mining of mineral deposits, generate more comprehensive 3D geological information than is generally obtainable in unmineralized rocks, and increasing amounts of this formerly proprietary data are being released into the public domain.

The greatly enhanced concentrations of metals and minerals found in mineral deposits, over normal rocks, are a result of transport, concentration and deposition at these keys sites by common Earth processes. Either these processes operated at greater rates (or greater efficiencies) than normal, or there were fortuitous combinations of processes acting in the right place and at the right time to bring about the formation of the deposit. This revolution from documenting mainly the descriptive aspects of mineral deposits (in order to recognize the next one better) to trying to understand processes and derive genetic models for how the mineralization formed has gathered pace dramatically over the last 30 years. This revolution has been picked up and driven by the most perceptive Earth scientists who have recognized the potential for using evidence preserved in mineral deposits to probe more fundamental questions about Earth history and the evolution of the Earth system with time.

This volume contains papers presented at the Geological Society's Fermor Flagship Meeting, entitled *World Class Mineral Deposits and Earth Evolution*, held at Cardiff University and the National Museum and Gallery of Wales from 18–21 August 2003. The aim of the 2003 Fermor Meeting was to bring together geologists from academia and industry to highlight the importance of mineral deposits in their own right and in understanding the many and varied links between mineral deposits and Earth system science.

The first two chapters deal with perhaps the longest-running and most fundamental process on the Earth, namely the accretion of extra-terrestrial material to form our planet that has continued from the Hadaean to the present day. What is less well known is that many of the 150 or so impact craters that have been recognized thus far contain valuable mineral or hydro-carbon resources. The opening paper by **Grieve** reviews the key aspects of the impact process and crater formation, and how the formation of impact breccias and impact melts can lead to the development of mineral deposits and trap sites for migrating hydrocarbons. The value of the resources that are extracted is truly astonishing; US$18 billion annually, from North American impact structures alone! This is followed by a study of gold mineralization in the Witwater-srand Basin of South Africa by **Hayward et al.** Debate has raged recently over whether the enormous amounts of gold in the basin are of placer or hydrothermal origin. The giant Vrede-fort impact crater formed in the centre of the Witwatersrand Basin and heat generated by the impact almost certainly affected the gold-bearing rocks. **Hayward et al.** present miner-alogical evidence to suggest that the gold, although modified, was primarily of placer origin. The impact event only produced a short-lived phase of brittle deformation and small-scale remobilization of gold.

The next group of papers covers the role of mineral deposits in constraining models of tectonic evolution on different scales. **De Wit & Thiart** present a statistical analysis of the metal distributions in the Archaean cratons and post-Archaean rocks of the former continent of Gondwanaland. Their analysis reveals that not only are Archaean cratons more richly endowed in metals than younger rocks (i.e. mineral diversity has apparently decreased with time), but that each Archaean craton also carries its own distinctive metal signature. These metal signatures appear to have been inherited close to the time that the craton separated from the mantle and reflect mantle heterogeneity as well as the tectonic and magmatic processes involved in craton formation. The reasons why the distributions of mineral deposits vary with time are examined in greater depth by **Groves et al.** They conclude that the temporal distribution of each mineral-deposit type is a function of formation and preservational processes. The most fundamental geodynamic control is exerted by the change from the formation of positively buoyant lithospheric mantle in the Archaean and Proterozoic to negatively buoyant lithospheric mantle in the Phanerozoic. Redox-sensitive sedimentary mineral deposits are most strongly affected by long-term oxidation by the atmosphere–hydrosphere–biosphere system.

Harcouët *et al.* place constraints on the evolution of temperature during the Eburnean orogeny in the Ashanti Belt of Ghana using finite-element thermal modelling. In order to satisfy the observed thermobarometric regime, they conclude that an anomalously high mantle heat flow (at least three times the present value) must have been in operation. Such a thermal anomaly may explain the widespread development of gold mineralization in the Ashanti Belt. This theme is expanded by **Leahy** *et al.* who evaluate the distribution of giant gold deposits using a plate-tectonic framework. They propose a new six-fold geodynamic classification system for gold deposits that emphasizes subduction and crustal accretion zones. **Leahy** *et al.* conclude that the distribution of giant gold deposits is controlled by fluid access to regional gold sources and is ultimately a function of the amount of oceanic crust (the principal source for gold) that is consumed during successive orogenic episodes.

The idea that the deep-seated source rocks (often lower crustal rocks) determine the metal composition and sulphur-isotopic ratio of mineral deposits is explored further by **Lowry** *et al.*, who consider the potential for terrane discrimination using mineral deposits. They describe significant differences in the sulphur isotope signatures and metal contents of mineral deposits from different terranes making up Northern Britain, and show that these differences are most probably related to the major basement blocks that were amalgamated during the Caledonian orogeny. The most exciting use of this approach comes when mineralization styles for the British terranes are compared with mineral deposits of similar age, and comparable terranes, in Eastern Canada, as similar patterns are evident. This compositional inheritance suggests that mineral deposit signatures can constrain models of terrane accretion, even where the orogenic zone has been rifted apart in more recent times.

The Uralide orogenic belt is one of the world's great metallogenic provinces and contains mineral deposits associated with pre-, syn-, and post-collisional events during formation of the orogen. **Herrington** *et al.* present an analysis of the Uralide tectonic framework, using different classes of mineral deposit, to constrain the formational settings of the different tectonic blocks. The recognition of major north–south trending strike–slip faults and thrusts suggests that instead of multiple collided magmatic arcs there may only be two arcs, separated by the continental sliver of the East Mugodzhar Precambrian massif, and accretionary wedges of the Transuralian zone. **Herrington** *et al.* suggest that newly recognized strike–slip faults can be traced from the Polar Urals to the Tien Shan for more than 4000 km, approximately along the collision zone between the two arc systems. These studies illustrate that apparently parochial studies of mineral deposits can stimulate fundamental questions about regional tectonic settings and can lead to conclusions of much wider significance.

Sediment-hosted mineral deposits occupy a special niche in studies of the Earth System because, if they are truly syndepositional, they may preserve direct evidence for the state of the atmosphere and hydrosphere at the time the deposit formed. One of the most exciting recent discoveries in the Archaean rock record is the presence of mass-independent sulphur-isotope fractionations in volcanogenic massive sulphide deposits and banded iron formations. **Farquhar & Wing** review the evidence for these isotope fractionations. They describe the extent of the effect in the Archaean rock record, compared to younger rocks, and conclude that the fractionation may have occurred via ultraviolet photolysis of sulphur dioxide in the atmosphere and transfer of elemental sulphur to the Earth's surface. If this is correct, the implications of this discovery for the Earth's early atmosphere are profound; the Earth's early atmosphere would have lacked a UV shield (like the modern ozone layer) and possessed very low concentrations of free oxygen.

Iron deposits are particularly important indicators of redox conditions in seawater and sedimentary porewaters. Modern and ancient euxinic sediments are often enriched in iron that is highly reactive with dissolved sulphide, compared to continental margin and deep-sea sediments. **Raiswell & Anderson** outline a model where this iron enrichment arises from mobilization of dissolved iron from anoxic pore waters into overlying seawater, followed by transport into deep-basin environments and precipitation as iron sulphides in sediments. The addition of reactive iron to deep-basin sediments is determined by the magnitude of the diffusive iron flux, the export efficiency of recycled iron from the shelf, the ratio of source area to basin sink area and the extent to which reactive iron is trapped in the deep basin.

The discovery of life around modern deep-sea hydrothermal vents has led to the suggestion that ancient VMS and SEDEX deposits may also contain the fossils of organisms living on the vents when they were active, and that such environments may have been the warm oases where life on Earth first developed. Biological activity produces recognizable shifts in carbon and sulphur isotopes that may leave a fingerprint of ancient life in the early rock record. **Grassineau et al.** carried out a stable isotope study of cherts, iron formations and massive sulphides and unmineralized rocks in the 3.8 Ga Isua greenstone belt (Greenland) and the 2.7 Ga Belingwe greenstone belt (Zimbabwe). Their data suggest that recognizable isotope signatures of biological origin exist in both greenstone belts. They attempted to estimate the degree of change in biological activity over the billion years that separates the two settings. **Grassineau et al.** suggest that early life at Isua was most likely present in transitory, short-lived, settings whereas a billion years later at Belingwe, the biological carbon and sulphur cycles were in full operation, with the development of well-established algal mat communities.

The normal processes of erosion, transport, sorting and grading of sediment can also lead to some spectacular mineral deposits, none more so than the giant diamond (mega) placers of the SW African coast. The paper by **Bluck et al.** provides the first comprehensive synthesis of the tectonic and sedimentary factors that lead to the formation of the Orange River and Namaqualand mega-placers. Their study indicates that formation of a diamond mega-placer requires the interaction of several key factors that may extend back over large periods of geological time. These are: first, an adjacent craton hosting diamondiferous kimberlites and secondary alluvial deposits that may be remobilized; second, a drainage system that encompasses as much of the craton as possible and that focuses the supply of diamonds to a limited point; and third, a high energy regime at the terminal placer site that removes the fine grained sediment accompanying the diamonds. **Bluck et al.** describe how the tectonic and geomorphological evolution of southern Africa led to this fortunate combination of events and ultimately the formation of some of the most valuable diamond deposits ever discovered.

Over the last 40 years, many innovative analytical techniques have been developed by mineral deposits researchers. Increasingly sophisticated and micro-analytical techniques are being applied to hydrothermal mineral deposits to obtain direct information about the compositions and P/T conditions of the mineralizing fluids. The final paper by **Heinrich et al.** describes how the direct analysis of metals in individual fluid and melt inclusions from minerals in porphyry Cu–Au–Mo deposits is now achievable using laser ablation ICP mass spectrometry. Their study shows that a feature such as the economically important ratio of Au to Cu is inherited from the magmatic source and that bulk grade of different porphyry deposits is optimized when a large influx of magmatic fluids are cooled through 420–320 °C over a restricted flow volume.

As stated at the outset, the economic value of mineral deposits is self-evident. What this volume illustrates is that there is an accompanying body of research that is aimed at understanding long-term Earth processes and that mineral deposits are unique and vital probes into the functioning of the Earth system. Mineral deposit studies contribute to a much wider range of fundamental, and regional, research questions than may appear obvious at first. The range of contributions in this volume illustrates this link clearly. For most 'economic' geologists it has been self-evident that mineral deposits can contribute intellectual as well as monetary wealth to society. However, many within the wider geological community, and funding organizations, are less aware of this than they should be. Mineral deposits and the interlinking processes that formed them have always been at the centre of Earth system science and the more people with different backgrounds and ideas that work on them, the greater their contribution can be. We hope that this volume will inspire more novel research on these wonders of nature.

I. McDonald
A. J. Boyce
I. B. Butler
R. J. Herrington
D. A. Polya

Acknowledgements

The following people, in no particular order, kindly acted as reviewers for the papers in this volume: Jamie Wilkinson, Hartwig Frimmel, Gary Stevens, Tony Fallick, Richard Herrington, Simon Bottrell, Jack Middleburg, Tim Lyons, Iain McDonald, Uwe Reimold, Richard Davies, Adrian Boyce, P. Gallagher, P. E. J. Pitfield, Chris Stanley, Fanus Viljoen, Kevin Leahy, Gus Gunn, Clive Rice, Steve Grimes, John Dulles and Steve Kessler. Their contribution is gratefully acknowledged.

Finally, we would like to thank all those who helped make the 2003 Fermor Meeting such a resounding success. The commercial sponsors: Thermo Electron Spectroscopy, Rio Tinto plc, Anglo American plc, SRK Consulting, the British Geological Survey and Goldfields International. And the supporting organizations who funded many of the keynote speakers: The Geological Society, The Applied Mineralogy Group of the Mineralogical Society, the Institute of Materials, Minerals and Mining, the Society of Economic Geologists, the Society for Geology Applied to Mineral Deposits. And most importantly of all, the many delegates from around the world who attended and participated in an exciting exchange of science and ideas.

Economic natural resource deposits at terrestrial impact structures

RICHARD A. F. GRIEVE

Earth Sciences Sector, Natural Resources Canada, Ottawa, Ontario, Canada K14 0E4

Abstract: Economic deposits associated with terrestrial impact structures range from world-class to relatively localized occurrences. The more significant deposits are introduced under the classification: progenetic, syngenetic or epigenetic, with respect to the impact event. However, there is increasing evidence that post-impact hydrothermal systems at large impact structures have remobilized some progenetic deposits, such as some of the Witwatersrand gold deposits at the Vredefort impact structure. Impact-related hydrothermal activity may also have had a significant role in the formation of ores at such syngenetic 'magmatic' deposits as the Cu–Ni–platinum-group elements ores associated with the Sudbury impact structure. Although Vredefort and Sudbury contain world-class mineral deposits, in economic terms hydrocarbon production dominates natural resource deposits found at impact structures. The total value of impact-related resources in North America is estimated at US$18 billion per year. Many impact structures remain to be discovered and, as targets for resource exploration, their relatively invariant, but scale-dependent properties, may provide an aid to exploration strategies.

Natural impact craters are the result of the hypervelocity impact of an asteroid or comet with a planetary surface and involve the virtually instantaneous transfer of the considerable kinetic energy in the impacting body to a spatially limited, near-surface volume of a planet's surface. Impact is an extraordinary geological process involving vast amounts of energy, and extreme strain rates, causing immediate rises in temperature and pressure that produce fracturing, disruption and structural redistribution of target materials. This is followed by a longer period of time during which the target rocks readjust to the 'local' structural and lithological anomaly that constitutes the resultant impact structure and re-equilibrate from the thermal anomaly that is the result of shock metamorphism. Currently, around 170 individual terrestrial impact structures or small crater fields have been recognized, with the discovery rate of around five new structures per year. A listing of known terrestrial impact structures and some of their salient characteristics (location, size, age, etc.) is maintained by the Planetary and Space Science Centre at the University of New Brunswick and can be found at: http://www.unb.ca/passc/ ImpactDatabase/index.html

Some economic deposits of natural resources occur within specific impact structures or are, in someway, impact-related. Masaitis (1989, 1992) noted approximately 35 known terrestrial impact structures that have some form of potentially economic natural resource deposits. In a review of the economic potential of terrestrial impact structures, Grieve & Masaitis (1994) reported that there were 17 known impact structures that have produced some form of economic resources. This contribution represents an update of their review. In the intervening 10 years, there has been clarification of both the nature and relation of the resource to the specific impact event and a greater understanding of impact processes and the character of specific impact structures. As with Grieve & Masaitis (1994), this contribution is not comprehensive with respect to all natural resources related to terrestrial impact structures and does not consider those structures that have been or are being exploited as a source of aggregate, lime or stone for building material (e.g. Ries, Rouchechouart), are a source of groundwater or serve as reservoirs for hydroelectric power generation (e.g. Manicouagan, Puchezh-Katunki). However, the economic worth of these types of natural resources at terrestrial impact structures can be considerable. For example, the hydroelectric power generated by the Manicouagon reservoir is of the order of 4500 GWh per year, sufficient to supply power to a small city and worth approximately US$200 million per year.

The examples considered here are the resource deposits directly related to the impact structure, through structural disturbance and/or brecciation of the target rocks, impact heating and/or hydrothermal activity, and the formation of a structural or topographic trap. This contribution follows the logic and the terminology of Grieve & Masaitis (1994); namely, deposits are considered in the order progenetic, syngenetic, epigenetic. The most significant development in

From: McDonald, I., Boyce, A. J., Butler, I. B., Herrington, R. J. & Polya, D. A. (eds) 2005. *Mineral Deposits and Earth Evolution.* Geological Society, London, Special Publications, **248**, 1–29. 0305-8719/$15.00

viewing natural resources deposits at terrestrial impact structures, since Grieve & Masaitis (1994), is the recognition of a greater role for impact-related, post-impact hydrothermal activity in impact events (e.g. McCarville & Crossey 1996; Ames *et al.* 1998, 2005; Osinski *et al.* 2001; Naumov 2002). The remobilization of progenetic deposits has blurred some of the separation between progenetic and syngenetic deposits, and hydrothermal deposits are considered as a continuation of syngenetic processes, not as epigenetic as in Grieve & Masaitis (1994).

Characteristics of terrestrial impact structures

Morphology

On most planetary bodies, impact structures are recognized by their characteristic morphology and morphometry. Detailed appearance, however, varies with crater diameter. With increasing diameter, impact structures become proportionately shallower and develop more complicated rims and floors, including the appearance of central peaks and interior rings. Impact craters are divided into three basic morphologic subdivisions: simple craters, complex craters, and basins (Dence 1972; Wood & Head 1976).

Simple impact structures have the form of a bowl-shaped depression with an upraised rim (Fig. 1). At the rim, there is an overturned flap of ejected target materials, which displays inverted stratigraphy, with respect to the original target materials. Beneath the floor is a lens of brecciated target material that is roughly parabolic in cross-section. This breccia lens consists of allochthonous material. In places, the breccia lens may contain highly shocked (even melted) target materials. Beneath the breccia lens, parautochthonous, fractured rocks define the walls and floor of what is known as the true crater. Shocked rocks in the parautochthonous materials of the true crater floor are confined to a small central volume at the base.

With increasing diameter, simple craters display increasing evidence of wall and rim collapse and evolve into complex craters. Complex impact structures on Earth are found with diameters greater than 2 km in layered, sedimentary target rocks but not until diameters of 4 km or greater in stronger, more homogeneous, igneous or metamorphic, crystalline target rocks (Dence 1972). Complex impact structures are characterized by a central topographic peak or peaks, a broad, flat floor, and

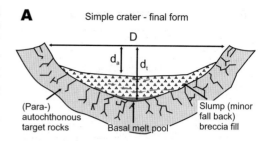

Fig. 1. (**A**) Schematic cross-section of a terrestrial simple impact structure. No vertical exaggeration. D is diameter and d_a and d_t are apparent and true depth, respectively. (**B**) Oblique aerial photograph of the 1.2 km diameter terrestrial simple impact structure, Barringer, Arizona, USA.

terraced, inwardly slumped and structurally complex rim areas (Fig. 2). The broad flat floor is partially filled by a sheet of impact melt rock and/or polymict allochthonous breccia. The central region is structurally complex and, in large part, occupied by a central peak, which is the topographic manifestation of a much broader and extensive area of structurally uplifted rocks that occurs beneath the centre of complex craters. Grieve & Therriault (2004) and Melosh (1989), respectively, provide further details of observations of terrestrial crater forms and cratering mechanics at simple and complex structures.

There have been claims that the largest known terrestrial impact structures have multiring forms, e.g. Chicxulub (Sharpton *et al.* 1993), Sudbury, Canada (Stöffler *et al.* 1994; Spray & Thompson 1995), and Vredefort (Therriault *et al.* 1997). Although some of their geological and geophysical attributes form annuli, it is not clear whether these correspond, or are related in origin, to the obvious topographical rings observed in lunar multi-ring basins (Spudis

A   Complex structure - final form

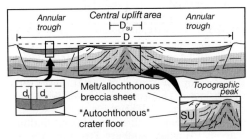

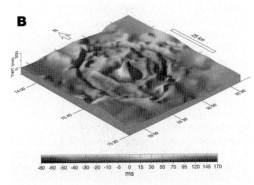

Fig. 2. (**A**) Schematic cross-section of a terrestrial complex impact structure. No vertical exaggeration. Abbreviations are as in Fig. 1, with SU as the vertical amount of structural uplift, and D_{su} as the diameter of the structural uplift, at its base. (**B**) 3D topography of a relatively uneroded complex impact structure, with a central peak and a possible peak ring, as illustrated by the residual two-way travel time just above the impact horizon at the Mjølnir impact structure, in the Barents Sea. View is from the SW, at an angle of 30° above the horizon. Note top of central peak has been eroded off. Vertical exaggeration is 20 times. Source: F. Tsikalas, University of Oslo.

1993; Grieve & Therriault 2000). Attempts to define morphometric relations, particularly depth–diameter relations, for terrestrial impact structures have had limited success, because of the effects of erosion and, to a lesser degree, sedimentation. The most recent empirical relations can be found in Grieve & Therriault (2004).

Geology of impact structures

Although an anomalous circular topographic, structural, or geological feature may indicate the presence of an impact structure on Earth, there are other geological processes that can produce similar features in the terrestrial environment. The burden of proof for an impact

origin in the terrestrial environment for a particular structure or lithology in the stratigraphic record generally lies with the documentation of the occurrence of shock-metamorphic effects.

On impact, the bulk of the impacting body's kinetic energy is transferred to the target by means of a shock wave. This shock wave imparts kinetic energy to the target, which leads to the formation of a crater and the ejection of target materials. It also increases the internal energy of the target materials, which leads to the formation of so-called shock-metamorphic effects. Details of the physics of shock wave behaviour and shock metamorphism can be found in Melosh (1989) and Langenhorst (2002). Shock metamorphism is the progressive breakdown in the structural order of minerals and rocks and requires pressures and temperatures well above the pressure–temperature field of endogenic terrestrial metamorphism (Fig. 3). Minimum shock pressures required for the production of diagnostic shock-metamorphic effects are 5–10 GPa for most silicate minerals. Strain rates produced on impact are of the order of 10^6 s^{-1} to 10^9 s^{-1} (Stöffler & Langenhorst 1994), many orders of magnitude higher than typical tectonic strain rates (10^{-12} s^{-1} to 10^{-15} s^{-1}; e.g. Twiss & Moores 1992), and shock-pressure duration is measured in seconds, or less, in even the largest impact events (Melosh 1989). Endogenic geological processes do not reproduce these physical conditions. They are unique to impact and, unlike endogenic terrestrial metamorphism, disequilibrium and metastability are common phenomena in shock metamorphism. Shock-metamorphic effects are well described in papers by Stöffler (1971, 1972, 1974), Stöffler

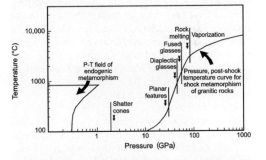

Fig. 3. Logarithmic plot of shock pressure (GPa) against post-shock temperature (°C) range of shock metamorphism for granitic rocks, with pressure ranges of some specific shock metamorphic effects indicated. Shown for comparison is the pressure–temperature range of endogenic terrestrial metamorphism.

& Langenhorst (1994), Grieve *et al.* (1996), French (1998), Langenhorst & Deutsch (1994), Langenhorst (2002) and others. They are discussed here only in general terms, as they relate to the recognition of impact materials in the terrestrial environment.

Impact melting. During shock compression, there is considerable pressure–volume work but pressure release occurs adiabatically. Heating of the target rocks occurs, as not all the pressure–volume work is recovered upon pressure release and the excess is manifest as irreversible waste heat. Above 60 GPa, the waste heat is sufficient to cause whole-rock melting and, at higher pressures, vaporization (Melosh 1989). Impact melted lithologies occur as glass particles and bombs in crater ejecta (Engelhardt 1990), as dykes within the crater floor and walls, as glassy to crystalline pools and lenses within the breccia lenses of simple craters, or as coherent, central sheets lining the floor of complex structures (Fig. 4).

The final composition of impact-melt rocks depends on the wholesale melting of a mix of target rocks, in contrast to partial melting and/or fractional crystallization relationships that occurs in endogenous igneous rocks. The composition of impact-melt rocks is, therefore, characteristic of the target rocks and may be reproduced by a mixture of the various country rock types in their appropriate geological proportions. Such parameters at $^{87}Sr/^{86}Sr$ and $^{143}Nd/^{144}Nd$ ratios of impact-melt rock also reflect the pre-existing target rocks (Jahn *et al.* 1978; Faggart *et al.* 1985). In general, even

Fig. 4. Approximately 150 m high cliffs of impact melt rock at the edge of the inner plateau at Manicouagan impact structure.

relatively thick impact-melt sheets are chemically homogeneous over radial distances of kilometres. In large impact structures, and where the target rocks are not homogeneously distributed, this observation may not hold true in detail, such as for Manicouagan, Canada (Grieve & Floran 1978), Chicxulub (Kettrup *et al.* 2000), and Popigai (Kettrup *et al.* 2003). Differentiation is not a characteristic of relatively thick coherent impact-melt sheets, with the exception of the extremely thick *c.* 2.5 km, Sudbury Igneous Complex, Sudbury Structure, Canada (Ariskin *et al.* 1999; Therriault *et al.* 2002).

Enrichments above target rock levels in siderophile and platinum-group elements (PGE) and Cr have been identified in some impact-melt rocks and ejecta. These are due to an admixture of up to a few percent of meteoritic material from the impacting body. In some melt rocks, the relative abundances of the various siderophiles have constrained the composition of the impacting body to the level of meteorite class (Palme *et al.* 1979; McDonald *et al.* 2001). In other melt rocks, no geochemical anomaly has been identified. This may be due to the inhomogeneous distribution of meteoritic material within the impact-melt rocks and sampling variations (Palme *et al.* 1981), or to differentiated impacting bodies, such as basaltic achondrites. More recently, high precision Cr, Os and He-isotopic analyses have been used to detect meteoritic material in the terrestrial environment (e.g. Koeberl *et al.* 1994; Koeberl & Shirley 1997; Peucker-Ehrenbrink 2001; Farley 2001).

Fused glasses and diapletic glasses. Shock-fused minerals are characterized morphologically by flow structures and vesiculation. Peak pressures required for shock melting of single crystals are in the order of 40–60 GPa (Stöffler 1972, 1974). Under these conditions, the minerals in the rock melt independently and selectively after the passage of the shock wave. Conversion of framework silicates to isotropic, dense, glassy, but not fused, phases occurs at peak pressures and temperatures, well below their normal melting point. These 'diaplectic glasses' require peak pressures of 30–45 GPa for feldspar and 35–50 GPa for quartz in quartzo-feldspathic rocks (e.g. Stöffler & Hornemann 1972; Stöffler 1984). Diaplectic glass has the same morphology as the original mineral crystal (Fig. 5), and a lower density than the crystalline form from which it is derived (but higher than thermally melted glasses of equivalent composition; e.g. Stöffler & Hornemann 1972; Langenhorst & Deutsch 1994).

Fig. 5. Some shock metamorphic effects. (**A**) Shatter cones at Gosses Bluff impact structure. (**B**) Photomicrograph of planar deformation features (PDFs) in quartz in compact sandstone from Gosses Bluff impact structure. Crossed polars, width of field of view 0.4 mm. (**C**) Photomicrograph of quartz (centre, higher relief) with biotite (darker grey, upper right) and feldspar (white, bottom) in a shocked granitic rock from Mistastin impact structure. Plane light, width of field of view 1.0 mm. (**D**) Photomicrograph as in (c) but with crossed polars. The biotite is still birefringent but the quartz and feldspar are isotropic, as they have been metamorphosed to diaplectic glasses by the shock wave, although retaining their original crystalline shapes.

High-pressure polymorphs. Shock can result in the formation of metastable polymorphs, such as stishovite and coesite from quartz (Chao *et al.* 1962; Langenhorst 2002) and cubic and hexagonal diamond from graphite (Masaitis 1993; Langehorst 2002). Coesite and diamond are also products of high-grade metamorphism but the paragenesis and, more importantly, the geological setting are completely different from those of impact events. The high-pressure polymorphs of quartz (i.e. stishovite and coesite) have only rarely been produced by laboratory shock-recovery experiments (cf. Stöffler & Langenhorst 1994). In terrestrial impact structures in crystalline targets, these polymorphs generally occur in small or trace amounts as very fine-grained aggregates and are formed by partial transformation of the host quartz. In porous quartz-rich target lithologies, however, they may be more abundant. For example, coesite may constitute 35% of the mass of highly shocked

Coconino sandstone at Barringer (Kieffer 1971). Further details on the characteristics of coesite and stishovite are given by Stöffler & Langenhorst (1994).

Planar microstructures. The most common documented shock-metamorphic effect is the occurrence of planar microstructures in tecto-silicates, particularly quartz (Fig. 5; Hörz 1968). The utility of planar microstructures in quartz reflect the ubiquitous nature of the mineral and the stability of the quartz and microstructures themselves, in the terrestrial environment, and the relative ease with which the microstructures can be documented. Recent reviews of the shock metamorphism of quartz, with an emphasis on the nature and origin of planar microstructures in experimental and natural impacts, are given in Stöffler & Langenhorst (1994), Grieve *et al.* (1996), and Langenhorst (2002). Planar deformation features (PDFs) in

minerals are produced under pressures of
c. 10–35 GPa. Planar fractures (PFs) form
under shock pressures ranging from
c. 5–35 GPa (Stöffler 1972; Stöffler & Langenhorst 1994).

Shatter cones. The only known large-scale diagnostic shock effect is the occurrence of shatter
cones (Dietz 1968). Shatter cones are unusual,
striated, and horse-tailed conical fractures (Fig.
5), ranging from millimetres to metres in length,
produced by the passage of a shock wave
through rocks (e.g. Sagy *et al.* 2002). Shatter
cones have been initiated most commonly in
rocks that experienced moderately low shock
pressures, 2–6 GPa, but have been observed in
rocks that experienced c. 25 GPa (Milton 1977).
These conical striated fracture surfaces are best
developed in fine-grained, structrually isotropic
lithologies, such as carbonates and quartzites.
Generally, they are found as individual or
composite groups of partial to complete cones
in rocks below the crater floor, especially in the
central uplifts of complex impact structures, and
rarely in isolated rock fragments in breccia
units, indicating the shatter cones formed before
the material was set in motion by the cratering
flow-field.

Geophysics of impact structures

Geophysical anomalies over terrestrial impact
structures vary in their character and, in
isolation, do not provide definitive evidence for
an impact origin. Interpretation of a single
geophysical data set over a suspected impact
structure may be ambiguous (e.g. Hildebrand *et
al.* 1991; Sharpton *et al.* 1993). However, when
combined with complementary geophysical
methods and the existing database of other
known impact structures, a more definite assessment can be made (e.g. Ormö *et al.* 1999). Since

potential-field data are available over large
areas, with almost continuous coverage, gravity
and magnetic observations have been the
primary geophysical indicators used for evaluating the occurrence of possible terrestrial impact
structures. Reflection seismic data, although
providing much better spatial resolution of
subsurface structure (e.g. Morgan *et al.* 2002a)
are used less often, because datasets are generally less widely available. Electrical methods
have been used even less commonly (e.g.
Henkel 1992). Given the lack of specificity of
the geophysical attributes of terrestrial impact
craters, they are not discussed here. The most
recent synthesis of the geophysical character of
terrestrial impact structures is Grieve &
Pilkington (1996).

Characteristics of natural resource deposits

The location and origin of economic natural
resource deposits in impact structures are
controlled by several factors related to the
impact process and the specific nature of the
target. The types of deposits are classified
according to their time of formation relative to
the impact event: progenetic, syngenetic and
epigenetic (Table 1). Progenetic economic
deposits are those that originated prior to the
impact event by purely terrestrial concentration
mechanisms. The impact event caused spatial
redistribution of these deposits and, in some
cases, brought them to a surface or near-surface
position, from where they can be exploited.
Syngenetic deposits are those that originated
during the impact event, or immediately afterwards, as a direct result of impact processes.
They owe their origin to energy deposition from
the impact event in the local environment,
resulting in phase changes and melting.
Hydrothermal deposits, where the heat source

Table 1. *Genetic groups of natural resource deposits at terrestrial impact structures*

Genetic group	Principal mode of origin	Types of known deposits
Progenetic	Brecciation	Building stone, silica
	Structural displacement	Iron, uranium, gold
Syngenetic	Phase transitions	Impact diamond
	Crustal melting	Cu, Ni, PGE, glass
	Hydrothermal activity	Lead, zinc, uranium, pyrite, gold, zeolite, agate
Epigenetic	Sedimentation	Placer diamond and tektites
	Chemical and biochemical sedimentation	Zeolite, bentonite, evaporites, oil shale, diatomite, lignite, amber, calcium phosphate
	Fluid flow	Oil, natural gas, fresh and mineralized water

was a direct result of the impact event, are also considered to be syngenetic. This differs from Grieve & Masaitis (1994). Epigenetic deposits result from the formation of an enclosed topographic basin, with restricted sedimentation, or the long-term flow of fluids into structural traps formed by the impact structure.

Progenetic deposits

Progenetic economic deposits in impact structures craters include iron, uranium, gold and others (Tables 1 & 2). In many cases, the deposits are relatively small. Only the larger and more active deposits are considered here.

Table 2. *Deposits and/or indications of natural resources in terrestrial impact structures*

Structure	Location (lat.; long.)	Diameter (km)	Economic material	Genetic type of economic deposit
*Ames, USA	36°15′N; 98°10′W	16	Oil, gas	Epigenetic
*Avak, USA	71°15′N; 156°38′W	12	Gas	Epigenetic
*Barringer, USA	35°02′N; 111°01′W	1.2	Earth silica	Progenetic
Beyenchime-Salaatin, Russia	71°50′N; 123°30′W	8	Pyrite	Epigenetic
*Calvin, USA	41°45′N; 85°57′	8.5	Oil	Epigenetic
Boltysh, Ukraine	48°45′N; 32°10′W	24	Oil shale	Epigenetic
*Carswell, Canada	58°27′N; 109°30′W	39	Uranium	Progenetic/ Syngenetic
*Charlevoix, Canada	47°32′N; 70°18′W	54	Ilmenite	Progenetic
*Chicxulub, Mexico	21°20′N; 89°30′W	180	Oil, gas	Epigenetic
*Crooked Creek, USA	37°50′N; 91°23′W	7	Lead, zinc	Syngenetic
Decaturville, USA	37°54′N; 92°43′W	6	Lead, zinc	Syngenetic
Ilyinets, Ukraine	49°06′N; 29°12′E	4.5	Agate	Epigenetic
Kaluga, Russia	54°30′N; 36°15′W	15	Mineral water	Epigenetic
Kara, Russia	69°05′N; 64°18′E	65	Diamond, zinc	Syngentic
Logoisk, Belarus	54°12′N; 27°48′E	17	Amber, calcium phosphate	Epigenetic
*Lonar, India	19°59′N; 76°31′E	1.8	Various salts	Epigenetic
*Marquez, USA	31°17′N; 96°18′W	22	Oil, gas	Epigenetic
Morokweng, S. Africa	26°28′S; 23°32′E	80	Ni-oxides, sulphides, silicates	Syngenetic
Obolon, Ukraine	49°30′N; 32°55′E	15	Oil shale	Epigenetic
Popigai, Russia	71°30′N; 111°00′E	100	Diamond	Syngenetic
Puchezh-Katunki, Russia	57°00′N; 43°35′E	80	Diamond, zeolite	Syngenetic
Ragozinka, Russia	58°18′N; 62°00′E	9	Diatomite	Epigenetic
*Red Wing, USA	47°36′N; 103°33′W	9	Oil, gas	Epigenetic
*Ries, Germany	48°53′N; 10°37′E	24	Diamond, lignite, bentonite, moldavites	Syngenetic/ Epigenetic
*Rotmistrovka, Ukraine	49°00′N; 32°00′E	2.7	Oil shale	Epigenetic
*Saint Martin, Canada	51°47′N; 98°32′W	40	Gypsum, anhydrite	Epigenetic
*Saltpan, South Africa	25°24′S; 28°50′E	1.1	Various salts	Epigenetic
Serpent Mound, USA	39°02′N; 83°24′W	8	Lead, zinc	Syngenetic
*Siljan, Sweden	61°02′N; 14°52′E	55	Lead, zinc, oil	Syngenetic/ Epigenetic
*Sierra Madera, USA	30°36′N; 102°55′W	13	Gas	Epigenetic
Slate Islands, Canada	48°40′N; 87°00′W	30	Gold	Progenetic
*Steen River, Canada	59°31′N; 117°37′W	25	Oil, gas	Epigenetic
*Sudbury, Canada	46°36′N; 81°11′W	250	Copper, nickel, PGE, Diamond	Syngenetic
*Ternovka, Ukraine	48°08′N; 33°31′E	12	Iron, uranium	Progenetic/ Syngenetic
Tookoonooka, Australia	27°00′S; 143°00′E	55	Oil	Epigenetic
*Viewfield, Canada	49°35′N; 103°04′W	2.5	Oil	Epigenetic
*Vredefort, South Africa	27°00′S; 27°30′E	300	Gold, uranium	Progenetic/ Syngenetic
Zapadnaya, Ukraine	49°44′N; 29°00′E	4	Diamond	Syngenetic
*Zhamanshin, Kazakhstan	48°24′N; 60°58′E	13.5	Bauxite, impact glass	Progenetic/ Syngenetic

*Resources exploited currently or in the past. Above listing does not include structures that are a source of fresh water, various building materials, or are used as hydroelectric reservoirs or the impact-related Cantarell oil field in Mexico.

Iron and uranium at Ternovka. Iron and uranium ores occur in the basement rocks of the crater floor and in impact breccias at the Ternovka or Terny structure, Ukraine (Table 2; Nikolsky *et al.* 1981, 1982). The structure is 375 ± 25 Ma old according to Nikolsky (1991) and was formed in a Lower Proterozoic fold belt. It is a complex crater from which erosion has removed more than 700 m, including most of the allochthonous impact lithologies, and has exposed the floor of the structure (Krochuk & Sharpton 2002). The central uplift is, in part, brecciated and injected by dykes of impact-melt rock up to 20 m wide. The annular trough contains remnants of allochthonous breccia, with patches and lenses of suevite. The present diameter of the structure is 10–11 km and its original diameter may have been 15–18 km. However, there is a smaller estimate for the original diameter of *c.* 8 km by Krochuk & Sharpton (2003), who also reported a younger, single whole rock $^{39}Ar/^{40}Ar$ date of 290 ± 10 Ma from an impact-melt rock.

The iron ores at Ternovka have been exploited through open pit and underground operations for more than fifty years. The ores are the result of hydrothermal and metasomatic action, which occurred during the Lower Proterozoic, on ferruginous quartzites (jaspilites) and some other lithologies, producing zones of albitites, aegirinites, and amphibole–magnetite and carbonate–hematite rocks, along with uranium mineralization. Post-impact hydrothermal alteration led to the remobilization of some of the uranium mineralization and the formation of veins of pitchblende. The production of uranium ceased in 1967 but iron ore was extracted, until recently, from two main open pits: Annovsky and Pezromaisk. The total reserves at the Pezromaisk open pit are estimated at 74 million tonnes, with additional reserves of lower grade deposits estimated at *c.* 675 million tonnes. Due to brecciation and displacement, blocks of iron ore are mixed with barren blocks. These blocks are up to hundreds of metres in dimension, having been rotated and displaced from their pre-impact positions. This displacement and mixing of lithologies causes difficulties in operation and evaluating the reserves but impact-induced fracturing aids in extraction and processing. Currently, mining operations are in a maintenance mode but may resume under new ownership (R. Krochuk, pers. comm. 2004).

Uranium at Carswell. The Carswell impact structure (Table 2) is located in northern Saskatchewan, Canada, approximately 120 km south of Uranium City. Details of various geological, geophysical and geochemical aspects of the Carswell structure, with heavy emphasis on the uranium ore deposits, can be found in Lainé *et al.* (1985). The Carswell structure is apparent in Shuttle Radar Topography as a two circular ridges, corresponding to the outcrop of the dolomites of the Carswell Formation (Fig. 6). The outer ridge is *c.* 39 km in diameter and is generally quoted as the diameter of the structure (e.g. Currie 1969; Innes 1964; Harper 1982). It represents a minimum original diameter. The outer ring of the structure is about 5 km wide and forms cliffs 65 m high of the Douglas and Carswell Formations. Interior to this, there is an annular trough, occupied by sandstones and conglomerates of the William River Subgroup of the Athabasca Group (Fig. 7). This trough is also approximately 5 km wide and rises to a core, *c.* 20 km in diameter, of metamorphic crystalline basement. The Carswell impact structure has been eroded to below the floor of the original crater.

The crystalline basement core consists of mixed feldspathic and mafic gneisses of the Earl River Complex, overlain by the more aluminous Peter River gneiss. Details of their mineralogy and chemistry can be found in Bell *et al.* (1985), Harper (1982), and Pagel & Svab (1985). The basement core is believed to have been uplifted by a minimum of 2 km. In a detailed structural study of the Dominique-Peter uranium deposit near the southern edge of the crystalline core, Baudemont & Fedorowich (1996) estimated that the amount of structural uplift in that area was in the order of 1.2 km. Surrounding the

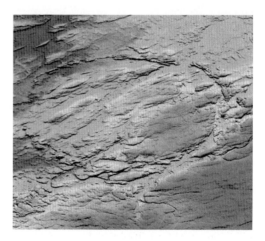

Fig. 6. Shuttle topographic radar image of digital topography over the Carswell impact structure (Table 2).

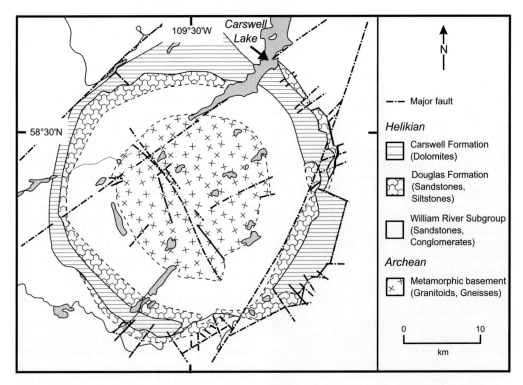

Fig. 7. Geological map of the Carswell impact structure, indicating uplifted crystalline basement core and down-faulted annulus of Carswell and Douglas Formations approximately corresponding to the circle in Fig. 14.

crystalline basement core are units of the unmetamorphosed Athabasca Group of sediments. The inner contact with the basement core is faulted and truncated in places and offset by radial faults. The outer contact of the Carswell Formation is also characterized by arcuate faulting, drag folding and local overturning of beds, as well as being offsets by radial faults (Fig. 7). There are several regional faults unrelated to the impact structure. The Carswell and Douglas Formations are the uppermost units of the Athabasca Group. Their outcrop is unique to the area and they owe their preservation to having been down-faulted at least 1 km to their present position (Harper 1982; Tona et al. 1985). Brecciation is common at Carswell and affects all lithologies. Currie (1969) used the term 'Cluff Breccias', after exposures near Cluff Lake, to include autochthonous monomict breccias, allochthonous polymict clastic breccias and impact melt rocks (Fig. 8). The latter two lithologies occur as dyke-like bodies and the relationships between the various breccias are locally complex (Wiest 1987).

The Athabasca Basin is the largest and richest known uranium-producing region in the world. Cumulative uranium production from the basin is approximately 1.5 billion pounds of uranium oxide, with a value of close to US$1.5 billion. Within the Carswell structure, the six known commercial uranium deposits occur in two main settings: at the unconformity between the Athabasca sandstone of the William River Subgroup and the uplifted crystalline basement core, and in mylonites and faults in the crystalline core. These deposits had grades between 0.3 and 6.8% uranium oxide and have produced close to US$70 million worth of uranium. The original uranium mineralization in the Athabasca Basin, and at Carswell, occurred during regolith development in the Precambrian, with later remobilization due to hydrothermal activity in response to thermotectonic events (Bell et al. 1985; Lainé 1986). The original commercial uranium deposit discovered at Carswell, the Cluff Lake D deposit, was a pre-existing or progenitic ore deposit that was brought to its present location by structural uplift in the Carswell impact event and subsequent erosion. The mineralization at

Fig. 8. Photomicrographs of lithologies at Carswell.
(**A**) Planar deformation features in quartz in
crystalline basement core. (**B**) Impact melt rock with
aphanitic matrix and acicular feldspar. Crossed polars,
fields of view 1 mm.

Cluff Lake D is also associated with shear zones
and faulting (Tona *et al.* 1985). At the time of
mining, it was the richest uranium ore body in
the world (Lainé 1986).

In their detailed structural study of
Dominique-Peter, the largest basement-hosted
deposit, Baudemont & Fedorowich (1996)
recognized four episodes of deformation. Two
of these were prior to mineralization, the third
episode related to mineralization and the final
episode related to the Carswell impact event.
They note that Carswell-related deformation
reactivated earlier faults associated with the
main mineralization. In addition, the paragenesis of the basement-hosted ore bodies contrasts
with the other unconformity-type deposits of
the Athabasca Basin and the Cluff D deposit at
Carswell, in that they are vein-type deposits,
related to fault zones reactivated by the
Carswell impact. 'Cluff Breccia' also commonly
occurs in the same fault structures as the
uranium mineralization. Baudemont &

Fedorowich (1996) found the association
'striking, albeit complexing'. Mineralized
material occurs within the Cluff Breccias, which
are, themselves, mineralized with veins of
coffinite (Lainé 1986).

It is not clear to what extent the Carswell
impact event was involved in remobilizing the
ores, beyond physical movement associated with
fault reactivation during structural uplift. The
basement-hosted ores are all associated with
extensive alteration, indicative of hydrothermal
fluid movement (Lainé 1986). At present, the
last of the known commercial uranium deposits
within the Carswell structure is in the mine-
decommissioning phase. However, there are
additional known exploration targets and reactivated faults, with pseudotachylite and/or 'Cluff
Breccias' and uranium mineralization, are
considered to be good future exploration drill
targets (Baudemont & Fedorowich 1996).

Gold and uranium of Vredefort. The Vredefort
structure, South Africa (Table 2), consists of an
uplifted central core of predominantly
Archaean granites (44 km in diameter)
surrounded by a collar of steeply dipping to
overturned Proterozoic sedimentary and
volcanic rocks of the Witwatersrand and
Ventersdrop Supergroup (18 km wide) and an
outer broad synclinorium of gently dipping
Proterozoic sedimentary and volcanic rocks of
the Transvaal Supergroup (28 km wide).
Younger sandstones and shales of the Karoo
supergroup cover the southeastern portion of
the structure. The general circular form with an
uplifted central core, the occurrence of
stishovite and coesite, as well as planar deformation features in quartz and shatter cones have all
been presented as evidence that the Vredefort
structure is the eroded remnant of a very large,
complex impact structure (e.g. Dietz 1961;
Hargraves 1961; Carter 1965; Manton 1965,
Martini 1978, 1991; Gibson & Reimold 2000).

The Witwatersrand Basin (Fig. 9) is the
world's largest goldfield, having supplied some
40% of the gold ever mined in the world. Since
gold was discovered there in 1886, it has
produced 47 000 tonnes of gold. The annual
Witwatersrand gold production for 2002 was
approximately 350 tonnes, or approximately
13.5% of the global gold supply, and current
reserve estimates are around 20 000 tonnes of
gold. The Vredefort impact event, occurred at
2023 ± 4 Ma. Based on the spatial distribution
of impact-related deformation and structural
features, Therriault *et al.* (1997) derived a self-
consistent, empirical estimate of the original
size of the Vredefort impact structure at

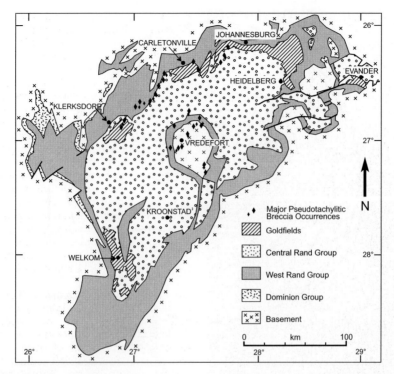

Fig. 9. Geological map of the Witwatersrand Basin, with partially obscuring Karoo Supergroup removed, indicating the central location of Vredefort.

between 225–300 km in diameter. A similar size estimate was derived by Henkel & Reimold (1998), based on potential field and reflection seismic data. These estimates effectively equate the spatial extent of the Vredefort impact structure to the entire Witwatersrand Basin (Fig. 10).

Independent of impact studies at Vredefort, structural analyses have identified a series of concentric anticlinal and synclinal structures related to Vredefort (e.g. McCarthy *et al.* 1986, 1990). The preservation of progenitic ores from erosion in these Vredefort-related structures (McCarthy *et al.* 1990) provided the emphasis for Grieve & Masaitis (1994). The origin of the gold in the Basin is still debated, with pure detrital and hydrothermal models and combinations of the two (e.g. Barnicoat *et al.* 1999; Minter 1999; Phillips & Law 2000). Gold with clear detrital morphological features (Minter *et al.* 1993), occurs with secondary, remobilized gold. This suggests that detrital gold was introduced into the Basin but that some gold was subsequently remobilized by hydrothermal activity. Grieve & Masaitis (1994) speculated that some remobilization of uranium may have occurred due to the Vredefort event but were equivocal as to the role of thermal activity

resulting from the spatially and temporally (2.05–2.06 Ga) close igneous event associated with the Bushveld Complex. Recent work has clarified the situation, with the Vredefort impact event forming both an important temporal marker and a critical element in the process of gold remobilization.

Two thermal or metamorphic events affected the rocks of the Basin. A regional amphibolite facies metamorphism predates the Vredefort impact event. However, a later, low pressure (0.2–0.3 GPa), post-impact event produced peak temperatures of 350 ± 50 °C in the Witwatersrand Supergroup to >700 °C in the centre of the crystalline core at Vredefort (Gibson *et al.* 1998). This post-impact metamorphism, which increased in intensity radially inwards, explains the previously documented progressive annealing of PDFs in the core rocks (Grieve *et al.* 1990) and is directly attributed to the combination of post-shock heating and the structural uplift of originally relatively deep-seated parautochthonous rocks during the Vredefort impact event (Gibson *et al.* 1998).

This post-impact metamorphism can be regarded as an integral part of the Vredefort impact event (Gibson & Reimold 1999). The

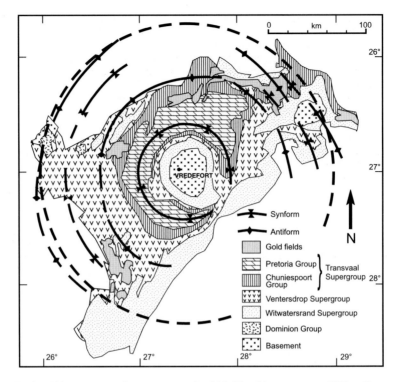

Fig. 10. Distribution of large concentric structures and gold fields, with respect to a 300 km diameter (dashed line) impact structure centred on Vredefort (Table 2).

heat engine that drove post-impact hydro-thermal activity was a result of the impact. Brecciation and the formation of pseudo-tachylite and other breccia dykes in the impact event also provided channels for fluid migra-tion. Reimold *et al.* (1999) coined the term 'autometasomatism' to describe the associated chlorite ± sericite alteration associated with the hydrothermal activity, in response to the thermal anomaly in the parautochthonous rocks of the basin and resulting from the Vredefort impact. This activity remobilized gold (and uranium) within impact-related structures and fractures, on all scales, that provided channels for fluid migration (Reimold *et al.* 1999). These fluids appear to have been originally meteoric and local in origin; gold was associated with chlorite and quartz veins, for example in the Ventersdrop Contact Reef (Frimmel *et al.* 1999). It is now evident the Vredefort impact event played a larger role in the genesis of the Witwatersrand Basin gold fields than simply preserving them from erosion by structural modification (McCarthy *et al.* 1990; Grieve & Masaitis 1994).

Syngenetic deposits

Syngenetic economic natural resources at impact structures include impact diamonds, Cu–Ni sulphides and platinum-group and other metals (Tables 1 & 2).

Impact diamonds. The first indication of impact diamonds was the discovery in the 1960s of diamond with lonsdaleite, a high-pressure (hexagonal) polymorph of carbon, in placer deposits, e.g. in the Ukraine, although their source was unknown (Cymbal & Polkanov 1975). In the 1970s, diamond with lonsdaleite was discovered in the impact lithologies at the Popigai impact structure. Since then, impact diamonds have been discovered at a number of structures, e.g. Kara, Lappajärvi, Puchezh-Katunki, Ries, Sudbury, Ternovka, Zapadnaya, and others (Gurov *et al.* 1996; Langenhorst *et al.* 1998; Masaitis 1993, 1998; Siebenschock *et al.* 1998).

Impact diamonds originate as a result of phase transitions from graphite, or crystalliza-tion from coal, and occur when their precursor carbonaceous lithologies were subjected to shock pressures > 35 GPa (Masaitis 1998). The diamonds from graphite in crystalline targets usually occur as paramorphs, with inherited crystallographic features (Masaitis *et al.* 1990; Val'ter *et al.* 1992) and as microcrystalline aggregates. At Popigai, these aggregates can reach 10 mm in size but most are 0.2–5 mm in size (Masaitis 1998). They consist of cubic diamond and lonsdaleite, with individual micro-crystals of 10^{-4} cm. The diamonds, generated from coal or other carbon in sediments, are generally porous and coloured.

Diamonds are most common as inclusions in impact-melt rocks and glass clasts in suevite breccias. For example at Zapadnaya, Ukraine, they occur in impact-melt dykes in the central uplift and in suevite breccias in the peripheral trough (Gurov *et al.* 1996). Zapadnaya, *c.* 3.8 km in diameter, 115 ± 10 Ma old, was formed in Proterozoic granite containing graphite (Gurov *et al.* 1985). At Popigai (Table 2), the allochthonous breccia filling the peripheral trough is

capped by diamond-bearing suevites and coherent bodies of impact-melt rocks (Fig. 11). The largest of these melt-rock bodies can be traced for 10–15 km along strike and is 500 m thick (Fig. 12; Masaitis *et al.* 1980). In the case of Popigai, the original source of the carbon is Archaean gneisses with graphite. The diamonds at Kara, Russia, also occur in impact-melt rocks. Kara, 65 km in diameter, 67 ± 6 Ma old, is located in a Palaeozoic fold belt, which contains Permian terrigenous sediments containing coal (Ezerskii 1982). Impact diamonds can also be found in strongly shocked lithic clasts in suevite breccias, e.g. at Popigai and Ries (Masaitis 1998; Siebenschock *et al.* 1998). In impact-melt rocks at structures with carbon-bearing lithologies, diamonds occur in relatively minor amounts, with provisional average estimates in the order of 10 ppb, although the cumulative volumes can be enormous. Although diamonds associated with known impact structures are not currently

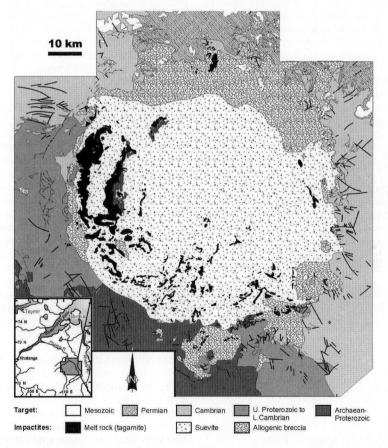

Fig. 11. Geological map of the Popigai impact structure (Table 2) indicating the distribution of impact melt rocks and suevite.

Fig. 12. Field photograph of outcrop of the sheet of diamond-bearing impact melt rocks, overlying allochthonous breccia, at the Popigai impact structure, Cliffs are approximately 150 m high.

exploited commercially, those produced by shock transformation of graphite tend to be harder and more resistant to breaking than the normal cubic diamonds from kimberlites.

At the Ries, Hough *et al.* (1995) have reported the only known occurrence of impact cubic diamond from impact lithologies, again suevite. They also reported the occurrence of SiC grains. The cubic diamonds are skeletal in appearance and they attributed their origin, and that of the SiC, to chemical vapour deposition in the ejecta plume over the impact site. Based on carbon isotope studies, they considered the source of the carbon and other elements to be from sedimentary rocks, including carbonates, which overlay the crystalline basement at the time of Ries impact and calculated that the Ries suevite might contain 7.2×10^4 tonnes of diamonds and SiC, in the proportion of 3:1. Other workers (e.g. Siebenschock *et al.* 1998) failed to find either cubic diamond or SiC at the Ries. Silicon carbide, however, has been reported from the Onaping Formation at the Sudbury impact structure (Masaitis *et al.* 1999). In this location, it is associated with impact diamonds that are a mixture of cubic diamond and hexagonal lonsdaleite, resulting not from chemical vapour deposition but from solid-state transformation by shock of precursor graphite.

Cu–Ni sulphides and platinum group metals at Sudbury. The Sudbury structure, Canada (Table 2), is the site of world-class Ni–Cu sulphide and platinum group metal ores. The pre-mining resources at Sudbury are estimated at 1.65×10^9 tonnes of 1.2% Ni and 1.1% Cu (Naldrett & Lightfoot 1993) and are associated with the Sudbury Igneous Complex (SIC). Sulphides were first noted at Sudbury in 1856. It was not until they were 'rediscovered' during

the building of the trans-Canada railway in 1883 that they received attention, with the first production occurring in 1886 (Naldrett 2003). By 2000, the Sudbury mining camp had produced 9.7 million tonnes of Ni, 9.6 million tonnes of Cu, 70 thousand tonnes of Co, 116 tonnes of Au, 319 tonnes of Pt, 335 tonnes of Pd, 37.6 tonnes of Rh, 23.3 tonnes of Ru, 11.5 tonnes of Ir, 3.7 thousand tonnes of Ag, 3 thousand tonnes of Se and 256 tonnes of Te (Lesher & Thurston 2002).

The most prominent feature of the Sudbury structure is the *c.* 30 × 60 km elliptical basin formed by the outcrop of the SIC, the interior of which is known as the Sudbury Basin (Figs 13 & 14). Neither the SIC nor the Sudbury Basin are synonymous with the Sudbury impact structure. The Sudbury impact structure includes the Sudbury Basin, the SIC and the surrounding brecciated basement rocks and covers a present area of > 15 000 km^2 (Giblin 1984*a,b*). From the spatial distribution of shock metamorphic features (e.g. shatter cones; Fig. 15) and other impact related attributes and by analogy with equivalent characteristics at other large impact structures, Grieve *et al.* (1991) estimated that the original crater rim diameter was 150–200 km. Central to these estimates is the assumption that the SIC was originally circular and, at its present level of erosion, was 60 km in diameter.

The elliptical shape of the Sudbury Basin is due to deformation caused by the Penokean Orogeny (Rousell 1984). The extent of this deformation was only recently appreciated. The results of a reflection seismic traverse across the Sudbury Basin, in the course of a LITHO-PROBE transect, indicated NW thrusting (Milkereit *et al.* 1992; Wu *et al.* 1995). Additional works (Cowan & Schwerdtner 1994; Hirt *et al.*

Fig. 13. Digital elevation image of the centre of the Sudbury impact structure (Table 2). The Sudbury Igneous Complex appears as a NE–SW orientated elliptical body c. 30 × 60 km in diameter, enclosing the smoother terrain of the Sudbury Basin. Traces of NNW trending faults can be clearly seen cutting the North Range of the Sudbury Igneous Complex. Arcuate structures associated with the superimposed younger Wanapitei impact structure are visible in the NW corner of the image. Source: V. Singhroy, Canada Centre for Remote Sensing.

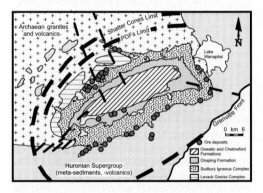

Fig. 14. Simplified regional geological map of the area of the Sudbury Igneous Complex, indicating major lithological Precambrian provinces (Archaean, Huronian, Grenville), Sudbury Igneous Complex and post-impact Whitewater Series of the Sudbury Basin. Dashed lines are prominent faults of varying age. Also indicated are the limits of the occurrence of shock metamorphic features (shatter cones, PDFs) and major ore deposits.

Fig. 15. Shatter cones in Huronian quartzite at the Sudbury impact structure.

suggested (e.g. Naldrett 2003; Tuchsherer & Spray 2002). There is some thermal disturbance of $^{40}Ar/^{39}Ar$ ages of Sudbury-related pseudo-tachylite north of the SIC, which suggest that there may be previously unrecognized post-impact Penokean metamorphism and possibly tectonic shortening in the north, in addition to the observed shortening in the south (Thompson et al. 1998). With the tectonic deformation and the considerable erosion, estimated to be c. 10 km (Schwarz & Buchan 1982), that has taken place at the Sudbury impact structure, it is difficult to constrain its original form. From its estimated original dimensions, it was most likely a peak-ring or a multi-ring basin (Stöffler et al. 1994; Spray & Thompson 1995).

Details of the geology of the Sudbury structure and the general area can be found in Dressler (1984a). In the simplest terms, the target rocks consisted of Archaean granite–greenstone terrain of the Superior Province of the Canadian Shield overlain, at the time of impact, by 5–10 km of Proterozoic Huronian metasediments, mostly arenaceous quartzites and wackes, and metavolcanics. The present outcrop pattern is illustrated in Figure 14, which

1993) indicated that there was a significant component of ductile deformation prior to the thrusting and brittle deformation. The original diameter of the SIC, at its present level of erosion, may have been 75–80 km and would have extended the original diameter of the final rim of the structure to the 250 km range (Deutsch & Grieve 1994; Stöffler et al. 1994). Even larger original diameters have been

also indicates the outcrop of Grenvillian gneisses to the SE, that were not present at the time of impact. North of the SIC, the presently exposed target rocks are Archaean granite–greenstones with a partial ring of down-dropped Huronian metasediments (Fig. 14). Closer to and immediately adjacent to the, so-called, North Range of the SIC, the relatively low grade granite–greenstone is replaced by the amphibolite facies rocks and, closest to the SIC, granulite facies Levack Gneiss Complex. These gneisses are complex, in detail, and have been thermally metamorphosed for a distance of > 1 km by the SIC (Dressler 1984b). The Levack gneisses formed at depths of 21–28 km (James et al. 1992). It is not certain that they were uplifted to their present position by the Sudbury impact event but uplift of this magnitude (c. 26 km) is presumed from an impact structure 250 km in diameter, based on empirical relations at other large impact structures (Grieve & Therriault 2004). Preliminary ^{40}Ar/^{39}Ar ages on hornblende separates from the Levack gneisses are consistent with uplift as a result of the Sudbury impact event (N. Wodicka, personnal communication, 2003). To the south and west, the footwall of the SIC is largely Huronian metasediments and metavolcanics (Fig. 14), but also includes Proterozoic granitic plutons.

The SIC has been subdivided into a number of phases or units. At the base, there is the Contact Sublayer (Fig. 16). This is the host to much of the sulphide mineralization at Sudbury. The sublayer has an igneous-textured matrix (Pattison 1979; Naldrett et al. 1984). Inclusions are of locally derived target rocks and mafic to ultramafic rocks. Recent U–Pb dating of these mafic inclusions indicates an age equivalent to that of the SIC, suggesting that it may represent materials that crystallized at an early stage from the SIC (Corfu & Lightfoot 1996). Generally included with the sublayer are the so-called Offset Dykes of the SIC (Naldrett et al. 1984). These dykes are most often radial to the SIC but some are concentric and they are hosts to Ni–Cu sulphide deposits. Some are extensive, e.g. the Foy Offset can be traced for c. 30 km from the North Range of the SIC and ranges from approximately 400 m in width near the SIC to 50 m at its most distal part (Grant & Bite 1984; Tuchscherer & Spray 2002).

Stratigraphically above the Sublayer lies the Main Mass of the SIC (Fig. 16), which is relatively, but not completely, clast free. For example, there are rare quartz clasts with partially annealed PDFs (Therriault et al. 2002). The contact relations between the sublayer and the Main Mass are apparently contradictory in places, with inclusions of Main Mass in the sublayer and vice versa (Naldrett 1984). Traditionally, the Main Mass has been divided into a number of facies: mafic norite, quartz-rich norite, felsic norite, quartz gabbro and granophyre, depending on the mineralogy. These facies are actually misnomers on the basis of the modal quartz, alkali feldspar and plagioclase proportions, as plotted on a Streckeisen diagram. Only the sublayer is gabbroic, the others fall in the quartz gabbro, quartz monzogabbro, granodiorite, and granite fields (Therriault et al. 2002). The nomenclature confusion arises from the fact that, like most impact melts of granitic/granodioritic composition, the mafic component of the melt crystallized as ortho-, and clino-pyroxene; although, there are also primary hydrous amphibole and biotite. Therriault et al. (2002), who made their observations on continuous cores rather than discontinuous outcrop, also demonstrated, on the basis of both mineralogy and geochemistry, that the contacts between the various facies of the SIC are gradational. The details of the mineralogy and geochemistry support a cogenetic source for the facies of the Main Mass of the SIC and fractional crystallization of a single batch of silicate liquid (e.g. Warner et al. 1998; Therriault et al. 2002).

The most comprehensive survey of the major element chemistry of the SIC is still Collins (1934), although more recent analyses of the Worthington Offset Dyke and Main Mass of the SIC can be found in Lightfoot & Farrow (2002) and Therriault et al. (2002), respectively. There is, of course, a massive proprietary geochemical database held by the mining companies in the area. The most unusual character of the average composition of the SIC is its high SiO_2 and K_2O, depletion in CaO and a low Na_2O/K_2O ratio compared to magmatic rocks of similar Mg number (Naldrett & Hewins 1984; Naldrett 1984). It also has an REE pattern, which is enriched in light rare earths (Kuo & Crockett 1979; Faggart et al. 1985), and is essentially that of the upper continental crust (Fig. 17; Faggart et al. 1985; Grieve et al. 1991). To account for these characteristics, earlier proponents of a magmatic origin for the SIC had to call upon massive crustal contamination of a mantle-derived magma (Kuo & Crockett 1979; Naldrett & Hewins 1984; Naldrett 1984). Strong arguments against such a proposal come from isotopic data and from thermal constraints. The $(^{87}Sr/^{86}Sr)^{T=1.85\,Ga}$ ratios for various lithologies of the SIC cluster at 0.707 but can range as high as 0.710 for granophyre samples (Gibbins &

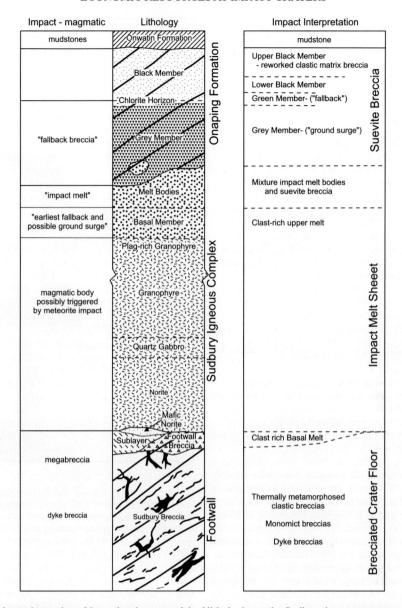

Fig. 16. Schematic stratigraphic section (not to scale) of lithologies at the Sudbury impact structure, with traditional nomenclature. Left-hand column is hybrid impact-magmatic interpretation of their genesis. Right-hand column is solely impact interpretation.

McNutt 1975; Hurst & Farhat 1977) or up to 0.7175 for the sublayer in the South Range (Faggart *et al.* 1985; Ostermann *et al.* 1996; Morgan *et al.* 2002*b*). The SIC is isotopically dated at 1.85 Ga (Krogh *et al.* 1984). These values are not compatible with a mantle-derived intrusion with a 1.85 Ga age. A similar argument holds for Nd isotopic systematics. Faggart *et al.* (1985) determined $Nd^{T=1.85\,Ga}$ values of −7 to

−8.8 for different units of the SIC that are typical for average upper continental crust at that time.

The conclusion that the SIC and its ores are not mantle-derived but crustal in composition is borne out by other isotopic studies (e.g. Fig. 18; Dickin *et al.* 1992, 1996, 1999; Walker *et al.* 1991). Cohen *et al.* (2000) analysed the Re–Os isotopes in the ultramafic inclusions in the sublayer of the

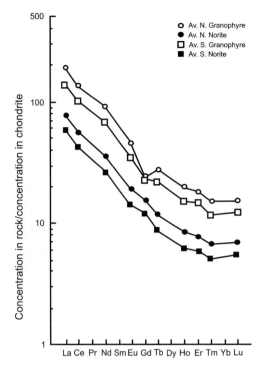

Fig. 17. Normalized REE abundances in phases of the Sudbury Igneous Complex lithologies of the North and South Ranges, indicating a crustal REE pattern for the Sudbury Igneous Complex. Source: Therriault *et al.* (2002).

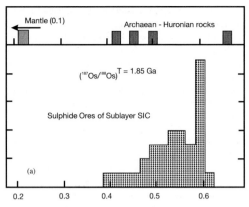

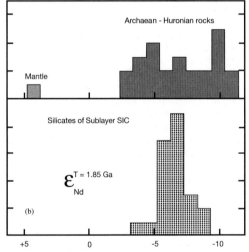

Fig. 18. Plot of (**a**) osmium and (**b**) neodymium isotopic data of ores and silicates in sublayer of the Sudbury Igneous Complex, compared to local crustal sources in the area of the Sudbury impact structure and the mantle at 1.85 Ga. Source: Faggart *et al.* (1985), Dickin *et al.* (1996, 1999).

SIC and concluded that they were also consistent with melting of pre-existing lithologies at 1.85 Ga. They also produced an imprecise crystallization age of 1.97 ± 0.12 Ga, which is within the error of 1.85 Ga crystallization age of the SIC, from a number of the inclusions. Recent, high precision Os isotope studies of sulphides from several mines have confirmed their crustal origin from a binary mixture of Superior Province and Huronian metasedimentary rocks (Morgan *et al.* 2002*b*).

There is a considerable geophysical database for the Sudbury impact structure. Most of the geophysical data are concentrated around the SIC and within the Sudbury Basin, because of the ore deposits, and a great deal of the data are proprietary. Most recently, the area around the SIC has been the subject of multidisciplinary geophysical studies, as part of a LITHOPROBE transect in the area. A north–south profile was completed in 1991 and additional data were acquired over parts of the South Range of the SIC in 1993. The north–south profile, which consisted of more than 100 km of conventional

and 40 km of high frequency Vibroseis seismic reflection data, has provided considerable insight into the structure of the SIC and the Sudbury Basin at depth (Milkereit *et al.* 1992; Wu *et al.* 1995; Boerner *et al.* 1999). These have been complemented with a variety of other geophysical data. The interpretation of these data can be found in *Geophysical Research Letters*, **21**, 1994. The combination of known ore deposits and reflection seismic data has also led Sudbury to be chosen as the first test site for the use of 3D reflection seismic survey methods for detecting massive sulphide deposits in crystalline rocks. The experiment concluded that such massive sulphide bodies produce a characteristic seismic scattering response and the technique has potential as a new exploration tool in crystalline rocks (Milkereit *et al.* 2002).

Recently, Farrow & Lightfoot (2002) have reviewed the nature of the ore deposits at Sudbury and placed their formation in an integrated time-sequence model. As with others (e.g. Naldrett 1984), they recognize: 'Contact' deposits associated with embayments at the base of the SIC and hosted by sublayer and footwall breccia; 'Offset' deposits associated with discontinuities and variations in thickness in the offset dykes; and Cu-rich footwall deposits. They also recognize a fourth deposit type associated with Sudbury Breccia. This is an acknowledgement that the Frood-Stobie deposit, which contained some 15% of the known mineralization and produced 600 million tonnes of ore, is not hosted in a traditional offset dyke but rather in Sudbury Breccia (Scott & Spray 2000).

The contact deposits consist of massive sulphides and are volumetrically the largest deposit-type, hosting approximately 50% of the known ore deposits. They include the Creighton and Whistler deposits and the North Range deposits of Levack and Coleman. The offset deposits include the Copper Cliff and Worthington Offsets in the South Range, which, along with the Frood-Stobie, contain approximately 40% of the known ores at Sudbury. The Cu-rich footwall deposits are volumetrically small relative to the contact deposits but are extremely valuable ore bodies, as they are relatively enriched in platinum group metals, in addition to copper. This type of deposit is hosted in the brecciated footwall of the SIC and is best known in the North Range, e.g. McCreedy East and West, Coleman, Strathcona and Fraser, where they occur as complex vein networks.

There is increasing realization that hydrothermal remobilization played a role in the genesis of the footwall deposits (e.g. Carter *et al.* 2001; Farrow & Watkinson 1997; Marshall *et al.* 1999; Molnar *et al.* 1997, 1999, 2001). Although Farrow & Lightfoot (2002) note the importance of this hydrothermal activity, they are equivocal as to the timing and suggest it may be related to regional metamorphic events. The fluids responsible for remobilization were saline, Cl-rich and oxidizing, and were initially at temperatures in excess of 300–400 °C. As Magyarosi *et al.* (2002) noted, the Sudbury impact event occurred before the peak of Penokean metamorphism in the area. Thus the regional metamorphism could not have been responsible for the hydrothermal activity that resulted in the remobilization of metals, particularly the copper and platinum

group metals. Given the growing body of knowledge in support of hydrothermal activity driven by the end result of large impact events, such as at Vredefort, and the emplacement of a massive post-impact hydrothermal system above and driven by the SIC (Ames *et al.* 1998, 2005), it is presumed that this was also the case beneath the SIC, with respect to the genesis of these secondary Cu and platinum group metal-rich ore deposits. The impact-induced hydrothermal system at Sudbury was recently modelled by Abramov & Kring (2004).

One feature common to all the major ore deposits is that they lie at the base or just beneath the SIC (Fig. 14). Earlier magmatic models of the origin of ores at Sudbury suggested that they resulted from the segregation of sulphides as an immiscible liquid, due to the assimilation of siliceous rocks by a basaltic magma, followed by gravitational settling and, later, fractional crystallization and, in some cases, remobilization (Naldrett *et al.* 1984; Morrison *et al.* 1994). The key difference, therefore, between endogenic magma with assimilation and the impact models is that a 'disequilibrium' composition, with respect to the expected equilibrium crystallization of endogenic silicate melts, was an original property of the SIC. That the metals in the ores, as well as the associated silicates, have an original crustal source is indicated by the Re–Os isotopic composition of the ores and the Nd–Sm composition of the silicates accompanying the ores (Fig. 18; Dickin *et al.* 1992). Although some details are still to be determined, recent work at Sudbury can mostly be fitted into the framework of the formation of a 200–250 km impact basin 1.85 Ga ago, with accompanying massive crustal melting producing a superheated melt of an unusual composition, which gave rise to immiscible sulphides that gravitationally settled, resulting ultimately in the present ore deposits. Complicating factors, but essential components of the evolutionary history, are the creation of a 'local' hydrothermal system resulting from the impact and the deformation by the Penokean orogeny that took place shortly after the impact.

Epigenetic deposits

Epigenetic deposits are due to the fact that impact structures can result in isolated topographic basins or locally influence underground fluid flow. Such deposits may originate almost immediately or over an extended period after the impact event and may include reservoirs of liquid and gaseous hydrocarbons, oil shales, various organic and chemical sediments, as well as flows of fresh and mineralized waters (Tables 1 & 2).

Hydrocarbons. Hydrocarbons occur at a number of impact structures. In North America, approximately 50% of the known impact structures in hydrocarbon-bearing sedimentary basins have commercial oil and/or gas fields.

The Ames structure (Table 2) is located in Oklahoma, USA, and is a complex impact structure about 14 km in diameter, with a central uplift, an annular trough, and slightly uplifted rim. It is buried by up to 3 km of Ordovician to Recent sediments (Carpenter & Carlson 1992). The structure was actually discovered in the course of oil exploration (Roberts & Sandridge 1992) and is the principal subject of a compilation of research papers (Johnson & Campbell 1997). The rim of the structure is defined by the structurally elevated Lower Ordovician Arbuckle dolomite; more than 600 m of Cambro-Ordovician strata and some underlying basement rocks are missing in the centre of the structure due to excavation. The entire structure is covered by Middle Ordovician Oil Creek shale, which forms both the seal and source for hydrocarbons and may have produced as much as 145 million barrels of oil (Curtiss & Wavrek 1997).

The first oil and gas discoveries were made in 1990 from an approximately 500 m thick section of Lower Ordovician Arbuckle dolomite in the rim (Fig. 19). Due to impact-induced fracturing and karsting, the Arbuckle dolomite in the rim of Ames has considerable economic potential. For example the 27-4 Cecil well, drilled in 1991, had drill stem flow rates of 3440 million cubic feet of gas and 300 barrels of oil per day (Roberts & Sandridge 1992). Wells drilled in the centre failed to encounter the Arbuckle dolomite and bottomed in granite breccia of the central uplift or, closer to the rim, and the granite–dolomite breccia. These central wells (Fig. 19) produce over half the daily production from Ames and include the famous Gregory 1-20, which is the most productive oil well from a single pay zone in Oklahoma at more than 100 000 barrels of oil per year (Carpenter & Carlson 1997). Gregory 1-20 encountered a *c.* 80 m section of granite breccia below the Oil Creek shale, with very effective porosity. A drill-stem test of the zone flowed at approximately 1300 barrels of oil per day, with a conservative estimate of primary recovery in excess of 5 million barrels from this single well (Donofrio 1998). Approximately 100 wells have been drilled at Ames, with a success rate of 50%. These wells produce more than 2500 barrels of oil and more than 3 million cubic feet of gas per day. Conservative estimates of primary reserves at Ames suggest they will exceed 25–50 million barrels of oil and 15–20 billion cubic feet of gas (Donofrio 1998; Kuykendall *et al.* 1997). Hydrocarbon production is from the Arbuckle dolomite, the brecciated granite and granite–dolomite breccia and is largely due to impact-induced fracturing and brecciation, which has resulted in significant porosity and permeability.

In the case of Ames, the impact not only produced the required structural traps but also the palaeoenvironment for the deposition of post-impact shales that provided oil and gas, upon subsequent burial and maturation (Curtiss & Wavrek 1997). There are similarities between the Ames crater shale and locally developed Ordovician shale in the Newporte structure (North Dakota), an oil-producing 3.2 km diameter impact crater (*c.* 120 000 barrels per year) in Precambrian basement rocks of the Williston Basin (Donofrio 1998). The Ames and Newporte discoveries have important implications for oil and gas exploration in crystalline rock underlying hydrocarbon-bearing basins. Donofrio (1981, 1997) first proposed the existence of such hydrocarbon-bearing impact craters and that major oil and gas deposits may occur in brecciated basement rocks.

At the Red Wing Creek structure in North Dakota, USA, hydrocarbons are also recovered from the rocks of the central uplift. In this case, the impact structure resulted in a structural trap but, unlike Ames, it is not responsible for the source of the oil. Red Wing Creek is a complex structure, approximately 9 km in diameter, with seismic records and drill-core data indicating a central peak in which strata have been uplifted by up to 1 km, an annular trough containing

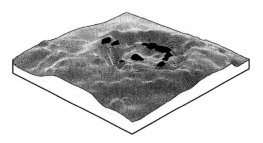

Fig. 19. Three-dimensional mesh diagram of residual structure on the post-impact upper Ordovician Sylvan shale at Ames impact structure (Table 2). View is to the NW at 25° elevation, with 20 times vertical exaggeration. Solid areas indicate where hydrocarbons are produced. Source: Carpenter & Carlson (1997).

crater-fill products, and a partially eroded structural rim (Brenan *et al.* 1975; Sawatsky 1977). As the result of a pronounced seismic anomaly, Shell Oil drilled the structure in 1965 on the NW flank of the central uplift. The drill hole indicated a structurally high and thickened Mississippian and Pennsylvanian section, compared to drill holes outside the structure. The well, however, was dry. In 1968, Shell drilled another hole to the NW in the annular trough. Here, the Mississippian was found to be structurally low compared to exterior. It was also dry and the structure, as a whole, was assumed to be dry. True Oil redrilled what was later recognized as the central uplift in 1972 and discovered *c.* 820 m of Mississippian oil column, with considerable high angle structural complexity and brecciation and a net pay of approximately 490 m. This is in contrast to the area outside the structure, which displays gentle dips and *c.* 30 m oil columns.

The large oil column is due to the structural repetition of the Mississippian Mission Canyon Formation in the central uplift (Brenan *et al.* 1975). The impact-induced porosity and permeability results in relatively high flow rates of more than 1000 barrels per day. Cumulative production in the 20 years since discovery is in excess of 12.7 million barrels of oil and 16.2 billion cubic feet of natural gas (Pickard 1994). Current production is restricted to about 300 000 barrels per year, to preserve unexploited reserves of natural gas. However, it is estimated that the brecciated central uplift contains more than 120 million barrels of oil and primary and secondary recoverable reserves may exceed 70 million barrels (Donofrio 1981, 1998; Pickard 1994). The natural gas reserves are estimated at 100 billion cubic feet. Virtually all the oil has been discovered within a diameter of 3 km, corresponding to the central uplift. Based on net pay and its limited aerial extent, Red Wing is the most prolific oil field in the United States, in terms of producing wells per area, with the wells in the central uplift having the highest cumulative productivity of all the wells in North Dakota.

The Avak structure is located on the Arctic coastal plain of Alaska, USA. The structure has been known for some time as a 'disturbed zone' in seismic data (Lantz 1981) and hydrocarbons were discovered in 1949 (Donofrio 1998). Only recently was evidence of shock metamorphism discovered in the form of shatter cones and planar deformation features in quartz (Kirschner *et al.* 1992). The structure itself has the form of a complex impact structure roughly 12 km in diameter. It is bounded by listric faults,

which define a rim area, and has an annular trough and central uplift. In the central uplift, the Lower–Middle Jurassic Kingak Shale and Barrow sand are uplifted more than 500 m from their regional levels. The central uplift has been penetrated by the Avak 1 well, which penetrated to a depth of 1225 m. Oil shows occur in Avak 1 but the well is not a commercial producer. Kirschner *et al.* (1992) suggest that pre-Avak hydrocarbon accumulations may have been disrupted and lost due to the formation of the Avak structure. There are, however, the South Barrow, East Barrow and Sikulik gas fields, which are post-impact and are related to the Avak structure. They occur outside the structure and are due to listric faults in the crater rim, which have truncated the Lower Jurassic Barrow sand and placed Lower Cretaceous Torok shales against the sand, creating an effective up-dip gas seal. The South and East Barrow fields are currently in production and primary recoverable gas is estimated at 37 billion cubic feet (Lantz 1981).

The Campeche Bank in the SE corner of the Gulf of Mexico is the most productive hydrocarbon producing area in Mexico. Oil and gas, from Jurassic source rocks, are recovered from breccia deposits at the Cretaceous–Tertiary (K/T) boundary. This area includes the world-class Cantarell oil field (Santiago-Acevedo 1980), which has produced close to 7 billion barrels of oil and 3 trillion cubic feet of gas, since discovery in 1974 to 1999. The bulk of production comes from the breccias at the K/T boundary. Primary reserves may range as high as 30 billion barrels of oil and 15 trillion cubic feet of gas. Production from the K/T boundary rocks is from up to 300 m of dolomitized limestone breccia, with a porosity of around 10%. Clasts of shocked quartz and plagioclase occur in the upper portion of the K/T breccia (Limon *et al.* 1994; Grajales-Nishimura *et al.* 2000). These breccias are the reservoir rocks for the hydrocarbons. The traps are Tertiary structural traps in the form of faulting and anticlinal structures. The seal to the reservoir rocks is an impermeable bentonitic bed, several tens of metres thick, which contains fragments of quartz and plagioclase with PDFs and some pristine impact melt glass fragments (Grajales-Nishimura *et al.* 2000). This bentonitic bed, with shocked materials, is considered to be altered ejecta materials from the K/T impact structure Chicxulub, which lies some 350 to 600 km to the NE.

Grajales-Nishimura *et al.* (2000) proposed the following sequence of events from the K/T lithologies. The main, hydrocarbon-bearing, breccias resulted from the collapse of the offshore carbonate platform resulting from seismic energy from the Chicxulub impact. This was followed by the deposition of K/T ejecta through slower atmospheric transport. The upper part of the ejecta deposit was later reworked by the action of impact-related tsunamis crossing the Gulf of Mexico. Subsequent dolomitization and Tertiary tectonics served to form the seal and trap for migrating Jurassic hydrocarbons, resulting in an oil field that produces more than 60% of Mexico's daily production and has reserves in excess of the entire onshore and offshore hydrocarbon reserves of the United States, including Alaska (Donofrio 1998). This oil field also accounts for the bulk of the current US$16 billion gross value of hydrocarbons produced from North American impact structures per year.

Other impact structures also produce hydrocarbons. For example, the 25 km diameter Steen River structure, Canada, produces oil from two wells on the northern rim. Oil and gas are produced from beneath the c. 13 km diameter Marquez and Sierra Madera structures, USA (Donofrio 1997, 1998). Newporte, which was noted earlier (Clement & Mayhew 1979), and Viewfield, Canada (Sawatsky 1977) also produce hydrocarbons. These are simple bowl-shaped craters. Viewfield has approximately 50

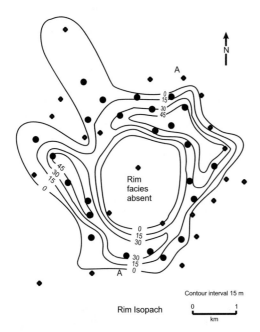

Fig. 20. Thickness of rim facies at Viewfield impact structure (Table 2). Black dots are hydrocarbon producing wells. Black dots with crosses are service or dry wells. Modified from Sawatzky (1977).

wells (Fig. 20) producing some 600 barrels of oil and 250 million cubic feet of gas per day. The recoverable reserves associated with Viewfield are estimated to be 10–20 million barrels of oil (Donofrio 1997, 1998). More than 500 000 barrels of oil have been produced since 1978 from the Calvin structure, USA, which is most likely an 8.5 km diameter complex impact structure (Milstein 1988).

Oil shales are known at Boltysh (25 km, 88 ± 3 Ma), Obolon (15 km, 215 ± 25 Ma) and Rotmistrovka (2.7 km, 140 ± 20 Ma) in the Ukraine (Masaitis *et al.* 1980; Gurov & Gurova 1991). They represent the unmatured equivalent of the hydrocarbon reserves at Ames. The most significant reserves are at Boltysh, where there are an estimated 4.5 billion tonnes (Bass *et al.* 1967). The oil shales are the result of biological activity involving algae in this isolated basin.

Concluding remarks

The total gross direct worth of natural resources from impact structures in North America alone is estimated at in excess of US$18 billion per year. Given the relatively small number of known impact structures, as a class of geological features, impact structures have considerable overall economic potential. There are areas of the world where the numbers of known impact structures are well below those expected from the known cratering rate, indicating that there are impact structures yet to be found. For example, the average cratering rate suggests that approximately 17 ± 8 structures with diameter of 20 km or more should have been formed in an area the size of Africa (approximately 30×10^6 km^2) in the last 100 Ma. The known impact record in Africa indicates no impact structures of the appropriate size and age. However, the c. 80 km diameter 145 Ma Morokweng impact structure in South Africa has approximately 500 ppm Ni in its impact-melt rocks. The nickel and other PGEs occur as Ni-oxides, Ni-sulphides and as high Ni content in silicates, e.g. 0.2% NiO in orthopyroxene (Hart *et al.* 2002; Koeberl & Reimold 2003). Unlike Sudbury, the metals appear to have come from the impacting body, which was an ordinary chondrite (McDonald *et al.* 2001). Although high Ni and PGE concentrations have been found in drill core associated with thin zones containing highly altered projectile fragments (Hart *et al.* 2002), the available evidence suggests that the impact melt failed to produce immiscible sulphides and, thus, large concentrations of metals in ore deposits.

The largest impact structures have the

greatest probability of having significant economic resources. These are the most energetic events; they affect the largest volumes of target rocks, have the largest post-impact hydrothermal systems and form the largest topographic basins. It is estimated that ten impact structures in the size range of Sudbury were formed on the Earth's land surface in the last 2 Ga. At present three are known: Chicxulub, Sudbury and Vredefort. Chicxulub is buried beneath 1 km of post-impact sediments and is unlikely, therefore, to have economically viable mineral deposits; although, it has exerted considerable control over the local hydrology (Perry *et al.* 2002) and resulted in a world-class oil field at Campeche. Both Sudbury and Vredefort are major mining camps, with world-class syngenetic and pro-genetic ore deposits, respectively.

Impact structures do have a general property that is an advantage in the exploration for natural resources. They have relatively fixed morphometric and structural relationships for a given diameter. Once a structure is known to be of impact origin, and its diameter established, it is possible to make considerable predictions as to the structural and lithological character of the structure as a whole. This scale-depending characteristic is generally lacking in most endogenic geological structures. The development of an exploration strategy based on these relationships is most notably illustrated, in hindsight, by the drilling for hydrocarbons at Ames. Similarly, one can only speculate how the level of current knowledge on the origin and evolution of Sudbury as an impact structure will guide future mineral exploration there.

For progenetic types of deposits, the central uplift area and annular trough of complex impact structures are the most promising targets. They result in environments where buried ore deposits are structurally brought close to the surface or near surface deposits are down-dropped and protected from erosion. Syngenetic deposits are less associated with the physical redistribution of lithologies and structural changes and more closely associated with the effects of shock metamorphism through phase changes. They are, thus, more likely to be concentrated in and around such lithologies as impact melt sheets and suevitic breccias. Secondary syngenetic associations are also to be expected due to hydrothermal processes in areas above and below such lithologies. Epigenetic deposits are most closely linked with the crater form itself; generally as an isolated basin with localized sedimentary and geochemical activity not present in the area or as a set of structures controlling the migration of fluids.

The brecciated and fractured rocks of the central uplift also provide an environment for the structural repetition of beds and increased porosity and permeability and are targets for hydrocarbon exploration.

Reviews of the manuscript by R. Davies, W.U. Reimold and A. Therriault are appreciated. Geological Survey of Canada Contribution 2004024.

References

ABRAMOV, O. & KRING, D.A. 2004. Numerical modeling of an impact-induced hydrothermal system at the Sudbury crater. *Journal of Geophysical Research*, **109**, E10007; doi: 10.1029/2003JE002213.

AMES, D.E., WATKINSON, D.H. & PARRISH, R.R. 1998. Dating of a regional hydrothermal system induced by the 1850 Ma Sudbury impact event. *Geology*, **26**, 447–450.

AMES, D.E., JONASSON, I.R., GIBSON, H.L. & POPE, K.O. 2005. Impact-generated hydrothermal system from the large Paleoproterozoic Sudbury crater, Canada. *In*: COCKELL, C., GILMOUR, I. & KOEBERL, C. (eds) *Biological Processes associated with Impact Events*. Impact Studies, Springer-Verlag, Berlin-Heidelberg (in press).

ARISKIN, A.A., DEUTSCH, A. & OSTERMANN, M. 1999. The Sudbury 'Igneous' Complex: Simulating phase equilibria and *in situ* differentiation for two proposed parental magmas. *In*: DRESLER, B.O. & SHARPTON, V.L. (eds) *Large Meteorite Impacts and Planetary Evolution II*. Geological Society of America Special Paper, **338**, 373–387.

BARNICOAT, A.C., YARDLEY, B.W.D., HENDERSON, I.H.C. & FOX, N.P.C. 1999. Discussion of detrital origin of hydrothermal gold by H.E. Frimmel. *Terra Nova*, **10**, 347–349.

BASS, YU. B., GALAKA, A.I. & GRABOVSKIY, V.I. 1967. [The Boltysh oil shales.] *Razvadka I Okhrana Nedr*, **9**, 11–15 [in Russian].

BAUDEMONT, D. & FEDOROWICH, J. 1996. Structural control of uranium mineralization at the Dominique Peter deposit, Saskatchewan, Canada. *Economic Geology*, **91**, 855–874.

BELL, K., CACCIOTTI, A.D. & SCHNESSL, J.H. 1985. Petrography and geochemistry of the Earl River Complex, Carswell structure, Saskatchewan – a possible Proterozoic komatiitic succession. *Geological Association Canada Special Paper*, **29**, 71–80.

BOERNER, D.E., MILKEREIT, B. & DAVIDSON, A. 1999. Geoscience impact: A synthesis of studies of the Sudbury Structure. *Canadian Journal of Earth Sciences*, **37**, 477–501.

BRENAN, R.L., PETERSON, B.L. & SMITH, H.J. 1975. The origin of Red Wing Creek Structure, McKenzie County, North Dakota. *Wyoming Geological Association Earth Science Bulletin*, **8**, 1–41.

CARPENTER, B.N. & CARLSON, R. 1992. The Ames impact crater. *Oklahoma Geological Survey*, **52**, 208–223.

CARPENTER, B.N. & CARLSON, R. 1997. The Ames meteorite impact crater. *In*: JOHNSON, K.S. & CAMPBELL, J.A. (eds) *Ames structure in northwest Oklahoma and similar features: Origin and petroleum production (1995 Symposium)*. Oklahoma Geological Survey Circular, **100**, 104–119.

CARTER, N.L. 1965. Basal quartz deformation lamellae – a criterion for recognition of impactites. *American Journal Science*, **263**, 786–806.

CARTER, W.M., WATKINSON, D.H. & JONES, P.C. 2001. Post-magmatic remobilisation of platinum-group elements in the Kelly Lake Cu–Ni sulfide deposit, Copper Cliff Offset, Sudbury. *Exploration Mining Geology*, **10**, 95–110.

CHAO, E.C.T., FAHEY, J.J., LITTLER, J. & MILTON, D.J. 1962. Stishovite, SiO_2, a very high pressure new mineral from Meteor Crater, Arizona. *Journal of Geophysical Research*, **67**, 419–421.

CLEMENT, J.H. & MAYHEW, T.E. 1979. Newporte discovery opens new pay. *Oil & Gas Journal*, **77**, 165–172.

COHEN, A.S., BURNHAM, O.M., HAWKESWORTH, C.J. & LIGHTFOOT, P.C. 2000. Pre-emplacement Re–Os ages for ultramafic inclusions in the sublayer of the Sudbury Igneous Complex, Ontario. *Chemical Geology*, **165**, 37–46.

COLLINS, W.H. 1934. Life-history of the Sudbury Nickel Irruptive, Part I, Petrogenesis. *Transactions of the Royal Society of Canada*, **28**, 123–177.

CORFU, F. & LIGHTFOOT, P.C. 1996. U–Pb geochronology of the sublayer environment, Sudbury Igneous Complex, Ontario. *Economic Geology*, **91**, 1263–1269.

COWAN, E.J. & SCHWERDTNER, W.M. 1994. Fold origin of the Sudbury Basin. *In*: LIGHTFOOT, P.C. & NALDRETT, A.J. (eds) *Proceedings of the Sudbury –Noril'sk Symposium*, Ontario Geological Survey Special Volume, **5**. Toronto, Ministry of Northern Development and Mines, 45–55.

CURRIE, K.L. 1969. Geological notes on the Carswell circular structure, Saskatchewan (74K). *Canadian Geological Survey of Canada Paper*, **67-32**, 1–60.

CURTISS, D.K. & WAVREK, D.A. 1997. The Oil Creek–Arbuckle Petroleum System, Major county, Oklahoma. *In*: JOHNSON, K.S. & CAMPBELL, J.A. (eds) *Ames structure in northwest Oklahoma and similar features: Origin and petroleum production (1995 Symposium)*. Oklahoma Geological Survey Circular, **100**, 240–258.

CYMBAL, S.N. & POLKANOV, YU. A. 1975. [*Mineralogy of titanium-zirconium placers of Ukraine.*] Nauk Press, Kiev [in Russian].

DENCE, M.R. 1972. The nature and significance of terrestrial impact structures. *24th International Geological Congress Section*, **15**, 77–89.

DEUTSCH, A. & GRIEVE, R.A.F. 1994. The Sudbury Structure: Constraints on its genesis from the probe results. *Geopyhsics Research Letters*, **21**, 963–966.

DICKIN, A.P., RICHARDSON, J.M., CROCKET, J.H., MCNUTT, R.H. & PEREDERY, W.V. 1992. Osmium isotope evidence for a crustal origin of platinum group elements in the Sudbury nickel ore, Ontario, Canada. *Geochimica et Cosmochimica Acta*, **56**, 3531–3537.

DICKIN, A.P., ARTAN, M.A. & CROCKET, J.H. 1996. Isotopic evidence for distinct crustal sources of North and South Range ores, Sudbury Igneous Complex. *Geochimica et Cosmochimica Acta*, **60**, 1605–1613.

DICKIN, A.P., NGUYEN, T. & CROCKET, J.H. 1999. Isotopic evidence for a single impact melting origin of the Sudbury Igneous Complex. *In*: DRESSLER, B.O. & SHARPTON, V.L. (eds) *Large Meteorite Impacts and Planetary Evolution II*. Geological Society of America Special Paper, **339**, 361–371.

DIETZ, R.S. 1961. Vredefort ring structure. Meteorite impact scar? *Journal of Geology*, **69**, 499–516.

DIETZ, R.S. 1968. Shatter cones in cryptoexplosion structures. *In*: FRENCH, B.M. & SHORT, N.M. (eds) *Shock Metamorphism of Natural Materials*. Mono Book Corporation, Baltimore, 267–285.

DONOFRIO, R.R. 1981. Impact craters: Implications for basement hydrocarbon production. *Journal of Petroleum Geology*, **3**, 279–302.

DONOFRIO, R.R. 1997. Survey of hydrocarbon-producing impact structures in North America: Exploration results to date and potential for discovery in Precambrian basement rock. *In*: JOHNSON, K.S. & CAMPBELL, J.A. (eds) *Ames Structure in Northwest Oklahoma and Similar Features: Origin and Petroleum Production (1995 Symposium)*. Oklahoma Geological Survey Circular, **100**, 17–29.

DONOFRIO, R.R. 1998. North American impact structures hold giant field potential. *Oil & Gas Journal*, **May**, 69–83.

DRESSLER, B.O. 1984a. General geology of the Sudbury area. *In*: PYE, E.G., NALDRETT, A.J. & GIBLIN, P.E. (eds) *The Geology and Ore Deposits of the Sudbury Structure*. Ministry of Natural Resources, Toronto, 57–82.

DRESSLER, B.O. 1984b. The effects of the Sudbury event and the intrusion of the Sudbury igneous complex on the footwall rocks of the Sudbury Structure. *In*: PYE, E.G., NALDRETT, A.J. & GIBLIN, P.E. (eds) *The Geology and Ore Deposits of the Sudbury Structure*. Ministry of Natural Resources, Toronto, 97–136.

ENGELHARDT, W.V. 1990. Distribution, petrography and shock metamorphism of the ejecta of the Ries crater in Germany – a review. *Tectonophysics*, **171**, 259–273.

EZERSKII, V.A. 1982. [Impact-metamorphosed carbonaceous matter in impactites.] *Meteoritika*, **41**, 134–140 [in Russian].

FAGGART, B.E. JR., BASU, A.R. & TATSUMOTO, M. 1985. Origin of the Sudbury complex by meteoritic impact: Neodymium isotopic evidence. *Science*, **230**, 436–439.

FARLEY, K.A. 2001. Extraterrestrial helium in seafloor sediments: Identification, characteristics and accretion rate over geologic time. *In*: PEUCKER-EHRENBRINK, B. & SCHMITZ, B. (eds) *Accretion of extraterrestrial matter throughout Earth's history*. Kluwer Academic/Plenum Publishers, New York, 179–204.

FARROW, C.E.G. & LIGHTFOOT, P.C. 2002. Sudbury

PGE revisited: Toward an integrated model. *In*: CABRI, L.J. (ed.) *The geology, geochemistry, mineralogy and benefication of platinum-group elements*. Canadian Institute of Mining and Metallurgy Special Volume, **54**, 273–297.

FARROW, C.E.G. & WATKINSON, G.H. 1997. Diversity of precious-metal mineralization in footwall Cu–Ni–PGE deposits, Sudbury, Ontario. Implications for hydrothermal models of formation. *Canadian Mineralogist*, **35**, 817–839.

FRENCH, B.M. 1998. *Traces of a Catastrophe: A Handbook of Shock Metamorphic Effects in Terrestrial Meteorite Impact Structures*. LPI Contribution, **No. 954**, Lunar and Planetary Institute, Houston TX.

FRIMMEL, H.E., HALLBAUER, D.K. & GARTZ, V.H. 1999. Gold mobilizing fluids in the Witwatersrand Basin: Composition and possible sources. *Mineralogy and Petrology*, **66**, 55–81.

GIBBINS, W.A. & MCNUTT, R.H. 1975. The age of the Sudbury Nickel Irruptive and the Murray granite. *Canadian Journal Earth Sciences*, **12**, 1970–1989.

GIBLIN, P.E. 1984*a*. History of exploration and development, of geological studies and development of geological concepts. *In*: PYE, E.G., NALDRETT, A.J. & GIBLIN, P.E. (eds) *The Geology and Ore Deposits of the Sudbury Structure*. Ministry of Natural Resources, Toronto, 3–23.

GIBLIN, P.E. 1984*b*. Glossary of Sudbury geology terms. *In*: PYE, E.G., NALDRETT, A.J. & GIBLIN, P.E. (eds) *The Geology and Ore Deposits of the Sudbury Structure*. Ministry of Natural Resources, Toronto, 571–574.

GIBSON, R.L. & REIMOLD, W.V. 1999. The significance of the Vredefort Dome for the thermal and structural evolution of the Wiwatersrand Basin, South Africa. *Mineralogy and Petrology*, **66**, 5–23.

GIBSON, R.L. & REIMOLD, W.V. 2000. Deeply exhumed impact structures; a case study of the Vredefort Structure, South Africa. *In*: GILMOUR, I. & KOEBERL, C. (eds) *Impacts and the early Earth*. Lecture Notes in Earth Science, **91**, 249–277.

GIBSON, R.L., REIMOLD, W.V. & STEVENS, G. 1998. Thermal–metamorphic signature of an impact event in the Vredefort Dome, South Africa. *Geology*, **26**, 787–790.

GRAJALES-NISHIMURA, J.M., CEDILLO-PARDO, E., ROSALES-DOMINGUEZ, C., *ET AL*. 2000. Chicxulub impact. The origin of reservoir and seal facies en the southeastern Mexico oil fields. *Geology*, **28**, 307–310.

GRANT, R.W. & BITE, A. 1984. Sudbury quartz diorite offset dikes. *In*: PYE, E.G., NALDRETT, A.J. & GIBLIN, P.E. (eds) *The Geology and Ore Deposits of the Sudbury Structure*. Ministry of Natural Resources, Toronto, 275–300.

GRIEVE, R.A.F. & FLORAN, R.J. 1978. Manicouagan impact melt, Quebec. II: Chemical interrelations with basement and formational processes. *Journal of Geophysical Research*, **83**, 2761–2771.

GRIEVE, R.A.F. & MASAITIS, V.L. 1994. The economic potential of terrestrial impact craters. *International Geology Review*, **36**, 105–151.

GRIEVE, R.A.F. & PILKINGTON, M. 1996. The signature of terrestrial impacts. *AGSO Journal of Australian Geology and Geophysics*, **16**, 399–420.

GRIEVE, R.A.F. & THERRIAULT, A.M. 2000. Vredefort, Sudbury, Chicxulub: Three of a kind? *Annual Review of Earth & Planetary Sciences*, **28**, 305–338.

GRIEVE, R.A.F. & THERRIAULT A.M. 2004. Observations at terrestrial impact structures: Their utility in constraining crater formation. *Meteorites and Planetary Science*, **39**, 199–216.

GRIEVE, R.A.F., CODERRE, J.M., ROBERTSON, P.B. & ALEXOPOULOS, J.S. 1990. Microscopic planar deformation features in quartz of the Vredefort structure: Anomalous but still suggestive of an impact origin. *Tectonophysics*, **171**, 185–200.

GRIEVE, R.A.F., STÖFFLER, D. & DEUTSCH, A. 1991. The Sudbury Structure: Controversial or misunderstood? *Journal of Geopyhsical Research*, **96**, 22 753–22 764.

GRIEVE, R.A.F., LANGENHORST, F. & STÖFFLER, D. 1996. Shock metamorphism of quartz in nature and experiment: II. Significance in geoscience. *Meteorites and Planetary Science*, **31**, 6–35.

GUROV, E.P. & GUROVA, E.P. 1991. [*Geological structure and composition of rocks in impact craters.*] Nauk Press, Kiev [in Russian].

GUROV, E.P., MELNYCHUK, E.V., METALIDI, S.V., RYABENKO, V.A. & GUROVA, E.P. 1985. [The characteristics of the geological structure of the eroded astrobleme in the western part of the Ukranian Shield.] Dopovidi Akad. *Nauk Ukrainskoi Radyanskoi Sotsialisychnoi Respubliky*, **Seriya B**, 8–11 [in Ukrainian].

GUROV, E.P., GUROVA, E.P. & RATISKAYA, R.B. 1996. Impact diamonds of the Zapadnaya crater; phase composition and some properties (abstract). *Meteoritics & Planetary Science*, **31 Supplement**, 56.

HARGRAVES, R.B. 1961. Shatter cones in the rocks of the Vredefort Ring. *Transactions and Proceedings Geological Society of South Africa*, **64**, 147–154.

HART, R.J., CLOETE, M., MCDONALD, I., CARLSON, R.W. & ANDREOLI, M.A.G. 2002. Siderophile-rich inclusions from the Morokweng impact melt sheet, South Africa: possible fragments of a chondritic meteorite. *Earth and Planetary Science Letters*, **198**, 49–62.

HARPER, C.T. 1982. Geology of the Carswell structure, central part. *Saskatchewan Geological Survey, Report*, **214**, 1–6.

HENKEL, H. 1992. Geophysical aspects of meteorite impact craters in eroded shield environment, with special emphasis on electric resistivity. *Tectonophysics*, **216**, 63–90.

HENKEL, H. & REIMOLD, W.U. 1998. Integrated geophysical modelling of a giant, complex impact structure: Anatomy of the Vredefort Structure, South Africa. *Tectonophysics*, **287**, 1–20.

HILDEBRAND, A.R., CAMARGO, A.Z., JACOBSEN, S.B., BOYNTON, W.V., KRING, D.A., PENFIELD, G.T. & PILKINGTON, M. 1991. Chicxulub crater: A possible Cretaceous–Tertiary boundary impact crater on the Yucatan Peninsula, Mexico. *Geology*, **19**, 867–871.

HIRT, A.M., LOWRIE, W., CLENDENEN, W.S. & KLIG-
FIELD, R. 1993. Correlation of strain and the
anisotropy of magnetic susceptibility in the
Onaping Formation: Evidence for a near-circular
origin of the Sudbury Basin. *Tectonophysics*, **225**,
231–254.

HÖRZ, F. 1968. Statistical measurements of deforma-
tion structures and refractive indices in experi-
mentally shock loaded quartz. *In*: FRENCH, B.M.
& SHORT, N.M. (eds) *Shock Metamorphism of
Natural Materials*. Mono Book Corporation,
Baltimore, 243–253.

HOUGH, R.M., GILMOUR, I., PILLIGER, C.T., ARDEN,
J.W., GILDES, K.W.R., YUAN, J. & MILLEDGE, H.J.
1995. Diamond and silicon carbide in impact melt
rock from the Ries impact crater. *Nature*, **378**,
41–44.

HURST, R.W. & FARHAT, J. 1977. Geochronologic
investigations of the Sudbury nickel irruptive and
the Superior Province granites north of Sudbury.
Geochimica et Cosmochimica Acta, **41**, 1803–1815.

INNES, M.J.S. 1964. Recent advances in meteorite
crater research at the Dominion Observatory,
Ottawa, Canada. *Meteoritics*, **2**, 219–241.

JAHN, B., FLORAN, R.J. & SIMONDS, C.H. 1978. Rb–Sr
isochron age of the Manicouagan melt sheet,
Quebec, Canada. *Journal of Geophysical
Research*, **83**, 2799–2803.

JAMES, R.S., PEREDERY, W. & SWEENY, J.M. 1992.
Thermobarometric studies on the Levack
gneisses – footwall rocks to the Sudbury Igneous
Complex (abstract). International Conference on
Large Meteorite Impacts & Planetary Evolution.
LPI Contribution No. **790**, 41.

JOHNSON, K.S. & CAMPBELL, J.A. (eds) 1997. *Ames
structure in northwest Oklahoma and similar
features: Origin and petroleum production (1995
Symposium)*. Oklahoma Geological Survey
Circular, **100**.

KETTRUP, B., AGRINIER, P., DEUTSCH, A. & OSTER-
MANN, M. 2000. Chicxulub impactites: Geochem-
ical clues to precursor rocks. *Meteoritics and
Planetary Science*, **35**, 1129–1158.

KETTRUP, B., DEUTSCH, A. & MASAITIS, V.L. 2003.
Homogeneous impact melts produced by a
heterogeneous target? Sr–Nd isotopic evidence
from the Popigai crater, Russia. *Geochimica et
Cosmochimica Acta*, **67**, 733–750.

KIEFFER, S.W. 1971. Shock metamorphism of the
Coconino sandstone at Meteor Crater, Arizona.
Journal of Geophysical Research, **76**, 5449–5473.

KIRSCHNER, C.E., GRANTZ, A. & MULLEN, W.W. 1992.
Impact origin of the Avak Structure and genesis
of the Barrow gas fields. *American Association of
Petroleum Geologists Bulletin*, **76**, 651–679.

KOEBERL, C. & REIMOLD, W.U. 2003. Geochemistry
and petrography of impact breccias and target
rocks from the 145 Ma Morokweng impact structure, South Africa. *Geochimica et Cosmochimica
Acta*, **67**, 1837–1862.

KOEBERL, C. & SHIREY, S.B. 1997. Re–Os isotope
systematics as a diagnostic tool for the study of
impact craters and ejecta. *Paleogeography, Paleo-
climatology, Paleoecology*, **132**, 25–46.

KOEBERL, C., REIMOLD, W.U., SHIREY, S.B. & LE
ROUX, F.G. 1994. Kalkkop crater, Cape Province,
South Africa: Confirmation of impact origin using
osmium isotope systematics. *Geochimica et
Cosmochimica Acta*, **58**, 1229–1234.

KROCHUK, R.V. & SHARPTON, V.L. 2002. Overview of
Terny astrobleme (Ukranian Shield) studies
(abstract). *Lunar and Planetary Science*, **XXXIII**,
1832 pdf.

KROCHUK, R.V. & SHARPTON, V.L. 2003. Morphology
of the Terny astrobleme based on field observations and sample analysis (abstract). *Lunar and
Planetary Science*, **XXXIV**, 1489 pdf.

KROGH, T.E., DAVIS, D.W. & CORFU, F. 1984. Precise
U–Pb zircon and baddeleyite ages for the
Sudbury area. *In*: PYE, E.G., NALDRETT, A.J. &
GIBLIN, P.E. (eds) *The Geology and Ore Deposits
of the Sudbury Structure*. Ontario Ministry of
Natural Resources, Toronto, 431–446.

KUO, H.Y. & CROCKETT, T.H. 1979. Rare earth
elements in the Sudbury nickel irruptive:
Comparison with layered gabbros and implications for nickel irruptive petrogenesis.
Economic Geology, **74**, 590–605.

KUYKENDALL, M.D., JOHNSON, C.L. & CARLSON, R.A.
1997. Reservoir characterization of a complex
impact structure: Ames impact structure,
northern shelf, Anadarko basin. *In*: JOHNSON, K.S.
& CAMPBELL, J.A. (eds) *Ames structure in northwest Oklahoma and similar features: Origin and
petroleum production (1995 Symposium)*.
Oklahoma Geological Survey Circular, **100**,
199–206.

LAINÉ, R.T. 1986. Uranium deposits of Carswell structure. *In*: EVANS, E.L. (ed.) *Uranium Deposits of
Canada*. Canadian Institute of Mining and Metallurgy Special Volume, **33**, 155–169.

LAINÉ, R., ALONSO, D. & SVAB, M. (eds) 1985. *The
Carswell Structure Uranium Deposits,
Saskatchewan*. Geological Association of Canada
Special Paper, **29**.

LANGENHORST, F. 2002. Shock metamorphism of some
minerals: Basic introduction and microstructural
observations. *Bulletin Czech Geological Survey*,
77, 265–282.

LANGENHORST, F. & DEUTSCH, A. 1994. Shock experiments on pre-heated α and β-quartz: I. Optical
and density data. *Earth and Planetary Science
Letters*, **125**, 407–420.

LANGENHORST, F., SHAFRANOVSKY, G. & MASAITIS,
V.L. 1998. A comparative study of impact
diamonds from the Popigai, Ries, Sudbury and
Lappajärvi craters (abstract). *Meteoritics and
Planetary Science*, **33 Supplement**, 90–91.

LANTZ, R. 1981. Barrow gas fields – N. Slope, Alaska.
Oil & Gas Journal, **79**, 197–200.

LESHER, C.M & THURSTON, P.C. (eds) 2002. A special
issue devoted to mineral deposits of the Sudbury
Basin. *Economic Geology*, **97**, 1373–1606.

LIGHTFOOT, P.C. & FARROW, C.E.G. 2002. Geology,
geochemistry and mineralogy of the Worthington
offset dike: A genetic model for offset dike
mineralization in the Sudbury Igneous Complex.
Economic Geology, **97**, 1419–1446.

LIMON, M., CEDILLO, E., QUEZADA, J.M., *ET AL.* 1994. Cretaceous-Tertiary boundary sedimentary breccias from southern Mexico: Normal sedimentary deposits or impact-related breccias? (abstract). *AAPG Annual Convention*, 199.

MAGYAROSI, Z., WATKINSON, D.H. & JONES, P.C. 2002. Mineralogy of Ni–Cu–Platinum Group Element sulfide ore in the 800 and 810 ore bodies, Copper Cliff South Mine, and *P-T-X* conditions during the formation of platinum-group minerals. *Economic Geology*, **97**, 147–1486.

MANTON, W.I. 1965. The orientation and origin of shatter cones in the Vredefort Ring. *New York Academy of Science Annals*, **123**, 1017–1049.

MARSHALL, D., WATKINSON, D.H., FARROW, C., MOLNAR, F. & FOUILLAC, A.M. 1999. Multiple fluid generations in the Sudbury Igneous Complex. Fluid inclusions, Ar, O, H, Rb and Sr evidence. *Chemical Geology*, **154**, 1–19.

MARTINI, J.E.J. 1978. Coesite and stishovite in the Vredefort Dome, South Africa. *Nature*, **272**, 715–717.

MARTINI, J.E.J. 1991. The nature, distribution and genesis of the coesite and stishovite associated with the pseudotachylite of the Vredefort Dome, South Africa. *Earth and Planetary Science Letters*, **103**, 285–300.

MASAITIS, V.L. 1989. The economic geology of impact craters. *International Geology Reviews*, **31**, 922–933.

MASAITIS, V.L. 1992. Impact craters: Are they useful? *Meteoritics*, **27**, 21–27.

MASAITIS, V.L. 1993. [Diamantiferous impactites, their distribution and petrogenesis.] *Regional Geology and Metallogeny*, **1**, 121–134 [in Russian].

MASAITIS, V.L. 1998. Popigai crater: Origin and distribution of diamond-bearing impactites. *Meteoritics and Planetary Science*, **33**, 349–359.

MASAITIS, V.L., DANILIN, A.I., MASHCHAK, M.S., RAIKHLIN, A.I., SELIVANOVSKAYA, T.V. & SHADENKOV, E.M. 1980. [*The Geology of Astroblemes.*] Nedra Press, Leningrad [in Russian].

MASAITIS, V.L., SHAFRANOVSKY, G.I., EZERSKY, V.A. & RESHETNYAK, N.B. 1990. [Impact diamonds from ureilites and impactites.] *Meteoritika*, **49**, 180–195 [in Russian].

MASAITIS, V.L., SHAFRANOVSKY, G.I. *ET AL.* 1999. Impact diamonds in suevite breccias of the Onaping Formation, Sudbury Structure, Ontario Canada. *In*: DRESSLER, B.O. & SHARPTON, V.L. (eds) *Large Meteorite Impacts and Planetary Evolution II.* Geological Society of America, Special Paper, **339**, 317–321.

MCCARTHY, T.S., CHARLESWORTH, E.G. & STANISTREET, I.G. 1986. Post-Transvaal structural features of the northern portion of the Witwatersrand Basin. *Transactions of the Geological Society of South Africa*, **89**, 311–323.

MCCARTHY, T.S., STANISTREET, I.G. & ROBB, L.J. 1990. Geological studies related to the origin of the Witwatersrand Basin and its mineralization: An introduction and strategy for research and exploration. *South African Journal of Geology*, **93**, 1–4.

MCCARVILLE, P. & CROSSEY, L.J. 1996. Post-impact hydrothermal alteration of the Manson impact structure. *In*: KOEBERL, C. & ANDERSON, R.R. (eds) *The Manson Impact Structure: Anatomy of an Impact Crater.* Geological Society of America Special Paper, **302**, 347–376.

MCDONALD, I., ANDREOLI, M.A.G., HART, R.J. & TREDOUX, M. 2001. Platinum-group elements in the Morokweng impact structure, South Africa: Evidence for impact of a very large ordinary chondrite projectile at the Jurassic–Cretaceous boundary. *Geochimica et Cosmochimica Acta*, **65**, 299–309.

MELOSH, H.J. 1989. *Impact Cratering: A Geologic Process.* Oxford University Press, New York.

MILKEREIT, B., GREEN, A. & SUDBURY WORKING GROUP, 1992. Deep geometry of the Sudbury structure from seismic reflection profiling. *Geology*, **20**, 807–811.

MILKEREIT, B., BOERNER, E.K., *ET AL.* 2002. Development of 3-D seismic exploration technology for deep nickel–copper deposits – A case history from the Sudbury basin, Canada. *Geophysics*, **65**, 1890–1899.

MILSTEIN, R.L. 1988. The Calvin 28 structure: Evidence for impact origin. *Canadian Journal of Earth Science*, **25**, 1524–1530.

MILTON, D.J. 1977. Shatter cones – an outstanding problem in shock mechanics. *In*: RODDY, D.J., PEPIN, R.O. & MERRILL, R.B. (eds) *Impact and Explosion Cratering.* Pergamon Press, New York, 703–714.

MINTER, W.E.L. 1999. Irrefutable detrital origin of Witwatersrand gold and evidence of eolian signatures. *Economic Geology*, **94**, 665–670.

MINTER, W.E.L., GOEDHART, M., KNIGHT, J. & FRIMMEL, H.E. 1993. Morphology of Witwatersrand gold grains from the Basal Reef: Evidence for their detrital origin. *Economic Geology*, **88**, 237–248.

MOLNAR, F., WATKINSON, D.H., JONES, P.C. & GALTER, I. 1997. Fluid inclusion evidence for hydrothermal enrichment of magmatic ore at the contact zone of Ni–Cu Platinum-group 4b deposit, Lindsley mine, Sudbury, Canada. *Economic Geology*, **92**, 674–682.

MOLNAR, F., WATKINSON, D.H. & EVEREST, J. 1999. Fluid-inclusion characteristics of hydrothermal Cu–Ni–PGE veins in granitic and metavolcanic rocks at the contact of the Little Stobie deposit, Sudbury, Canada. *Chemical Geology*, **154**, 279–301.

MOLNAR, F., WATKINSON, D.H. & JONES, P.C. 2001. Multiple hydrothermal processes in footwall units of the North Range, Sudbury Igneous Complex, Canada, and implications for the genesis of vein-type Cu–Ni–PGE deposits. *Economic Geology*, **96**, 1645–1670.

MORGAN, J., GRIEVE, R.A.F. & WARNER, M. 2002a. Geophysical constraints on the size and structure of the Chicxulub impact center. *In*: KOEBERL, C. & MACLEOD, K.G. (eds) *Catastrophic events and mass extinctions: Impact and beyond.* Geological Society of America Special Paper, **356**, 39–46.

MORGAN, J.W., WALKER, R.J., HORAN, M.F., BEARY, E.S. & NALDRETT, A.J. 2002b. ^{190}Pt-^{186}Os and ^{187}Re-^{187}Os systematics of the Sudbury Igneous Complex, Ontario. *Geochimica et Cosmochimica Acta*, **66**, 273–290.

MORRISON, G.G., JAGO, B.C. & WHITE, T.L. 1994. Footwall mineralization of the Sudbury Igneous Complex. *In*: LIGHTFOOT, P.C. & NALDRETT, A.J. (eds) *Proceedings of the Sudbury – Noril'sk Symposium*. Ontario Geological Survey Special Volume, **5**. Ministry Northern Development and Mines, Toronto, 57–64.

NALDRETT, A.J. 1984. Mineralogy and composition of the Sudbury ores. *In*: PYE, E.G., NALDRETT, A.J. & GIBLIN, P.E. (eds) *The Geology and Ore Deposits of the Sudbury Structure*. Ontario Geological Survey Special Volume, **1**. Ministry Natural Resources, Toronto, 309–325.

NALDRETT, A.J. 2003. From impact to riches: Evolution of geological understanding as seen at Sudbury, Canada. *GSA Today*, **13**, 4–9.

NALDRETT, A.J. & HEWINS, R.H. 1984. The main mass of the Sudbury Igneous Complex. *In*: PYE, E.G., NALDRETT, A.J. & GIBLIN, P.E. (eds) *The Geology and Ore Deposits of the Sudbury Structure. Ontario Geological Survey Special Volume*, **1**. Ministry Natural Resources, Toronto, 235–251.

NALDRETT, A.J. & LIGHTFOOT, P.C. 1993. *Ni–Cu–PGE ores of the Noril'sk region Siberia: A model for giant magmatic ore deposits associated with flood basalt*. Society of Economic Geologists Special Publication, **2**, 81–123.

NALDRETT, A.J., HEWINS, R.H., DRESSLER, B.O. & RAO, B.V. 1984. The contact sublayer of the Sudbury Igneous Complex. *In*: PYE, E.G., NALDRETT, A.J. & GIBLIN, P.E. (eds) *The Geology and Ore Deposits of the Sudbury Structure*. Ontario Geological Survey Special Volume, **1**. Ministry Natural Resources, Toronto, 253–274.

NAUMOV, M.V. 2002. Impact-generated hydrothermal systems: Data from Popigai, Kara and Puchezh-Katunki impact structures. *In*: PLADO, J. & PESONEN, L.J. (eds) *Impacts in Precambrian Shield*. Springer-Verlag, Berlin, 117–172.

NIKOLSKY, A.P. 1991. [*Geology of the Pervomaysk iron-ore deposit and transformation of its structure by meteoritic impact.*] Nedra Press, Moscow [in Russian].

NIKOLSKY, A.P., NAUMOV, V.P. & KOROBKO, N.I. 1981. [Pervomaysk iron deposit, Krivoj Rog and its transformations by shock metamorphism.] *Geologia rudnykh mestorozhdenii*, **5**, 92–105 [in Russian].

NIKOLSKY, A.P., NAUMOV, V.P., MASHCHAK, M.S. & MASAITIS, V.L. 1982. [Shock metamorphosed rocks and impactites of Ternorvka astrobleme.] *Transactions of VSEGEI*, **238**, 132–142 [in Russian].

ORMÖ, J., BLOMQVIST, G., STRUKELL, E.F.F. & TÖRNBERG, R. 1999. Mutually constrained geophysical data for evaluating a proposed impact structure: Lake Hummeln, Sweden. *Tectonophysics*, **311**, 155–177.

OSINSKI, G.R., SPRAY, J.G. & LEE, P. 2001. Impact-induced hydrothermal activity within Haughton impact structure, arctic Canada: Generation of a transient, warm, wet oasis. *Meteoritics and Planetary Science*, **36**, 731–745.

OSTERMANN, M., SCHÄRER, U. & DEUTSCH, A. 1996. Impact melt dikes in the Sudbury multi-ring basin (Canada): Implications from uranium–lead geochronology on the Foy Offset Dike. *Meteoritics and Planetary Science*, **31**, 494–501.

PAGEL, M. & SVAB, M. 1985. Petrographic and geochemical variations within the Carswell structure metamorphic core and their implications with respect to uranium mineralization. *Geological Association of Canada Special Paper*, **29**, 55–70.

PALME, H., GOEBEL, E. & GRIEVE, R.A.F. 1979. The distribution of volatile and siderophile elements in the impact melt of East Clearwater (Quebec). *Proceeding 10th Lunar and Planetary Science Conference*. Pergamon, New York, 2465–2492.

PALME, H., GRIEVE, R.A.F. & WOLF, R. 1981. Identification of the projectile at Brent crater, and further considerations of projectile types at terrestrial craters. *Geochimica et Cosmochimica Acta*, **45**, 2417–2424.

PATTISON, E.F. 1979. The Sudbury sublayer. *Canadian Mineralogist*, **17**, 257–274.

PERRY, E., VELAQUEZ-OLIMAN, G. & MARIN, L. 2002. The hydrogeology of the karst aquifer system of the northern Yucatan Peninsula, Mexico. *International Geology Review*, **44**, 191–221.

PEUCKER-EHRENBRINK, B. 2001. Iridium and osmium as tracers of extraterrestrial matter in marine sediments. *In*: PEUCKER-EHRENBRINK, B. & SCHMITZ, B. (eds) *Accretion of extraterrestrial matter throughout Earth's history*. Kluwer Academic/Plenum Publishers, New York. 163–178.

PHILLIPS, G.N. & LAW, J.D.M. 2000. Witwatersrand gold fields; geology, genesis and exploration. *Reviews in Economic Geology*, **13**, 459–500.

PICKARD, C.F. 1994. Twenty years of production from an impact structure, Red Wing Creek Field, McKenzie County, North Dakota. *The Contact*, **41**, 2–5.

REIMOLD, W.U., KOEBERL, C., FLETCHER, P., KILLICK, A.M. & WILSON, J.D. 1999. Pseudotachylite breccias from fault zones in the Witwatersrand Basin, South Africa: Evidence of autometasomatism and post-brecciation alteration processes. *Mineralogy and Petrology*, **66**, 25–53.

ROBERTS, C. & SANDRIDGE, B. 1992. The Ames hole. *Shale Shaker*, **42**, 118–121.

ROUSELL, D.H. 1984. Mineralization in the Whitewater group. *In*: PYE, E.G., NALDRETT, A.J. & GIBLIN, P.E. (eds) *The Geology and Ore Deposits of the Sudbury Structure*. Ontario Geological Survey Special Volume, **1**. Ministry Natural Resources, Toronto, 219–232.

SAGY, A., FINEBERG, J. & RECHES, Z. 2002. Dynamic fracture by large extra-terrestrial impacts as the origin of shatter cones. *Nature*, **418**, 310–313.

SANTIAGO-ACEVEDO, J. 1980. Giant oilfields of the Southern Zone-Mexico. *In*: HALBOUTY, M.T. (ed.) *Giant oilfield and gas fields of the decade:*

1968–1978. American Association of Petroleum Geologists Memoir, **30**, 339–385.

SAWATZKY, H.B. 1977. Buried impact craters in the Williston Basin and adjacent area. *In*: RODDY, D.J., PEPIN, R.O. & MERRILL, R.B. (eds) *Impact and Explosion Cratering*. Pergamon Press, New York, 461–480.

SCHWARZ, E.J. & BUCHAN, K.L. 1982. Uplift deduced from remanent magnetization; Sudbury area since 1250 Ma ago. *Earth and Planetary Science Letters*, **58**, 65–74.

SCOTT, R.G. & SPRAY, J.G. 2000. The South Range breccia belt of the Sudbury impact structure; a possible terrace collapse feature. *Meteoritics and Planetary Science*, **35**, 505–520.

SHARPTON, V.L., BURKE, K., ET AL. 1993. Chicxulub multiring impact basin: Size and other characteristics derived from gravity analysis. *Science*, **261**, 1564–1567.

SIEBENSCHOCK, M., SCHMITT, R.T. & STÖFFLER, D. 1998. Impact diamonds in glass bombs from suevite of the Ries crater, Germany (abstract). *Meteoritics and Planetary Science*, **33 Supplement**, 145.

SPRAY, J.G. & THOMPSON, L.M. 1995. Friction melt distribution in terrestrial multi-ring impact basins. *Nature*, **373**, 130–132.

SPUDIS, P.D. 1993. *The Geology of Multi-ring Impact Basins*. Cambridge University Press, Cambridge.

STÖFFLER, D. 1971. Progressive metamorphism and classification of shocked and brecciated crystalline rocks in impact craters. *Journal of Geophysical Research*, **76**, 5541–5551.

STÖFFLER, D. 1972. Deformation and transformation of rock-forming minerals by natural and experimental shock processes. I. Behavior of minerals under shock compression. *Fortschritte der Mineralogie*, **49**, 50–113.

STÖFFLER, D. 1974. Deformation and transformation of rock-forming minerals by natural and experimental shock processes. II. Physical properties of shocked minerals. *Fortschritte der Mineralogie*, **51**, 256–289.

STÖFFLER, D. 1984. Glasses formed by hypervelocity impacts. *Journal of Non-Crystalline Solids*, **7**, 465–502.

STÖFFLER, D. & HORNEMANN, U. 1972. Quartz and feldspar glasses produced by natural and experimental shock. *Meteoritics*, **7**, 371–394.

STÖFFLER, D. & LANGENHORST, F. 1994. Shock metamorphism of quartz in nature and experiment: 1. Basic observation and theory. *Meteoritics*, **29**, 155–181.

STÖFFLER, D., DEUTSCH, A., ET AL. 1994. The formation of the Sudbury Structure, Canada: Towards a unified impact model. *In*: DRESSLER, B.O., GRIEVE, R.A.F. & SHARPTON, V.L. (eds) Geological Society of America Special Paper, **293**, 303–318.

THERRIAULT, A.M., GRIEVE, R.A.F. & REIMOLD, W.U. 1997. Original size of the Vredefort structure: Implications for the geological evolution of the Witwatersrand Basin. *Meteoritics and Planetary Science*, **32**, 71–77.

THERRIAULT, A.M., FOWLER, A.D. & GRIEVE, R.A.F. 2002. The Sudbury Igneous Complex: A differentiated impact melt sheet. *Economic Geology*, **97**, 1521–1540.

THOMPSON, L.M., SPRAY, J.G. & KELLEY, S.P. 1998. Laser probe argon-40/argon-39 dating of pseudotachylite from the Sudbury Structure: Evidence for post impact thermal overprinting in the North Range. *Meteoritics and Planetary Science*, **33**, 1259–1269.

TONA, F., ALONSO, D. & SVAB, M. 1985. Geology and mineralization in the Carswell structure – A general approach. *Geological Association of Canada Special Paper*, **29**, 1–18.

TUCHSHERER, M.G. & SPRAY, J.G. 2002. Geology, mineralisation and emplacement of the Foy offset dike, Sudbury. *Economic Geology*, **97**, 1377–1398.

TWISS, R.S. & MOORES, E.M. 1992. *Structural Geology*. W.H. Freeman, New York.

VAL'TER, A.A., EREMENKO, G.K., KWASNITSA, V.N. & POLKANOV, YU. A. 1992. [*Shock metamorphosed carbon minerals.*] Nauk Press, Kiev [in Russian].

WALKER, R.J., MORGAN, J.W., NALDRETT, A.J., LI, C. & FASSETT, J.D. 1991. Re–Os isotope systematics of Ni–Cu sulfide ores, Sudbury Igneous Complex, Ontario: Evidence for a major crustal component. *Earth and Planetary Science Letters*, **105**, 416–429.

WARNER, S., MARTIN, R.F., ABDEL-RAHAM, A.F.M. & DOIG, R. 1998. Apatite as a monitor of fractionation, degassing and metamorphism in the Sudbury Igneous Complex, Ontario. *Canadian Mineralogist*, **36**, 981–999.

WIEST, B, 1987. *Diskordante Gangbreccien und strukturelle Merkmale im Untergrund komplexer Impaktstrukturen und ihre Bedeutung für die Kraterbildungsprozesse*. PhD thesis, Westfälische Wilhems Universität, Münster, Germany.

WOOD, C.A. & HEAD, J.W. 1976. Comparison of impact basins on Mercury, Mars and the Moon. *Proceeding 7th Lunar Science Conference*. Pergamon, New York, 3629–3651.

WU, J., MILKEREIT, B. & BOERNER, D.E. 1995. Seismic imaging of the enigmatic Sudbury structure. *Journal* of *Geophysical Research*, **100**, 4117–4130.

Gold mineralization within the Witwatersrand Basin, South Africa: evidence for a modified placer origin, and the role of the Vredefort impact event

C. L. HAYWARD[1,2], W. U. REIMOLD[2], R. L. GIBSON[2] & L. J. ROBB[3]

[1]*Department of Earth Sciences, University of Cambridge, Downing Street, Cambridge CB2 3EQ, UK (e-mail: chay02@esc.cam.ac.uk)*

[2]*Impact Cratering Research Group, School of Geosciences, University of the Witwatersrand , Private Bag 3, PO WITS 2050, Johannesburg, South Africa*

[3]*Economic Geology Research Institute-Hugh Allsopp Laboratory, School of Geosciences, University of the Witwatersrand, Private Bag 3, PO WITS 2050, Johannesburg, South Africa*

Abstract: The chemical composition of gold within the Archaean metasedimentary rocks of the Witwatersrand Supergroup displays significant heterogeneity at the micro-, meso- and regional scales. A detailed electron microbeam analytical and petrological study of the main auriferous horizons in the Central Rand Group throughout the Witwatersrand Basin indicates that gold has been remobilized late in the paragenetic sequence over distances of less than centimetres. Contemporaneous chlorite formation was strongly rock-buffered. Gold mobilization occurred under fluid-poor conditions at temperatures that did not exceed 350 °C. Widespread circulation of mineralizing fluids within the Central Rand Group is not supported by the gold and chlorite chemical data. Brittle deformation that affects most of the paragenetic sequence of the Central Rand Group late in its post-depositional history is followed by sequences of mineral growth and dissolution that appear throughout the Central Rand Group and have consistent textural relationships with gold. The consistent location within the paragenetic sequence, the wide regional and strati-graphic extent of the brittle deformation, together with mineral chemical and petrological data suggest that the Vredefort Impact Event (2.02 Ga) was the cause of this late deformation, and that post-impact fluid-poor metamorphism resulted in crystallization of a significant proportion of the gold on and within mineral grains that were deformed during this event.

The world-class Witwatersrand gold deposit in South Africa has produced some 50 000 tonnes of gold since mining began in the late 1880s (Robb & Robb 1998). The gold is mostly hosted by lower greenschist grade Archaean metasedi-mentary rocks of the Central Rand Group (CRG), the upper group of the 3.07–2.7 Ga Witwatersrand Supergroup, and by the over-lying Ventersdorp Contact Reef (VCR). Most of the gold is located within conglomeratic units, known locally as 'reefs'. Mining occurs in gold-fields that form the so-called 'Golden Arc' extending from east and SE of Johannesburg to the west and SW towards Welkom in the northern Free State Province (Fig. 1). Mining began around Johannesburg in outcrop and subcrop, and then proceded down-dip into deeper buried portions of the CRG. To date, these, the deepest mines in the world, have reached mining depths close to 4000 m. Current activity centres on the West Wits and

Klerksdorp goldfields, as well as the recently opened Target Mine at the northern edge of the Welkom goldfield, and in the more marginally economic Evander goldfield (Fig. 1). However, considerable exploration and evaluation activity is currently centred on the down-dip extensions of the Central Rand goldfield, where several extensive payshoots seem to continue (Stewart *et al.* in press). Good models for the mineraliza-tion are ever more important as existing reserves become depleted and new exploration extends beyond the currently exploited volumes of the CRG.

Despite prolonged and intensive study, the origin of the Witwatersrand gold and the mode of its incorporation into the CRG have remained controversial ever since the early twentieth century (e.g. Gregory 1908; Young 1917; Graton 1930). The controversy exists because of the apparent paradox of gold distribution being controlled mostly by primary

From: MCDONALD, I., BOYCE, A. J., BUTLER, I. B., HERRINGTON, R. J. & POLYA, D. A. (eds) 2005. *Mineral Deposits and Earth Evolution.* Geological Society, London, Special Publications, **248**, 31–58. 0305-8719/$15.00
© The Geological Society of London 2005.

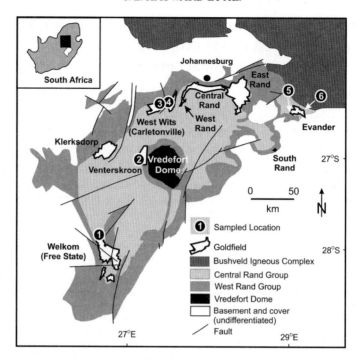

Fig. 1. Simplified geological map showing the subcrop of the Central Rand and West Rand Groups, the location of the Vredefort Dome and the goldfields, both active and inactive. Modified from Pretorius (1986). The locations that were sampled during this study are numbered as follows: 1, Target; 2, Venterskroon drill core; 3, East Driefontein; 4, Placer Dome; 5, Leslie; 6, Kinross.

sedimentological features within the CRG, and yet appearing to be of predominantly authigenic-hydrothermal origin in its textural associations within these metasedimentary rocks, including occurrences within base metal mineralized quartz veins (Frimmel 1997*a*; Robb & Robb 1998; Frimmel & Minter 2002). Textural evidence for gold being late in the paragenetic sequence has been widely reported and accepted by most researchers (e.g. Barnicoat *et al.* 1997; Robb & Robb 1998; Foya *et al.* 1999; Jolley *et al.* 2004). Current debate maintains the lines drawn during recent decades between two possible models for the gold mineralization: the modified placer and the hydrothermal models. The modified placer model advocates the incorporation of detrital gold, pyrite and uraninite as heavy minerals into the CRG, with subsequent hydrothermal remobilization during one or more local or regional metamorphic events. This explains the sedimentary control on ore grades and the clearly hydrothermal textures of the gold, and some other minerals. This is the model favoured by most current Witwatersrand workers (e.g. Els 1991; Robb *et al.* 1997; Frimmel & Minter 2002). Recently, additional

support for a detrital origin of gold and sulphides has come from the application of Re–Os dating, which gives ages for pyrite and the associated gold, of 3.03 Ga ± 0.021 Ga (Kirk *et al.* 2001, 2002) that predate the maximum age of the CRG (2.89 Ga; Robb *et al.* 1990). In contrast, the hydrothermal model proposes that the gold, pyrite and uraninite, as well as carbon, within the CRG were introduced together during a late-stage, basin-wide, hydrothermal fluid circulation event (e.g. Barnicoat *et al.* 1997; Phillips & Law 2000; Jolley *et al.* 2004). These hydrothermal fluids deposited gold preferentially within conglomeratic units apparently owing to greater permeability within these units and/or the occurrence of fracture and shear zones within them. The origins of these fluids are not well constrained. According to various authors (e.g. Barnicoat *et al.* 1997; Phillips & Law 2000), they may be either extrabasinal, having been derived during the craton-wide Ventersdorp magmatic event around 2.7 Ga, and/or intrabasinal, having been generated by metamorphic dehydration of the Central Rand Group and the underlying West Rand and Dominion Groups (Fig. 1) during Transvaal basin times (2.6–2.15 Ga Armstrong *et al.* 1991; Walraven 1997; Walraven & Martini 1995), or as a consequence of the subsequent Bushveld and Vredefort events at 2.06 and 2.02 Ga, respectively.

The two mineralization models have very different implications for mineral chemistry within the parageneses that occur in different parts of the CRG. Of special pertinence to efforts to discriminate between detrital and hydrothermal origins of the gold mineralization is the chemistry of the gold itself and that of the minerals that formed in association with gold, in particular phyllosilicates and sulphides. If the gold was originally detrital, some degree of variation in the composition among different populations of gold particles within the CRG may be expected to survive from the original variation present in the gold from the source region(s). The degree of preservation of such primary heterogeneity depends upon the degree to which any later fluids present were able to remobilize gold; remobilization by hydrothermal fluids of an originally chemically heterogeneous population of gold grains would have led to a reduction in this primary heterogeneity. In contrast, a relatively low degree of chemical heterogeneity is expected from gold that was deposited during a single episode of hydrothermal circulation, if the fluids were reasonably homogeneous chemically. Gold in the Witwatersrand reefs has many and varied textural associations with a range of authigenic minerals which include chlorite, pyrite, pyrrhotite, galena and muscovite as well as altered hydrocarbon, commonly referred to in the literature as 'carbon' or 'bitumen'. Chlorite is of particular importance, as it occurs in a large number of locations within the CRG and can be demonstrated through its textural associations to have formed contemporaneously with gold at many of these locations. The degree of chemical heterogeneity of chlorite associated with gold should indicate the degree to which the composition(s) of fluid(s) present during gold mobilization or recrystallization was/were rock-buffered or fluid-buffered, with consequent implications for the volume of fluid and the volume of rock within which auriferous fluids were circulating.

Mineral chemistry and textural relations should, thus, offer key insights into the formation and evolution of the Witwatersrand gold deposits. Mineral chemical data cannot, however, be interpreted in terms of deposit formation and evolution without significant ambiguity, unless accompanied by detailed textural and paragenetic information. Several electron microbeam analytical studies of gold grains *in situ* in thin section have been carried out on individual Witwatersrand depositional units and/or mines (Frimmel & Gartz 1997; Frimmel *et al.* 1993; Oberthür & Saager 1986; Saager 1969), and one on individual grains from a gold concentrate from a number of mines (von Gehlen 1983). However, most of the numerous analyses of Witwatersrand gold have been on 'bulk' samples or on gold grains isolated from their host rock by acid digestion, with consequent loss of paragenetic context (e.g. Reid *et al.* 1988; Hallbauer & Utter 1977). In other studies, the 'fineness' of the gold ($F = Au/(Au + Ag) \times 1000$) rather than the individual element concentrations was reported (Saager 1969). The loss of textural information, reporting of only fineness values, and the effect of averaging of gold compositions leads unavoidably to over-simplification of the data on gold composition and, thus, has implications for models regarding gold origin.

In the literature on the CRG and the overlying Ventersdorp Contact Reef there is, with the exceptions of the studies mentioned above, a general paucity of *in situ* microbeam analysis of petrologically well-characterized samples. Such analyses and petrological observations enable detailed examination of the validity of various aspects of the two current models of mineralization within the CRG (e.g. Frimmel & Gartz 1997). In particular, mineral chemical and paragenetic information will identify the scales and distribution of any chemical heterogeneity, the relative timing of mineral growth and dissolution, provide information on the types of fluids that may have existed in the rock at different times, and place the gold mineralization within a well-constrained petrogenetic framework that links the regional and mine-scale with the centimetre and microscopic scales.

Accordingly, the present study was initiated with the aim of gaining a detailed understanding of the composition of gold and key allogenic and metamorphic minerals (chlorite, chloritoid, muscovite, pyrite, pyrrhotite, chalcopyrite and uraninite) within a well-defined paragenetic framework from samples representative of as many of the surviving reefs in the CRG as possible. Samples were obtained underground from reefs and their hanging walls and footwalls, and from locations that span the preserved stratigraphic and geographic extent of the CRG (Figs 1 & 2). The regional distribution of samples combined with detailed mine geology and thin-section scale observation provides the link between large-scale features and the details of gold mobility and metamorphism of the CRG

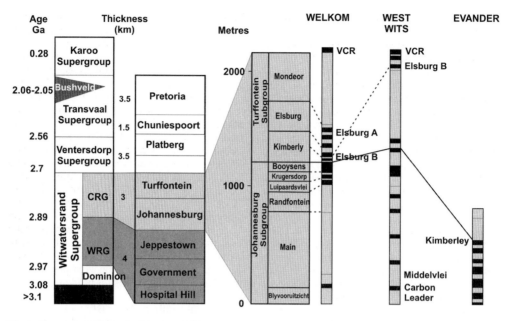

Fig. 2. Summary of the main features of the stratigraphy of the Central Rand Group and sampled levels within the Welkom, West Wits and Evander goldfields. Modified from Pretorius (1986) and Robb & Robb (1998).

at the millimetre to kilometre scales. It also enables regional validation and expansion of the findings of earlier, smaller scale studies on gold chemistry (e.g. Frimmel & Gartz 1997; Oberthür & Saager 1986). The first publication from this comprehensive study was centred on compositions of gold and some associated chlorite, and on initial interpretations of deformation that may be related to the Vredefort impact event that affected the basin at 2.02 Ga (e.g. Gibson & Reimold 2001), and its effect on gold distribution within the CRG.

Geological background

The Witwatersrand Basin is the structurally-preserved remnant of a presumably originally larger Archaean sedimentary basin, formed in an inferred foreland setting during collision of the Kaapvaal craton with other continental fragments to the NW between 3.07 and 2.71 Ga (Armstrong *et al.* 1991). However, the precise details of the tectonic setting of the basin and timing of events during Witwatersrand Supergroup deposition remain elusive (Catuneau 2001). The basin was filled progressively, first by bimodal volcanics as well as sediments of the Dominion Group (3086–3074 Ma; Armstrong *et al.* 1991), and then by the sediments of the Witwatersrand Supergroup (Fig. 2). Following

subsidence, either due to thermal collapse or the development of a foreland basin, the argillaceous and sandy sediments of the West Rand Group were deposited in an environment that evolved from epicontinental to tidally-dominated marine shelf (e.g. Phillips & Law 2000; Frimmel & Minter 2002 and references therein). Extensional tectonism gave way to compression, causing fracturing of the basement into discrete blocks (McCarthy 1994). Uplift of the hinterland and a progressive series of falls in sea level led to the formation of prograding alluvial fans and low-relief braid-plain environments depositing proximal gravel-dominated and distal sand-dominated sequences within topographically restricted areas between 2.91 and 2.71 Ga (Barton *et al.* 1989). These sediments comprise the Central Rand Group, from which most of the gold mined in the Witwatersrand Basin was derived.

Gold is present almost exclusively in the fluvial sediments. The locations of depository fans and channels have been mapped in all gold mines (e.g. McWha 1994) and these sedimentological features are used on a day-to-day basis in mining and exploration. Palaeocurrent determinations indicate that these sediments were derived from source areas to the north and west of the Witwatersrand Basin. A component of the mineralization occurs as allogenic grains, including pyrite, leucoxene (after ilmenite), chromite, uraninite, titanite, sphalerite, zircon, garnet, diamond and gold, amongst others (e.g. Feather & Koen 1975; Robb *et al.* 1997), and is restricted to gravel facies, deposited mainly as mature scour-based pebble lag and gravel bar deposits. In mine outcrop, heavy mineral concentration is controlled by the presence of bars and channels and the prevailing processes of winnowing, reworking, and erosion. The different conglomerate sequences vary widely because of differences in local source regions, topography and depositional environment (Robb & Robb 1998; McCarthy 1994; Smith & Minter 1980; Frimmel & Minter 2002). Most of the conglomerate volume does not contain significant (economic) gold, although at least some gold is present in most of the conglomerate units. Among the most productive of the conglomerate horizons is the Ventersdorp Contact Reef (VCR), for which there is evidence that supports formation through extensive and widespread episodes of erosion and reworking of underlying sediments that are crossed by the basal unconformity of the VCR (e.g. McCarthy 1994; McWha 1994; Krapez 1985). In places, over 2500 m of erosion occurred prior to the final formation of the VCR (Rust 1994). Deposition of the VCR was abruptly terminated by the rapid outpouring between 2714 and 2704 Ma (Armstrong *et al.* 1991) of 60 000 km^3 of Klipriviersberg Group tholeiitic flood basalts of the Ventersdorp Supergroup, which buried the VCR. The Witwatersrand Basin underwent episodic subsidence between 2700 and 2200 Ma during the deposition of the Ventersdorp Supergroup and, later, the formation of the Transvaal Basin (from c. 2.6 to 2.15 Ga). During early Transvaal times, much of the craton was covered by a shallow epicontinental sea in which the Chuniespoort Group dolomites formed. Slow subsidence and chemical sedimentation was followed by regional uplift and erosion in the north of the craton, and deposition of the Pretoria Group siliclastic rocks (Eriksson *et al.* 2001).

According to various workers, the CRG was affected by a series of regional metamorphic events related to tectonomagmatic events in the region of the Witwatersrand Basin and beyond (e.g. Frimmel 1994; Robb & Robb 1998; Phillips & Law 2000; Gibson & Jones 2002). Pre-Transvaal thrust-related metamorphism occurred at 2.64–2.71 Ga, and syn- and mid-Transvaal regional metamorphism is reported to have occurred at 2.55–2.58 Ga (Frimmel 1994, 1997b; Robb *et al.* 1997; Gartz & Frimmel 1999) and 2.2–2.3 Ga, respectively (Frimmel 1994;

Robb *et al.* 1997). Steady-state geotherms between 2.60 Ga (early Transvaal) and 2.06 Ga (late Transvaal) were calculated by Gibson & Jones (2002) at 15–20 °C km^{-1}. A further regional metamorphic episode was caused by the intrusion of the vast Bushveld Igneous Complex (BIC) (e.g. Gibson & Wallmach 1995; Frimmel 1997b; Robb *et al.* 1997) into the central region of the Transvaal Basin from approximately 2060–2050 Ma (Walraven & Hattingh 1993; Walraven *et al.* 1990; Walraven 1997), to the north and NE of the Witwatersrand Basin. This caused elevation of upper crustal geotherms to around 35–40 °C km^{-1}, and led to regional greenschist-facies metamorphism, with grades increasing towards the intrusion (Gibson & Stevens 1998). The peak metamorphic conditions attained within the CRG were reached following the intrusion of the BIC, with temperatures of 350 ± 50 °C and pressures of 0.15–0.3 GPa developed generally, as indicated by the reported coexistence of chloritoid and pyrophyllite (Phillips 1987, 1988).

Following the cooling of the BIC, the area suffered a major meteorite impact event at 2023 ± 4 Ma (e.g. Kamo *et al.* 1996; Gibson *et al.* 1997). This event excavated a transient crater of at least 100 km diameter and likely 30 km depth, which, in turn, collapsed to produce a broad, shallow impact basin of 250–300 km diameter (Henkel & Reimold 1998; Therriault *et al.* 1997). The Kaapvaal crust below and around the impact structure experienced widespread and strong macro- and micro-deformation. The deeply eroded crust now exposed in the Vredefort Dome, the central uplift feature of the impact structure in the geographic centre of the preserved extent of the Witwatersrand Basin (Fig. 1) exhibits evidence of shock pressures up to at least 30 GPa and post-impact temperatures of at least 900 °C in its central parts (Gibson 2002; Gibson & Reimold 2005). In addition, extensive macro-brecciation was caused by the Vredefort impact event, leading to the formation of pseudo-tachylitic breccias (Dressler & Reimold 2004; Reimold & Gibson 2005) in the central uplift and environs (Reimold & Colliston 1994; Reimold *et al.* 1999). Massive breccias are also known from bedding-parallel and normal faults in the wider Witwatersrand Basin (e.g. Killick *et al.* 1998; Reimold *et al.* 1999 and references therein). Brecciation and fracturing occurred throughout the Witwatersrand strata at that time. This brecciation event and associated microdeformation provides a convenient time marker in the evolution of the Witwatersrand basin and its gold deposits. Detailed regional

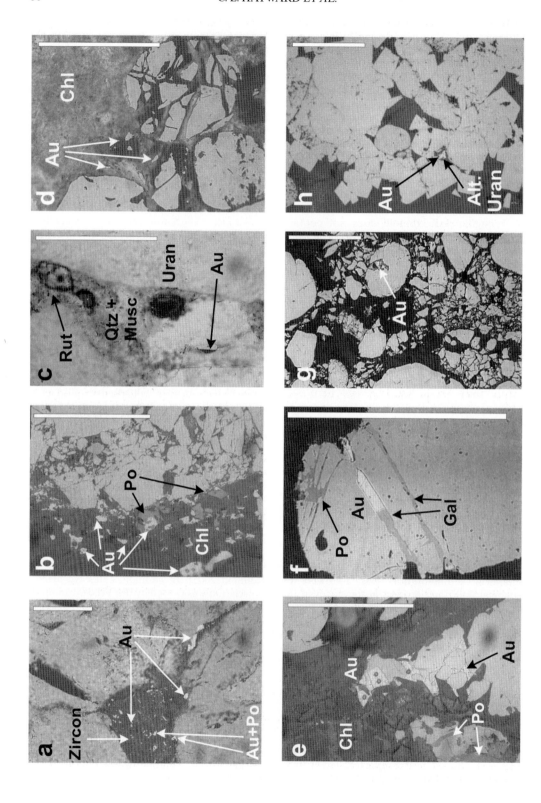

metamorphic studies have shown that the upper crust in the region of the Witwatersrand basin experienced a further, retrograde phase of greenschist-facies metamorphism immediately after the impact event (Gibson *et al.* 1998; Gibson & Reimold 2001; Reimold *et al.* 1999).

Analytical methods

Electron probe microanalysis (EPMA), although not the most sensitive technique available, has the highest resolution, down to 1–2 μm. This resolution is essential in attempts to analyse the widest possible range of gold grains in the Witwatersrand deposit, where gold is generally very fine-grained (<1 μm–10 μm are common sizes, but larger grains of gold occur).

EPMA was carried out at the Departments of Earth Sciences of the Universities of Cambridge and Manchester, using Cameca SX100 electron probe microanalysers. Data reduction employed the PAP correction routine (Pouchou & Pichoir 1987). Analyses on each SX100 used the same experimental conditions and the same set of calibration standards. Data from both instruments are therefore directly comparable. Prior to each gold analysis, analysing crystals were driven to Fe Kα, Si Kα, Ti Kα and Al Kα in order to eliminate beam overlap with surrounding minerals (pyrite, pyrrhotite, quartz, chlorite, rutile, muscovite). The beam was focused to the maximum degree, which at 20 kV in gold will produce X-rays from an excitation volume no greater than 2 μm in diameter. In each auriferous thin section, every grain of gold greater than 4 μm across was analysed and multiple analyses were made on all larger grains. Analyses with totals between 99.2 and 100.8 wt% are included in the dataset. For chlorite analysis, the beam was operated at 15 kV and a diameter of 7 μm, with beam currents of 10 nA for major elements and 40 nA for minor and trace elements. All chlorite analyses with (Na + K + Ca) >0.09 wt%

were rejected. For all analyses, counting time on the peak was twice that on each side of the background.

Cathodoluminescence imaging was done out at the University of Manchester using a CITL Mk 2 cold cathode instrument operated at 15 kV and 400 μA. The thin sections were cooled using a copper cold block to liquid nitrogen temperature (design by C. Hayward). Images were recorded using Fuji Provia 400 ASA 35mm slide film, push-processed to 1250 ASA. Average exposure times using this film were of the order of 3–8 seconds.

Results

Paragenetic relationships

The paragenetic relations of gold in the CRG are varied and complex. Gold occurs in association with chlorite, pyrite, pyrrhotite, bitumen, alteration products of uraninite, chalcopyrite, bravoite and galena (Table 1; Fig. 3). This observation contradicts the assertion of Jolley *et al.* (2004) that most gold is associated with hydrocarbon alone. From Table 1 it is clear that gold is most commonly associated with texturally late chlorite and quartz (VCR and Carbon Leader reefs, West Wits goldfield and Kimberley reef, Evander goldfield) (Fig. 3b, d, e, g), and with late pyrrhotite (Carbon Leader reefs, West Wits goldfield and Kimberley reef, Evander goldfield) (Fig. 3a, b, e), and occurs as overgrowths on fragmented and corroded pyrite (VCR and Carbon Leader reefs, West Wits goldfield and Kimberley reef, Evander goldfield) (Fig. 3b, d, e, f, g, h) and overgrowing altered uraninite grains (Carbon Leader, West Wits goldfield) (Fig. 3h). In contrast with the findings of Gartz & Frimmel (1999), Frimmel & Minter (2002) and England *et al.* (2002), gold was found only very rarely as inclusions within secondary, hydrothermal pyrite.

There is no relation between the mineralogical

Fig. 3. Examples of mineralogical and textural associations of gold in the CRG. Scale bar in all images is 100 μm long. (**a**) Kimberley reef, Leslie Mine (Evander). Gold ± pyrrhotite in fractures within quartz and zircon. Reflected + transmitted light. (**b**) VCR, E. Driefontein Mine (W. Wits). Gold plus pyrrhotite in chlorite overgrowing fragments of brittly-deformed pyrites. Reflected light. (**c**) Carbon Leader reef, E. Driefontein Mine (W. Wits), Gold interstitial between detrital quartz grains. Reflected + transmitted light. (**d**) Kimberley reef, Leslie Mine (Evander), gold with chlorite overgrowing corroded fragments of brittly-deformed pyrites. Reflected + transmitted light. (**e**) Carbon Leader reef, E. Driefontein Mine (W. Wits), gold and chlorite overgrowing and infilling corroded fragments of brittly-deformed pyrite. Reflected light. (**f**) Carbon Leader reef, E. Driefontein Mine (W. Wits), gold with galena infilling fractures in pyrite and cross-cutting earlier pyrrhotite-filled fractures. Reflected light. (**g**) Elsburg B reef, Target Mine (Welkom), gold and chlorite within cracks in brittly-deformed pyrite. (**h**) Elsburg B reef, Target Mine (Welkom), gold with altered uraninite between brittly-deformed rounded and subhedral pyrites and overgrown by euhedral pyrite. Po, pyrrhotite; Rut, rutile; Chl, chlorite; Qtz, quartz; Musc, muscovite; Uran, uraninite; Gal, galena.

Table 1. *Mineralogical associations of gold occurrences, analyses of which are presented in Figure 7*

West Wits, VCR

4	Occ. 1	Within late chlorite between slightly corroded fragments of pyrite.
	Occ. 2	Within late chlorite between quartz grains.
	Occ. 3	Overgrowing fragmented and corroded pyrite. Associated with late chlorite and quartz.
5	Occ. 1	Within late chlorite and overgrowing fragmented pyrite and corroded quartz.
	Occ. 2	Within late chlorite and overgrowing fragmented pyrite.
	Occ. 3	Within late chlorite and overgrowing corroded fragmented pyrite. One grain within a corroded quartz grain, others within embayments in quartz grain margins.

West Wits, Carbon leader

1a	Occ. 1	Between quartz grains (interstitial) with muscovite.
	Occ. 2	Overgrowing fragmented and corroded pyrite.
1b	Occ. 2	Within alteration products of uraninite.
	Occ. 3	Within and overgrowing altered uraninite and associated with pyrrhotite.
	Occ. 4	Associated with pyrrhotite. Overgrowing fragmented and corroded pyrite.
	Occ. 6	Between quartz grains (interstitial) with muscovite and overgrowing fragmented and corroded pyrite.
	Occ. 7	Associated with pyrrhotite and overgrowing fragmented and corroded pyrite.
	Occ. 8	Within quartz and overgrowing fragmented pyrite.
	Occ. 9	Within quartz and altered uraninite and overgrowing fragmented and corroded pyrite.
	Occ. 10	Within quartz and grain boundaries between quartz and/or muscovite, and overgrowing fragmented and corroded pyrite.
	Occ. 11	Between quartz grains (interstitial), together with muscovite.
	Occ. 12	Between quartz grains (interstitial), together with muscovite and pyrrhotite.
	Occ. 13	Overgrowing fragmented and corroded pyrite, associated with altered uraninite and muscovite.
	Occ. 14	Between quartz grains (interstitial), and overgrowing breakdown products of uraninite.
4b	Occ. 1	Overgrowing muscovite and altered uraninites, together with pyrrhotite.
	Occ. 2	Within and overgrowing altered uraninites and corroded pyrites, together with pyrrhotite.
	Occ. 5	Interstitial between quartz grains and overgrowing altered uraninite.
	Occ. 6	Within a corroded, rounded pyrite grain
	Occ. 7	Interstitial between quartz grains (possibly recrystallized detrital grains?)
	Occ. 9	Within chloritoid
	Occ. 10	Overgrowing fragmented and corroded pyrite.
	Occ. 13	Overgrowing corroded pyrite.
	Occ. 14	Overgrowing fragmented and corroded pyrite.
	Occ. 15	With pyrrhotite in a fracture within quartz.
	Occ 16	Within fractures in a fragmented and corroded pyrite.
4c	Occ. 1	Intergrown with pyrrhotite within a corroded pyrite.
	Occ. 2	With muscovite, interstitial between quartz grains (possibly recrystallized detrital grains?).
	Occ. 3	With muscovite, interstitial between quartz grains, and adjacent to but not overgrowing, pyrite (possibly recrystallized detrital grains?).
	Occ. 4	With muscovite, interstitial between quartz grains (possibly recrystallized detrital grains?).
	Occ. 5	With muscovite, interstitial between quartz grains (possibly recrystallized detrital grains?).
	Occ. 6	Overgrowing corroded pyrite.
	Occ. 7	With muscovite, interstitial between quartz grains (possibly recrystallized detrital grains?).
	Occ. 8	With muscovite, interstitial between quartz grains (possibly recrystallized detrital grains?).
	Occ. 9	With muscovite, interstitial between quartz grains
	Occ. 10	A seam within an altered uraninite grain, truncated by the margins of that grain.
	Occ. 11	With muscovite, interstitial between quartz grains (possibly recrystallized detrital grains?), and overgrowing altered uraninite grains.

Evander, Kimberley

3c	Occ. 1	With chlorite within quartz.
	Occ. 2	With chlorite and pyrrhotite interstitial between quartz grains.
4a	Occ. 1	Intergrown with chalcopyrite
	Occ. 2	With muscovite, interstitial between quartz grains (possibly recrystallized detrital grains?).
	Occ. 3	With chlorite, interstitial between quartz grains (possibly recrystallized detrital grains?).
	Occ. 4	Interstitial between quartz grains (possibly recrystallized detrital grains?).
	Occ. 5	Associated with bravoite and overgrowing a corroded pyrite and within fracture in quartz grains.
	Occ. 6	Within quartz with minor chlorite.
	Occ. 7	Within quartz with minor chlorite.

Table 1. (*Continued*)

2b	Occ. 1	Between fragmented and corroded rounded pyrites, later enclosed by euhedral pyrite overgrowth.
	Occ. 3	Overgrowing fragmented and corroded pyrites and overgrowing different zones of the enclosing, euhedral pyrite overgrowth.
	Occ. 4	Overgrowing fractured and corroded pyrite.
	Occ. 5	With muscovite, interstitial between quartz grains

or textural association of the gold and Ag or Hg content. It is difficult to be certain about the morphologies of every gold grain when working only with thin sections. However, of the more than 300 gold grains observed in the present study, almost all displayed characteristics that are best interpreted as deriving from hydrothermal crystallization of the gold. These include intergrowth of gold with chlorite, gold infilling fractures within sulphide and silicate minerals, the presence of crystal faces on some gold grains and the growth of gold against faces of metamorphic minerals (Fig. 3). More than 70% of the gold observed is located within fractured minerals or overgrowing fragments of fractured grains, in particular pyrite, but also zircon, arsenopyrite and quartz (Fig. 3). No grains displayed morphologies characteristic of detrital deposition, such as disc-like particles with overturned edges, and spheroidal and toroidal forms as reported, for example, by Frimmel (1997*a*) and Minter *et al.* (1993). Some gold in the Carbon Leader reef occurs within the quartz–muscovite matrix that is interstitial between detrital quartz grains. Although this gold currently has the appearance of hydrothermally crystallized gold (straight margins boundaries, crystal faces), the location of the gold within interstices between detrital grains together with the diversity of chemical composition of some groups of these grains (see below) could be suggestive of an originally detrital deposition of the gold.

Chlorite occurs in a number of textural associations within the CRG. In the studied samples, it occurs in rounded areas, which may represent replaced clasts, with or without quartz, rutile and pyrite. It also occurs commonly in veins and brittle shear zones, and in strain shadows with or without quartz. Chlorite is widespread in the matrix between quartz grains, where it occurs with muscovite, pyrite, pyrrhotite, rutile, bitumen and alteration products of uraninite. The chlorite associated with gold is present within fractures, in the interstices between quartz grains, and in brittle microshears. This chlorite and the associated gold appear to have formed contemporaneously (e.g. gold intergrown with and overgrowing chlorite grains, occurrences of gold and chlorite ± quartz infilling fractures in fragmented mineral grains, with or without pyrrhotite (Fig. 3)). This chlorite is generally undeformed, relatively coarse-grained, and overgrows and/or cross-cuts earlier chlorite-defined fabric(s).

Pyrrhotite, which is commonly intergrown with the gold (Fig. 3a, e) and is therefore interpreted as having crystallized contemporaneously, is sometimes also observed overgrowing the gold grains, with well-formed pyrrhotite grains growing outwards from the margins of the enclosed gold grains (Fig. 3b). This suggests that pyrrhotite formation continued after gold crystallization was completed.

Mineral chemistry

Gold chemistry

The Witwatersrand gold grains contain highly variable Ag and Hg contents. The data for Ag and Hg contents obtained in this study, shown by goldfield, mine and reef name, are presented in Figure 4. They demonstrate considerable chemical heterogeneity in gold composition on the regional scale. In general, there is a greater degree of chemical similarity amongst gold mined from different stratigraphic levels within a single goldfield, than there is between gold mined from different goldfields. Most gold from the Evander goldfield has high Ag and low Hg, most from the Target mine has low Ag and Hg, and most gold from the West Wits goldfield has low Ag and high Hg. However, the VCR, Elsburg 'B' and Carbon Leader reefs have Ag and Hg contents in gold that overlap the ranges of other reefs in different goldfields, for example the VCR gold from the West Wits goldfield has Ag and Hg contents similar to the Kimberley reef of the Evander goldfield, and the Carbon Leader has similar concentrations of these elements to gold from the Elsburg 'B' reef in the Welkom goldfield. In addition to Ag and Hg, Cu,

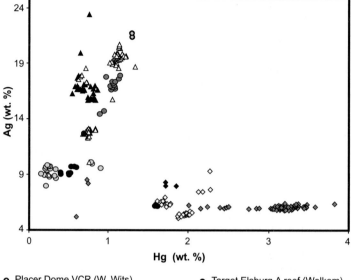

Fig. 4. Silver and mercury contents of gold analysed in the present study.

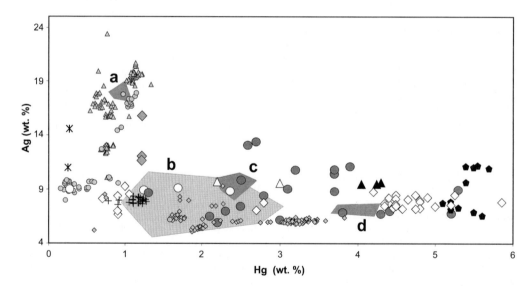

Fig. 5. Silver and mercury contents of gold analysed in the present and in previous studies. The data from the present study are shown with small grey symbols without differentiation between different reefs and goldfields. See Fig. 4 for this information.

Co, As and Te were also detected in some grains. Preliminary X-ray mapping and back-scattered electron imagery (Hayward, unpublished data) indicate that these elements are contained within the gold, or as inclusions below the resolution of the electron probe microanalyser (approximately 1 μm for X-ray maps and 0.02 μm for the backscattered electron images).

In Figure 5, the present data are plotted together with EPMA analyses from other sources (von Gehlen 1983; Oberthür & Saager 1986; Reid *et al.* 1988; Frimmel *et al.* 1993; Frimmel & Gartz 1997). Few of these earlier data are listed numerically in these publications, and values have been assigned to data points by graphical measurement of the chemical variation plots in the various published figures and where this was impossible, fields of data are indicated by shaded areas. The current and previously published data sets are in broad agreement with the generalized statements about Ag and Hg contents made above. However, the addition of the earlier data in most cases increases the scatter of data for individual reefs and goldfields. In addition to scatter caused by chemical heterogeneity of the gold, some must be expected because of the differences in experimental conditions applied during the collection of the different data sets. Also plotted on Figure 5 are data from analyses of gold from potential source region lithologies or analogous lithologies (von Gehlen 1983). The Barberton vein-type gold has low Hg and Ag contents in the middle of the range of values known for Witwatersrand gold. The Barberton gold is reasonably similar in composition to that of the Elsburg reefs in the Target goldfield, and the Kimberley reef at Evander. It would be premature to make any judgements about potential sources of gold on the basis of so few data.

From the gold composition data (Fig. 4), the variations of Ag and Hg content in individual thin sections (ΔAg and ΔHg, Fig. 6a), and in grains large enough for multiple analyses to have been made (δAg and δHg, Fig. 6b) have been calculated. The values of ΔAg and ΔHg indicate variability in gold composition on the centimetre scale, and as they represent individual thin sections, may represent minimum values for the variability in gold chemistry in the whole sample (see the discussion below of the examples from the Carbon Leader reef of the East Driefontein Mine). The level of variation within single thin sections can be high, approaching 5 wt% for Ag and 3 wt% for Hg, for some samples in the present study, and exceeding 5 wt% for Hg in one sample of VCR

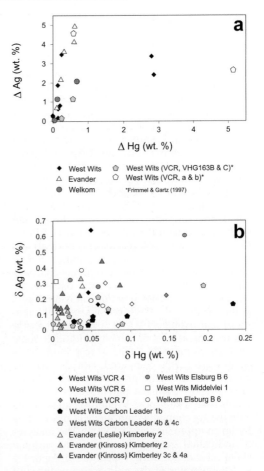

Fig. 6. (**a**) Variation in Ag and Hg contents of all analysed gold individual thin sections. (**b**) Variation in Ag and Hg contents of individual gold grains. The thin section number is indicated at the end of each entry in the legend.

from the Klerksdorp goldfield (Frimmel & Gartz 1997). In general, the values of ΔAg and ΔHg reflect the abundance of Ag and Hg in the gold from the same location. Samples from the Carbon Leader Reef of the West Wits goldfield have high ΔHg (approx. 2.5 wt%) and fall among the samples with the largest ΔAg (approx. 3.2 wt%), whereas the VCR at Klerksdorp (Frimmel & Gartz 1997) has ΔHg in excess of 5 wt%. Most VCR samples from the West Wits goldfield have comparatively lower variation (ΔAg <0.8 wt% and ΔHg <0.25 wt%). However, those from the West Wits VCR and Elsburg B reefs have some of the largest ΔAg values (1.8 and 3.5 wt% respectively).

Values for δAg and δHg which reflect

variability in gold composition on the 5–100 μm scale, are approximately an order of magnitude lower than those for ΔAg and ΔHg and fall mostly within 0.35 wt% for Ag and 0.1 wt% for Hg. Variation in the Ag and Hg contents within individual grains can, however, exceed 0.6 wt% Ag and 0.15 wt% Hg (Fig. 6b). The link between the variability in Ag and Hg contents and the concentrations of these elements present in gold from any given locality is less clear than that between concentration and ΔAg and ΔHg. Previously, it had been reported that gold in the CRG was homogenous at the scale of individual grains (Oberthür & Saager 1986; Reid *et al.* 1988; Frimmel *et al.* 1993; Frimmel & Gartz 1997). The variability measured in the present study is far greater than the precision of EPMA (2σ precision of ± 1% (relative) is normal) and is, therefore, interpreted to indicate real chemical variation within gold grains analysed. Unfortunately, there is very little data on the chemical heterogeneity within individual gold grains. Reid *et al.* (1988) considered a variability of approximately 0.4 wt% Ag and 0.6 wt% Hg to represent homogeneity, although some analyses within gold grains were poorer in Hg by at least 1 wt% compared with the average content for the same grain. Frimmel & Gartz (1997) report chemical inhomogeneity in gold grains from a late (post-Vredefort) quartz vein, with the Ag content within single grains varying by up to approximately 1.5 wt%. The analyses of individual gold grains reveal no rational chemical zonation with respect to Au, Ag or Hg and the margins of the gold grains or those of associated minerals.

There is a decrease in the amount of heterogeneity in gold composition from the regional scale to smaller scales of mine outcrop, to single thin section and even individual gold grains. Similar observations have been made on samples from smaller-scale studies: VCR at East and West Driefontein (West Wits) and from Vaal Reefs (Klerksdorp) (Saager 1969; Reid *et al.* 1988; Frimmel & Gartz 1997), Steyn Reef, St Helena Mine (Welkom), VCR, Western Deep Levels (West Wits), Carbon Leader, Blyvooruitzicht (West Wits) and VRC at Kloof (West Rand) (von Gehlen 1983), VCR, Elsburg, Gold Estates, 'C', MB3, MB4, Vaal, MB5, Livingston-Johnston and Commonage Reefs (Klerksdorp) (Utter 1979).

Detailed petrological and mineral chemical studies: the micrometre to hundred metre scales

Heterogeneity in composition between gold grains that are separated by less than a few hundred micrometres is observed in many, but not all, thin sections. The results from two VCR and two Carbon Leader locations from East Driefontein mine (West Wits) and two Kimberley reef locations from the Kinross Mine (Evander) are shown in Figure 7. These examples were chosen because the mineralized units represent horizons that have previously been interpreted as sedimentologically-controlled ore bodies (Krapez 1985; Engelbrecht *et al.* 1986; Tweedie 1986) and, in the cases of the VCR and Kimberley reefs, have been the subject of significant hydrothermal overprint. In most samples, the majority of analyses cluster into one or more compact groups with respect to Ag and Hg content. In the Carbon Leader and Kimberley reef samples there are also individual grains with compositions that deviate significantly from those of the main groups of analyses.

Gold composition is heterogeneous within most thin sections studied at scales between one or two centimetres and tens to hundreds of micrometres. The location of each occurrence of gold is shown on the thin sections in Figure 7, together with their Ag and Hg contents. Groups of analyses enclosed by black rings on these plots are from single grains. In the following discussion, an 'occurrence' of gold refers to a single isolated grain or localized group of grains within an area of one or two square millimetres. In almost all cases, petrographic observations demonstrate that a single occurrence comprises gold in a single paragenetic association (e.g. within fractures in pyrite, interstitial between quartz grains, within late-forming chlorite, overgrowing alteration products of uraninite, see Fig. 3).

Two of the VCR samples (Nos 4 & 5; Fig. 7a, b) were collected from stratigraphically adjacent depositional units within the same panel, at the East Driefontein Mine, West Wits goldfield, with a stratigraphic separation of 35 cm. The gold in each of these samples has distinct and non-overlapping Ag and Hg contents, with an average difference between each sample of approximately 1 wt% in Ag and 0.35 wt% in Hg (Fig. 8). Sample 7, from a panel in the VCR several hundred metres away forms another, distinct group in terms of Ag and Hg content (Fig. 8). Samples 1 and 4 from the Carbon Leader reef were collected from neighbouring panels and multiple thin sections were studied from each sample (1a, 1b & 4b, 4c; Fig. 7c, d). Most gold in these samples has approximately 6 wt% Ag and 3–3.4 wt% Hg. There are small groups of analyses from both sample locations with approximately 8.1–8.5 wt% Ag and 0.7 wt% Hg, and with approximately 6 wt% Ag and 2–2.5 wt% Hg. A similar pattern of variation in the gold composition is seen in some Kimberley reef samples (Fig. 7e, f). Gold in two samples from adjacent parts of the same panel mostly belongs to one range of compositions (approximately 19–20 wt% Ag and 1.1 wt% Hg), but a smaller group of gold analyses from both samples with lower Ag and Hg was identified, and also individual grains of gold whose composition fall outside the main ranges of Ag and Hg contents were observed. At scales between hundreds of metres and millimetres (i.e. between different panels within one mine and locations within individual thin sections), gold compositions are heterogeneous.

In VCR thin section 4 (East Driefontein mine, West Wits goldfield; Fig. 7a) the analyses of some of the individual gold grains form chemically distinct groups, whose internal chemical variability (0.2–0.6 wt% Ag and 0.5–0.7 wt% Hg) is significantly lower than that of the gold in the thin section as a whole and is, in some cases, comparable to the differences in Ag and Hg contents between grains. Individual gold grains have Ag and Hg contents that cluster into distinct groups. However, grains from the different occurrences overlap so that it is not certain whether the groupings from individual grains represent truly distinct compositions of these grains, or subsets of an indivisible range of compositions for the whole thin section. This uncertainty is greater in section 5, in which the ranges of composition for each occurrence and individual gold grains mutually overlap.

The Carbon Leader reef shows the greatest degree of compositional variation with respect to Ag and Hg for the data of the present study and also that of Frimmel & Gartz (1997). The heterogeneity in gold composition present on the scale of tens of kilometres (between the West Wits and Klerksdorp goldfields) continues down to scales from millimetres to micrometres. Figures 4, 5, 7c, d show the compositions of individual occurrences and grains of gold within the pairs of thin sections (1a, b and 4b, c) from each of two hand specimens examined in this study from different panels within the Carbon Leader reef of the East Driefontein Mine, West Wits goldfield. In thin section 1b (Fig. 7c), the analyses of a grain of gold of approximately $100 \times 25 \, \mu m$ size (Occurrence 2) indicate the maximum range of composition within a single grain in this thin section: $\delta Ag = 0.19 \, wt\%$, $\delta Hg = 0.13 \, wt\%$. Other single grains in this section have δAg and δHg values in the general ranges of 0.17–0.09 wt% and 0.13–0.09 wt% respectively. The gold in Occurrence 3 in the same thin section is within and overgrowing an altered uraninite grain. The various gold grains have similar ranges of Ag contents of 5.9–6.1 wt%, but highly variable Hg contents. The occurrences of gold labelled in Figure 7c have respective Hg contents of 2.8, 2.2 and 3.3 wt%, a total range of 1.1 wt%. A grain approximately 400 μm away from the altered uraninite (Occurrence 4), which overgrows a fragmented and corroded pyrite and is also associated with pyrrhotite has a similar Ag content and a Hg content of 3.1 wt%. In Occurrence 6, gold overgrows a corroded pyrite grain within quartz and muscovite. The Ag and Hg content of this gold are 6.01–6.1 wt% and 2.4–2.5 wt%, respectively. An isolated grain within the quartz and muscovite matrix and about 100 μm away has similar Ag values but a Hg content of 3.2 wt%. In Occurrences 8 and 9, as noted above, grains separated by 150–300 μm from other gold grains have radically different Ag and Hg contents. The gold in Occurrences 3, 6, 7, 8 and 9 is all within a few square millimetres and occurs with fine-grained, fragmented rounded pyrite, uraninite and zircon between large quartz clasts many millimetres across. Within this small area, there is a range of 3.35 wt% in Ag content and 2.7 wt% in Hg. In addition, the data demonstrate that significant differences in gold chemistry occur over distances of less than 100 μm. In Occurrences 1, 10, 11 and 12 single grains and/or groups of grains that appear paragenetically coeval have ranges of Ag and Hg contents of up to 0.23 wt% and 0.14 wt%, respectively. There is a similar range in composition within the larger single grains to that between closely-spaced groups of smaller grains. Thin section 1a, from the same hand specimen as 1b, has similar values for Hg but higher Ag values (6.3–6.5 wt%) than most grains from section 1b. Only two analyses from this section were possible, because of the fine grain size of the gold; thus it is not certain whether all the gold is relatively high in Ag, or whether other grains have compositions that would plot within the mass of data for section 1b.

In thin sections b and c from sample 4 the gold has a range of Ag contents between 5.8 and 6.4 wt%. In section 4c, although individual grains may differ in their Ag content by up to 0.6 wt%, the ranges of Ag contents of grains for which multiple analyses were possible is from

a

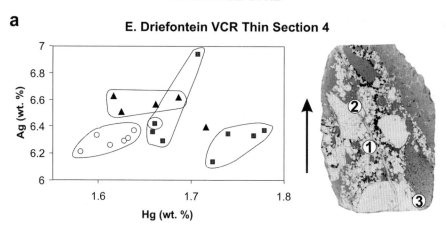

E. Driefontein VCR Thin Section 4

b

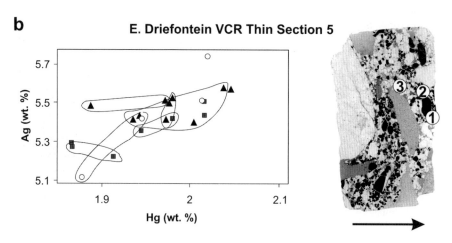

E. Driefontein VCR Thin Section 5

c

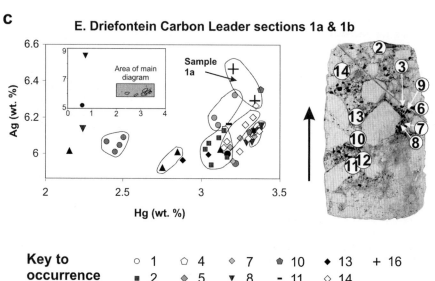

E. Driefontein Carbon Leader sections 1a & 1b

Key to occurrence numbers

○ 1   ⬠ 4   ◇ 7   ⬠ 10   ◆ 13   + 16
■ 2   ◆ 5   ▼ 8   - 11   ◇ 14
▲ 3   ● 6   ● 9   ● 12   ▽ 15

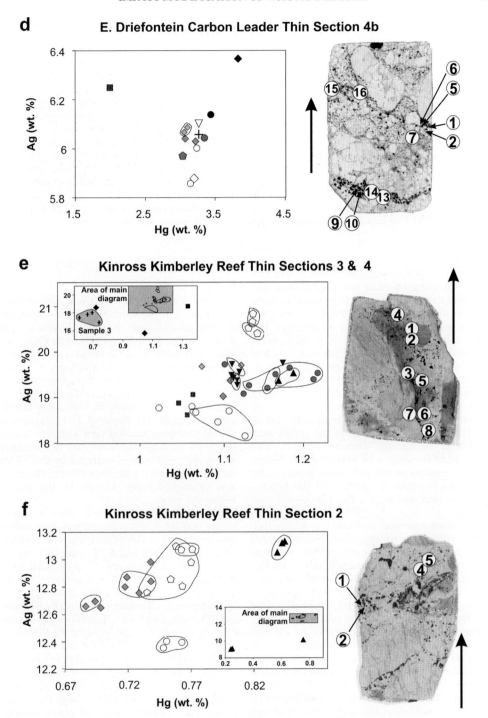

Fig. 7. Locations of gold occurrences in thin sections and the Ag and Hg contents for each occurrence. Black rings enclosing groups of points on the chemical plots indicate groups of analyses from single grains of gold. Occurrences of gold are marked on the images of the thin sections by circled numbers. The black arrow adjacent to the images of the thin sections indicates stratigraphically 'up' on the sample. Thin sections are all of the order of 4–5.5 cm in height.

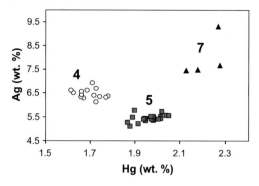

Fig. 8. Silver and mercury contents of Samples 4, 5 and 7 from the VCR of East Driefontein Mine, West Wits goldfield.

sample 2b has two grains that differ strongly from the main group of analyses; one with lower Ag and Hg and one with lower Ag. Looking in more detail at each sample, Occurrence 5 in thin section 4a has wide variation in Ag and Hg contents. Two grains only 75 μm apart have Ag contents that differ by 3.8 wt% and Hg by only 0.06 wt%. Another grain approximately 550 μm away has a Hg content approximately 0.4 wt% lower than these two grains and Ag contents that differ by approximately 1 wt% and 2.9 wt%. As with other reefs, there is greater similarity of chemical composition within individual gold grains and (usually) within single occurrences of gold grains than there is between different grains and occurrences.

0.04 wt% up to 0.27 wt%, or almost half of this range. In common with other gold from the Carbon Leader reef, it is the Hg content in these thin sections that enables the best discrimination between different gold occurrences. The total range of Hg content in section 4c is 0.54 wt% (excluding the analyses with very low Hg contents), whereas the ranges of Hg content for individual grains is 0.07–0.19 wt%. The relatively narrow ranges of Hg content of individual gold grains compared with the overall range within the whole thin section suggest the possibility that the compositions of the grains may represent chemically discrete objects. The question of the degree of heterogeneity within possible sources of the gold must, however, be addressed in order to understand the heterogeneities observed within the CRG gold (see below).

The final example is from the Kimberley (UK9) reef of the Kinross mine, Evander goldfield, the closest operating mine to the presently preserved extent of the Bushveld intrusion. The variation in gold compositions mirrors that observed for the VCR and Carbon Leader reefs of the West Wits goldfield (Fig. 7e, f). The data are from three thin sections taken from three samples, two from an underground site (3c and 4a) and one from a different location, the closest possible to the Bushveld intrusion (2b). Most data from each of the three samples form fairly compact groups in terms of Ag and Hg content, and each sample is distinct. Each of the three groups has outliers, as also seen in the VCR and Carbon Leader samples above. One gold grain from sample 4a has Ag and Hg closely similar to gold in sample 3c. An analysis from the same occurrence (5) has over 2 wt% less Ag than most other gold in that sample. Occurrence 2 in

Composition of chlorite associated with gold

Chlorite from all mineralogical and textural associations in the auriferous thin sections and in those samples from the footwalls and hanging walls was analysed by EPMA. Chemical variation in chlorite is useful in two respects. First, it gives an indication of temperature of formation from differences in the occupancy of tetrahedral sites by aluminium (Kranidiotis & MacLean 1987; Cathelineau 1988; Zang & Fyfe 1995). Second, the scale(s) and range(s) of chlorite composition, together with petrological observations, may help to indicate the degree to which composition was buffered by that of the surrounding rock, or whether chlorite chemistry was influenced more strongly by the fluid(s) within the rock.

All the chlorites analysed occur in assemblages that are Al-saturated (that is, they include muscovite and/or chloritoid). The compositions of chlorite associated with gold vary significantly. Chlorite from the VCR of East Driefontein Mine falls within the centre of the compositional range of clinochlore, compared with the more aluminous and slightly less iron-rich compositions of clinochlores from the Elsburg B reef of Target Mine or the Kimberley reef of Kinross Mine. The VCR chlorite and that of the Target Mine have relatively restricted ranges of Fe/(Fe + Mg), and variable Al(IV) (Fig. 9). However, chlorite from the Kinross Mine has a range of Fe/(Fe + Mg) that equals that of the chlorite from all other sampled locations, and has the same range of Al(IV) contents. A limited number of analyses from each goldfield plot near the compositional range of sudoite; these were reported previously from the CRG (Frimmel *et al.* 1993; Zhou & Phillips

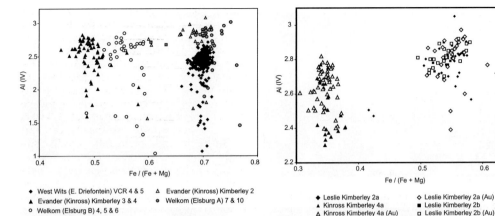

Fig. 9. Composition of chlorite formed contemporaneously with late gold crystallization, from samples from the West Wits, Evander and Welkom goldfields. The thin section number is indicated at the end of each entry in the legend.

Fig. 10. Compositions of chlorite formed contemporaneously with gold crystallization (labelled 'Au' in the legend) and chlorite cross-cut by this chlorite from the Kimberley reef, Kinross Mine, Evander goldfield. The thin section number is indicated at the end of each entry in the legend.

1994). In most hand specimens, there is little variation in the composition of the chlorite, suggesting homogeneity at the 10–30 cm scale.

At the centimetre scale there is almost no variation in chlorite composition. At most locations sampled in this study, chlorite that formed at the same time as gold is chemically indistinguishable from other, probably earlier, generations of chlorite. For example, three samples from the Kimberley reef of the Evander goldfield, one from Kinross Mine and two from the Leslie Mine show complete overlap between chlorites from all textural associations in the analysed thin sections, whether in chlorite-replaced mineral grains, veins or microshears, or late-formed chlorite associated with gold. In many thin sections, all generations and associations of chlorite are compositionally indistinguishable (Fig. 10). The observations presented in Figure 10 are repeated across the CRG. The variability of chlorite composition on the mine outcrop scale (tens of centimetres) and above, coupled with

the homogeneity of chlorite composition on the millimetre–centimetre scale indicate strong control on chlorite compositions by the surrounding rock, i.e. a rock-buffered system. In each of the sampled locations, groups of analyses of single chlorite grains at different distances to pyrite and pyrrhotite indicate that there is no correlation in any of the current samples between the Fe/(Fe + Mg) ratio of the chlorites measured and the proximity of the chlorite to iron sulphides, such as that observed in Frimmel (1997a).

The temperatures derived for the current data set using tetrahedral aluminium occupancy from the formula of Zang & Fyfe (1995) give formation temperatures for the chlorite that cluster around 230–315 °C. These temperatures are consistent with previous calculations of formation temperature derived from chlorite chemistry (e.g. Frimmel 1997b; Gartz & Frimmel 1999). The temperatures calculated from chlorite chemistry (Table 2) incorporate a

Table 2. *Temperatures (in degrees celsius) of chlorite formation derived from EPMA data using various published thermometric calculation functions**

Goldfield	Zang & Fyfe (1995)	Kranidiotis & MacClean (1987)	Cathlineau (1988)
Evander	173–323	218–387	221–466
Welkom	116–314	146–384	137–454
W. Wits	110–284	144–357	141–406

* The lower range of values, just above 110 °C, lie outside the stability range for tri-octahedral chlorite. The main cluster of data points give lower limits of temperature of approximately 145 °C (Zang & Fyfe 1995; Kranidiotis & MacClean 1987) and 195 °C (Cathlineau 1988).

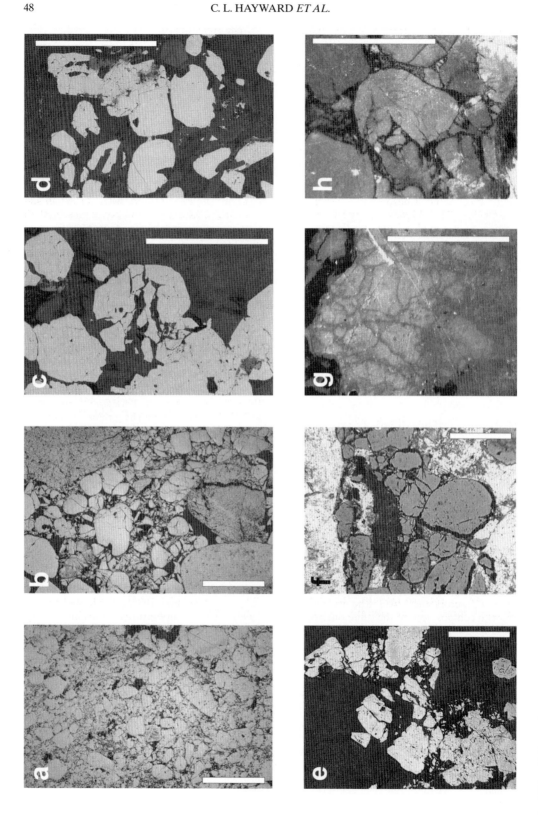

correction made to account for the relatively high iron contents (Zang & Fyfe 1995) of the Witwatersrand chlorites (Fe/(Fe + Mg) generally greater than 0.58). Other chlorite thermometers give unrealistically high temperatures, for example the thermometer of Cathelineau (1988) gives apparent formation temperatures that exceed 450 °C. Temperatures significantly in excess of 400 °C are not supported by the observed chlorite–quartz–muscovite ± chloritoid ± pyrophyllite metamorphic assemblage of the CRG (Phillips 1987; Frimmel *et al.* 1993; Phillips & Law 1994; Gartz & Frimmel 1999), in which there is no widespread development of andalusite. The highest temperatures are recorded for samples from the Kinross mine, Evander, from East Driefontein (Middelvlei reef), West Wits, and from the Target mine, Welkom. Most chlorites associated with gold in these locations indicate temperatures in the range 260–320 °C. The results from the VCR at East Driefontein fall mostly into the range 230–280 °C. All locations have groups of analyses that correspond to lower temperatures of around 210–230 °C. The temperatures do not correlate with proximity of sample locations to either the Bushveld Igneous Complex or the Vredefort Dome.

Brittle mineral deformation in the CRG: post-deformation paragenesis and crystallization of gold

The CRG has been affected by many brittle deformation events during its long history, as is evident from the numerous cross-cutting and variably healed fractures and fluid inclusion trains in quartz clasts, and larger scale brittle shears and fractures at various scales (see also Barnicoat *et al.* 1997; Jolley *et al.* 2004). A notable feature of the CRG samples studied is the widespread evidence for brittle deformation very late in the paragenetic sequence that affected much of the mineral assemblage, including detrital quartz, zircon, pyrite, chromite, uraninite and arsenopyrite clasts, and

metamorphic quartz, pyrite and chloritoid (Figs 3 & 11). The spaces between the fragments are filled most commonly with quartz and/or chlorite but, in localized areas of some samples, fragmented pyrite is 'cemented' by significant later pyrite overgrowth (Fig. 11), in addition to quartz and/or chlorite. Mineral deformation was observed in the form of fracturing and fragmentation, sometimes very severe, of quartz pebbles and smaller clasts in quartzite, some generations of quartz overgrowths, most of the compact, rounded pyrite, a significant proportion of the euhedral pyrite and clasts undergoing pyrite-replacement, chromite, arsenopyrite, zircon and uraninite. This late deformation is present at all locations sampled in the present study and appears to affect the entire mineral assemblage that was present prior to the latest episodes of mineral growth within the samples studied. The intensity of the late deformation is not related to relative proximity, or otherwise, of grains to small- to medium-scale brittle deformation features such as faults or brittle shear zones. The fragments are distributed in thin section in such a way that indicates that the shattering was multidirectional. No consistent sense of shear on fragments, or consistent spatial relationship with intergranular microfaults was noted in thin section. This contrasts with the observations of predominantly sub-horizontal fracturing made by Barnicoat *et al.* (1997). Furthermore, significant fracturing at high angles to bedding was visible within many of the thin sections examined in the present study.

The severity of the late deformation and the spatial distribution of fragments of the deformed grains appear to be related to the mineralogy and sizes of the grains prior to deformation and to local fabric, specifically the relative positions of nearest and in places, the second- or third-nearest neighbour grains. For example in conglomerates that comprise coarse quartz pebbles in mutual contact with an interstitial medium/fine-grained quartz–phyllosilicate matrix, the quartz pebbles are fractured, but the matrix quartz grains are less severely affected. Small grains (quartz, sulphide, zircon,

Fig. 11. Examples of brittle deformation of minerals from the different reefs and goldfields. Scale bar in all images is 500 μm long. (**a**) & (**b**) Elsburg reef, Target Mine (Welkom). Pyrite with quartz and late chlorite. Both thin sections are from the same sample, and yet show different degrees of post-deformation pyrite crystallization (see text for discussion). Reflected light; (**c**) VCR, E. Driefontein Mine (W. Wits), Pyrite and quartz. Reflected + transmitted light. (**d**) E. Driefontein Mine, Meddelvlei reef (W. Wits). Pyrite (rounded and subhedral and zircon (below scale bar) plus quartz and chloritoid. Reflected + transmitted light. (**e**) Venterskroon, Kimberley reef. Pyrite and quartz. Reflected light. (**f**) Venterskroon, Kimberley reef. Pyrite and zircons (upper centre) plus quartz. Reflected + transmitted light. (**g**) Carbon Leader reef, E. Driefontein Mine (W. Wits). Quartz. Cathodoluminescence. (**h**) Leslie Mine, Kimberley reef (Evander). Quartz. Cathodoluminescence.

etc.) adjacent to large quartz clasts are fragmented against these clasts. In pyrite-rich layers,
the pyrites and other minerals (especially
arsenopyrite, chromite and zircon) are
commonly very severely fragmented (Fig. 11).
The degree of brittle deformation is far less
severe where these minerals occur within or
adjacent to fine-grained quartz grains, especially
where they occur together with phyllosilicates,
and are not in these types of textural associations (i.e. where they do not occur within
concentrations of similar grains or adjacent to
large quartz clasts); such mineral grains may be
essentially undeformed. The brittle deformation
is commonly highly localized, with, for example,
almost pulverized zircon grains within half a
millimetre of intact detrital zircon grains.

The effects of brittle deformation are
observed to be relatively rare in the latest
minerals to form within the CRG. These are late
quartz cements, chlorite, and minor pyrrhotite
and pyrite, and in some locations also calcite,
chalcopyrite and bravoite. A number of
sequential growth and/or dissolution events that
involve quartz, chlorite, pyrrhotite and pyrite
are observed in most auriferous samples
studied. There is a consistent relationship
between the spatial occurrence of gold with
respect to the late brittle deformation and these
sequences of post-deformation mineral growth
and dissolution. Gold occurs very commonly
within fractures, or among the resulting fragments. Approximately 70% of the gold
observed from all goldfields studied occurs in
this textural setting and much of this gold is
associated with the latest and least deformed, or
undeformed chlorite, pyrrhotite and quartz
growth. Other gold appears related to fractures
and other features that are not obviously
connected with the latest fabrics, reflecting
earlier events that influenced gold distribution.
These observations indicate the existence of a
temporal and paragenetic relationship between
the late brittle deformation, gold crystallization
and late quartz, chlorite and pyrrhotite formation, which occurs throughout the entire
geographic and stratigraphic ranges sampled
(noting that sampling locations were chosen to
be representative of the entire CRG). Gold has
been mobilized after widespread brittle deformation, in fluids that caused: variable dissolution of pyrite and deposition of pyrrhotite; no
pyrite dissolution; and in other locations, pyrite
growth. The variability in the sequences of late
mineral growth and dissolution that occur
within the CRG suggests that there were local
differences in the conditions experienced
following the brittle deformation, and that
uniform conditions of post-deformation metamorphism are not applicable in detail throughout the Witwatersrand Basin.

Discussion

Gold chemistry shows a strong geographically-
influenced variation, with gold from specific
goldfields seemingly displaying a broad
chemical 'signature' with respect to Ag and Hg
contents. Individual reefs from different
locations have different gold compositions
(compare the VCR from the Placer Dome and
East Driefontein mines and Elsburg reef from
the Placer Dome and Target mines, Fig. 4).
However, there are important examples of gold
from certain reefs that have compositions that
are atypical of such goldfield 'signatures' (for
example West Wits Elsburg reef and VCR gold
with Evander-type 'signatures' and West Wits
Carbon Leader reef gold with Welkom-type
'signatures'). No systematic variation in gold or
chlorite chemistry with respect to proximity to
the Bushveld Igneous Complex has been
observed in the present study, nor has the proximity of the intrusion to the Evander goldfield
caused homogenization across that goldfield.
The original extent of the Bushveld Complex
remains unclear, and the related possible
thermal influence on the CRG and its gold
similarly obscure. Presumed outliers of the
Bushveld Complex north of the Vredefort
Dome (Losberg, Coetzee & Kruger 1989) and
post-Transvaal sills within the Potchefstroom
syncline surrounding the Vredefort dome
(Cawthorn *et al.* 1981) indicate that the thermal
effects of the intrusions may not have been
localized only within the northeastern portion
of the remaining CRG metasediments.

The regional chemical heterogeneity of gold
is most simply explained by the incorporation of
compositionally varied gold from eroded source
regions during the deposition of the CRG. It is
premature to invoke the similarities between
gold compositions of Archaean greenstone-
derived gold and some CRG gold too strongly
(Fig. 5), because of the paucity of data on gold
from potential source region lithologies. It is
difficult to explain the observed variation in
gold compositions at scales between regional
and sub-millimetre through models that argue
for the introduction of gold into the CRG by
regional circulation of hydrothermal and/or
metamorphic fluids. Gold mineralization mechanisms in which large volumes of metamorphic
or extra-basinally derived fluids carry gold
through major stratigraphically-controlled
pathways have previously been advocated

(Phillips 1987; Phillips & Law 1994; Barnicoat *et al.* 1997; Jolley *et al.* 2004). To account for the presence of gold of different compositions within single mine outcrops and thin sections, chemically different fluids would need to have deposited gold, presumably at different times, within the same volume of rock. It is unlikely that multiple episodes of fluid ingress would all have been gold-forming, and that they could have occured without influencing the composition of gold already present. Furthermore, since the same types of chemical heterogeneities are observed in all auriferous samples included in the present study, this would mean that multiple auriferous fluids would have been required in all parts of the CRG, which seems improbable.

Chemical homogeneity of gold within the CRG can be demonstrated in most of the samples only at scales smaller than centimetres and, in many examples at less than tens to hundreds of micrometres, even in reefs such as the VCR, where significant hydrothermal activity is widely reported to have occurred (Zhao *et al.* 1994; Frimmel & Gartz 1997). The observed variation in gold chemistry suggests that homogenization of gold composition occurred over distances of, at most, only millimetres and more often at a scale comparable to that of the individual gold grains themselves. The present EPMA data indicate the presence of chemical heterogeneity within individual gold grains, which is at variance with earlier findings (Saager 1969; Utter 1979; von Gehlen 1983; Frimmel & Gartz 1997, 1983), in which individual grains of gold were reported to be chemically homogenous. This observation is problematic in the context of experimental studies on the rates of chemical homogenization of gold that indicate that periods less than thousands to tens of thousands of years are sufficient to homogenize gold grains of sizes similar to those within the CRG at temperatures of 300 °C and less than 10 Ma at 200 °C (Czamanske *et al.* 1973; Frimmel *et al.* 1993). Heterogeneity of Ag and Hg in gold has previously been reported only from post-Vredefort quartz veins (Frimmel 1997a). The chemical heterogeneity measured within individual gold grains implies that metamorphism of detrital gold was neither sufficiently prolonged nor sufficiently high in temperature to cause complete chemical re-equilibration. An average geotherm of 15–20 °C km^{-1}, calculated for the Witwatersrand Basin between 2.60 and 2.05 Ga (Gibson & Jones 2002) indicates that the CRG, which reached maximum burial depths of approximately 7 km, did not reach temperatures much in excess of 140–170 °C prior to the intrusion of the

Bushveld Complex and/or the regional metamorphism that followed the Vredefort impact. On the basis of these data, achievement of the widely reported temperatures in excess of 300 °C (e.g. Frimmel 1997a; Drennan *et al.* 1999; Frimmel & Gartz 1997) outside of a transient thermal event would have required unrealistic burial depths of around 15 km for the CRG. We conclude from this that the optimal conditions for the development of temperatures of up to 320 °C calculated from chlorite that crystallized contemporaneously with a significant proportion of the CRG gold would have been met when the CRG rocks were at their deepest levels of burial (late Transvaal) and then they were affected by transient thermal events such as the Bushveld magmatism and/or the Vredefort impact.

However, the preservation of regional to sub-millimetre variation (i.e. at scales greater than those of single grains) in gold compositions throughout the CRG suggests that in many locations, chemical homogenization at the centimetre scale and above has never occurred during the post-depositional history of the CRG. This in turn implies that transportation of gold during metamorphic remobilization events, evident from the hydrothermal textures of most of the gold grains, has never been over distances greater than millimetres to centimetres in most of the volume of the CRG (noting the exceptions of gold within late-stage veins; Frimmel 1997a). This is in agreement with previous findings (e.g. Frimmel & Gartz 1997; Frimmel & Minter 2002). For some gold occurrences, for example in the Carbon Leader reef (Table 1), the location of gold grains between detrital grains of quartz, together with the chemical diversity of populations of gold grains that occur in similar textural settings within the samples, may indicate *in situ* recrystallization of an original detrital grain (Fig. 3c) rather than a remobilization of gold particles. Such conclusions fit well with both the observed chemical variations within chlorite associated with gold, and the common examples of localized fluid flow encountered within the studied CRG samples (e.g. centimetre-scale localization of pyrite dissolution and/or overgrowth, highly localized distribution of alteration surrounding small-scale bedding-parallel shear zones), and similar observations made in other parts of the CRG (e.g. Reimold *et al.* 1999). Chemical diversity of gold grain populations, together with the chemical homogeneity of different generations of chlorite within the same, localized, volumes of rock both argue for small volumes of rock-buffered fluids and highly restricted fluid flow pathways. This is contrary to models involving large quantities of fluid circulating widely within the CRG and transporting major quantities of gold in solution, as for example proposed by Stevens *et al.* (1997) and Phillips *et al.* (1990). Even where there is evidence of brittle shearing, the gold in the shear zones appears not to be chemically homogenized (e.g. Fig. 7a). This observation is also at odds with models that advocate major gold (re)mobilization along fracture networks (e.g. Jolley *et al.* 2004), as gold deposition from fluids moving in such networks would lead to a higher degree of homogeneity of gold composition at the micrometre up to metres scales than is evident from any of the samples analysed in the present study and in previous ones that included *in situ* microanalysis of gold (Figs 4, 5 & 6; Saager 1969; Oberthür & Saager 1986; Frimmel *et al.* 1993; Frimmel & Gartz 1997).

Analyses of gold from the present study and previous work (von Gehlen 1983; Oberthür & Saager 1986; Reid *et al.* 1988; Frimmel & Gartz 1997) show that the gold chemistry of the VCR is the most diverse of all Witwatersrand reefs. The Ag and Hg contents of VCR gold overlap with the analytical data from all other reefs and all goldfields (Fig. 5) and appear to support the interpretation of the VCR as an erosional feature derived from underlying CRG units (McCarthy 1994). The VCRs compositionally diverse gold is reflected in the chemical diversity derived from varied source regions during erosion of earlier auriferous CRG units. These findings are contrary to those of Frimmel (1997a), who reported relatively restricted chemical variation of VCR gold because of greater fluid–rock interaction in this reef compared with other reefs. The data from this study indicate that the degree of variation in Au, Ag and Hg contents of gold grains in the VCR samples is not markedly lower than those observed in the Kimberley reef of the Evander (Hg) goldfield or for samples of Elsburg reef from Target Mine (Hg and Ag).

The association between gold and pyrrhotite, and the corrosion of pyrite in the Kimberley reef of the Evander goldfield and the Carbon Leader reef of the West Wits goldfield, suggest the possibility of desulphidation of pyrite, forming pyrrhotite during prograde metamorphism (Craig & Vokes 1993). The transportation of gold as a sulphide complex is supported by fluid inclusion compositions and experimental data on gold solubility (Boer *et al.* 1995; Stevens *et al.* 1997; Robb & Meyer 1991).

The Vredefort impact event is widely acknowledged as having played a role in the history of the gold mineralization within the CRG (Robb *et al.* 1997; Reimold *et al.* 1999). However, the impact has not been considered in detail in many of the models of CRG gold mineralization (Robb & Meyer 1995; Barnicoat *et al.* 1997; Phillips & Law 1994, 2000), although the preservation of the auriferous metasedimentary rocks of the CRG has been attributed to the deformation caused during and immediately after the impact (the central dome and rim syncline; McCarthy *et al.* 1990; Henkel & Reimold 1998). The 300 km diameter final crater suggested by Therriault *et al.* (1997) would have encompassed the locations of all goldfields, with the possible exception of the Evander goldfield (Fig. 12). Thus, the goldfields would have lain within the area covered by the crater (Fig. 12), and beneath a blanket of ejecta and collapse breccia, with those areas nearer to the centre of the impact structure possibly also lying under impact melt. The almost universal presence within the CRG of brittle deformation of mineral grains very late in the paragenetic sequence, the consistent expression of the deformation in thin section, and the apparent lack of consistent associated directional deformation such as shearing and independence from localized features (such as faults and brittle shear zones) all point towards a single, regional-scale event having caused the deformation. Furthermore, since the deformation affects all but the latest mineral growth it probably occurred relatively late in the paragenetic sequence, but prior to the final thermal pulse. The late-stage brittle deformation, if caused by the Vredefort impact as indicated by our detailed petrographic studies, provides a time marker within the CRG that is far more widespread than the pseudotachylitic breccias that are used within the CRG to distinguish between pre- and post-impact mineral growth and deformation events (Reimold *et al.* 1999). This is consistent with the impact scenario in which catastrophic deformation is followed by thermal relaxation of shock-heated rocks and impact melts (Gibson *et al.* 1998). Finally, the occurrence of almost three quarters of the gold observed in the present study within brittle deformation features that formed late in the paragenetic sequence, and in association with relatively undeformed chlorite and pyrrhotite, *inter alia* indicates that some gold deposition post-dated the last major deformation episode(s) affecting the CRG.

Shock deformation, in the form of shatter cones, is seen today at a maximum of approximately 60 km from the centre of the Vredefort Dome (Wieland *et al.* 2003; Fig. 12). The impact would have been experienced as an elastic wave

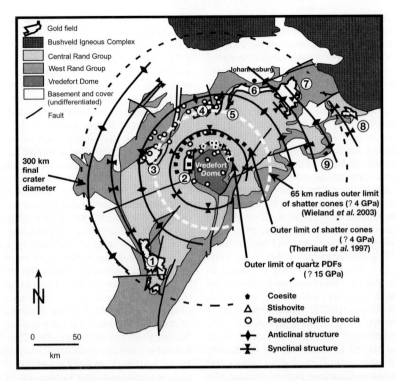

Fig. 12. Location of the goldfields with respect to the known distribution of impact indicators. The goldfields are numbered: 1, Welkom; 2, Venterskroon; 3, Klerksdorp; 4, West Wits; 5, West Rand; 6, Central Rand; 7, East Rand; 8, Evander; 9, South Rand. Locations of anticlinal and synclinal features from McCarthy *et al.* (1990); pseudotachylitic breccia occurrences, outer limits of shatter cones and planar deformation features in quartz, and the location of a 300 km diameter final crater from Therriault *et al.* (1997); maximum radius from the centre of the Vredefort Dome at which shatter cones were identified in the northern Vredefort collar (represented by a circle at this radius of 65 km), Wieland *et al.* (2003); locations of coesite and stishovite from Gibson (2002).

at the radial distances of the goldfields (e.g. Melosh 1989). Modelling of the Vredefort impact (Therriault *et al.* 1997; Henkel & Reimold 1998; Turtle & Pierazzo 1998; Turtle *et al.* 2003) and of the Sudbury impact in Canada (Ivanov & Deutsch 1999) indicates that shock pressures of 2 GPa would have extended as far as 70->100 km from the centre of the impact structure. However, the calculations of Therriault *et al.* (1997) assumed shatter cones at a maximum radius of only 38 km from the impact centre, incorporating the known distribution of shatter cones at that time. The later extension of the radial distribution of shatter cones (Wieland *et al.* 2003) may well have implications for the final size of the Vredefort crater. For conditions most relevant to the calculated size and speed of the Vredefort impactor in various numerical simulations (approximately 10-20 km diameter and 20 kms⁻¹), it is reasonable to presume that the relatively low pressures required to cause

brittle deformation in quartz, sulphides and zircon (of the order of tens to hundreds of MPa; Melosh 1989) would have extended radially to distances significantly in excess of 100 km, thus encompassing the locations within the CRG in which the bulk of the goldfields are now situated.

Computer simulations of Vredefort-scale impacts (Turtle & Pierazzo 1998; Turtle *et al.* 2003) do not yet extend beyond the collapse of the transient crater. Thus, the final distribution of shock pressure and post-shock temperature, beneath the impact basin is not yet constrained by such numerical modelling. The CRG in the areas now exploited for gold would have experienced multiple episodes of brittle deformation (other than those that pre-date the impact), not only from the distal effects of the impact shock wave, but also from the faulting and thrusting that resulted from the collapse of the crater walls and rebound of the crater floor, and later modification and adjustment of the rock volume below and around the final crater structure.

The large number of different mineralogical associations of gold in various locations sampled in this study (for a sub-set of the total sample suite, see Table 1 and Fig. 3) indicate that, rather than a single fluid common to all gold in the CRG, there were local differences in the conditions during gold remobilization/recrystallization. This is supported by the variability of gold chemistry among closely-spaced populations of grains and the uniformity of chlorite composition, which indicates strong rock buffering of the fluids. Thus, the data argue against large volumes of fluids, and strongly suggest that those fluids that were present were unable to transport gold efficiently and/or that they were restricted in their ability to travel within their host rocks. This observation holds even for units such as the VCR, which is widely regarded as having been the reef most severely affected by hydrothermal activity (Zhao *et al.* 1994; Frimmel & Gartz 1997). Rock-buffering of fluids within the VCR has been reported also by Frimmel & Gartz (1997), and Gartz & Frimmel (1999) have argued for there having been only relatively minor quantities of fluid associated with gold remobilization events within the VCR. In the CRG reefs, a similar scenario is proposed, despite the ubiquitous microscopic fracturing (Frimmel & Gartz 1997; this study). If large volumes of meteoric water did penetrate the fractured impact crater basement to the levels occupied by the CRG, the evidence suggests that they must have been channelled in a network of larger fractures, brittle shear zones and other larger-scale deformation features. An alternative explanation is that immediately after the impact only small volumes of fluids managed to penetrate to more than the 5 km depth at which the reefs were located (McCarthy *et al.* 1990; Gibson *et al.* 1998; Henkel & Reimold 1998). Fluid volumes derived internally from dehydration of the CRG are also likely to have been low, given that the rocks had already experienced dehydration during the slightly higher-grade pre-impact peak metamorphic event. It seems, therefore, that the post-Vredefort hydrothermal activity within the CRG was relatively fluid-poor.

Conclusions

A variety of data from both this and previous studies indicate that the presently observed disposition of the gold in the CRG is the result of very limited mobilization of gold in small volumes of fluids that were rock-buffered within

small volumes of rock, and not the result of significant gold mobilization in voluminous, widely circulating hydrothermal fluids, as proposed by proponents of the hydrothermal model. The chemical heterogeneity of gold at scales from regional down to hundreds of micrometres, and of associated chlorite from tens of kilometres down to tens of centimetres, and chemical homogeneity at the centimetre scale indicate that hydrothermal remobilization and chemical homogenization of gold occurred only over small distances (from centimetres to less than a millimetre) throughout most of the CRG during its post-depositional history. The present data argue for the detrital incorporation of gold grains into the CRG sediments, with chemical diversity occurring because of primary heterogeneity in gold that was inherited from the source regions, rather than introduction of gold in large-scale hydrothermal fluid circulations. Although fluids have clearly moved through the extensive fracture networks observed within the CRG at the macroscopic to microscopic scales, these have not mobilized the majority of the gold present. The Vredefort impact is considered the most likely cause of the latest brittle deformation, which is observed almost universally in the samples of the CRG included in the present study. The final crater, currently estimated at around 300 km in diameter (Therriault *et al.* 1997; Henkel & Reimold 1998), would have included most of the volume of the CRG now being mined for gold. The post-impact regional metamorphism caused minor remobilization of gold throughout the CRG, but did not initiate a major gold-transporting hydrothermal event.

We are grateful to the personnel of the mines in the Witwatersrand Basin included in this study for permission to sample on their properties, for their logistical and geological support. Thin section preparation was carried out at the Council for Geoscience, Pretoria with the assistance of Odette Smith and Thinus Cloeter. Funding for the work has been gratefully received from the National Research Foundation of South Africa and the Research Office of the University of the Witwatersrand. We thank Dave Plant, University of Manchester, for supplying standard materials for EPMA. The useful comments made by reviewers H. Frimmel, I. McDonald and G. Stevens are gratefully acknowledged. This is the University of the Witwatersrand Impact Cratering Research Group Contribution No. 78.

References

ARMSTRONG, R.A., COMPSTON, W., RETIEF, E.A., WILLIAMS, I.S. & WLEKE, H.J. 1991. Zircon ion-microprobe studies bearing on the age and evolution of the Witwatersrand triad. *Precambrian Research*, **53**, 243–266.

BARNICOAT, A.C., HENDERSON, I.H.C., KNIPE, R.J., ET AL. 1997. Hydrothermal gold mineralisation in the Witwatersrand Basin. *Nature*, **386**, 820–824.

BARTON, E.S., COMPSTON, W., WILLIAMS, I.S., BRISTOW, J.W, HALLBAUER, D.K. & SMITH, C. 1989. Provenance ages for the Witwatersrand Supergroup and Ventersdorp Contact Reef: constraints from ion microbe U–Pb ages of detrital zircons. *Economic Geology*, **84**, 2012–2019.

BOER, R.H., REIMOLD, W.U. & KESLER, S.E. 1995. Conditions of gold remobilisation in the Ventersdorp Contact Reef, Witwatersrand Basin. *Abstracts, Centennial Geocongress*, Geological Society of South Africa, Johannesburg, **2**, 696.

CATHELINEAU, M. 1988. Cation site occupancy in chlorites and illites as a function of temperature. *Clay Minerals*, **23**, 471–485.

CATUNEAU, O. 2001. Flexural partitioning of the late Archaean Witwatersrand foreland system, South Africa. *Sedimentary Geology*, **141–142**, 95–112.

CAWTHORN, R.G., DAVIES, G., CLUBLEY-ARMSTRONG, A. & McCARTHY, T.S. 1981. Sills associated with the Bushveld Complex, South Africa: an estimate of parental magma composition. *Lithos*, **14**, 1–15.

COETZEE, H. & KRUGER, F.J. 1989. The geochronology and Sr- and Pb-isotope geochemistry of the Losberg Complex and the southern limit of the Bushveld Complex magmatism. *South African Journal of Geology*, **92**, 37–41.

CRAIG, J.R. & VOKES, F.M. 1993. The metamorphism of pyrite and pyritic ores: an overview. *Mineralogical Magazine*, **57**, 3–18.

CZAMANSKE, G.K., DESBOROUGH, G.A. & GOFF, F.E. 1973. Annealing history limits for inhomogenous, native gold grains as determined from Au–Ag diffusion rates. *Economic Geology*, **68**, 1275–1288.

DRENNAN, G.R., BOIRON, M.-C., CATHLINEAU, M. & ROBB, L.J. 1999. Characteristics of post-depositional fluids in the Witwatersrand Basin. *Mineralogy and Petrology*, **66**, 83–108.

DRESSLER, B.O. & REIMOLD, W.U. 2004. Order or chaos? Origin and mode of emplacement of breccias in floors of large impact structures. *Earth-Science Reviews*, **67**, 1–54.

ELS, B.G. 1991. Placer formation during progradational fluvial degredation: the late Archaen Middelvlei gold placer, Witwatersrand, South Africa. *Economic Geology*, **86**, 261–277.

ENGELBRECHT, C.J., BAUMBACH, G.W.S., MATTHYSEN, J.L. & FLETCHER, P. 1986. The West Wits Line. *In*: ANHAEUSSER, C.A. & MASKE, S. (eds) *Mineral Deposits of Southern Africa, Volume 1*. Geological Society of South Africa, Johannesburg, 599–648.

ENGLAND, G.L., RASMUSSEN, B., KRAPEZ, B. & GROVES, D.I. 2002. Palaeoenvironmental significance of rounded pyrite in siliciclastic sequences of the Late Archaean Witwatersrand Basin: oxygen-deficient atmosphere or hydrothermal evolution? *Sedimentology*, **49**, 1122–1156.

ERIKSSON, P.G., ALTERMANN, W., CATUNEANU, O., VAN DER MERWE, R. & BIMBY, A.J. 2001. Major influences on the evolution of the 2.67–2.1 Ga Transvaal basin, Kapvaal craton. *Sedimentary Geology*, **141–142**, 205–231.

FEATHER, C. & KOEN, G.M. 1975. The mineralogy of the Witwatersrand reefs. *Minerals Science and Engineering*, **7**, 189–224.

FOYA, S.N., REIMOLD, W.U., PRZYBYLOWICZ, W.J. & GIBSON, R.L. 1999. PIXE microanalysis of gold–pyrite associations from the Kimberley Reefs, Witwatersrand Basin, South Africa. *Proceedings of ICNMTA-98, Nuclear Instruments and Methods in Physics Research B*, **158**, 588–592.

FRIMMEL, H.E. 1994. Metamorphism of Witwatersrand gold. *Exploration and Mining Geology*, **3**, 357–370.

FRIMMEL, H.E. 1997a. Detrital origin of hydrothermal Witwatersrand gold – a review. *Terra Nova*, **9**, 192–197.

FRIMMEL, H.E. 1997b. Chlorite thermometry in the Witwatersrand Basin: constraints on the Palaeozoic geotherm in the Kapvaal Craton, South Africa. *Journal of Geology*, **105**, 601–615.

FRIMMEL, H.E. & GARTZ, V.H. 1997. Witwatersrand gold particle chemistry matches model of metamorphosed, hydrothermally altered placer deposits. *Mineralium Deposita*, **32**, 523–530.

FRIMMEL, H.E. & MINTER, W.E.L. 2002. *Recent developments concerning the geological history and genesis of the Witwatersrand gold deposits*, South Africa. Society of Economic Geologists Special Publication, **9**, 17–45.

FRIMMEL, H.E., LE ROEX, A.P., KNIGHT, J. & MINTER, W.E.L. 1993. A case study of the postdepositional alteration of the Witwatersrand Basal reef gold placer. *Economic Geology*, **88**, 249–265.

GARTZ, V.H. & FRIMMEL, H.E. 1999. Complex metasomatism of an Archaean placer in the Witwatersrand Basin, South Africa: the Ventersdorp Contact Reef – a hydrothermal aquifer? *Economic Geology*, **94**, 689–706.

GIBSON, R.L. 2002. Impact-induced melting of Archaean granulites in the Vredefort Dome, South Africa. I: anatexis of metapelitic granulites. *Journal of Metamorphic Geology*, **20**, 57–70.

GIBSON, R.L. & JONES, M.Q.W. 2002. Late Archaean to Palaeoproterozoic geotherms in the Kapvaal craton, South Africa: constraints on the thermal evolution of the Witwatersrand Basin. *Basin Research*, **14**, 169–181.

GIBSON, R.L. & REIMOLD, W.U. 2001. *The Vredefort Impact Structure, South Africa: The scientific evidence and a two-day excursion guide*. Council for Geoscience, Memoir, **92**, 111p.

GIBSON, R.L. & REIMOLD, W.U. 2005. Shock pressure distribution in the Vredefort impact structure, South Africa. *In*: KENKMANN, T., HÖRZ, F. & DEUTSCH, A. (eds) *Large Meteorite Impacts and Planetary Evolution III*. Geological Society of America Special Paper 384.

GIBSON, R.L. & STEVENS, G. 1998. Regional metamorphism due to anorogenic intracratonic magmatism. *In*: TRELOAR, P.J. & O'BRIEN, P.J. (eds) *What Drives Metamorphism and Metamorphic*

Reactions? Geological Society, London, Special Publications, **138**, 121–135.

GIBSON, R.L. & WALLMACH, T. 1995. Low pressure–high temperature metamorphism in the Vredefort Dome, South Africa: anticlockwise pressure-temperature path followed by rapid decompression. *Geological Journal*, **30**, 319–331.

GIBSON, R.L., ARMSTRONG, R.A. & REIMOLD, W.U. 1997. The age and thermal evolution of the Vredefort impact structure: A single-grain U–Pb zircon study. *Geochimica et Cosmochimica Acta*, **283**, 241–262.

GIBSON, R.L., REIMOLD, W.U. & STEVENS, G. 1998. Thermal–metamorphic signature of an impact event in the Vredefort dome, South Africa. *Geology*, **26**, 787–790.

GRATON, L.C. 1930. Hydrothermal origin of the Rand gold deposits, Part 1. *Economic Geology*, **25**, Supplement to Number 3. 185 p.

GREGORY, J.W. 1908. The origin of the gold in the Rand Banket. *Geological Society of South Africa Transactions.* Annexure, **17**, 2–41.

HALLBAUER, D.K. & UTTER, T. 1977. Geochemical and morphological characteristics of gold particles from recent river deposits and the fossil placers of the Witwatersrand. *Mineralium Deposita*, **12**, 293–306.

HENKEL, H. & REIMOLD, W.U., 1998. Integrated geophysical modeling of a giant, complex impact structure: anatomy of the Vredefort Structure, South Africa. *Tectonophysics*, **287**, 1–20.

IVANOV, B.A. & DEUTSCH, A. 1999. Sudbury impact event: Cratering mechanics and thermal history. *In*: DRESSLER, B.O. & SHARPTON, V.L. (eds) *Large Meteorite Impacts and Planetary Evolution II.* Geological Society of America, Special Paper, **339**. Boulder, Colorado.

JOLLEY, S.J., FREEMAN, S.R., BARNICOAT, A.C., *ET AL.* 2004. Structural controls on Witwatersrand gold mineralisation. *Journal of Structural Geology*, **26**, 1067–1086.

KAMO, S.L., REIMOLD, W.U., KROGH, T.E. & COLLISTON, W.P. 1996. A 2.023 Ga age for the Vredefort impact event and a first report of shock metamorphosed zircons in pseudotachylitic breccias and granophyre. *Earth and Planetary Science Letters*, **144**, 369–387.

KILLICK, A.M., THWAITES, A.M., GERMS, G.J.G. & SCHOCH, A.E. 1998. Pseudotachylite associated with a bedding-parallel fault zone between the Witwatersrand and Ventersdorp Supergroups, South Africa. *Geologische Rundschau*, **77**, 329–344.

KIRK, J., RUIZ, J., CHESLEY, J., TITLEY, S. & WALSHE, J. 2001. A detrital model for the origin of gold and sulfides in the Witwatersrand Basin based on Re–Os isotopes. *Geochimica et Cosmochimica. Acta*, **65**, 2149–2159.

KIRK, J., RUIZ, J., WALSHE, J. & ENGLAND, G. 2002. A major Archaen gold and crust-forming event in the Kaapvaal Craton, South Africa. *Science*, **297**, 1156–1158.

KRANIDIOTIS, P. & MACLEAN, W.H. 1987. Systematics of chlorite alteration at the Phelps Dodge massive sulphide deposit, Matagami, Quebec. *Economic Geology*, **82**, 1898–1911.

KRAPEZ, B. 1985. The Ventersdorp Contact placer: a gold-pyrite placer of stream and debris-flow origins from the Archaean Witwatersrand Basin of South Africa. *Sedimentology*, **32**, 223–234.

MCCARTHY, T.S. 1994. The tectonosedimentary evolution of the Witwatersrand Basin with special reference to the occurrence and character of the Ventersdorp Contact Reef. *South African Journal of Geology*, **97**, 247–259.

MCCARTHY, T.S., STANISTREET, I.G. & ROBB, L.J. 1990. Geological studies related to the origin of the Witwatersrand Basin and its mineralization – an introduction and a strategy for research and exploration. *South African Journal of Geology*, **93**, 1–4.

MCWHA, M. 1994. The influence of landscape on the Ventersdorp Contact Reef at Western Deep Levels South Mine. *South African Journal of Geology*, **97**, 319–331.

MELOSH, H.J. 1989. *Impact Cratering: A Geological Process.* Oxford University Press, New York.

MINTER, W.E.L., GOEDHART, M., KNIGHT, J. & FRIMMEL, H.E. 1993. Morphology of Witwatersrand gold grains from the Basal reef: Evidence for their detrital origin. *Economic Geology*, **88**, 237–248.

OBERTHÜR, T. & SAAGER, R. 1986. Silver and mercury in gold particles from the Proterozoic Witwatersrand placer deposits of South Africa: metallogenic and geochemical implications. *Economic Geology*, **81**, 20–31.

PHILLIPS, G.N. 1987. Metamorphism of the Witwatersrand goldfields: conditions during peak metamorphism. *Journal of Metamorphic Geology*, **5**, 307–322.

PHILLIPS, G.N. 1988. Widespread fluid infiltration during metamorphism of the Witwatersrand goldfields: generation of chloritoid and pyrophyllite. *Journal of Metamorphic Geology*, **6**, 311–332.

PHILLIPS, G.N. & LAW, D.M. 1994. Metamorphism of the Witwatersrand goldfields: a review. *Ore Geology Reviews*, **9**, 1–31.

PHILLIPS, G.N. & LAW, J.D.M. 2000. Witwatersrand goldfields: geology, genesis and exploration. *Society of Economic Geology Reviews*, **13**, 439–500.

PHILLIPS, G.N., KLEMD, R. & ROBERTSON, N.S. 1988. Summary of some fluid inclusion data from the Witwatersrand Basin and some surrounding granitoids. *Geological Society of India Memoir*, **11**, 59–65.

PHILLIPS, G.N., LAW, J.D.M. & MYERS, R.E. 1990. The role of fluids in the evolution of the Witwatersrand Basin. *South African Journal of Geology*, **93**, 54–69.

POUCHOU, J.L. & PICHOIR, F. 1987. Basic expression of 'PAP' computation for quantitative EPMA. *In*: BROWN, J.D. & PACKWOOD, R.H. (eds) *Proceedings of the 11th ICXOM*, University of Western Ontario, London, Ontario, 249–253.

POUJOL, M., ROBB, L.J., RESPAUT, J.-P. & ANHAEUSSER, C.A. 1996. 3.07–2.97 Ga greenstone belt formation in the northeastern Kapvaal craton: implications for the origin of the Witwatersrand Basin. *Ecomomic Geology*, **91**, 1455–1461.

PRETORIUS, D.A. 1986. The Witwatersrand Basin, surface and subsurface geology and structure (Map). *In*: ANHAEUSSER, C.R. & MASKE, S. (eds) *Mineral Deposits of Southern Africa I*. Geological Society of South Africa. Back Pocket.

REID, A.M., LE ROEX, A.P. & MINTER, W.E.L. 1988. Composition of gold grains in the Vaal placer, Klerksdorp, South Africa. *Mineralium Deposita*, **23**, 211–217.

REIMOLD, W.U. & COLLISTON, W.P. 1994. Pseudotachylites of the Vredefort Dome and the surrounding Witwatersrand Basin, South Africa. *In*: DRESSLER, B.O., GRIEVE, R.A.F. & SHARPTON, V.L. (eds) *Large Meteorite Impacts and Planetary Evolution*. Geological Society of America Special Paper **293**, Geological Society of America, Boulder Colorado, 177–196.

REIMOLD, W.U. & GIBSON, R.L. 2005. 'Pseudotachylites' in large impact structures. *In*: KOEBERL, C. & HENKEL, H. *Impact Tectonics*. Springer, Berlin, 1–53.

REIMOLD, W.U., KÖBERL, C., FLETCHER, P., KILLICK, A.M. & WILSON, J.D. 1999. Pseudotachylitic breccias from fault zones in the Witwatersrand Basin, South Africa: evidence of autometasomatism and post-brecciation processes. *Mineralogy and Petrology*, **66**, 25–53.

ROBB, L.J. & MEYER, F.M. 1991. A contribution to the debate concerning epigenetic versus syngenetic mineralisation processes in the Witwatersrand Basin. *Economic Geology*, **86**, 396–401.

ROBB, L.J. & MEYER, F.M. 1995. The Witwatersrand Basin, South Africa: geological framework and mineralization processes. *Ore Geology Reviews*, **10**, 67–94.

ROBB, L.J. & ROBB, V.M. 1998. Gold in the Witwatersrand Basin. *In*: WILSON, M.G.C. & ANHAEUSSER, C.R. (eds) *The Mineral Resources of South Africa*. Council for Geoscience Handbook 16. Council for Geoscience, Pretoria. 294–349.

ROBB, L.J., DAVIS, D.W. & KAMO, S.L. 1990. U–Pb ages on single detrital zircon grains from the Witwatersrand Basin, South Africa: constraints on the age of sedimentation and the evolution of the granites adjacent to the basin. *Journal of Geology*, **98**, 311–328.

ROBB, L.J., CHARLESWORTH, E.G., DRENNAN, G.R., GIBSON, R.L. & TONGU, E.L. 1997. Tectonometamorphic setting and paragenetic sequence of the Au–U mineralisation in the Archaean Witwatersrand Basin, South Africa. *Australian Journal of Earth Sciences*, **44**, 353–371.

RUST, I.C. 1994. A note of the Ventersdorp Contact Reef: a gravel in transit. *South African Journal of Geology*, **97**, 238.

SAAGER, A. 1969. The relationship of silver and gold in the Basal reef of the Witwatersrand System, South Africa. *Mineralium Deposita*, **4**, 93–113.

SMITH, N.D. & MINTER, W.E.L. 1980. Sedimentological controls on gold and uranium in two Witwatersrand palaeoplacers. *Economic Geology*, **75**, 1–14.

STEVENS, G., BOER, R. & GIBSON, R. 1997. Metamorphism, fluid-flow and gold mobilisation in the Witwatersrand Basin: towards a unifying model. *South African Journal of Geology*, **100**, 363–375.

STEWART, R.A., REIMOLD, W.U. & CHARLESWORTH, E.G. 2004. Tectonosedimentary model for the Central Rand Basin, South Africa: Implications for exploration. *South African Journal of Geology*, **107**, 603–618.

THERRIAULT, A.M., GRIEVE, R.A.F & REIMOLD, W.U. 1997. Original size of the Vredefort structure: Implications for the geological evolution of the Witwatersrand Basin. *Meteoritics and Planetary Science*, **32**, 71–77.

TURTLE, E.P. & PEIRAZZO, E. 1998. Constraints on the size of the Vredefort impact crater from numerical modelling. *Meteoritics and Planetary Science*, **33**, 483–490.

TURTLE, E.P., PIERAZZO E. & O'BRIEN, P.O. 2003. Numerical modelling of impact heating and cooling of the Vredefort impact structure. *Meteoritics and Planetary Science*, **38**, 293–303.

TWEEDIE, E.B. 1986. The Evander Goldfield. *In*: ANHAEUSSER, C.A. & MASKE, S. (eds) *Mineral Deposits of Southern Africa, Volume 1*. Geological Society of South Africa, Johannesburg, 705–730.

UTTER, T. 1979. The morphology and silver content of gold from the Upper Witwatersrand and Ventersdorp Systems in the Klerksdorp goldfield, South Africa. *Economic Geology*, **74**, 27–44.

VON GEHLEN, K. 1983. Silver and mercury in single gold grains from the Witwatersrand and Barberton, South Africa. *Mineralium Deposita*, **18**, 529–534.

WALRAVEN, F. 1997. Geochronology of the Rooiberg Group, Transvaal Supergroup, South Africa. *Economic Geology Research Unit Information Circular*, **316**. University of the Witwatersrand, Johannesburg.

WALRAVEN, F. & HATTINGH, E. 1993. Geochronology of the Nebo Granite, Bushveld Complex. *South African Journal of Geology*, **96**, 31–42.

WALRAVEN, F. & MARTINI, J. 1995. Zircon Pb-evaporation age determinations of the Oak Tree Formation, Chuniespoort Group, Transvaal Sequence: implications for Transvaal–Griquualand West basin correlations. *South African Journal of Geology*, **98**, 58–67.

WALRAVEN, F., ARMSTRONG, R.A. & KRUGER, F.J. 1990. A chronostratigraphic framework for the north–central Kapvaal Craton, the Bushveld Complex and the Vredefort structure. *Tectonophysics*, **171**, 23–48.

WIELAND, F., REIMOLD, W.U. & GIBSON, R.L. 2003. New evidence related to the formation of shatter cones: with special emphasis on structural observations in the collar of the Vredefort Dome, South Africa. *3rd International Conference on Large Meteorite Impacts, August 2003, Noerdlingen/Germany*, CD-ROM, Abstract # 4008, 2pp.

YOUNG, R.B. 1917. *The Banket: A study of the auriferous conglomerates of the Witwatersrand and the associated rocks*. Gurney and Jackson, London.

ZANG, W. & FYFE, W.S. 1995. Chloritization of the hydrothermally altered bedrock at the Igarapé Bahia gold deposit, Carajás, Brazil. *Mineralium Deposita*, **30**, 30–38.

ZHAO, B., MEYER, F.M., ROBB, L.J. & MCWHA, M. 1994. A preliminary study of alteration associated with the Ventersdorp Contact Reef at Western Deep Levels South Mine, Witwatersrand Basin, South Africa. *South African Journal of Geology*, **97**, 348–356.

ZHOU, T. & PHILLIPS, G.N. 1994. Sudoite in the Archaean Witwatersrand Basin. *Contributions to Mineralogy and Petrology*, **116**, 352–359.

Metallogenic fingerprints of Archaean cratons

MAARTEN DE WIT[1] & CHRISTIEN THIART[2]

CIGCES, and Departments of Geological[1] and Statistical[2] Sciences, University of Cape Town, Rondebosch 7701, South Africa (e-mail: maarten@cigces.uct.ac.za; thiart@stats.uct.ac.za)

'The duty of the geologist and the prospector is in fact to deliver the goods'
(Sir Lewis L. Fermor, 1951, sixth geological President of the Institution of Mining and Metallurgy)

Abstract: Archaean cratons are fragments of old continents that are more richly endowed with mineral deposits than younger terrains. The mineral deposits of different cratons are also diversely enriched with useful (to humankind) chemical elements. Cratons are therefore mineral-diversity hotspots that represent regional geochemical heterogeneities in the early Earth, evidence for which remains encoded on each craton as unique metallogenic 'fingerprints'. Some of the younger cratons (<3.0 Ga, e.g. Superior Province, Yilgarn and Zimbabwe) have strong Au, Cu, Pb and Zn imprints. Older (>3.0 Ga) cratons, however, are remarkably enriched in siderophile elements such as Ni, Cr, PGE, in both their crustal and mantle sections (e.g. Pilbara and Kaapvaal Cratons). Still other Archaean cratons are relatively enriched in Sn, W, U and Th (e.g. Amazonian, Leo-Man, Ntem and South China Cratons). How most of these fragments of old continents inherited their rich and diverse metallogenic characteristics is unresolved. Their dominant metallogenic inventories were formed near the time of their separation from the mantle; thereafter the inherited metals were frequently remobilized and redistributed during subsequent tectono-metamorphic, magmatic and erosion-deposition processes (e.g. tin in South America; platinum and gold in Southern Africa). Because different cratons are likely to represent only small remnants of once much larger and probably varied Archaean continents, part of the total metal inventories of Archaean continents must have been recycled back into the mantle. Using six selected element groups from our extensive in-house GIS database of Gondwana mineral deposits, we derive the metallogenic fingerprints of 11 Archaean cratons of the southern hemisphere, and compare these against metallogenic fingerprints of the same elements in younger crust of three continents (Africa, Australia and South America). We confirm that the mineral deposit density and diversity of Earth's continental lithosphere has decreased with time. We conclude that metallogenic elements were transferred more efficiently from the mantle to the continental lithosphere in the Archaean and/or that subsequently (<2.5 Ga) recycling of these elements (mineral deposits) back into the mantle became more effective.

The resilience of cratons as archives of Archaean processes is well established. Archaean cratons (>2.5 Ga) are underlain by relatively thin crust (c. 30–40 km), and thick mantle lithosphere (up to c. 250–300 km: de Wit 1998; James et al. 2001; Shirez et al. 2002; Stankiewicz et al. 2002; Fouch et al. 2004). Near surface, the crust of most of these cratons is well endowed with concentrations of metallic elements useful and economic to humankind (hereafter referred to as mineral deposits). Cratons are distinct in that their mineral deposits contain different mixtures of metallic elements from craton to craton. For example the Kaapvaal Craton of South Africa is known to be relatively enriched in gold and platinum group elements (PGE), the Zimbabwe Craton and the Yilgarn Craton of Australia in gold and tungsten, the São Fransisco Craton in gold and base-metals (Cu/Pb/Zn), and the Amazonian Craton in gold and tin (Groves et al. 1987; Wilsher et al. 1993; Wilsher 1995; de Wit et al. 1999; Thiart & de Wit 2005).

We assume that most mineral deposits of Archaean cratons reflect geochemical concentration processes that operated during the formation of the cratons. In most cases this can be verified if the relative age of the mineral deposits is geologically well constrained and/or reliably dated radiometrically. In other cases it may not always be obvious if these mineral enrichments are inherited from Archaean times or if they were added to the cratons at a later stage. For example, the platinum deposits of the

From: McDONALD, I., BOYCE, A. J., BUTLER, I. B., HERRINGTON, R. J. & POLYA, D. A. (eds) 2005. *Mineral Deposits and Earth Evolution.* Geological Society, London, Special Publications, **248**, 59–70. 0305-8719/$15.00

Kaapvaal Craton are mostly found to be associated with the igneous rocks of the Bushveld Complex that intruded the centre of this craton in the Palaeoproterozoic (2.05–2.06 Ga; Cawthorne & Walraven 1998; Eglington & Armstrong 2004). Thus these elements may have been added from the asthenospheric mantle at that time. However, the PGE geochemistry (and that of other siderophile elements like Ni, Cr, Au) of lithospheric mantle xenoliths found in kimberlites across the Kaapvaal craton (McDonald *et al.* 1995; Hart *et al.* 1997), as well as that of Archaean mid-lower crust (Hart *et al.* 2004) and of Archaean mineral deposits in the greenstone belts of this craton (Tredoux *et al.* 1989), all suggest that both the lithospheric mantle and crust of this craton were already rich in PGE in Archaean times. Indeed the PGE in the igneous rocks of the Bushveld Complex may have been inherited from the underlying depleted Archaean mantle lithosphere (McDonald *et al.* 1995; Maier & Barnes 2004). Similarly, most of the extraordinary enrichment of the Kaapvaal Craton in gold occurs in Archaean sedimentary rocks of the Witwatersrand Basin. Even though some of the gold may have been introduced and/or remobilized at a later time (Frimmel & Minter 2002), gold and osmiridium grains produce Re–Os isochron ages and Re depletion model ages that range from 3.5–2.9 Ga (Hart & Kinloch 1989; Kirk *et al.* 2001). Thus, we can assume that much of the concentration of precious elements in the continental lithosphere of the Kaapvaal and other cratons occurred during their formation in the Archaean (Groves *et al.* 1987; Tredoux *et al.* 1989; McDonald *et al.* 1995).

We set out to verify that different Archaean cratons have distinct metallogenic patterns, which we refer to as their metallogenic 'fingerprints'. These fingerprints may reveal something fundamental about the formation of cratons and Earth's earliest continents. Then we address the question of whether a unit of Archaean continental crust is more enriched in mineral deposits than younger crust. We compare our results from Archaean cratons with those from younger crust (<2.5 Ga) at different scales: first at a continental scale (e.g. South America, Australia, Africa), and then at a supercontinental scale using our database of the mineral deposits of Gondwana. This reveals important information about evolution of continental crust. However, our analyses are limited to the present southern hemisphere, because our mineral database is confined only to continental fragments of the former supercontinent Gondwana (see below).

Establishing metallogenic fingerprints of different cratons using selected elements from their mineral deposits allows us to address a spectrum of controversial tectonic models that compare early Earth processes with those of the present. For example on the modern plate tectonic Earth, specific elements in mineral deposits (reflecting different mineral deposit type) characterize specific plate tectonic environments (Sawkins 1990; Windley 1991). As a corollary, therefore, cratonic mineral deposit diversity may also be used to test for plate tectonic processes in the Archaean (Herrington *et al.* 1997). In addition, variations in the total concentration of these elements in continental crust of different ages may be used to test for changes in the rates of these processes over time, and the mineral diversity patterns of cratons may help us to decipher the recycling history of Earth's continental lithosphere. Similarly, mineral-diversity patterns of cratons and younger lithosphere may help to test reconstructions of past supercontinents (e.g. de Wit *et al.* 1999).

Taking fingerprints of Archaean cratons and younger crust using selected elements from their mineral deposits

Database

The geological and mineral deposit data used for this study are incorporated in a GIS database, called GO-GEOID, housed at our centre (CIGCES). This database is restricted to the main continental fragments of Gondwana; its geological component is based on the geological map of Gondwana (de Wit *et al.* 1988), whereas the mineral database was constructed using open access literature sources (Wilsher 1995) and thereafter regularly updated at the CIGCES. This has been described in detail elsewhere (Wilsher *et al.* 1993; de Wit *et al.* 1999; Thiart & de Wit 2000; de Wit *et al.* 2004)

Eleven elements were selected for our analyses, and these were divided into six element groups according to their geochemical affinities (e.g. lithophile, chalcophile, siderophile), as well as their relative abundance in the database. The six element groups, and the total number of deposits in which these groups occur on each craton (Fig. 1) are tabulated in Table 1a. In total there are just over 6000 deposits spread over twelve identified cratons (Table 1a). There are eleven Gondwana cratons (seven in Africa, two in South America and two

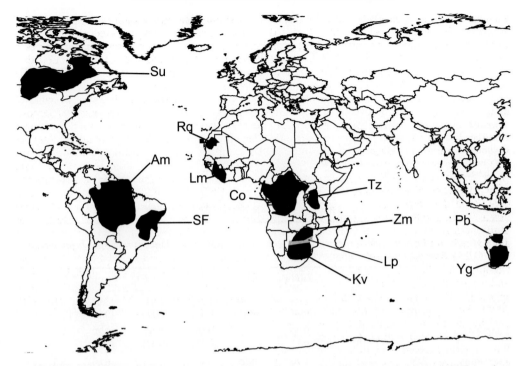

Fig. 1. Cratons examined in this study. Kv, Kaapvaal; Lp, Limpopo; Zm, Zimbabwe; Co, Congo; Tz, Tanzania; Lm, Leo-Man; Rq, Requibath; Su, Superior; SF, São Fransisco; Am, Amazonia; Yg, Yilgarn; Pb, Pilbara.

in Australia) for which we have sufficient mineral deposit data in our Gondwana databank. For comparison, we also selected one Canadian craton. The data for the latter are from the Geological Survey of Canada (Kirkham *et al.* 1994, 2002; Jenkins *et al.* 1997; Eckstrand & Good 2000; Kirkham & Dunne 2000, 2002; Eckstrand *et al.* 2002). Because their data structure is significantly different from our

Gondwana database, we used only the data from the Superior Province in this study for contingency analyses (see below).

Methods and results

Two simple, well-established statistical methods are used to normalize and evaluate our data across three scales (viz. Archaean cratons

Table 1a. *Number of mineral deposits of selected element groups across Gondwana and the Superior province of Canada*

	Au	CrNiPgeTi	CuZnPbBa	SnSb	W	UThREE	Cra Total
Kaapvaal	894	241	152	96	5	77	1465
Limpopo	0	40	33	0	8	9	90
Zimbabwe	625	114	106	81	305	15	1246
Congo	53	2	24	1	0	17	97
Tanzania	7	0	3	1	0	5	16
Leo-Man	47	22	11	10	2	9	101
Requibath	0	5	21	0	3	4	33
Superior	162	134	52	3	0	0	351
Amazonia	434	38	109	645	8	13	1247
São Fransisco	435	127	462	149	6	66	1245
Pilbara	7	6	11	1	0	4	29
Yilgarn	97	50	17	1	0	11	176
Dep Total	2761	779	1001	988	337	230	6096

Table 1b. *Chi square (χ^2, similarity measure) values between cratons and element groups. The higher the value, the stronger the association*

	Au	CrNiPgeTi	CuZnPbBa	SnSb	W	UThRee
Kaapvaal	80	16	33	84	71	9
Limpopo	41	71	23	15	2	9
Zimbabwe	7	13	48	72	809	22
Congo	2	9	4	14	5	49
Tanzania	0	2	0	1	1	32
Leo-Man	0	6	2	3	2	7
Requibath	15	0	45	5	1	6
Superior	0	177	1	51	19	13
Amazonia	30	92	45	971	54	25
São Fransisco	30	7	325	14	57	8
Pilbara Craton	3	1	8	3	2	8
Yilgarn	4	34	5	27	10	3

Table 1c. *Natural log of the spatial coefficient (ρ_{ij}) with the approximate standard error ($s(\ln\rho_{ij})$, below and in brackets)) of element groups within cratons*

	Au	CrNiPgeTi	CuZnPbBa	SnSb	W	UThREE	Cra Tot	Area in km²
Kaapvaal	1.79 (0.03)	1.75 (0.06)	1.04 (0.08)	0.59 (0.10)	−1.29 (0.45)	1.83 (0.11)	5.03 (0.02)	6.60E + 05
Limpopo		1.37 (0.16)	0.93 0.17		0.60 (0.35)	1.10 (0.33)	1.20 (0.10)	1.59E + 05
Zimbabwe	2.06 (0.04)	1.63 (0.09)	1.30 (0.10)	1.05 (0.11)	3.45 (0.06)	0.82 (0.26)	6.60 (0.03)	3.52E + 05
Congo	−2.29 (0.14)	−4.30 (0.71)	−2.06 (0.20)	−5.23 (1.00)		−0.94 (0.24)	0.13 (0.08)	2.31E + 06
Tanzania	−2.37 (0.38)		−2.20 (0.58)	−3.29 (1.00)		−0.22 (0.45)	0.11 (0.22)	3.33E + 05
Leo-Man	−0.43 (0.15)	0.08 (0.21)	−0.87 (0.30)	−0.95 (0.32)	−1.48 (0.71)	0.40 (0.33)	0.88 (0.08)	3.20E + 05
Requibath	(0.45)	−0.76 (0.22)	0.42		−0.43 (0.58)	0.24 (0.50)	0.47 (0.15)	1.68E + 05
Superior	−1.33 (0.08)	−0.26 (0.09)	−1.46 (0.14)	−4.30 (0.58)			0.23 (0.05)	2.73E + 06
Amazonia	−0.59 (0.05)	−1.76 (0.16)	−0.96 (0.10)	0.83 (0.04)	−2.48 (0.35)	−1.61 (0.28	0.66 (0.03)	3.48E + 06
São Fra.	0.69 (0.05)	0.73 (0.09)	1.77 (0.05)	0.65 (0.08)	−1.49 (0.41)	1.29 (0.12)	2.48 (0.03)	9.64E + 05
Pilbara	−1.58 (0.38)	−0.47 (0.41)	−0.11 (0.30)	−2.50 (1.00)		0.35 (0.50)	0.39 (0.18)	1.51E + 05
Yilgarn	−0.37 (0.10)	0.23 (0.14)	−1.10 (0.24)	−3.92 (1.00)		−0.06 (0.30)	0.53 (0.07)	6.25E + 05

(Fig. 1), continents and super-continents), as described briefly below.

Scale 1, method 1: Testing associations between element groups and Archaean cratons (>2.5 Ga). Observed frequencies (or deposit counts/ craton) are given in a contingency table (Table 1a); and from these, expected frequencies are determined by multiplying marginal frequencies, and dividing these by the total number of observed deposits. To standardize data in a contingency table, observed frequencies (O) and expected (E) frequencies are used to calculate a Chi-square value:

$$\chi_{ij}^2 = \frac{(O_{ij} - E_{ij})^2}{E_{ij}},$$

where i and j indicate the corresponding row or column of the contingency table (Table 1a).

The Chi-square value is sometimes referred to as the similarity measure (or the degree of similarity) or the chi-square distance (Davis 2002), and is effectively a measure of the degree (or tightness) of association between the corresponding data sets. In our case, high values of this coefficient indicate a close association between a craton and element groups (Table 1b). The similarity measure provides a 'ranking' order that can be used qualitatively to test for relative associations between different mineral occurrences and specific cratons. For example it is clear from the data in Table 1b that there is a relatively strong association between the Zimbabwe Craton and tungsten (W), and that this association is one of the strongest observed for all the cratons.

The disadvantage of the Chi-square values is that they do not contrast the different sizes of the cratons. To normalize to a unit craton area we use a spatial coefficient as described below.

Scale 1, method 2: Spatial coefficients of element groups and Archaean cratons: cratonic fingerprints. To compare the association between different element groups and cratons directly, we first normalized the data per unit area of crust, and then derived a measure of spatial association, which we termed the spatial coefficient, ρ_{ij},

$$\rho_{ij} = \frac{N(C_i \cap D_j)/N(D_j)}{A(C_i)/A(C_\bullet)},$$

in which $A(C_i)$ is the area of the ith craton (C_i), and $A(C_\bullet)$ is the total area of all the cratons in the study $(A(C_\bullet) = \Sigma A(C_i))$. $N(D_j)$ is the total number of deposits in the jth mineral group (D_j). $N(C_i \cap D_j)$ represents the number of deposits of group D_j in craton C_i.

This normalized spatial measure represents the proportion of deposits (say gold – j) of all the jth deposits (in cratons) that occur in the specified craton (i) per unit area of all twelve cratons. The spatial coefficient (ρ_{ij}) ranges in value from 0 to infinity, and it is equal to 1 if there is no spatial association between a craton and an element group. For values of $\rho_{ij} > 1$, there is a positive association between mineral j and craton i; $\rho_{ij} < 1$ indicates a negative association. All the negative associations are compressed in the range from 0 to 1. Positive associations fall

in the range of 1 to infinity. To eliminate the skewness in ρ_{ij} values, we derive the natural log of ρ_{ij} (Table 1c). This represents a symmetric value around 0: positive associations are greater than 0, and negative associations are less than 0.

In a similar study, Mihalasky & Bonham-Carter (2001) evaluated the spatial association between lithodiversity and mineral deposits in Nevada, USA, and from this derived a relationship from between $\ln\rho_{ij}$ and a 'weights of evidence' (WofE) measure (Bonham-Carter 1996; Thiart & de Wit 2000). $\ln\rho_{ij}$ and WofE converge when the WofE measure becomes infinitely small. Thus the standard error calculated for WofE can be substituted as an approximate standard error for $\ln\rho_{ij}$ (Mihalasky & Bonham-Carter 2001; Bonham-Carter 1996, Chapter 9). The approximate standard error of $\ln\rho_{ij}$, is given by:

$$s(\ln\rho_{ij}) = \sqrt{\frac{1}{N(C_i \cap D_j)}}$$

These approximate standard errors are given in Table 1c (in brackets and below, the actually natural log of the spatial coefficients) and as vertical error bars in Figure 2.

Figure 2 expresses the diversity amongst cratons for each set of elements in the six element groups and provides 12 cratonic (natural log) spatial coefficient values for each set of elements. The last chart in Figure 2 represents the natural log of the spatial coefficient for the total number of element groups (e.g. sum of all six groups) for each craton; we refer to these as their metallic 'fingerprints'.

From Figure 2, Zimbabwe shows the strongest (natural log) spatial coefficient for five element groups, and the sum of all six groups (total): metaphorically a strong 'fingerprint'. The Tanzanian and Congo Cratons represent the lowest values (for the individual element groups as well as their totals). The latter may be due to the low number of discovered mineral deposits on these cratons (and thus in our database), and in turn emphasizes the need to incorporate into our analyses a value that reflects the past exploration history of the areas under consideration (see below).

There is a distinct difference in the metallogenic fingerprints of some cratons that may have had a common history in the past. For example, the Kaapvaal and Pilbara Cratons may have been part of the same craton (Vaalbara) in the Meso- and Neo-Archaean (Zegers *et al.* 1998; Strik 2004). Yet Figure 2 displays a significant difference in their metallogenic fingerprints. It is not clear why this is so. Perhaps because more

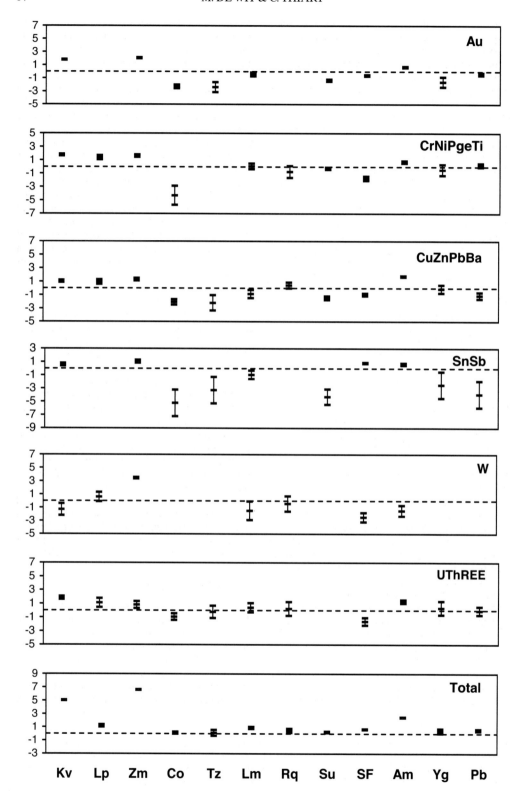

than half the Pilbara is covered by the thick sequence of late Archaean–early Palaeoproterozoic sediments and volcanics of the Mount Bruce Supergroup (Trendall 1983), a number of large deposits may be hidden. However, at present, the total number of deposits in our database for the Kaapvaal Craton is almost fifteen times greater than those for the Pilbara Craton. Either the Kaapvaal Craton is indeed mineralized to a much greater degree than the Pilbara, or a number of deposits are yet to be discovered in the Pilbara Craton. Alternatively, the Vaalbara model for an Archaean supercontinent may be wrong.

A similar observation may be made about the differences between the fingerprints of the Congo Craton and the São Francisco Craton. Several studies suggest, but have not proven, a former connection between these two cratons. However, the most robust data point to the possible amalgamation of these two cratons into a single crustal segment only in Ebunian times (c. 2.0 Ga; Ledru et al. 1994), and therefore do not negate a separate history for these two cratons in Archaean times.

Clearly more data are needed to test the above discrepancies further. It also emphasizes that major mineralizing pulses on the cratons during post-Archaean events (e.g. the Bushveld Complex of the Kaapvaal Craton) are significant factors that need to be incorporated better into our analysis. These are issues beyond the scope of the present paper.

Scale 2: Spatial association between element groups and post-Archaean crust on a continental scale. The analyses described above were repeated for three large continental fragments (Africa, South America and Australia) of the former supercontinent Gondwana, to enable direct comparison of the metallogenic fingerprints of cratons (>2.5 Ga; 'old' crust) with mineralization in younger crust (<2.5 Ga; 'young' crust). The data from the Superior Province were omitted from further analysis.

The mineral deposit data for the old cratons and younger crust are summarized in Table 2a; the similarity association measures (χ^2) in Table 2b; the natural log of the spatial coefficients are given in Table 2c and illustrated in Figure 3 with the standard error bars.

For Africa, the old crust (cratons) forms a stronger association with the six element groups (and for the total of all six groups) than does its younger crust. For Australia the old crust (cratons) forms a stronger association with the three groups Au, CrNiPGE, CuZnPbBa, as well as the total of all element groups than does its younger crust, but for the three groups SnSb, W and UThREE we do not observe any significant differences between old crust and younger crust. South America follows a similar trend, except for the W and UThREE element groups, that are concentrated to greater degree in younger crust. We have previously explored the Gondwana–tin association between West Africa and South America, to confirm their combined pre-Gondwana history, and to question a frequently advocated fit between them and North America in support of models of a proposed Mesoproterozoic supercontinent Rodinia (de Wit et al. 1999).

Scale 3: Comparing the spatial association between elements groups and post-Archaean crust on a supercontinental-Gondwana-scale. The above analysis is repeated using all the data of the cratons and the three continents combined. This allows us to compare and contrast the mineral inventory of all old (Archaean) crust with that of younger crust, as it would have been in Gondwana times (c. 200–500 Ma). To a certain extent this should buffer the concentration of recent mineral deposits related to one specific present day plate tectonic environment (such as subduction below South America) that may otherwise skew the analyses when considering a single continent. The data are summarized in Table 3a–c, and plotted in Figure 4. From this we conclude that Archaean crust is mineralized to a significantly greater degree than younger crust, except for the strong lithophile element group UThREE .

Discussion and conclusions

There is a greater concentration of mineral deposits in Archaean crust compared to younger crust. However, we have also shown that a significant mineral diversity exists in the crust of different cratons: each craton appears to have a unique metallogenic fingerprint. Such differences resemble variations in more recent mineralization between continents that can be

Fig. 2. Mineral diversity of selected elements of Archaean cratons, as shown by plotting the (natural log, *Y*-axis) spatial coefficient ($\ln p_{ij}$) of the six element groups (and the total of these groups (lower most plot)) of twelve craton. Acronyms for cratons as in Figure 1. Confidence intervals = 2 × standard error. Note how each craton has a unique combination of elements that we term their 'Metallogenic fingerprint'.

Table 2a. *Number of mineral deposits of selected element groups of cratonic crust (old) and younger crust (young) of three continents: Africa (Afr), South America (SAm) and Australia (Aus)*

	Au	CrNiPGE	CuZnPbBa	SnSb	W	UThREE	Cra Tot
Afr old	1626	424	350	189	323	136	3048
Afr young	499	91	768	142	119	491	2110
SAm old	869	165	571	794	14	79	2492
SAm young	797	119	999	236	828	671	3650
Aus old	104	56	28	2	0	15	205
Aus young	70	2	15	22	5	157	271
Deposit total	3965	857	2731	1385	1289	1549	11776

Table 2b. *Chi square (χ^2, similarity measure) values between element groups and old or younger crust of three continents: Africa (Afr), South America (SAm) and Australia (Aus)*

	Au	CrNiPGE	CuZnPbBa	SnSb	W	UThREE
Afr old	350	184	180	80	0	175
Afr young	63	25	159	45	54	164
SAm old	1	1	0	856	245	189
SAm young	152	81	27	87	460	76
Aus old	18	113	8	20	22	5
Aus young	5	16	36	3	21	413

Table 2c. *Natural log of the spatial coefficient (ρ_{ij}) with standard errors (in brackets) of element groups within cratons (old) and younger crust (young) of three continent: Africa (Afr), South America (SAm) and Australia (Aus). The total area of the younger crust is 3.73×10^7, and that of the cratons is 9.52×10^6 km^2*

	Au	CrNiPgeTi	CuZnPbBa	SnSb	W	UThREE	Cra Tot	Area in km^2
Afr old	1.50 (0.02)	1.68 (0.05)	0.33 (0.05)	0.40 (0.07)	1.00 (0.06)	−0.05 (0.09)	1.04 (0.02)	4.31E + 06
Afr young	−1.25 (0.04)	−1.42 (0.10)	−0.45 (0.04)	−1.46 (0.08)	−1.56 (0.09)	−0.33 (0.05)	−0.90 (0.02)	2.06E + 07
SAm old	0.84 (0.03)	0.71 (0.08)	0.79 (0.04)	1.80 (0.04)	−2.17 (0.27)	−0.62 (0.11)	0.80 (0.02)	4.44E + 06
SAm young	−0.13 (0.04)	−0.50 (0.09)	0.47 (0.03)	−0.29 (0.07)	1.03 (0.03)	0.64 (0.04)	0.30 (0.02)	1.07E + 07
Aus old	0.46 (0.10)	1.37 (0.13)	−0.48 (0.19)	−2.44 (0.71)		−0.54 (0.26)	0.05 (0.07)	7.75E + 05
Aus young	−1.98 (0.12)	−4.01 (0.71)	−3.15 (0.26)	−2.09 0.21	−3.50 (0.45)	−0.24 (0.08)	−1.72 (0.06)	6.01E + 06

Fig. 3. Mineral deposit concentration of selected elements in cratonic crust (old, O) and younger crust (young, Y; grey background) of three separate continents (Afr, Africa; Sam, South America; Aus, Australia), as shown by plotting the (natural log, Y-axis) spatial coefficient ($\ln\rho_{ij}$) of the element groups (and the total of these groups (lower most plot)) against the corresponding crust. Note that for each continent, concentration of mineral deposits across its cratons is, in general, greater than that in its younger crust. This confirms a decrease in mineral deposit density and diversity in continental crust younger than Archaean crust. A notable exception to this is for W (tungsten) and the lithophite elements UThREE of South America, and possibly tin and antimony (SnSb) in Australia. There is not sufficient data from cratonic Australia for robust analysis.

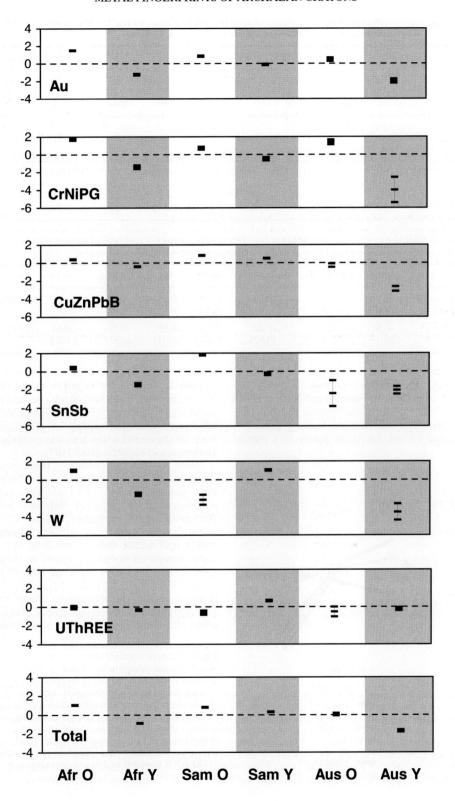

Table 3a. *Number of mineral deposits of selected element groups on all cratons (Old) and younger crust (Young) of three continents combined as a supercontinent*

	Au	CrNiPGE	CuZnPbBa	SnSb	W	UThREE	Area Total
Old	2761	779	1001	988	337	230	6096
Young	1204	78	1730	397	952	1319	5680
Total	3965	857	2731	1385	1289	1549	11776

Table 3b. *Chi square (χ^2, similarity measure) values between element groups and cratons (Old) or younger crust (Young) of three continents combined as a supercontinent*

	Au	CrNiPGE	CuZnPbBa	SnSb	W	UThREE
Old	228	123	110	142	135	366
Young	218	117	105	135	129	348

Table 3c. *Natural log of the spatial coefficient (ρ_{ij}) with standard error (in brackets) of element groups in all cratons (Old) and younger crust (Young) of three continents combined as a supercontinent*

	Au	CrNiPgeTi	CuZnPbBa	SnSb	W	UThREE	Cra Tot	Area in km^2
Old	1.17	1.31	0.54	1.25	0.25	−0.31	0.88	9.520E + 06
	(0.02)	(0.04)	(0.03)	(0.03)	(0.05)	(0.07)	(0.01)	
Young	−0.84	−1.17	−0.20	−1.01	−0.08	0.07	−0.44	3.732E + 07
	(0.03)	(0.07)	(0.02)	(0.05)	(0.03)	(0.03)	(0.01)	

linked to different plate tectonic environments (e.g. oceanic arcs, continental subduction zones etc.). This provides some support for the view that plate tectonic and ore forming processes operated on the Archaean Earth in a similar fashion to today. For example many of the NeoArchaean cratons have strong Au and base-metal signatures that fit with mineralization of subduction–accretion models as proposed for some of these cratons on tectonic grounds (e.g. Herrington *et al.* 1997; Stott 1997).

There is robust evidence that cratonic lithosphere is more enriched in mineral deposits than younger crust (<2.5 Ga). The greater concentration of mineral deposits in the Archaean may represent more efficient mineralization processes, perhaps related to higher heat and/or volatile loss from the early Earth compared to today (Abbott & Hoffman 1984; Abbott *et al.* 1994; de Wit & Hart 1993; de Wit & Hynes 1995; Pollack 1997; de Wit 1998), in which case our results may be interpreted to reflect greater 'partition coefficients' of selected elements between cratonic crust and (depleted)

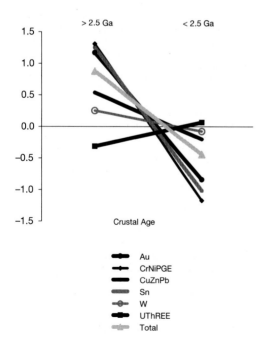

Fig. 4. Relative concentration of mineral deposits of all six element groups in the cratons (>2.5 Ga) and younger crust (<2.5 Ga) of the Gondwana continents [Africa–South America–Australia] combined. Five of the six element groups are concentrated to a greater degree in the cratonic crust compared to that of younger crust. This suggests a total loss of mineral deposit mass in Earth's preserved continental crust over time. Only the strong lithophile element group UThREE is concentrated to a greater degree in younger crust.

mantle during the formation of Archaean lithosphere.

We cannot rule out the possibility that the higher concentration of Archaean mineral deposits reflects the greater preservation potential of cratons (and their mineral deposits) relative to younger continents, in which case the great mineral wealth of cratons may be merely a consequence of greater rates of recycling of young continental crust relative to that of old Archaean crust preserved in cratons. This process may possibly be reflected by the fact that the strong lithophile group UThREE is the only group that has similar concentrations in both Archaean and post-Archaean crust.

To summarize, Earth's crust appears to signal a loss of mineral deposit mass and diversity per unit-area through time. However, as a final word of caution, our analysis ignores factors such as the infrastructure and exploration history of the study regions. Intuitively, we might expect that greater accessibility and political stability would increase exploration and discovery rates; and mining activity. Mineralization in some cratonic areas (and younger terrains) may not be adequately represented in our database and analyses. We address this issue elsewhere (Thiart & de Wit 2005).

This work is supported through funds of the South African National research Foundation (NRF). We are grateful to Graham Bonham-Carter and Fritz Agterberg for their interest and encouragement in our work, and for hosting CT during a research visit to Canada. We thank J. J. Wilkinson and R. Herrington for constructive reviews.

References

ABBOTT, D.H. & HOFFMAN, S.E. 1984. Archaean plate tectonics revisited, 1. Heat flow, spreading rate, and age of subducting lithosphere and their effects on the origin and evolution of continents. *Tectonics*, **3**, 429–448.

ABBOTT, D.H., BURGESS, L., LINGHI, J. & SMITH, W.H.F. 1994. An empirical thermal history of the Earth's upper mantle. *Journal of Geophysical Research*, **99**, 13 835–13 850.

BONHAM-CARTER, G.F. 1996. *Geographic Information Systems for Geoscientists: Modelling with GIS*. Pergamon/Elsevier Science Publications.

CAWTHORNE, R.G. & WALRAVEN, F. 1998. Emplacement and crystallization time for the Bushveld Complex. *Journal of Petrology*, **39**, 1669–1687.

DAVIS, J.C. 2002. *Statistics and Data Analysis in Geology*, 3rd edn. John Wiley & Sons.

DE WIT, M.J. 1998. On Archaean granites, greenstones, cratons and tectonics: does the evidence demand a verdict? *Precambrian Research*, **91**, 181–226.

DE WIT, M.J. & HART, R.A. 1993. Earth's earliest continetal lithosphere, hydrothermal flux, and crustal recycling. *Lithos*, **30**, 309–336.

DE WIT, M.J. & HYNES, A. 1995. The onset of interaction between the hydrosphere and oceanic crust, and the origin of the first continental lithosphere. *In*: COWARD, M. & RIES, A. (eds) *Early Precambrian Processes*. Geological Society, London, Special Publications, **95**, 1–9.

DE WIT, M.J., JEFFERY, M., BERGH, H. & NICOLAYSEN, L. 1988. *Geological map of sectors of Gondwana reconstructed to their disposition ~150 Ma, scale 1:10.000.000*. American Association Petroleum Geologists, Tulsa, Oklahoma, USA.

DE WIT, M.J., THIART, C., DOUCOURE, C.M. & WILSHER, W. 1999. Scent of a supercontinent: Gondwana's ores as chemical tracers – tin, tungsten and the Neoproterozoic Laurentia–Gondwana connection. *Journal of African Earth Sciences*, **28**, 35–51.

DE WIT, M.J., THIART, C., DOUCOURÉ, C.M., MILESI, J.P., BILLA, M., BRAUX, C. & NICOL, N. 2004. *The 'Gondwana Metal-Potential' GIS, A geological and metallogenic synthesis of the Gondwana supercontinent at 1:10 million scale*. Developed & published by CIGCES (South Africa) and BRGM (France) http://www.gondwana.brgm.fr; http://www.uct.ac.za/dept/cigces/gondwana

ECKSTRAND, O.R., GOOD, D.J. & GALL, Q. 2002. *Ni–PGE–Cr Deposits*. World Minerals Geoscience Database Project, Mineral Resources Division, Geological Survey of Canada, unpublished database under revision.

ECKSTRAND, O.R., & GOOD, D.J. (compilers) 2000. *World Distribution of Nickel Deposits*. Geological Survey of Canada, Open File 3791a, 3 diskettes.

EGLINGTON, B. & ARMSTRONG, R.A. 2004. The Kaapvaal craton and adjacent orogens, southern Africa: A geochronology database and overview of the development of the craton. *South African Journal of Geology*, **107**, 13–32.

FERMOR, L.L. 1951. The mineral deposits of Gondwanaland (presidential address). *Transactions of the Institution of Mining and Metallurgy*, London, **536**, 421–465.

FRIMMEL, H.E. & MINTER, W.E.L. 2002. *Recent developments concerning the geological history and genesis of the Witwatersrand gold deposits, South Africa*. Society of Economic Geologists, Special Publication, **9**, 17–45.

FOUCH, M.J., JAMES, D.E., VANDECAR, J.C., VAN DER LEE, S. & THE KAAPVAAL SEISMIC GROUP. 2004. Mantle seismic structure beneath the Kaapvaal and Zimbabwe Cratons. *South African Journal of Geology*, **107**, 33–44.

GROVES, D.I., HO, S.E., ROCK, N.M.S., BARLEY, M.E. & MUGGERIDGE, M.T. 1987. Archaean cratons, diamonds and platinum: evidence for long lived crust mantle systems. *Geology*, **15**, 801–805.

HART, R.J., DE WIT, M.J. & TREDOUX, M. 1997. Refractory trace elements in diamonds: further clues to the origin of the ancient cratons. *Geology*, **25**, 1143–1146.

HART, R.J., MCDONALD, I., TREDOUX, M., ET AL. 2004.

New PGE and Re/Os isotope data from lower crustal sections of the Vredefort dome and a reinterpretation of its 'crust on edge' profile. *South African Journal of Geology*, **107**, 173–184.

HART, S.R. & KINLOCH, E.D. 1989. Osmium isotope systematics in Bushveld and Witwatersrand ore deposits. *Economic Geology*, **84**, 1651–1655.

HERRINGTON, R.J., EVANS, D.M. & BUCHANAN, D.L. 1997. Metallogenic aspects. *In*: DE WIT, M.J. and ASHWAL, L.D. (eds) *Greenstone Belts*. Oxford University Press, Oxford, 176–220.

JAMES, D.E., FOUCH, M., VANDECAR, J. & VAN DER LAAN, S. 2001. Tectospheric structure beneath southern Africa. *Geophysical Research Letters*, **28**, 2485–2488.

JENKINS, C.L., VINCENT, R., ROBERT, F., POULSEN, K.H., GARSON, D.F. & BLONDÉ, J.A. 1997. *Index-level Database for Lode Gold Deposits of the World*. Geological Survey of Canada, Open File 3490, diskette.

KIRK, J., RUIZ, J., CHESLEY, J., TITLEY, S. & WALSHE, J. 2001. A detrital model for the origin of gold and sulfides in the Witwatersrand basin based on Re–Os isotopes. *Geochimica et Cosmochimica Acta*, **65**, 2149–2159.

KIRKHAM, R.V. & DUNNE, K.P.E. 2000. *World Distribution of Porphyry, Porphyry-Associated Skarn, and Bulk-Tonnage Epithermal Deposits and Occurrences*. Geological Survey of Canada, Open File 3792a, diskette.

KIRKHAM, R.V. & DUNNE, K.P.E. 2002. *Sediment-Hosted Copper Deposits*. World Minerals Geoscience Database Project, Mineral Resources Division, Geological Survey of Canada, unpublished database under revision.

KIRKHAM, R.V., CARRIÈRE, J.J., LARAMÉE, R.M. & GARSON, D.F. (compilers) 1994. *Global Distribution of Sediment-Hosted Stratiform Copper Deposits and Occurrences*. Geological Survey of Canada, Open File 2915b, Map, report, and diskette.

KIRKHAM, R.V., CARRIÈRE, J.J., RAFER, A. & BORN, P. 2002. *Sediment-Hosted Copper Deposits*. World Minerals Geoscience Database Project, Mineral Resources Division, Geological Survey of Canada, unpublished database under revision.

LEDRU, P., JOHAN, V., MILESI, J-P., TEGYEY, M. 1994. Markers of the last stages of the Paleoproterozoic collision: evidence for a 2 Ga continent involving circum-South Atlantic provinces. *Precambrian Research*, **69**, 169–191.

MAIER, W.D. & BARNES, S-J. 2004. Pt/Pd and Pd/Ir ratios in mantle-derived magmas: a possible role for mantle metasomatism. *South African Journal of Geology*, **107**, 333–340.

MCDONALD, I., DE WIT, M.J., SMITH, C.B., BIZZI, L. & VILJOEN, K.S. 1995. The geochemistry of the platinum-group elements in Brazilian and Southern African Kimberlites. *Geochemica et Cosmochimica Acta*, **59**, 2883–2903.

MIHALASKY, M.J. & BONHAM-CARTER, G.F. 2001. Lithodiversity and its spatial association with metallic mineral sites, Great Basin of Nevada. *Natural Resources Research*, **10**, 209–226.

POLLACK, H.N. 1997. Thermal characteisitcs of the Archaean. *In*: DE WIT, M.J. & ASHWAL, L.D. (eds) *Greenstone Belts*. Oxford University Press, Oxford, 223–232.

SAWKINS, F.J. 1990. *Mineral Deposits in relation to Plate Tectonics*. 2nd edition. Springer-Verlag, Berlin.

SHIREX, S.B., HARRIS, J.W., RICHARDSON, S.H., *ET AL.* 2002. Diamond genesis, seismic structure, and the evolution of the Kaapvaal–Zimbabwe Craton. *Science*, **297**, 1683–1686.

STANKIEWICZ, J., CHEVROT, S., VAN DER HILST, R.D. & DE WIT, M.J. 2002. Crustal thickness, discontinuity depth and upper mantle structure beneath southern Africa: constraints from body wave conversions. *Physics of Earth and Planetary Interiors*, **130**, 235–251.

STOTT, G.M. 1997. The Superior Province, Canada. *In*: DE WIT, M.J. & ASHWAL, L.D. (eds) *Greenstone Belts*. Oxford University Press, Oxford, 480–507.

STRIK, G.H.M.A. 2004. *Paleomagnetism of late Archaean flood basalt terrains*. PhD thesis, University of Utrecht, Netherlands, 149pp.

TRENDALL, A.F. 1983. The Hamersley basin. *In*: TRENDALL, A.F. & MORRIS, R.C. (eds) *Iron-formation: facts and problems*. Elsevier, Amsterdam, 69–129.

THIART, C. & DE WIT, M.J. 2000. Linking Spatial Statistics to GIS: exploring potential gold and tin models of Africa. *South African Journal of Geology*, **103**, 215–230.

THIART, C. & DE WIT, M.J. 2005. Metallogenic fingerprints of Archaean cratons and changing patterns of mineralisation during Earth evolution. *In*: KESLER, S. & OHMOTO, H. (eds) *Evolution of the Earth's early atmosphere, hydrosphere and biosphere: constraints from ore deposits*. Geological Society of America Special Memoir (in press).

TREDOUX, M., DE WIT, M.J., HART, R.J., ARMSTRONG, R.A., LINDSAY, N. & SELLSCHOP, J.P.F. 1989. Platinum-group elements in a 3.5 Ga nickel–iron occurrence: possible evidence of a deep mantle origin. *Journal of Geophysical Research*, **94**, (B1), 795–813.

WILSHER, W. 1995. *The distribution of selected mineral deposits across Gondwana with geodynamic implications*. Unpublished PhD thesis, University of Cape Town. 258 pp.

WILSHER, W., HERBERT, R., WULLSCHLEGER, N., NAICKER, I., VITALI, E. & DE WIT, M.J. 1993. Towards intelligent spatial computing for the Earth Sciences in South Africa. *South African Journal of Science*, **89**, 315–323.

WINDLEY, B.F. 1991. *The Evolving Continents*. 2nd edn. John Wiley, New York.

ZEGERS, T.E., DE WIT, M.J., DANN, J. & WHITE, S.H. 1998. Vaalbara, Earth's oldest supercontinent: a combines structural, geochronological and paleomagnetic test. *Terra Nova*, **10**, 250–259.

Controls on the heterogeneous distribution of mineral deposits through time

DAVID I. GROVES[1], RICHARD M. VIELREICHER[1], RICHARD J. GOLDFARB[2]
& KENT C. CONDIE[3]

[1]*Centre for Global Metallogeny, School of Earth and Geographical Sciences,
The University of Western Australia, Crawley, WA 6009, Australia
(e-mail: dgroves@segs.uwa.edu.au)*

[2]*United States Geological Survey, Box 25046, MS 964, Denver Federal Center, Denver,
Colorado 80225-0046, USA*

[3]*Department of Earth and Environmental Science, New Mexico Institute of Mining and
Technology, Soccorro, New Mexico 87801, USA*

Abstract: Mineral deposits exhibit heterogeneous distributions, with each major deposit type showing distinctive, commonly unique, temporal patterns. These reflect a complex interplay between formational and preservational forces that, in turn, largely reflect changes in tectonic processes and environmental conditions in an evolving Earth. The major drivers were the supercontinent cycle and evolution from plume-dominated to modern-style plate tectonics in a cooling Earth. Consequent decrease in the growth rate of continental crust, and change from thick, buoyant sub-continental lithospheric mantle (SCLM) in the Precambrian to thinner, negatively buoyant SCLM in the Phanerozoic, led to progressive decoupling of formational and preservational processes through time. This affected the temporal patterns of deposit types including orogenic gold, porphyry and epithermal deposits, volcanic hosted massive sulphide (VHMS), palaeoplacer Au, iron oxide, copper gold (IOCG), platinum group elements (PGE), diamond and probably massive sulphide SEDEX deposits. Sedimentary mineral deposits mined for redox-sensitive metals show highly anomalous temporal patterns in which specific deposit types are restricted to particular times in Earth history. In particular, palaeoplacer uranium, banded iron formation (BIF) and BIF-associated manganese carbonates that formed in the early Precambrian do not reappear in younger basins. The most obvious driver is progressive oxidation of the atmosphere, with consequent long-term changes in the hydrosphere and biosphere, the latter influencing the temporal distribution and peak development of deposits such as Mississippi Valley types (MVT), hosted in biogenic sedimentary rocks.

Economic geologists have long been fascinated by the heterogeneous distribution of mineral deposits, both in space in terms of metallogenic provinces and in time in terms of metallogenic epochs (e.g. Stanton 1972). Numerous studies of mineral deposits over the past 20 to 30 years have led to refinement of their classification into relatively well-defined types, such that their distribution in space and time can be defined. At the same time, enormous improvements in the robustness, accuracy and precision of mineral chronometers (e.g. SHRIMP U–Pb zircon and phosphate studies; Rasmussen *et al.* 2001; Compston *et al.* 2002; Vielreicher *et al.* 2003; Ar/Ar mica and amphibole studies: Renne *et al.* 1998; Re–Os in molybdenite and other sulphides and gold: Stein *et al.* 1997; Kirk *et al.* 2001) have allowed temporal peaks in the distribution of mineral deposits to become progressively better defined. Thus, although temporal patterns of deposit distribution have been defined broadly in the past (e.g. Meyer 1981, 1988), it is only in the last decade that details of these temporal patterns have been deciphered. Even now, the lack of geochronological data hampers definition of the temporal patterns for several mineral-deposit types. As discussed by Barley & Groves (1992), some periodic changes in the temporal distribution of specific mineral-deposit types may be due to long-term periodic supercontinent formation and breakup, with certain deposits (e.g. porphyry Cu–Au, epithermal Au–Ag, orogenic Au, VHMS) formed preferentially in convergent margins during supercontinent assembly, whereas others (e.g. iron oxide copper–gold; sediment-hosted uranium) formed during intracratonic extension (e.g. Mitchell & Garson

From: McDONALD, I., BOYCE, A. J., BUTLER, I. B., HERRINGTON, R. J. & POLYA, D. A. (eds) 2005. *Mineral Deposits and Earth Evolution.* Geological Society, London, Special Publications, **248**, 71–101. 0305-8719/$15.00
© The Geological Society of London 2005.

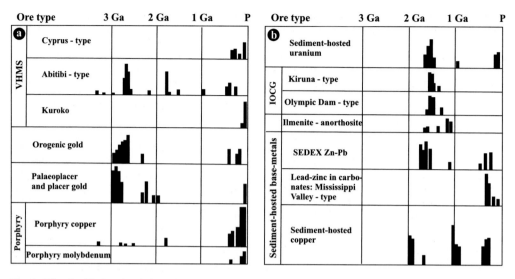

Fig. 1. Historical figure showing distribution through time of selected mineral deposits. (**a**) Orogenic convergent margin settings; (**b**) anorogenic or continental-basin settings. Peaks in abundance of anorogenic and continental-basin metal deposits appear to correspond to: extensive Mesoproterozoic continents; Neoproterozoic continent; and Pangea. Adapted from Barley & Groves (1992), with some deposit types renamed to better reflect the text in this paper, but no new deposit types added.

1981; Sawkins 1984; Kerrich *et al.* 2000). These differences are well illustrated in Figure 1, adapted from Barley & Groves (1992). However, the distribution of most mineral-deposit types is more complicated than this. Some are restricted to one or more, very specific periods in Earth history, whereas others occur periodically throughout the rock record (Fig. 1). The temporal patterns contain both formational and preservational components.

In a broad sense, the heterogenous temporal distribution of mineral deposits can be ascribed to both temporal changes in the processes that produce the deposits, or the depositional environments in which they form, and to the preservational potential of the deposit-hosting environments. In turn, temporal changes in ore-deposit forming processes can be ascribed to: the secular decrease in global heat flow in a cooling Earth; long-term changes in tectonic processes; and the evolution of the atmosphere–hydrosphere–biosphere system. Tectonic changes may be a consequence of a cooling Earth and may also induce relatively short-term fluctuations on long-term evolution of the atmosphere–hydrosphere system. Changes in tectonic regime may, in addition, affect the preservation potential of terranes in which the mineral deposits formed. There are thus feedback loops in dynamic evolving global systems.

In this paper, the temporal distribution of

several important mineral-deposit types is considered in terms of an evolving Earth. Length constraints make an exhaustive review of all mineral deposit types impossible, so those that best illustrate evolving Earth processes are selected for review. Deposits of some of the abundant elements, such as bauxites, chromitites and titanium and vanadium oxide deposits, are not discussed, nor are ores of the granitophile elements such as tin, tungsten, tantalum–niobium, beryllium and others. A brief review of the evidence for long-term change in tectonic processes, and related change in the nature of the lithosphere as a result of a cooling Earth, is followed by a discussion of the formational and/or preservational patterns of a number of hydrothermal and magmatic deposits of metals that are least influenced by redox conditions. A brief review of the various models for atmosphere–hydrosphere and biosphere evolution then leads into a discussion of mainly sedimentary rock-hosted deposits of redox-sensitive metals whose temporal patterns may reflect changes in the atmosphere–hydrosphere–biosphere system. The relative roles of tectonic and environmental changes on the formational and preservational patterns of mineral deposits are highlighted. The major features of the various deposit types discussed in this paper are summarized in Table 1 in order of description of their secular variation.

The paper revolves around the generation and interpretation of the temporal distribution of various ore deposit types. Numerous figures are constructed to facilitate the discussion. There are a number of critical problems in generating these temporal distributions:

1. it is impossible to include all deposits within each deposit type as they are too numerous, and hence a cutoff that captures > 90% of production has been applied;
2. in some places, particularly for central Asian countries, Russia and China, production and resource data are incomplete or vary between papers or databases, some of which are in construction but not yet commercially available;
3. even for well-described world-class to giant deposits, total tonnages (or Moz) of metal may vary by a factor of two, dependent on whether only production, or production-plus-reserves are included, or a total resource (includes inferred and probable resources) is given;
4. for some deposit types, databases are available, but confidential; and
5. the absolute age of many deposits is unknown or debated, and thus time ranges have to be given. Therefore, the figures reflect the temporal distribution of mineral deposits but are indicative rather than definitive in terms of quantification of resources.

Tectonic evolution in a cooling Earth

Since the advent of modern plate-tectonic theory, there has been a significant literature on the relationship between mineral deposits and their global tectonic setting (e.g. Mitchell & Garson 1981; Sawkins 1984; Titley 1993; Kesler 1997; Barley et al. 1998; Kerrich et al. 2000; Goldfarb et al. 2001a,b; Blundell 2002). The nature of tectonic processes through time must be defined if the temporal evolution of mineral deposits is to be understood. The brief review given below is summarized from a more exhaustive discussion by Groves et al. (2005).

Juvenile crustal growth, mantle plumes and the supercontinent cycle

The evolution of the Earth is inextricably linked to the progressive decay of heat production, lowering mantle temperatures and viscosity, and thickening of the sub-continental lithosphere (e.g. Pollack 1986). Some consider this to be a more-or-less continuous evolutionary process (e.g. Pollack 1997), but since the early 1990s, it has been more widely accepted, through the distribution of U–Pb zircon ages, coupled with Nd and Hf isotopic data, that there has been episodic growth of juvenile continental crust (e.g. Gastil 1960; Taylor & McLennan 1985; Stein & Hoffman 1994; Condie 1998, 2000). In this model, juvenile continental growth commenced before 3.0 Ga, with two major peaks at c. 2.7 Ga and at c. 1.9 Ga, with smaller peaks at c. 2.8, 2.5, 1.7, 0.48, 0.28 and 0.1 Ga (Fig. 2a). Ages from Archaean cratons suggest that the first supercontinent(s), probably small relative to later analogues, could have formed during frequent collisions and suturing of continental blocks and oceanic terranes between 2750 and 2650 Ma (e.g. Aspler & Chiarenzelli 1998). This correlates with the Late Archaean peak in juvenile crust production at about 2700 Ma, suggesting a connection between supercontinent formation and production of juvenile continental crust, although the nature of this connection is unclear (e.g. Hoffman 1988; Murphy & Nance 1992). Similarly, a new supercontinent appears to have formed between 1900 and 1800 Ma (Hoffman 1989; Rogers 1996; Condie 2002a; Pesonen et al. 2003), the rapid growth of Pangea occurred between 480 and 250 Ma, and a new supercontinent has begun to form over the last 150 Ma, further suggesting that increased production rates of juvenile crust correlate with formation of supercontinents (Condie 2004).

Major peaks in juvenile crust formation are interpreted as having been caused by mantle overturn events that gave rise to a large number of mantle plumes in less than 100 million years (Condie 1998, 2000; Isley & Abbott 1999). Using several plume proxies, such as flood basalts, high-Mg lavas including komatiites, giant dyke swarms, and layered intrusions, Abbott & Isley (2002) defined 36 mantle plume events, each lasting 10 million years, in the last 3.8 billion years (Fig. 2b). Several plume proxy peaks, particularly at c. 2.75, 1.8, 0.25 and 0.1 Ga, are close to calculated peaks in juvenile crust production, but it is evident that, if plume-proxy abundance reflects a mantle plume event, not all events correlate with increased rates of production of continental crust (Condie 2004).

For this reason, Condie (2004) suggests that there are two types of mantle plume events. The first is composed of long-lived (>200 Ma), shielding mantle plume events (Condie 2004) in which shielding of a large supercontinent (e.g. Trubitsyn et al. 2003) causes mantle upwelling beneath it (Lowman & Jarvis 1996), followed by

Table 1. *Summary of major features of deposit groups discussed in text*

Deposit type	Metal association	Tectonic setting	Geological setting	Major controls	P–T conditions	Giant/world-class examples (in age order)	Occurrence in geological time	Most-endowed periods	Key references
A Ni-Cu sulphide	Ni – Cu – PGE ± Co ± Au	craton margin: mantle plume?	intercratonic rift zones	major fault zones and intersections	>800 °C; 30–150 MPa	Mt Keith, Australia; Sudbury, Canada; Jinchuan, China; Noril'sk, Russia	Archaean to Phanerozoic	Proterozoic; Late Palaeozoic; Mesozoic	Lesher (1989); Naldrett (1997, 1999, 2002)
B Porphyry Cu-Au	Cu – Au – Ag ± Mo± Pb ± Sn ± Te ± W ± Zn	convergent margin	continental arc; island arc	oxidized, I-type, subvolcanic, alkalic intrusion	200–600 °C; 50–100 MPa	Almalyk, Uzbekistan; Bingham, USA; Baja de la Alumbrera, Argentina; Grasberg, Indonesia	Archaean to Recent	Mid-Mesozoic to Recent	Sillitoe (1997, 2000)
C Epithermal Au-Ag	Au – Ag – Sb – Hg ± Cu ± Pb ± Te ± W ± Zn	convergent margin	continental arc; island arc	subvolcanic, calc-alkalic intrusion	generally <300 °C; 5–20 MPa	Pueblo Viejo, Dominican Republic; Cripple Creek, USA; Porgera and Ladolam, Papua New Guinea	Phanerozoic	Cenozoic	White et al. (1995); Cooke & Simmons (2000)
D Orogenic Au	Au – Ag –As – Te – Bi ± Cu ± Pb ± Sb ± W ± Zn	convergent margin	fore-arc; accretionary wedge	shear zones; anticlinal fold hinges	250–650 °C; 50–600 MPa	Timmins, Canada; Golden Mile, Australia; Ashanti, Ghana; Bendigo, Australia; Mother Lode, California	Archaean to Phanerozoic	Late Archaean; Palaeoproterozoic; Mesozoic	Goldfarb et al. (2001a); Groves et al. (1998, 2003)
E Intrusion-related Au	Au – Ag – Te – Bi – As – Cu ± Mo ± Pb ± Sb ± Sn ± W ± Zn	convergent margin	near craton; back-arc	reduced alkalic granitoids along faults	350–500 °C; 150–300 MPa	Fort Knox, Alaska	Mainly Phanerozoic	Late Mesozoic	Thompson & Newberry (2000); Lang & Baker (2001)
F VHMS	Cu – Pb – Zn ± Ag ± Au	variable plate extension settings	mid-ocean ridge; back arc; accretionary terranes	seafloor convection zones	250–400 °C; 10–50 MPa	Kidd Creek and Flin Flon, Canada; Maikain, Kazakhstan; Rio Tinto, Spain; Windy Craggy, Canada	Archaean to Recent	Early Phanerozoic	Large (1992); Barrie & Hannington (1999); Large et al. (2002)
G Palaeoplacer and placer Au–U	Au ± U	foreland basin	fluvial marine environments	braided river system	Earth surface conditions	Witwatersrand, South Africa; Tarkwa, Ghana; Nome, Alaska	Archaean to Recent	Late Archaean; Early Proterozoic; Recent	Henley & Adams (1979); Frimmel & Minter (2002); Frimmel et al. (in press)
H Fe oxide Cu-Au	Cu – Au – U ± As ± Co ± Mo ± W	near-craton margin	intracratonic extensional zones	shear/fault zone; alkalic magmatism	200–500 °C; >150 MPa	Salobo, Brazil; Palabora, South Africa; Olympic Dam, Australia; Candelaria, Chile	Archaean to Phanerozoic	Late Archaean; Early to Mid-Proterozoic; Late Mesozoic	Hitzman et al. (1992); Hitzman (2000); Williams & Skirrow (2000); Haynes (2002)
I PGE in layered intrusion	PGE – Ni ± Cu	within craton; mantle plume?	intracratonic rift zones	giant intrusions	800–1000 °C; 100–300 MPa	Stillwater, USA; Great Dyke, Zimbabwe; Bushveld, South Africa	Archaean to Palaeoproterozoic	Late Archaean; Early Proterozoic	Cawthorn (1999); Stribrny et al. (2000); Cawthorn et al. (2002)

Table 1. (*Continued*)

Deposit type	Metal association	Tectonic setting	Geological setting	Major controls	P–T conditions	Giant/world-class examples (in age order)	Occurrence in geological time	Most-endowed periods	Key references
J Diamond	Carbon	cratons with early Pe SCLM	intracratonic extensional zones	kimberlite, lamproite intrusions	1200–1500 °C; 50–70 GPa	Mir, Russia; Ekati, Canada; Premier, South Africa; Argyle, Australia; Orapa, Botswana	Late Archaean to Recent	Late Proterozoic; Cenozoic	Gurney et al. (1999); Navon (1999); Richardson et al. (2004)
K Unconformity-related U	U ± Au ± Ag ± As ± Co ± Mo ± Ni±Se	intracratonic	sedimentary basin	major unconformities	<300 °C; <150 MPa	Key Lake, Canada; Jabiluka, Australia	Mesoproterozoic	Mesoproterozoic	Wilde et al. (1989); Plant et al. (1999)
L Sandstone-type U	U ± Mo ± V	intracratonic	terrestrial sedimentary basin	redox fronts; shale-sandstone contacts	~100 °C; <100 MPa	Ambrosia Lake, USA	Phanerozoic	Mesozoic; Tertiary	Adams (1991); Plant et al. (1999)
M BIF (enriched Fe)	Fe (low P)	intracratonic; passive margin; platforms	deep, sediment-starved basin	faults, folds, dykes (for enriched Fe)	marine conditions for BIF: >100 °C; <100 MPa for enriched Fe	Mt Tom Price and Mt Whaleback, Australia; Carajas Province, Brazil	Archaean to Phanerozoic	Palaeoproterozoic	Trendall & Morris (1983); Klein & Beukes (1992); Gross (1993)
N Minette-type (oolitic) ironstone	Fe	passive continental margin?	shallow marine environment	supergene leaching?	marine conditions	Birmingham District, Alabama, USA Lorraine, France	Phanerozoic	Silurian; Jurassic	Alling (1947); Melon (1962); Gross (1996)
O Mn-rich iron formation	Mn	as for BIF	open-shelf environment	deep-water environments	marine conditions	Kalahari deposits, South Africa	Archaean to Palaeoproterozoic	Earliest Palaeoproterozoic	Roy (1981, 1988, 2000)
P Oolitic oncolitic and shale-hosted Mn	Mn	passive continental margin	open-shelf environment	shallow-marine to deltaic environments	marine conditions	Molango, Mexico; Nikopol, Ukraine	Palaeoproterozoic to Tertiary	Palaeoproterozoic; Mesozoic	Gurvich (1981); Frakes & Bolton (1984)
Q MVT	Pb – Zn – Ag ± Co ± Cu ± Ni	peripheral to foreland basins	platformal carbonate sequences	faults and hydrothermal karst	80–150 °C; <30 MPa	Esker, Canada; East Tennessee and Tri-State, USA; Mehdiabad, Iran	Proterozoic to Phanerozoic	Mid- to Late-Palaeozoic; Late Mesozoic	Sangster (1996); Leach et al. (2001); Bradley & Leach (2003)
R SEDEX	Pb – Zn ± Cu ± Ag ± As ± Bi ± Sb	intracratonic rift zones	rift-cover sequences in sedimentary basins	faults associated with half-grabens	50–300 °C; 30–60 MPa	Broken Hill and HYC, Australia; Howards Pass, Canada; Red Dog, Alaska; Jinding, China	Proterozoic to Phanerozoic	Mid-Proterozoic; Palaeozoic	Lydon (1996); Large et al (1998); Cooke et al. (2000)

Letters A–R provide key for reference in text. SCLM, sub-continental lithospheric mantle.

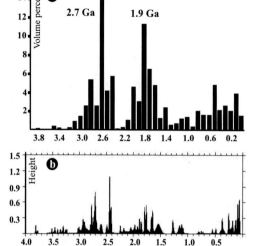

Fig. 2. (**a**) Frequency distribution of juvenile continental crust based on a total volume of continental crust of 7.177×10^9 km³. Juvenile crust ages are U/Pb zircon ages used in conjunction with Nd isotope data and lithological associations. Note, *y*-axis shows volume percent growth. Modified after Condie (1998, 2000). (**b**) Distribution of mantle plume events deduced from time series analysis of plume proxies from Abbott & Isley (2002). Peak height depends on the number of plume proxies and the errors of the age, the latter set at 5 million years.

the development of numerous mantle plumes within the upwelling, and fragmentation of the supercontinent. In contrast, events that are associated with peaks in juvenile crust production, for example at *c.* 2.7 and 1.9 Ga, are short-lived (<100 Ma), catastrophic mantle plume events (Condie 2004) whose exact cause is unclear. Crustal growth associated with catastrophic mantle plume events is mainly by addition of arc components to the continents, either by accretion or collision. In this scenario, the mineral deposits shown in Figure 1a (including both Abitibi-type VHMS and orogenic gold deposits) would broadly correspond to a possible catastrophic mantle plume event at 2.7 Ga and those in Figure 1b may correlate with shielding mantle plume events.

Transition from Archaean to modern-style plate tectonics

If the Earth has cooled steadily with time (e.g. Pollack 1986; Vlaar 2000; Martin & Moyen 2002), it is theoretically possible that plate

tectonics would have been different in the Archaean (e.g. Fyfe 1978). For example more melt would have been produced at ocean ridges from a hotter Archaean mantle and, consequently, oceanic crust would have been thicker (e.g. Bickle 1990), possibly 20 km or more than the present crust (Sleep & Windley 1982). Considerations of the density and thickness distribution of modern and likely Archaean lithosphere and crust (Davies 1992) suggest that Archaean oceanic plates would be ready to subduct before they reached neutral buoyancy, whereas the time required to reach neutral buoyancy at present is less than the subduction age (e.g. Sprague & Pollack 1980; Parsons 1982). The crossover point (Davies 1992), between about 2.5 and 2.0 Ga, was presumably when buoyant plate tectonics (de Wit 1998), possibly with numerous smaller plates (e.g. Pollack 1997), began to evolve into modern plate tectonics through an intermediate phase in the Proterozoic.

Thus, the Archaean and probably Palaeoproterozoic were dominated by various types of plume events that induced the growth of voluminous juvenile continental crust. The abundance of komatiites at this time attests to the intensity of plume activity (e.g. Campbell *et al.* 1989; Pirajno 2000). Although there have been arguments against the concept (e.g. Hamilton 1998), the overall similarity of volcanic and volcanosedimentary successions and deformation sequences and patterns of Archaean greenstone belts to those in modern convergent settings (e.g. de Wit *et al.* 1987; Barley *et al.* 1998), and the recognition of fossil subduction zones in the Abitibi Belt of Canada (e.g. Ludden & Hynes 2000), suggest that early Precambrian tectonics was dominated by a form of plume-influenced, modified plate tectonics. The contrasts in volcanic and intrusive suites can be ascribed to shallow subduction of hot, young oceanic lithosphere in the early Precambrian versus subsequent steeper subduction of cooler, older oceanic lithosphere, as summarized, for example, by Kerrich *et al.* (2000).

Evolution of sub-continental lithospheric mantle

The nature of the lithospheric mantle below the continental crust was also influenced by the greater heat flow and increased mantle plume activity in the early Earth. Geological evidence for this is provided by contrasts between the broadly equidimensional Archaean and Palaeoproterozoic cratons, with anomalously large volumes of granitoids, and the highly elongate orogens of the Mesoproterozoic to Recent. Each craton has its own distinct metallogenic associations, irrespective of the fact that the mineralization may be pre-, syn-, or post-cratonization (de Wit & Thiart 2003). This suggests that each craton is unique, not only in terms of its early orogenic stages, but also in terms of its sub-continental lithospheric mantle (hereafter SCLM) root. It is thus imperative to understand the evolution of the SCLM and the role it plays in the temporal patterns shown by mineral deposits.

Various studies (e.g. O'Reilly & Griffin 1996; Poudjom Djomani *et al.* 2001; Griffin *et al.* 2003) demonstrate that there are progressive changes in the mineralogical composition of Archaean to Phanerozoic SCLM, that the Archaean lithospheric geotherms were lower than those in the Phanerozoic, and that typical thicknesses of lithosphere varied from 250–180 km in the Archaean through 180–150 km in the Proterozoic to 140–60 km in the Phanerozoic (Fig. 3). Calculated mean densities (at 20 °C) range from about 3.31 Mg m^3 for Archaean SCLM as against about 3.35 and 3.36 Mg m^3 for Proterozoic and Phanerozoic SCLM, respectively. The driving force for progressive change with time is interpreted to be decreasing mantle plume activity, and hence decreasing deep high-degree partial melting, with associated residues and/or cumulates (Griffin *et al.* 2003). The secular evolution of SCLM implicates broadly synchronous formation of continental crust and its lithospheric root, and suggests that they have been linked throughout their subsequent history (Griffin *et al.* 2003; Sleep 2003). This explains not only some of the distinctive deposit associations within cratons (Groves *et al.* 1987), but also the unique metallogenetic associations of each craton (de Wit & Thiart 2003).

From a tectonic viewpoint, the most important aspect is the secular change in the mean density, and hence buoyancy, of the SCLM. Archaean SCLM would be buoyant relative to asthenosphere under any reasonable geological conditions, stress and viscosity would be reduced, and it could not be delaminated by gravitational processes, rather only by rifting and replacement by more fertile asthenosphere (Poudjom Djomani *et al.* 2001). Thus, mineral deposits formed in the late Archaean or thereafter should have enhanced preservational potential. Proterozoic SCLM is estimated to be moderately buoyant and unlikely to be delaminated, whereas Phanerozoic SCLM should be negatively buoyant and readily delaminated (Poudjom Djomani *et al.* 2001). Therefore, mineral deposits sited within progressively younger SCLM should have progressively lower preservational potential.

Summary

Under the influence of a cooling Earth, tectonic processes evolved from mantle-plume dominated, 'buoyant' style of plate tectonics to modern-style plate tectonics. Coincident with this was a trend from rapid and voluminous formation of juvenile continental crust, with major peaks at *c.* 2.7 and 1.9 Ga, which were associated with catastrophic mantle plumes and the development of equidimensional Archaean–Palaeoproterozoic cratons, to less voluminous crustal growth and the evolution of highly elongate orogens. Progressive decrease in mantle plume intensity and frequency with time led to progressive evolution of thinner, less-buoyant SCLM. Juvenile continental crust and its SCLM root were linked through their subsequent history, strongly influencing both their metallogenic and preservational potential. In particular, formational and preservational processes would have been linked in the early Earth but progressively decoupled in the Proterozoic and Phanerozoic. Temporal patterns are presented below for several, relatively redox-insensitive, mineral deposits whose

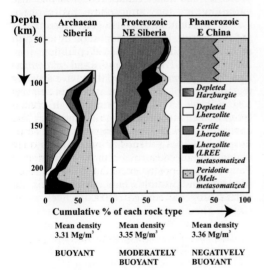

Fig. 3. Depth and depth variation in composition of selected Archaean, Proterozoic and Phanerozoic lithospheric sections, showing changes in both depth and composition of sub-continental mantle lithosphere with time. Mean densities (at 20 °C) and relative buoyancy also shown. Modified from Griffin *et al.* (2003).

secular variation is thought to be related to a cooling Earth and consequent changes in tectonic processes. The secular evolution of epigenetic gold deposits is discussed in more detail by Groves *et al.* (2005).

Secular variations in mineral deposits related to a cooling Earth

The most obvious example of a secular change in mineral deposits directly related to a cooling Earth is shown by magmatic iron–nickel–copper sulphide deposits, which commonly have PGE credits (Table 1A). These deposits show a secular pattern in which the oldest, mainly Archaean, deposits are komatiite-associated Fe–Ni–Cu deposits, with low PGE grades, whereas the post-Archaean deposits are hosted in mafic–ultramafic intrusions and have moderate to significant PGE credits (Fig. 4). Komatiites tend to be restricted to the period of maximum mantle plume activity in the early Earth, as they represent high degrees of mantle partial melting (Arndt *et al.* 2002), although Mesozoic examples are recorded on Gorgona Island as a result of plume activity (Storey *et al.* 1991; Arndt *et al.* 1997). The komatiite-associated deposits include massive to matrix Fe–Ni–Cu sulphide ores at the base of komatiitic flow sequences (e.g. Kambalda, Western Australia) and disseminated Fe–Ni–Cu sulphide ores within thicker komatiitic flows or sub-volcanic sills (e.g. Perseverance, Western Australia). Ground melting by the exceedingly high temperature lavas appears to have been important in the genesis of the massive sulphides (e.g. Lesher 1989). In contrast, large post-Palaeoproterozoic deposits are massive Ni–Cu sulphide ores at or near the base of

layered mafic–ultramafic sills that are sited on the outside of craton margins (e.g. Naldrett 1997). These include, amongst others, the giant Sudbury, Jinchuan and Noril'sk deposits (Fig. 4). The intrusions may have high-MgO parent magmas (e.g. Hawkesworth *et al.* 1995), but they are far less magnesian than the komatiitic melts, reflecting lower heat-flux in a cooling Earth.

In addition to the gross change from komatiitic to non-komatiitic parentage, the deposits show a very striking pattern with giant deposits restricted to the Proterozoic and late Palaeozoic (Table 1). This temporal pattern relates to periods of continental rifting, probably related to shielding mantle plume events and following supercontinent coalescence (e.g. Noril'sk in geometric centre of Laurasia), but predating extensive rift volcanism or ocean creation, with coincident meteorite bombardment of a suitable tectonic setting in the case of Sudbury (cf. Naldrett 2002).

Secular mineral deposit distribution related to changes in tectonic style

Epigenetic gold deposits

Epigenetic gold deposits potentially reflect secular changes in tectonic processes because most formed below the influence of surficial processes, and hence cannot reflect any secular variation in the atmosphere–hydrosphere system. In addition, although gold deposition is redox sensitive, it may be deposited by a shift to more oxidized or more reduced conditions (e.g. Mikucki 1998). *Porphyry Cu–Au* and *epithermal-type Au–Ag deposits* (Table 1B,C) have a strong tectonic control in convergent margin settings (e.g. Sillitoe 1997), but form at high crustal levels (<3 km to surface) in arc and back arc environments with high to extreme uplift rates. Porphyry Cu–Au deposits are thus mostly younger than Mesozoic, although significant Palaeozoic deposits occur in Central Asia, Mongolia and possibly China, particularly in the Altaid orogenic collage of Yakubchuck *et al.* (2002). Their temporal distribution is shown in Figure 5a together with the sources of data. Epithermal-type Au–Ag deposits are mostly Cenozoic to Tertiary with some in the Mesozoic and rare examples in the Palaeozoic (Fig. 5b). A temporal pattern similar to these also characterizes porphyry Mo deposits (Fig. 1a). The older examples of porphyry systems appear to be selectively preserved by accretion or collision of hosting arcs with continental blocks. The virtual absence of porphyry Cu–Au deposits from the late Palaeoproterozoic to the mid-Palaeozoic,

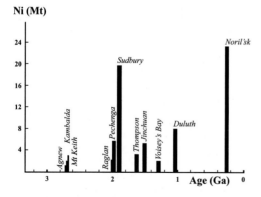

Fig. 4. Distribution of major nickel–copper sulphide deposits with time. Major sources are Naldrett (1999, 2002).

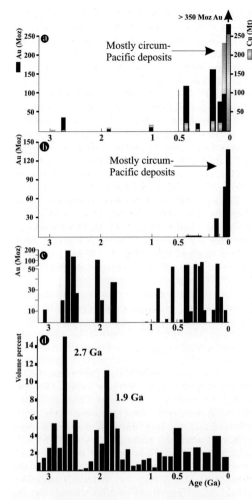

Fig. 5. Temporal distribution of epigenetic gold ± copper ± silver deposits compared with the temporal evolution of juvenile continental crust. (**a**) Porphyry Cu–Au deposits; data from Kirkham & Sinclair (1996); Deshbanduh Sikka & Nehru (1997); Sillitoe (1997, 2000); Perello *et al.* (2001); Camus (2002); Yakubchuck *et al.* (2002) and Chiaradia *et al.* (2004). (**b**) Epithermal Au–Ag deposits; data from White *et al.* (1995); Sillitoe (1997); Cooke & Simmons (2000) and Chiaradia *et al.* (2004); (**c**) Orogenic gold deposits: adapted from Goldfarb *et al.* (2001a) and Groves *et al.* (2003); see Figure 6b for major gold provinces. (**d**) Temporal evolution of continental crust growth; from Condie (2000).

but their appearance both subsequent to that period, as well as in the Archaean, is an interesting temporal pattern discussed further below.

In contrast to porphyry and epithermal deposits, orogenic gold deposits (Table 1D), although formed in the same convergent margin

settings, were deposited in a wide range of crustal environments (3–20 km depth: Groves 1993) during the compressional to trans-pressional deformation that stabilized their host orogens (Groves *et al.* 1998). The deposits, formed over 3.4 billion years of Earth history (Goldfarb *et al.* 2001a,b), are abundant and geographically widespread and, hence, are potentially excellent markers of temporal changes in tectonic processes.

As discussed by Goldfarb *et al.* (2001a,b), orogenic gold deposits of all ages record orogen-wide fluxes of deeply-sourced auriferous fluids as part of the orogenic process in convergent margins, almost certainly in response to high thermal fluxes due to anomalous plate geometries (e.g. Landefeld 1988; Haeussler *et al.* 1995; Griffin *et al.* 1998; Qiu & Groves 1999; Wyman *et al.* 1999) during global-scale tectonic events. As such, their temporal pattern (Fig. 5c) should reflect associated global events, such as periods of crustal growth (Fig. 5d).

The earliest orogenic gold deposits formed in the Middle Archaean in the Pilbara and Kaapvaal cratons, between about 3.4 (e.g. Zegers *et al.* 2002) and 3.1 Ga (e.g. deRonde *et al.* 1991). The ages of later deposits define two major Precambrian peaks at 2.75–2.55 Ga (centred at *c.* 2.65 Ga) and 2.1–1.75 Ga, a marked scarcity of deposits between 1.7 Ga and 600 Ma, and a more cyclic distribution from about 600 Ma to 50 Ma (Fig. 5c). The earliest orogenic gold deposits correspond broadly to the earliest significant growth of continental crust (Fig. 5c,d). The Archaean and Palaeo-proterozoic peaks reflect major episodic growth of Precambrian continental crust, centred at *c.* 2.7 Ga and 1.9 Ga, although Palaeoprotero-zoic orogenic gold peaks flank the latter crustal growth peak. Phanerozoic orogenic gold deposits show a more continuous, shorter-wavelength distribution, similar to that of Phanerozoic orogenic events and associated periods of crustal growth and supercontinent coalescence. The secular evolution of orogenic gold deposits thus reflects the evolutionary trend from strongly episodic, plume-influenced plate tectonics in the Archaean to modern-style plate tectonics discussed above. Gold produc-tion from early Precambrian orogenic gold provinces temporally related to the catastrophic mantle plume events of Condie (2004) is equiv-alent to superior to that of younger provinces (Fig. 5c), despite their antiquity, illustrating the critical role that thermal energy levels play in the formation of giant orogenic gold provinces.

The obvious anomaly is the overall scarcity of

orogenic gold deposits between about 1.7 Ga and 600 Ma (Fig. 5c) despite the record of crustal growth during that period. The only likely exception is the Olympiada deposit whose probable age is *c.* 820 Ma (Safonov 1997) although it may be as young as 600 Ma (Konstantinov *et al.* 1999). This hiatus suggests the influence of another factor in determining the temporal pattern of preserved orogenic gold deposits. Based on the secular changes in tectonic processes discussed above, the rarity of Mesoproterozoic to Neoproterozoic orogenic gold deposits may be explained in terms of preservational potential of hosting terranes. Deposits embedded in the central portions of Archaean (e.g. Kalgoorlie Terrane, Western Australia) and Palaeoproterozoic (e.g. Ashanti Belt, Ghana) cratons would have been underlain by relatively buoyant SCLM, which would have been difficult to delaminate. Hence, there would have been a very high chance of preservation except adjacent to craton margins, where later orogeny might cause uplift and erosion. In contrast, modern-style, highly elongate orogenic belts along the margins of the cratons, with their negatively buoyant SCLM and, in some cases, subsequently overprinting orogenic belts, would be much more prone to uplift and erosion. Thus it is most likely that Mesoproterozoic to Neoproterozoic orogenic gold deposits did form during major continental-crust forming events from 1.7 Ga to 600 Ma, but that progressive erosion of the elongate thin orogens down to their roots removed most of these deposits.

However, the temporal gap in orogenic gold deposits is almost exactly matched by the so-called Proterozoic glacial gap between Palaeoproterozoic and Neoproterozoic snowball events in the snowball Earth model (e.g. Hoffman & Schrag 2002, fig. 12). Snowball events appear to develop in the early stages of supercontinent breakup because this increases the area of continental margins globally, creates smaller, 'wetter continents', and the potential for burial of enhanced organic carbon and related drawdown of atmospheric CO_2 (e.g. Hoffman 1999): the best documented example relates to the fragmentation of Rodinia post-750 Ma (e.g. Hoffman 1999; Meert & Powell 2001). The similarities in the period of the glaciation gap and the dearth of orogenic gold deposits raises the alternative possibility that there was limited subduction and crustal growth at this time due to supercontinent stability. However, the supercontinent that formed towards the end of the Palaeoproterozoic must have been fragmented, even if incompletely (e.g. Condie 2002*b*), prior to the growth of Rodinia between *c.* 1.3–1.2 and 1.0 Ga (e.g. Meert & Powell 2001) which was achieved by a series of continental collisions, grouped under the term Grenvillian. Rare greenschist facies domains within the largely high metamorphic-grade Grenvillian belts contain, albeit relatively small (up to 1 Moz Au resources), orogenic gold deposits, for example in the Sunsas belt of the SW Amazonian Craton in Brazil (e.g. Geraldes *et al.* 1997). Hence, it appears that orogenic events conducive to the evolution of orogenic gold deposits did exist between *c.* 1.7 and 0.6 Ga, and that the dearth of deposits is caused primarily by lack of preservation.

The return of abundant orogenic gold deposits at *c.* 600 Ma (with exception of the potentially older Olympiada deposit) suggests that this is a broad-scale preservational threshold in modern-style orogenic belts. The abundance of gold placers spatially associated with many Phanerozoic orogenic gold provinces reflects their progressive erosion, with the lack of world-class orogenic gold deposits in rocks younger than *c.* 50 Ma suggesting that 50 million years is the minimum period to uplift and expose a major orogenic gold province.

It is concluded that world-class orogenic gold deposits formed by similar processes during major orogenic, crust-forming events throughout Earth history. However, the change from a plume-influenced plate tectonic style to a modern tectonic style induced changes in lithosphere buoyancy, particularly of the SCLM, which strongly influenced the preservational potential of terranes of different age. Hence, the temporal pattern (Fig. 5c) is largely preservational, resulting from secular change in the tectonic process, rather than one caused by a fundamental change in formational process.

During the past decade, there has been interest in a second type of gold deposit generated from aqueous-carbonic hydrothermal systems. These have been termed 'intrusion-related gold deposits' (e.g. Sillitoe 1991; Sillitoe & Thompson 1998; Thompson *et al.* 1999). Most recently, Lang & Baker (2001) have summarized the features of those deposits that contain gold as the only economic metal (Table 1E). There is general acceptance that the Fort Knox deposit and various, small deposits and prospects in the Tombstone gold belt of the Yukon, and at Timbarra, NSW, Australia are intrusion-related (Groves *et al.* 2003). Others have been classified as both orogenic and intrusion-related gold deposits by different authors, but only the examples cited above meet all the criteria of Lang & Baker (2001). It is thus impossible to define their temporal distribution except to note that the undoubted deposits of this type are of late Phanerozoic age. They are therefore not discussed further, other than to record that they seem to occur in deformed and metamorphosed shelf sequences adjacent to cratonic margins in association with small stocks, sills and dykes of unusual granitoid suites of mixed mantle and crustal parentage (e.g. Hart *et al.* 2004), well inboard from the convergent margins where orogenic gold deposits are possibly forming at the same time.

Volcanic-hosted massive sulphide deposits

The volcanic-hosted massive sulphide (VHMS) deposits (Table 1F) are commonly subdivided based upon geographic settings (e.g. Meyer 1981; Fig. 1a) or metal ratios (e.g. Large 1992). However, they are essentially a distinct and recognizable type of deposit formed at or below the seafloor by circulating hot seawater (Barrie & Hannington 1999). Their pattern through time is shown in Figure 6a, and the source of data is shown in the figure caption. Modern examples of VHMS systems form at seafloor spreading ridges (e.g. East Pacific Rise), and also in back-arc basins (e.g. the Lau Basin). They are accreted into the convergent margin in similar tectonic environments to those in which orogenic gold deposits are generated (e.g. Goldfarb 1997), or may even form in the margins themselves (e.g. Solomon & Quesada 2003).

Thus, VHMS deposits should show a broadly similar preservational temporal pattern to that of orogenic gold deposits if the model outlined above is correct. The oldest VHMS deposits in the eastern Pilbara and the Barberton terranes at c. 3.5–3.25 Ga (e.g. Vearncombe *et al.* 1995) correspond broadly with the oldest orogenic gold events globally (e.g. Zegers *et al.* 2002) and with the earliest record of formation of significant continental crust (Fig. 2). Major peaks in VHMS deposits at c. 2.7 Ga and 1.9 Ga also correspond to peaks in both orogenic gold deposits and crustal growth (Figs 2 & 6a, b), possibly coincident with mantle plume events at these times. Similarly, VHMS deposits have formed and been incorporated into continental crust almost continuously since the latest Neoproterozoic, with examples on most continents, although there are distinct peaks in the early to middle Palaeozoic and Mesozoic (e.g. Titley 1993), as there are also for orogenic gold deposits. The VHMS deposits, although recognized to a minor degree between c. 1.75 Ga and 600 Ma are, like orogenic gold deposits, markedly under-represented in this age range. Again, it appears most likely that the VHMS deposits were mainly lost from the geological record as their hosting linear belts above relatively buoyant SCLM were uplifted and deeply eroded to their roots during that period. The youngest mined VHMS deposits are of Tertiary age, corresponding broadly to the youngest exposed orogenic gold deposits.

It appears that VHMS deposits formed in back-arc basins or, less commonly, accretionary terranes during orogenic events throughout Earth history, but as was the case for orogenic gold deposits, their secular distribution has a large preservational component due to changes in the lithosphere caused by the evolution of global tectonic processes.

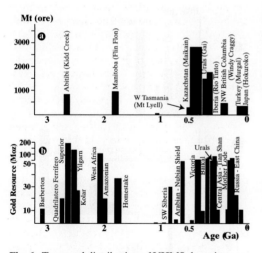

Fig. 6. Temporal distribution of VHMS deposits compared with that of orogenic gold deposits. (**a**) VHMS deposits, data from Gibson & Kerr (1993), Smith *et al.* (1993), Eremin & Dergachov (1994), Franklin (1996), Poulsen & Hannington (1996), Leistel *et al.* (1998), Prokin & Buslaev (1999), Barrie *et al.* (1999), Hannington *et al.* (1999), Pan & Xie (2001), and Relvas *et al.* (2002). (**b**) Orogenic gold deposits, adapted from Goldfarb *et al.* (2001a) and Groves *et al.* (2003).

Placer and palaeoplacer gold deposits

Most giant placer gold deposits were deposited in foreland basins in Mesozoic–Recent convergent margins surrounding the Pacific Rim (e.g. New Zealand, California, Alaska) through erosion of Palaeozoic–Mesozoic orogenic gold deposits (e.g. Henley & Adams 1979; Edwards & Atkinson 1986; Goldfarb *et al.* 1998). Different processes may have operated in NE Russia and in Siberia where melting of permafrost, with subsequent volume changes and release of organic-rich surface waters, may have liberated significant amounts of gold (A. Yakubchuck, pers. comm. 2004) from primary orogenic gold deposits (e.g. Yakubchuck & Edwards 1999). Most placer deposits were mined from Recent river systems and beaches, although some Tertiary palaeoplacers were preserved by overlying volcanic and volcaniclastic rocks (e.g. Ballarat and Bendigo, Australia), and some may be as old as Palaeozoic (e.g. Timan Range, Russia; A. Yakubchuck, pers. comm. 2004).

Palaeoplacer gold deposits older than the Tertiary are rare (Fig. 7a), yet the giant Late Archaean Witwatersrand deposits (Table 1G) represent the largest gold province globally. Although both modified placer (e.g. Minter *et al.* 1993) and hydrothermal (e.g. Phillips & Myers 1989; Barnicoat *et al.* 1997) models are proposed, the bulk of the evidence, particularly the pre-sedimentation Re–Os ages of both gold and associated rounded pyrites, mainly *c.* 3.0 Ga (e.g. Kirk *et al.* 2001, 2002), implicates the Witwatersrand gold ores as modified palaeoplacers (Frimmel *et al.* in press). If this model is correct, the gold would have been deposited in the Central Rand Group in a retroarc foreland basin setting (Kositcin & Krapez 2004) similar to the modern depositories of much of the

placer gold. Extreme environmental conditions on early Earth included a potentially more acidic and chemically aggressive atmosphere, the lack of vegetation and widespread organisms, and hence the predominance of braided stream environments and potential for effective wind erosion and sorting (e.g. Minter 1999). This can potentially explain the giant size of the Witwatersrand deposits, because many modern giant placers owe their origin to tectonic, erosional and sedimentary processes that

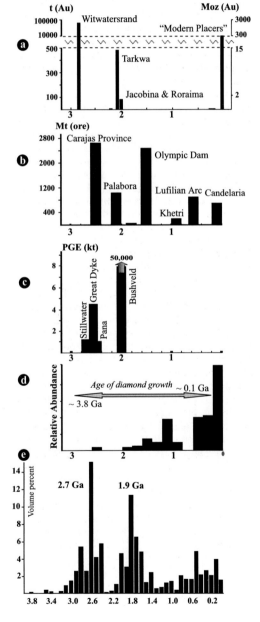

Fig. 7. Temporal distribution of palaeoplacer and placer gold deposits, iron oxide copper–gold deposits, PGE deposits in layered intrusions and primary diamond deposits, compared with secular evolution of juvenile continental crust. (**a**) Palaeoplacer and placer gold deposits; data from Boyle (1979), Bache (1987), Goldfarb *et al.* (1998), Yakubchuck & Edwards (1999), Milesi *et al.* (2002) and Frimmel *et al.* (in press). (**b**) Iron oxide copper–gold deposits with significant Cu–Fe sulphides and gold; data from Williams & Skirrow (2000), Nisbet *et al.* (2000) and Haynes (2002) and references therein. (**c**) PGE deposits in large layered intrusions; data from Stribrny *et al.* (2000) and Levine & Wilburn (2002). (**d**) Primary diamond deposits; data from Janse (1984), Jaques *et al.* (1984), Macnae (1995), Viljoen *et al.* (1996), Leckie *et al.* (1997), Schaerer *et al.* (1997), Chalapathi *et al.* (1998), Berryman *et al.* (1999), Carlson *et al.* (1999), Graham *et al.* (1999*a*), Graham *et al.* (1999*b*), Graham *et al.* (1999*c*), Kornilova *et al.* (1999), Menzies *et al.* (1999), Richardson *et al.* (1999), Sanders (1999), Digonnet *et al.* (2000), Moralev & Glukhovsky (2000), Sage (2000), Agashev *et al.* (2001), Shanker *et al.* (2001), Soni *et al.* (2001), Zhang *et al.* (2001), Brueckner *et al.* (2002), Burgess *et al.* (2002), Fang *et al.* (2002), Kaminsky *et al.* (2002), Spetsius *et al.* (2002). (**e**) Juvenile continental crust growth, from Condie (2000).

combine to produce super-effective sorting and detrital gold concentration from seemingly insignificant gold sources in the hinterland. Based on the Os isotopic compositions of the Witwatersrand gold (Kirk *et al.* 2002), the deposits could have been derived from a dispersed mantle-derived greenstone component of vanished granitoid–greenstone terranes in the hinterland to the Central Rand Basin (Frimmel *et al.* in press). Alternatively, they could have been derived from gold deposits equivalent in age to, or younger than, the orogenic Barberton gold deposits in the proposed hinterland of the Witwatersrand Basin (e.g. deposits in the Amalia–Kraaipan, Murchison, Pietersburg and Giyani belts).

An important question is why these deposits are preserved in the Late Archaean, yet absent from the geological record again until the Tertiary (Fig. 7a). The Witwatersrand almost certainly owes its preservation to its location on old SCLM (e.g. Shirey *et al.* 2003), where buoyancy protected it from subsequent destruction. The host Kaapvaal Craton is not alone in having very old SCLM (e.g. Richardson *et al.* 2004); the antiquity of its earliest well-preserved basins (*c.* 3.0–2.8 Ga) and the remarkable preservation of its Late Archaean to Mesoproterozoic and younger sedimentary and volcanic basins are evidence of its long-term stability and high preservational potential. The Re–Os age of the detrital gold and pyrite in the Witwatersrand palaeoplacers is *c.* 3.0 Ga, about 300 Ma older than the first significant crust-forming event at *c.* 2.7 Ga (Fig. 2), implicating old SCLM, as do detrital diamonds in the Witwatersrand rocks. The detrital gold is interpreted as having been derived from vanished terranes in the upthrusted margins of relatively small continental blocks to the north and SW to south (Schmitz *et al.* 2004), which would have been susceptible to uplift and erosion despite potentially buoyant SCLM.

The other significant gold palaeoplacer provinces, although two orders of magnitude smaller than the Witwatersrand, are Palaeoproterozoic in age (Fig. 7a), with Tarkwa in Ghana being the largest at *c.* 2.1 Ga (Pigois *et al.* 2003), and Jacobina (*c.* 2.0 ± 0.1 Ga) and Roraima (*c.* 1.96 Ga) representing smaller provinces (Frimmel *et al.* in press). These deposits, like the Witwatersrand ores, lie close to a peak in juvenile crust formation (Fig. 2), with Tarkwa predating the major peak in crust formation. These deposits appear to be preserved because of the buoyancy of the underlying Palaeoproterozoic SCLM, developed during the orogenic events in which the primary gold deposits formed.

Thus, placers and palaeoplacers appear to display a temporal pattern largely dictated by preservation. Presumably, their formation was linked to uplift and erosion of gold-enriched volcanic sequences or orogenic gold provinces throughout Earth history, but their pre-Tertiary survival was dictated by their preservation in buoyant early Precambrian SCLM.

Iron oxide copper–gold deposits

The iron oxide copper–gold deposit type (IOCG, Table 1H; e.g. Hitzman *et al.* 1992) includes a wide variety of deposits that are rich in iron-oxide minerals, including essentially sulphide-free phosphorus-, fluorine- and REE-rich deposits that formed in a variety of tectonic settings (cf. Hitzman 2000). However, only those deposits with significant Cu, Fe and Au resources, with the base metals contained in sulphide phases, are considered below. There is direct (e.g. Groves & Vielreicher 2001) and indirect (e.g. Campbell *et al.* 1998) geological evidence for an association of giant Archaean to Mesoproterozoic examples with alkaline magmatism, and recent dating in the Carajas mineral province of Brazil (e.g. Requia *et al.* 2003; Groves *et al.* 2004; Tallarico *et al.* 2005) confirms this. All significant Archaean to Mesoproterozoic examples are sited within about 100 km of the margin of Archaean cratons (e.g. Palabora, South Africa; Carajas, Brazil; Olympic Dam, South Australia) or close to the boundary of Archaean and Palaeoproterozoic lithosphere (e.g. Cloncurry, Queensland, Australia).

The temporal pattern of economically significant IOCG deposits (Fig. 7b) shows major peaks in the latest Archaean (*c.* 2.57 Ga: e.g. Carajas deposits), Palaeoproterozoic (*c.* 2.05 Ga: e.g. Palabora) and Mesoproterozoic (*c.* 1.59 Ga: e.g. Olympic Dam), which are significantly offset from the main periods of crustal growth at *c.* 2.7 Ga and 1.9 Ga (Fig. 2). The earliest Carajas deposits are sited at the eastern margin of the largest A-type granite province on Earth in a region of Late Archaean platformal sequences on the Amazonian craton, rather than in greenstone belts (e.g. Grainger *et al.* 2002). The Olympic Dam deposit formed at *c.* 1.59 Ga, very soon after amalgamation of Archaean and Palaeoproterozoic blocks to form proto-Australia at *c.* 1.75–1.70 Ga (e.g. Betts *et al.* 2002), in association with outpouring of the Gawler volcanic rocks and emplacement of associated intrusions of alkaline affinity. The location of giant deposits near craton margins suggests that associated alkaline magmatism was derived by small degrees of partial melting of SCLM, which had been previously metasomatized during plume-related cratonization events in which incompatible element-rich (e.g. K, Th, U, Au, LREE) low-degree partial melts may have stalled in the SCLM. This explains the offset between their ages and the ages of cratonization or crustal growth, and the lack of examples earlier than about 2.6 Ga. Thus, there is a major tectonic control on the formation of large IOCG deposits, but their temporal pattern (Fig. 7b) also has a preservational component, with giant Precambrian examples of the deposit type selectively formed and preserved in buoyant SCLM. There are Neoproterozoic (*c.* 850 Ma) deposits ascribed to the IOCG group at Khetri in India, which are associated with A-type granitoids, including syenites (Knight *et al.* 2002). Deposits ascribed to the IOCG group are also recorded from the Neoproterozoic Fulilian intracratonic rift basin of Zambia (e.g. Nisbet *et al.* 2000). Some large deposits contain 0.6–0.9% Cu but negligible gold (<0.1 g t^{-1} Au) and some contain limited iron oxides (Hitzman 2001). There are then no economic examples with significant copper and gold grades younger than Olympic Dam until the >450 million tonnes Candelaria deposit was formed at *c.* 115 Ma (Mathur *et al.* 2002). Spatially and temporally related subalkaline to alkaline metaluminous granitic plutons have Sr, Nd, and Pb isotopic ratios that suggest derivation of parent magmas from a subduction-modified metasomatized mantle source (Marschik *et al.* 2003a,b).

Platinum group element deposits in large layered intrusions

Platinum group element (PGE) deposits in large layered intrusions (Table 1I) are mined primarily for their PGEs, rather than as in most magmatic nickel–copper–(PGE) sulphide deposits, where PGEs are a by-product (Vermaak 1995). The largest known deposits of this type are in the Bushveld Complex and Great Dyke of southern Africa and the Stillwater Complex of the USA, all sited within cratons rather than on craton margins or between cratons, in contrast to the nickel–copper–sulphide deposits and to some smaller PGE deposits, for example the Palaeoproterozoic Pana deposits in the Kola Peninsular (Levine & Wilburn 2003) and the Partimo–Pennikat–Suharko deposits in northern Finland (Iljina & Hanski 2002), which lie on craton margins. Such deposits in large layered intrusions require a source for the PGE-enriched mafic magma (e.g. Cawthorn 1999) and thick buoyant lithosphere to support and preserve the large volumes of dense basic magma. Archaean SCLM has the required parameters for producing such PGE deposits. It is anomalously thick and buoyant (Fig. 3), and the high-degree melting that took place to produce such lithosphere also had the potential to produce PGE-fertile lithosphere via the concentration of dense PGE-rich sulphide residues in depleted mantle (Hamlyn & Keays 1986; Keays 1995). It is expected, therefore, that these PGE deposits would first appear in the geological record (Fig. 7c) after the main *c.* 2.7 Ga crust-forming event (Fig. 2), probably during the first significant intracratonic magmatism following cratonization. The deep crustal levels of emplacement of the host intrusions and the post-formational stability of the cratons, which mitigate against erosion and exposure of the commonly, centrally-sited, deep PGE deposits, particularly in cratons that have not experienced the relatively recent uplift of the Kaapvaal Craton, can readily explain the lack of economic post-Proterozoic deposits.

Primary diamond deposits

As summarized by Navon (1999), diamonds form naturally under asthenospheric conditions of high pressure and low geothermal gradient, such as those that existed below Archaean, or Palaeoproterozoic, SCLM. They are transported to the upper crust by highly volatile kimberlite and lamproite magmas. Published formational ages of diamonds are somewhat controversial (Navon 1999), but Nd model ages of mineral inclusions that are as old as *c.* 3.3 Ga are recorded (e.g. Richardson *et al.* 1984); some diamonds may be as old as *c.* 3.8 Ga (Sano *et al.* 2002). This indicates that conditions conducive to diamond formation did occur at the time of earliest recorded crustal growth (Fig. 2), as also indicated by rare detrital diamonds in the *c.* 2.9–2.84 Ga Central Rand Group of the Witwatersrand Basin (Frimmel *et al.* in press).

However, the earliest significant primary diamond deposits in kimberlites coincide almost exactly with the earliest PGE deposits in layered intrusions (Table 1J; compare Fig. 7c & d), which suggests that suitable conditions for both formation and preservation were first met in the Late Archaean to Early Proterozoic. However, superior diamond deposits are associated with funnel-shaped parts of kimberlite or lamproite pipes at high crustal levels, extruding to the surface (e.g. Argyle pipe, Australia: Boxer *et al.* 1989; Lorenz *et al.* 1999), and they are also susceptible to deep weathering due to their unstable high-Mg mineral assemblages. Hence, unlike the PGE deposits in layered intrusions, they show a classic preservational pattern with maximum abundance in the late Phanerozoic (Fig. 7d).

Evolution of the atmosphere–hydrosphere–biosphere system

In terms of secular variations in mineral deposits, the evolution of the atmosphere is important because atmospheric pO_2 controls the stability of ore minerals of redox-sensitive metals in near-surface environments, and also controls climate, which in turn controls the nature and severity of weathering. Weathering can also affect the evolution of mineral deposits: for example the Witwatersrand palaeoplacers discussed above. The atmosphere also provides a long-term buffer on the redox state of at least the shallow parts of the hydrosphere, although mantle plume events clearly cause more episodic redox shifts in the hydrosphere along with high-stands of global sea level (e.g. Larson 1991; Titley 1993). Several papers dealing with the evolution of the atmosphere–hydrosphere–biosphere system are included within this volume. Hence, only a brief summary of the controversy regarding early Earth atmospheric evolution is presented here. This is followed by a description of the temporal evolution of mineral deposits of redox-sensitive metals, because their distribution clearly contributes to the debate. Sedimentary rock-hosted deposits of Cu, Zn, and Pb are discussed separately.

Ohmoto *et al.* (2001) suggested that the history of oxygenation of the atmosphere and hydrosphere can be divided into four stages:

1   in a pre-cyanobacteria stage, the atmosphere and hydrosphere were essentially oxygen free;
2   the shallow hydrosphere was oxidized by cyanobacteria, the atmosphere pO_2 remained below 0.001 present atmospheric level (PAL), and the deep hydrosphere remained anoxic;
3   the atmosphere became oxygenated, with increase in pO_2 to 0.1–0.5 PAL, but the deep hydrosphere remained anoxic; and
4   the deep hydrosphere became oxygenated, pO_2 exceeded 0.5 PAL, and the entire hydrosphere became oxic, except for local anoxic basins.

It is the timing of these stages that has become so controversial in the past 10–15 years. Holland (1994, 1999) has summarized the case for the view that O_2 was absent or present in very low abundance in the early atmosphere. Then, at about 2.3 Ga, it rose rapidly in the 'Great Oxidation Event' until, by about 2.0 Ga, pO_2 was at least 0.1–0.15 PAL. Ohmoto *et al.* (2001) termed this the Cloud-Walker-Holland-Kasting (CWHK) model in which stages 2, 3 and 4 above are reached at about 2.7 Ga, 2.3–2.2 Ga and 0.6 Ga. Farquar *et al.* (2000) reported that mass-independent fractionation of sulphur isotopes occurred before *c.* 2.4 Ga, but not significantly after 2.0 Ga, supporting the concept of the 'Great Oxidation Event' in the period between *c.* 2.4 and 2.0 Ga. On the other hand, Ohmoto (1997) summarized the evidence for a model (the Dimroth-Kimberley-Ohmoto [DKO] model of Ohmoto *et al.* 2001) in which the transition from stage 1 to 4 occurred within 50 million years at *c.* 4.0 Ga; that is, pO_2 reached 0.5 PAL very early in Earth history. The two models are shown schematically in Figure 8.

In terms of the biosphere, early life, probably as thermophyllic chemotropic prokaryotes, evolved in Middle Archaean submarine hydrothermal systems (e.g. Rasmussen 2000; Russell 2003). In the Proterozoic, life became more diversified with the evolution of the eukaryotes. Metazoans, the first animal fossils, are widely accepted to have existed at *c.* 600 Ma, but may be older than 1200 Ma (Rasmussen *et al.* 2002). Organic material was confined to marine environments until the early to middle Palaeozoic, with terrestrial plants becoming abundant in the late Palaeozoic and Mesozoic, eventually producing some of the worlds great coalfields (cf. Craig *et al.* 1988).

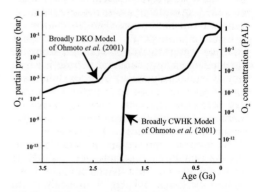

Fig. 8. Contrasting models for the evolution of the atmosphere, adapted from Solomon & Sun (1997). PAL, present atmospheric level.

The progressive evolution of life, such as sulphur-, silica- and carbonate-secreting organisms, led to changes in ocean chemistry. The organisms fixed these elements and compounds in biogenically-produced minerals rather than in hydrothermal products such as cherts and highly silicified and/or carbonated sedimentary and volcanic rocks that were abundant in Middle Archaean greenstone belts (e.g. Buick & Dunlop 1990). Specific consequences of the evolution of the biosphere were important for the deposition of mineral deposits. For example, marine carbon flux increased due to the proliferation of life, providing widespread sources of reductant for oxidized mineralizing fluids, and the evolution of organisms such as corals changed both the chemistry (from dolomite to calcite) and porosity and permeability of platformal carbonate rocks, and hence their suitability as both aquifers and reactive host rocks for the deposition of hydrothermal mineral deposits.

The temporal distribution of mineral deposits of redox-sensitive metals may contribute to the debate on the evolution of the atmosphere–hydrosphere–biosphere system. The mineral deposits of uranium, iron and manganese are used below as examples of temporal distributions that can contribute to the debate.

Mineral deposit distribution potentially related to changes in atmosphere–hydrosphere–biosphere

Uranium deposits

Uranium generally occurs in two oxidation states: U^{6+} is stable under oxidizing conditions, whereas U^{4+} is stable under reducing conditions. The most common ore mineral of U^{4+} is uraninite, which is unstable under modern oxidizing near-surface conditions. The formation of hydrothermal deposits is favoured when oxidizing waters scavenge uranium from the surrounding environment and deposit it where they become reduced at a redox boundary. Some important uranium deposits are associated with granitic magmatism (e.g. Rössing in Namibia), but only the deposits hosted in sedimentary sequences are considered below.

The earliest representatives of the deposit types of uraninite are palaeoplacers, which are restricted to the Late Archaean and Early Proterozoic (Table 1G; Fig. 9a), mainly in the Witwatersrand, South Africa, and Elliot Lake, Canada, deposits (e.g. Pretorius 1981). Models for the origin of the uraninite grains have

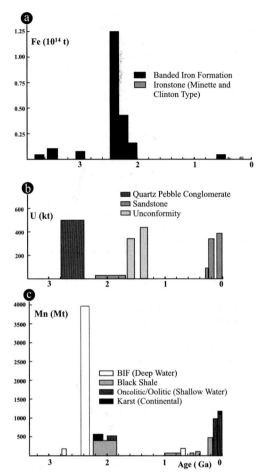

Fig. 9. Temporal distribution of sediment-hosted deposits of redox-sensitive metals. (**a**) uranium deposits; data from Dahlkamp (1993), Bell (1996) and Ruzicka (1996). (**b**) Iron deposits; data from Jensen and Bateman (1979) and Klemm (2000). (**c**) Manganese deposits; data from Schaefer *et al.* (2001) and Mikhailov (2004).

ranged from placer to hydrothermal but, certainly in the Witwatersrand, they appear to be detrital because of the variation in U/Th ratios even between adjacent grains (Frimmel *et al.* in press) and the fact that they predate early oil migration (England *et al.* 2001). Detrital uraninite is also widespread in Archaean sandstones and conglomerates (e.g. Rasmussen & Buick 1999), but is very rare in younger sedimentary rocks (Ohmoto *et al.* 2001).

The Mesoproterozoic (Fig. 7a) is dominated by unconformity-related uranium deposits (Table 1K), such as those of the Athabasca Basin in Saskatchewan, Canada, and the East

Alligator River area of the Northern Territory of Australia (e.g. Marmont 1987; Needham 1988). They are sited at redox boundaries on unconformities separating Palaeoproterozoic basement from overlying Mesoproterozoic sedimentary sequences. The deposits clearly formed where oxidized fluids, carrying uranyl complexes, were reduced by carbonaceous material of marine origin in the sedimentary rocks (e.g. Wilde *et al.* 1989). Phanerozoic uranium deposits belong to the sandstone-type class, best developed in the USA and Central Asia in the Mesozoic and Tertiary (e.g. Adams 1991). In these deposits, oxidized groundwater carrying uranyl complexes flowed through conglomerate or coarse-grained sandstone aquifers. Uranium was deposited by reaction with terrestrial organic material.

Thus, the sedimentary rock hosted uranium deposits show a remarkable evolutionary secular pattern from extensive Archaean to Palaeoproterozoic palaeoplacers, through unconformity-related deposits influenced by marine carbonaceous material, to sandstone-type deposits (Table 1L), influenced by terrestrial organic matter following the evolution of widespread terrestrial plants.

Iron deposits

Iron has two oxidation states, Fe^{2+} and Fe^{3+}. In hydrothermal mineral deposits, iron is transported as Fe^{2+} and deposited via redox reactions involving oxidation to Fe^{3+}. Iron has been mined from large magmatic–hydrothermal magnetite–apatite orebodies of Kiruna (Sweden) type (Fig. 1b) and from smaller skarn deposits, but only the sedimentary deposits are discussed here.

The most economically important iron deposits are those related to banded iron formations (BIFs), rhythmically layered sedimentary rocks comprising alternating layers of magnetite (less commonly siderite) and fine-grained chert (Table 1M). The highest grade iron ores are low-phosphorus hematitic deposits related to hydrothermal upgrading of BIF in structurally prepared sites (e.g. Barley *et al.* 1999; Taylor *et al.* 2001). Thin, discontinuous Algoma-type BIF was deposited in some of the oldest known Archaean greenstone belts (e.g. Isua, Greenland) from about 3.75 Ga, but they and the thicker Superior-type BIFs are most abundant in the period *c.* 2.6–2.0 Ga (Fig. 9b). They are remarkably rare to absent after *c.* 2.0 Ga.

The most voluminous BIFs appear to have been deposited in intracratonic, passive margin or platform basins during high sea-level stands

in the Late Archaean and Palaeoproterozoic (Simonson & Hassler 1997). The iron and silica appear to have been derived mainly from hydrothermal vents on the seafloor (Klein & Beukes 1992; Barley *et al.* 1999) related to mantle plume activity (Isley & Abbott 1999). Superplume events can account for several aspects of BIF deposition. The enhanced submarine volcanism and hydrothermal venting associated with ocean ridge and oceanic plateau volcanism during a mantle plume event may be the source of iron and silica, and the resultant elevated sea level could provide extensive shallow marine basins along stable continental platforms necessary to deposit and preserve BIF. Most authors argue that the proliferation of BIF in the Archaean and Proterozoic was due to anoxic oceans. Oxygenated surface waters for oxidation of widespread Fe^{2+} created by cyanobacterial photosynthesis were limited (e.g. Kump *et al.* 2001), assuming a less-oxygenated atmosphere at that time. There are counter arguments (e.g. Ohmoto 2003), and it is also possible that mantle plume events caused widespread oceanic anoxia as in the Phanerozoic (e.g. Titley 1993), but virtual restriction of BIF to the Archaean and Palaeoproterozoic remains one of the most remarkable temporal patterns shown by mineral deposits globally.

Similarly, the contrast with younger sedimentary iron ores is striking. These ores, typified by the Clinton-type deposits of the USA and minette-type ores of Europe (Table 1N), are oolitic beds of hematite–siderite–chamosite intercalated with sandstones and shales (e.g. Melon 1962). They formed predominantly in the early Palaeozoic and Mesozoic, in shallow shelf and estuarine sequences, in neritic, oxygenated to euxinic and restricted environments. Hematite is abundant and magnetite absent. This deposit group was important in the early days of the industrial revolution in Europe and the USA, but is dwarfed by the iron ores derived from BIF.

Manganese deposits

Manganese occurs normally in a variety of oxidation states, including Mn^{2+}, Mn^{3+} and Mn^{4+}. As for iron, manganese is soluble as Mn^{2+} with manganese minerals deposited via oxidation reactions. Manganese deposits show a similar temporal pattern to iron deposits (Fig. 9c). The dominant deposits in terms of global resources are manganiferous BIF, with the major deposits sited in the Kalahari manganese field in the Palaeoproterozoic Transvaal Group of South Africa (Astrup &

Tsikos 1998). The major ores are manganese carbonate (braunite–kutnahorite) sedimentary rocks that are interbedded with iron formation. There is considerable controversy over the genesis of these deposits (summarized by Astrup & Tsikos 1998), but a model equivalent to that for BIF appears most likely considering their restricted age range and close association with BIF (Table 1O).

Again, as for the sedimentary iron deposits, the deeper water, early Palaeoproterozoic, manganese carbonate deposits are succeeded by shallower water sedimentary manganese deposits (Table 1P), variously classed as black shale and oncolitic/oolitic deposits (Schaefer *et al.* 2001). These first appear in the Palaeoproterozoic in South Africa, Ghana, Gabon and Brazil, and are also common in the Neoproterozoic to Tertiary (e.g. Nikopol ores, Ukraine). Schaefer *et al.* (2001) suggested that the switchover in the type of sedimentary manganiferous ores was due to the migration of Mn-oxidizing pelagic micro-organisms into shallow water environments to access dissolved Mn^{2+} from continental runoff rather than from volcanic exhalations as in earlier oceans. Thus, they imply a biogenic control, perhaps coupled with a change in oxygen levels in the atmosphere–hydrosphere system.

Summary

The temporal patterns for mineral deposits of redox-sensitive metals (Fig. 9) are distinctive in that, in each case, different mineral-deposit types occupy different time periods in Earth history. Extensive arguments have been presented for and against an evolving atmosphere–hydrosphere–biosphere system based on individual lines of evidence. However, when the temporal patterns are considered together, and the principle of Ockham's Razor is applied, they favour a model involving progressive oxidation of the atmosphere–hydrosphere system, with increasing oxidation and decreasing CO_2 partial pressures in the atmosphere in the early Palaeoproterozoic (cf. Krupp *et al.* 1994).

Secular distribution of base-metal deposits in sedimentary rocks

Base-metal deposits in sedimentary rocks are treated separately because they represent a complex group of deposits, with variable classification schemes and apparently complex temporal controls that probably include components related to secular changes in

tectonic processes, preservational potential of SCLM, and atmosphere–hydrosphere–biosphere conditions. Mississippi Valley type (MVT) deposits are considered first because they are a well-defined group whose tectonic controls are relatively well understood. This contrasts with the group of deposits broadly classed as SEDEX (e.g. Lydon 1996) or shale-hosted, which probably contain several groups of deposits that formed in variable tectonic settings, in different sedimentary basinal settings, and potentially at different times ranging from synsedimentation through diagenesis to syn–early deformation. They are included as a broad group primarily for completeness of coverage of temporal trends of important mineral deposit types showing distinctive temporal patterns.

MVT deposits

The MVT deposits, essentially lead–zinc ores commonly with accessory barite or fluorite and hosted by platformal carbonate sequences (Table 1Q), are, unlike SEDEX deposits, demonstrably epigenetic (Sangster 1990). Their temporal distribution (Fig. 10a) contrasts markedly with that of SEDEX deposits (Fig. 10b) in that, although present, they are relatively rare and rather small in the Proterozoic (e.g. Kesler & Reich in press), with most large deposits having formed in the Phanerozoic,

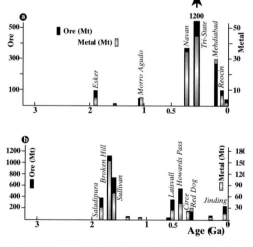

Fig. 10. Temporal distribution of sediment-hosted base-metal deposits. (**a**) Mississippi Valley type Pb – Zn ± Ba ± F deposits: data from Sangster (1996) and Leach *et al.* (2001); (**b**) SEDEX Zn – Pb – Cu ± Ag ± Au deposits in the sense of Lydon (1996): data from Lydon (1996) and Large *et al.* (2002).

particularly between the Devonian and Permian.

Leach *et al.* (2001) demonstrated, through a seminal synthesis that involved a compilation of all palaeomagnetic and isotopic dating, that the migration of the fluids that deposited MVT deposits was not a natural consequence of the evolution of their hosting basins. Instead, most MVT deposits formed in districts where the hosting basins, and particularly the platformal carbonates, had a hydrological connection to distant orogenic belts. Thus, there is a paucity of deposits related to breakup of supercontinents such as Rodinia and Pangea. Instead, the most important periods of MVT formation were the Devonian to Permian during the assembly of Pangea (75% of all MVT metal production) and the Cretaceous to Tertiary during accretion of microplates along the western margin of North America and between Africa and Eurasia. Although the deposits are sited in platformal sequences, ore fluid migration was related to convergent-margin tectonics (cf. Bethke & Marshak 1990). The precise mechanism for fluid movement is debated, but Leach *et al.* (2001) concluded that topographically-driven fluid migration was important.

The requirement for large amounts of precipitation to form regional fluid flow systems and for saline ore fluids generated from evaporated seawater (e.g. Kesler *et al.* 1995; Viets *et al.* 1996) suggests a palaeoclimatic control, with basins that formed at low latitudes being more favourable sites for MVT deposits (Leach *et al.* 2001). The significant increase in the number of MVT deposits from the Devonian onwards suggests also a biospheric control. Coralline limestone reefs are the favoured depositional sites over older, dominantly dolomitic, carbonates due to their enhanced porosity and permeability. Potentially of equal significance was the assembly of Pangea, particularly the North America segment, with uplifted orogenic belts along Pangean suture zones driving migration of ore fluids through structurally-prepared vast carbonate platforms to produce the largest MVT districts on Earth (Leach *et al.* 2001).

SEDEX deposits

Lydon (1996) defined a SEDEX deposit (Table 1R) as a massive sulphide deposit formed in a sedimentary basin by the submarine venting of hydrothermal fluids and whose principal minerals are sphalerite and galena, with $(Zn \times 100)/(Zn+Pb)$ from 15 to 100. The SEDEX deposits account for about 40% and 60% of global zinc and lead production (with significant silver as a by-product), respectively, and for greater percentages of global resources, but there are only some 25 mined deposits, among them (if Lydon's 1996 classification is accepted) Broken Hill, Cannington, Mount Isa, McArthur River (HYC) and Century in Australia, Sullivan in Canada and Red Dog in USA (Lydon 1996; Large *et al.* 1998, 2002). The SEDEX deposits normally occur within rift-cover sequences in sedimentary basins that are controlled by tectonic subsidence associated with major intracratonic or epicratonic rift systems, although Sullivan occurs in a rift-fill sequence (Lydon 1996). Sedimentary-hosted copper deposits of Zambian type are not discussed here as modern published literature is scarce, with much new data held confidential in a recent AMIRA Project carried out by CODES, University of Tasmania and Colorado School of Mines.

The deposit type has a temporal pattern (Fig. 10b) with two broad peaks in the Palaeo-Meso-proterozoic, mainly Australian deposits, and in the lower to middle Palaeozoic, mainly deposits in western Canada, NW USA (Alaska) and western Europe. The onset of the Mesoproterozoic does not appear to coincide with any major unidirectional changes in environmental conditions. However, the formation of the giant Australian deposits at *c.* 1650–1600 Ma (summarized in Cooke *et al.* 2000) follows shortly after the final assembly of Proto-Australia at *c.* 1750–1700 Ma (e.g. Betts *et al.* 2002), part of a global supercontinent assembly event centred at approximately 1850–1800 Ma and involving collisions in Laurentia, Baltica and Siberia (Condie 2000). This was arguably the first supercontinent assembly of major crustal fragments, providing a foundation for the sedimentary rift basins that host the SEDEX deposits (Solomon & Groves 1994). In Palaeoproterozoic Australia, this also coincided with a time of high heat flow through the crust, including the emplacement of voluminous high-heat producing granitoids (e.g. Wyborn *et al.* 1987) whose post-cooling radiogenic heat has been implicated in the genesis of the SEDEX deposits (Solomon & Heinrich 1992). The evolution of these basins probably provided, for the first time, the suitable conjunction of basin size and thickness, lithostratigraphy, including widespread evaporites, permeability, redox and structural architecture for the development of the regional hydrothermal systems required to produce the giant lead–zinc deposits. Distant collisional events at the margins of the proto-Australian cratonic blocks were accompanied by episodic extension and basin formation, remnants of which are preserved, for example in the extraordinary base-metal rich McArthur Basin and the Mt Isa Inlier (e.g. O'Dea *et al.* 1997; Betts *et al.* 1998, 2002; Jackson *et al.* 2000). Reactivation of deep extensional faults appears to be important in SEDEX formation (e.g. Large *et al.* 2002). The deposits lie close to the sutures along which the older cratonic fragments (West Australian, South Australian and North Australian craton) were assembled. Remarkably, the Mesoproterozoic SEDEX deposits lie close to the margin of later Grenvillian orogenic belts (Lydon 1996, fig. 6.1–8A), suggesting that these original sutures became fundamental intracratonic structures along which the cratons perforated in the Neoproterozoic.

The early to middle Palaeozoic SEDEX deposits show a similar distribution on a continental reconstruction for the Late Devonian (Lydon 1996, fig 6.1–8B), suggesting that similar tectonic controls on basin evolution and mineralization were in place during both major periods of SEDEX deposition. The Palaeozoic examples must be linked in some way to the assembly of Pangea.

Synthesis

Mineral deposits show heterogeneous temporal distributions, with each specific group of deposits having a distinctive, commonly unique, temporal pattern that relates to the conjunction of formational and preservational processes in an evolving Earth. Only nickel–copper sulphide deposits show a direct relationship to a cooling Earth, with komatiite-hosted ores restricted to the Archaean and Palaeoproterozoic.

Some deposit types, for example epithermal gold–silver deposits and, to a lesser extent, porphyry copper–gold or molybdenum deposits, show almost purely preservational patterns with virtual restriction to the late Phanerozoic, because they form at high crustal levels in arc settings in convergent margins with high uplift rates and rapid erosion. Other deposits, such as orogenic gold and VHMS deposits that formed throughout Earth history during major crust-forming events, show mixed formation and preservation patterns. These relate to secular changes in tectonic processes from strongly plume-influenced buoyant-style plate tectonics in the latest Archaean and early Palaeoproterozoic to modern-style plate tectonics from the Neoproterozoic to present. The consequences of this changeover included evolution from thick, buoyant SCLM beneath broadly equidimensional Archaean–Palaeoproterozoic cratons,

which preserved the orogenic gold and VHMS deposits, to thinner, negatively buoyant SCLM beneath elongate Phanerozoic orogenic belts that were more susceptible to uplift and erosion. This led to non-preservation of the orogenic gold and VHMS deposits in Mesoproterozoic and Neoproterozoic belts. Thus, crustal and metallogenic formation and preservation processes were linked in the early Earth but became decoupled in the Mesoproterozoic to Recent. Giant gold palaeoplacers were also selectively preserved in foreland basins underlain by thick early Precambrian SCLM, with a major hiatus and then the reappearance of large gold placers in the Tertiary to Recent from erosion of orogenic gold deposits in the hinterland to foreland basins.

Yet other types of mineral deposits seem to require the prior existence of thick, stable, early Precambrian SCLM, with low geothermal gradients and incompatible element-rich metasomatism of marginal SCLM, for formation. Hence, these make their first appearance in the Late Archaean or Palaeoproterozoic. These include primary diamond deposits, PGE deposits in layered intrusions and IOCG deposits. However, they show drastically different temporal patterns, with giant PGE deposits and IOCG deposits mainly restricted to the Precambrian due to their relatively deep crustal level of emplacement, or protection by cover rocks. In contrast, economic diamond deposits became progressively more abundant towards the late Phanerozoic because of their high crustal level of emplacement and susceptibility to weathering of hosting kimberlites and lamproites.

A schematic diagram showing the preservational environments of mineral deposits whose temporal patterns are related to tectonic processes is presented in Figure 11. Furthermore, Figure 12 shows the location of mineral deposit types in terms of their tectonic environment in relation to proximity to plate margins and, potentially, a supercontinent cycle.

First-order controls on sedimentary rock-hosted lead–zinc deposits are more problematic. The SEDEX deposits first formed in the early Mesoproterozoic, following the assembly of the first giant supercontinent and related generation of large rift basins with appropriate structural and sedimentary architecture, and again during the assembly of Pangea: precise processes are equivocal. The MVT deposits formed in carbonate platforms with a direct connection to distal orogenic belts. The more important examples show a more restricted temporal distribution, probably due to the

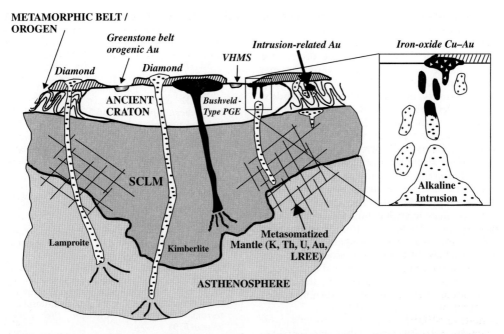

Fig. 11. Schematic diagram showing the preservational environments of mineral deposit types with a strong tectonic control (commonly strong plume component) as related to Archaean SCLM. Diagram adapted from Groves *et al.* (1987) and Kerrich *et al.* (2000).

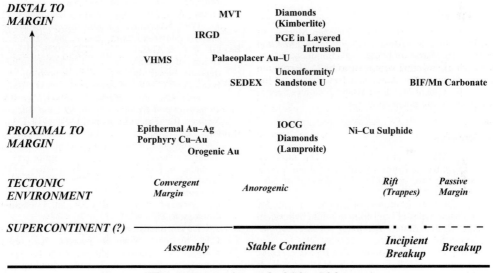

Time (approximately 300 - 500 m.a.)

Fig. 12. Location of major mineral deposit types discussed in the text in terms of their tectonic environment and proximity to plate margins or sutures at the time of formation broadly with respect to the supercontinent cycle. Note that the length of time shown for each stage of the supercontinent cycle is schematic (particularly for incipient breakup) to allow deposit types to be shown. Note also that some deposit types shown as forming at different times in the supercontinent cycle may actually form broadly at the same absolute time (e.g. Ni–Cu sulphides v. orogenic gold and VHMS deposits), particularly in the early Precambrian when there were probably no extensive supercontinents. IOCG, iron oxide Cu–Au deposit; MVT, Mississippi Valley type deposit; IRGD, intrusion-related gold deposit; VHMS, volcanic-hosted massive sulphide deposit.

requirement of permeable, reactive limestone reef sequences, rather than dolomitic carbonates, combined with suitable climatic conditions at low latitudes and topographic drivers for long-distance fluid flow, which were best met during the assembly of Pangea.

Sedimentary deposits of the more redox-sensitive resources, such as uranium, iron and manganese, show remarkable temporal patterns in which deposit types that dominate production from Archaean and early Palaeoproterozoic basins (i.e. palaeoplacer uranium, BIF and BIF-associated manganese carbonate deposits) are essentially absent in younger basins. However, they are replaced by different deposit types in these younger environments, including unconformity-related and sandstone-type uranium and oolitic iron and manganese deposits. Although individual lines of evidence are equivocal, these temporal patterns would be expected for progressive secular change in the atmosphere–hydrosphere–biosphere system during Earth evolution.

We are grateful to colleagues at the Centre for Global Metallogeny, The University of Western Australia, particularly N. Vielreicher, at the USGS, particularly D. Leach, and to R. Kerrich, University of Saskatchewan, and J. Hronsky, WMC Resources, for their contributions and discussions on this topic. This paper was inspired by the research of 'Chuck' Meyer and stimulated by the Fermor Flagship Meeting in Cardiff in 2003. The paper was significantly improved by the extensive and incisive comments of A. Yakubchuck, together with reviews by R. Herrington and C. Stanley. This paper is a contribution to the Centre for Global Metallogeny and Tectonics Special Research Centre publication #303.

References

ABBOTT, D.H. & ISLEY, A.E. 2002. The intensity, occurrence and duration of superplume events and eras over geological time. *Journal of Geodynamics*, **34**, 265–307.

ADAMS, S.S. 1991. Evolution of genetic concepts for principal types of sandstone uranium deposits in the United States. *Economic Geology Monograph*, **8**, 225–248.

AGASHEV, A.M., WATANABE, T., BYDAEV, D.A., POKHILENKO, N.P., FOMIN, A.S., MAEHARA, K. & MAEDA, J. 2001. Geochemistry of kimberlites from the Nakyn Field, Siberia; evidence for unique source composition. *Geology*, **29**, 267–270.

ALLING, H.L. 1947. Diagenesis of the Clinton hematite ores of New York. *Geological Society of America Bulletin*, **58**, 991–1018.

ARNDT, N.T., KERR, A.C. & TARNEY, J. 1997. Dynamic melting in plume heads; the formation of Gorgona komatiites and basalts. *Earth and Planetary Science Letters*, **146**, 289–301.

ARNDT, N.T., LEWIN, E. & ALBAREDE, F. 2002. Strange partners; formation and survival of continental crust and lithospheric mantle. *In*: FOWLER, C.M.R., EBINGER, C.J. & HAWKESWORTH, C.J. (eds) *The Early Earth; Physical, Chemical and Biological Development*. Geological Society, London, Special Publications, **199**, 91–103.

ASPLER, L.B. & CHIARENZELLI, J.R. 1998. Two Neoarchean supercontinents? Evidence from the Paleoproterozoic. *Sedimentary Geology*, **120**, 75–104.

ASTRUP, J. & TSIKOS, H. 1998. Manganese. *In*: WILSON, M.G.C. & ANHAEUSSER, C.R. (eds) *The Mineral Resources of South Africa*. Council for Geoscience Handbook **16**, 450–460.

BACHE, J.J. 1987. *World Gold Deposits: A Geological Classification*. Studies in Geology: Elsevier, New York.

BARLEY, M.E. & GROVES, D.I. 1992. Supercontinent cycles and the distribution of metal deposits through time. *Geology*, **20**, 291–294.

BARLEY, M.E., KRAPEZ, B., GROVES, D.I. & KERRICH, R. 1998. The Late Archaean bonanza: metallogenic and environmental consequences of the interaction between mantle plumes, lithospheric tectonics and global cyclicity. *Precambrian Research*, **91**, 65–90.

BARLEY, M.E., PICKARD, A.L., HAGEMANN, S.G. & FOLKERT, S.L. 1999. Hydrothermal origin for the 2 billion year old Mount Tom Price giant iron ore deposit, Hamersley Basin, Western Australia. *Mineralium Deposita*, **34**, 784–789.

BARNICOAT, A.C., HENDERSON, I.H.C., KNIPE, R.J., ET AL. 1997. Hydrothermal gold mineralization in the Witwatersrand basin. *Nature*, **386**, 820–824.

BARRIE, C.T. & HANNINGTON, M.D. 1999. Classification of volcanic-associated massive sulfide deposits based on host-rock composition. *In*: BARRIE, T.C. & HANNINGTON, M.D. (eds) *Volcanic-Associated Massive Sulfide Deposits: Progresses and Examples in Modern and Ancient Settings*. Reviews in Economic Geology, **8**, 1–11.

BARRIE, T.C., HANNINGTON, M.D. & BLEEKER, W. 1999. The giant Kidd Creek volcanic-associated massive sulphide deposits, Abitibi Subprovince, Canada. *In*: BARRIE, T.C. & HANNINGTON, M.D. (eds) *Volcanic-Associated Massive Sulfide Deposits: Progresses and Examples in Modern and Ancient Settings*. Reviews in Economic Geology, **8**, 247–260.

BELL, R.T. 1996. Sandstone Uranium. *In*: ECKSTRAND, O.R., SINCLAIR, W.D. & THORPE, R.I. (eds) *Geology of Canadian Mineral Deposit Types*. Geological Survey of Canada, Geology of Canada, **8**, 212–219.

BERRYMAN, A.K., STIEFENHOFER, J., SHEE, S.R., WYATT, B.A. & BELOUSOVA, E.A. 1999. The discovery and geology of the Timber Creek Kimberlites, Northern Territory, Australia. *In*: GURNEY, J.J., GURNEY, J.L., PASCOE, M.D. & RICHARDSON, S.H. (eds) *Proceedings of the VIIth International Kimberlite Conference*, Cape Town, South Africa, **1**, 30–39.

BETHKE, C.M. & MARSHAK, S. 1990. Brine migration

across North America – the plate tectonics of groundwater. *Annual Reviews of Earth and Planetary Sciences*, **18**, 228–315.

BETTS, P.G., LISTER, G.S. & O'DEA, M.G. 1998. Asymmetric extension of Middle Proterozoic lithosphere, Mount Isa Inlier, Queensland, Australia. *Tectonophysics*, **296**, 293–316.

BETTS, P.G., GILES, D., LISTER, G.S. & FRICK, L.R. 2002. Evolution of the Australian lithosphere. *Australian Journal of Earth Sciences*, **49**, 661–695.

BICKLE, M.J. 1990. Mantle evolution. *In*: HALL, R.P. & HUGHES, D.J. (eds) *Early Precambrian Basic Magmatism*. Blackie and Sons, Glasgow, 111–135.

BLUNDELL, D.J. 2002. The timing and location of major ore deposits in an evolving orogen: the geodynamic context. *In*: BLUNDELL, D.J., NEUBAUER, F. & VON QUANT, A. (eds) *The Timing and Location of Major Ore Deposits in an Evolving Orogen*. Geological Society, London, Special Publications, **204**, 1–12.

BOXER, G.L., LORENZ, V. & SMITH, C.B. 1989. The geology and volcanology of the Argyle (AK1) lamproite diatreme, Western Australia. *In*: ROSS, J., JAQUES, A.L., FERGUSON, J., GREEN, D.H., O'REILLY, S.Y., DANCHIN, R.V. & JANSE, A.J.A. (eds) *Kimberlite and Related Rocks*. Geological Society of Australia Special Publication, **14**, 140–152.

BOYLE, R.W. 1979. *The Chemistry of Gold and its Deposits*. Geological Survey of Canada Bulletin, **280**, 584 pp.

BRADLEY, D.C. & LEACH, D.L. 2003. Tectonic controls of Mississippi Valley-type lead–zinc mineralization in orogenic forelands. *Mineralium Deposita*, **38**, 652–667.

BRUECKNER, H.K., CARSWELL, D.A. & GRIFFIN, W.L. 2002. Paleozoic diamonds within a Precambrian peridotite lens in UHP gneisses of the Norwegian Caledonides. *Earth and Planetary Science Letters*, **203**, 805–816.

BUICK, R. & DUNLOP, J.S.R. 1990. Evaporitic sediments of Early Archaean age from the Warawoona Group, North Pole, Western Australia. *Sedimentology*, **37**, 247–277.

BURGESS, R., LAYZELLE, E., TURNER, G. & HARRIS, J.W. 2002. Constraints on the age and halogen composition of mantle fluids in Siberian coated diamonds. *Earth and Planetary Science Letters*, **197**, 193–203.

CAMPBELL, I.H., GRIFFITH, R.W. & HILL, R.I. 1989. Melting in an Archaean mantle plume: heads it's basalt, tails it's komatiites. *Nature*, **339**, 697–699.

CAMPBELL, I.H., COMPSTON, D.M., RICHARDS, I.P., JOHNSON, J.P. & KENT, A.R. 1998. Reviews of the application of isotopic studies to the genesis of Cu–Au mineralization at Olympic Dam and Au mineralization at Porgera, Tennant Creek district and Yilgarn Craton. *Australian Journal of Earth Sciences*, **45**, 201–218.

CAMUS, F. 2002. The Andean porphyry system. *In*: COOKE, D.R. & PONGRATZ, J. (eds) *Giant Ore Deposits: Characteristics, Genesis and Exploration*. University of Tasmania, Hobart, CODES Special Publication, **4**, 5–22.

CARLSON, J.A., KIRKLEY, M.B., THOMAS, E.M. & HILLIER, W.D. 1999. Recent Canadian kimberlite discoveries. *In*: GURNEY, J.J., GURNEY, J.L., PASCOE, M.D. & RICHARDSON, S.H. (eds) *Proceedings of the VIIth International Kimberlite Conference*, Cape Town, South Africa, **1**, 81–89.

CAWTHORN, R.G. 1999. Platinum-group element mineralization in the Bushveld Complex – a critical reassessment of geochemical models. *South African Journal of Geology*, **102**, 268–281.

CAWTHORN, R.G., MERKLE, R.K.W. & VILJOEN, M.R. 2002. Platinum-group-element deposits in the Bushveld Complex, South Africa. *In*: CABRI, L.J. (ed.) *The Geology, Geochemistry, Mineralogy and Mineral Benefication of Platinum-Group-Elements*. Canadian Institute of Mining Special Volume, **54**, 389–429.

CHALAPATHI, R.N.V., GIBSON, S.A., PYLE, D.M. & DICKIN, A.P. 1998. Contrasting isotopic mantle sources for Proterozoic lamproites and kimberlites from the Cuddapah Basin and eastern Dharwar Craton; implication for Proterozoic mantle heterogeneity beneath Southern India. *Journal of the Geological Society of India*, **52**, 683–694.

CHIARADIA, M., FONTBOTE, L. & BEATE, B. 2004. Cenozoic continental arc magmatism and associated mineralization in Ecuador. *Mineralium Deposita*, **39**, 204–222.

COMPSTON, W., WRIGHT, A.E. & TOGHILL, P. 2002. Dating the late Precambrian volcanicity of England and Wales. *Journal of the Geological Society of London*, **159**, 323–339.

CONDIE, K.C. 1998. Episodic continental growth and supercontinents: a mantle avalanche connection? *Earth and Planetary Science Letters*, **163**, 97–108.

CONDIE, K.C. 2000. Episodic continental growth models: afterthoughts and extensions. *Tectonophysics*, **322**, 153–162.

CONDIE, K.C. 2002a. Continental growth during a 1.9 Ga superplume event. *Journal of Geodynamics*, **34**, 249–264.

CONDIE, K.C. 2002b. Breakup of a Paleoproterozoic supercontinent. *Gondwana Research*, **5**, 41–43.

CONDIE, K.C. 2004. Supercontinents and superplume events: distinguishing signals in the geologic record. *Physics of the Earth and Planetary Interiors*, **146**, 319–332.

COOKE, D.R. & SIMMONS, S.F. 2000. Characteristics and genesis of epithermal gold deposits. *In*: HAGEMANN, S.G. & BROWN, P.E. (eds) *Gold in 2000*. Reviews in Economic Geology, **13**, 221–244.

COOKE, D.R., BULL, S.W., LARGE, R.R. & McGOLDRICK, P.J. 2000. The importance of oxidized brines for the formation of Australias Proterozoic stratiform sediment-hosted Pb–Zn (SEDEX) deposits. *Economic Geology*, **95**, 1–17.

CRAIG, J.R., VAUGHAN, D.J. & SKINNER, B.J. 1988. *Resources of the Earth*. Prentice Hall, New Jersey, 71–88.

DAHLKAMP, F.J. 1993. *Uranium Ore Deposits*. Springer Verlag, Berlin.

DAVIES, G.F. 1992. On the emergence of plate tectonics. *Geology*, **20**, 963–966.

DE RONDE, C.E.J., KAMO, S., DAVIS, D.W., DE WIT, M.J.
& SPOONER, E.T.C. 1991. Field, geochemical and
U–Pb isotopic constraints from hypabyssal felsic
intrusions within the Barberton greenstone belt,
South Africa; implications for tectonics and the
timing of gold mineralization. *Precambrian
Research*, **49**, 261–280.

DESHBANDUH, S. & NEHRU, C.E. 1997. Review of
Precambrian porphyry Cu±Mo±Au deposits with
special reference to Malanjkhand porphyry
copper deposits, Madhya Pradesh, India. *Journal
of the Geological Society of India*, **49**, 239–288.

DE WIT, M.J. 1998. On Archean granites, greenstones,
cratons and tectonics: does the evidence demand
a verdict? *Precambrian Research*, **91**, 181–226.

DE WIT, M.J. & THIART, C. 2003. Metallogenic scents
of Archaean cratons: changing patterns of mineralization during earth evolution. *Transactions of
the Institution of Mining and Metallurgy*, **112**,
B114–116.

DE WIT, M.J., ARMSTRONG, R., HART, R.J. & WILSON,
A.H. 1987. Felsic igneous rocks within the
3.3–3.5 Ga Barberton greenstone belt – high
crustal level equivalents of the surrounding
tonalite trondhjemite terrain, emplaced during
thrusting. *Tectonics*, **6**, 529–549.

DIGONNET, S., GOULET, N., BOURNE, J., STEVENSON, R.
& ARCHIBALD, D. 2000. Petrology of the
Abloviak aillikite dykes, New Quebec; evidence
for a Cambrian diamondiferous alkaline province
in northeastern North America. *Canadian
Journal of Earth Sciences*, **37**, 517–533.

EDWARDS, R. & ATKINSON, K. 1986. *Ore Deposit
Geology and its Influence on Mineral Exploration*. Chapman and Hall, London, 175–214.

ENGLAND, G.L., RASMUSSEN, B., KRAPEZ, B. &
GROVES, D.I. 2001. The origin of uraninite,
bitumen nodules and carbon seams in Witwatersrand gold–uranium–pyrite ore deposits, based
on a Permo-Triassic analog. *Economic Geology*,
96, 1907–1920.

EREMIN, N.I. & DERGACHOV, A.L. 1994. Kuroko-type
volcanogenic massive sulphide deposits in the
Rudny Altai province, Russia and Kazakhstan.
*Proceedings of the Ninth Quadrennial IAGOD
Symposium*, Beijing, China, August 12–14,
361–380.

FANG, W., HU, R., SU, W., XIAO, J., QI, L. & JIANG, G.
2002. On emplacement ages of lamproites in
Zhenyuan County, Guizhou Province, China.
Chinese Science Bulletin, **47**, 874–880.

FARQUAR, J., BAO, H. & THIEMENS, M. 2000. Atmospheric influence of Earth's earliest sulphur cycle.
Science, **289**, 756–758.

FRAKES, L.A. & BOLTON, B.R. 1984. Origin of
manganese giants: sea-level change and anoxic-
oxic history. *Geology*, **12**, 83–86.

FRANKLIN, J.M. (1996) Volcanic-associated massive
sulphide base metal. *In*: ECKSTRAND, O.R.,
SINCLAIR, W.D. & THORPE, R.I. (eds) *Geology of
Canadian Mineral Deposit Types*. Geological
Survey of Canada, Geology of Canada, **8**,
158–183.

FRIMMEL, H.E. & MINTER, W.E.L. 2002. Recent
developments concerning the geological history
and genesis of the Witwatersrand gold deposits,
South Africa. Society of Economic Geologists,
Special Publication, **9**, 17–45.

FRIMMEL, H.E., GROVES, D.I., KIRK, J., RUIZ, J.,
CHESLEY, J. & MINTER, W.E.L. in press. The
formation and preservation of the Witwatersrand
goldfields, the largest gold province in the world.
Economic Geology, 100th Anniversary Volume.

FYFE, W.S. 1978. The evolution of the Earth's crust:
Modern plate tectonics to ancient hot spot
tectonics? *Chemical Geology*, **23**, 89–114.

GASTIL, R.G. 1960. The distribution of mineral dates
in time and space. *American Journal of Science*,
258, 1–35.

GERALDES, M.C., FIGUEIREDO, B.R., TASSARINI,
C.C.G. & EBERT, H.D. 1997. Middle Proterozoic
vein-hosted gold deposits in the Pontese Lacerda
region, southwestern Amazonian Craton, Brazil.
International Geology Reviews, **39**, 438–448.

GIBSON, H.L. & KERR, D.J. 1993. Giant volcanic-
associated massive sulfide deposits: with
emphasis on Archean examples. *In*: WHITING,
B.H., HODGSON, C.J. & MASON, R. (eds) *Giant
Ore Deposits*. Society of Economic Geologists
Special Publication, **2**, 319–348.

GOLDFARB, R.J. 1997. *Metallogenic evolution of
Alaska*. Economic Geology Monograph, **8**, 4–34.

GOLDFARB, R.J., PHILLIPS, G.N. & NOCKLEBERG, W.J.
1998. Tectonic setting of synorogenic gold
deposits of the Pacific Rim. *Ore Geology
Reviews*, **13**, 185–218.

GOLDFARB, R.J., GROVES, D.I. & GARDOLL, S. 2001*a*.
Orogenic gold and geologic time: A global
synthesis. *Ore Geology Reviews*, **18**, 1–75.

GOLDFARB, R.J., GROVES, D.I. & GARDOLL, S. 2001*b*.
Rotund versus skinny orogens: well-nourished or
malnourished gold? *Geology*, **29**, 539–542.

GRAHAM, I., BURGESS, J.L., BRYAN, D., RAVENSCROFT,
P.J., THOMAS, E.M., DOYLE, B.J., HOPKINS, R. &
ARMSTRONG, K.A. 1999*a*. Exploration history
and geology of the Diavik Kimberlites, Lac de
Gras, Northwest Territories, Canada. *In*: GURNEY,
J.J., GURNEY, J.L., PASCOE, M.D. & RICHARDSON,
S.H. (eds) *Proceedings of the VIIth International
Kimberlite Conference*, Cape Town, South Africa,
1, 262–279.

GRAHAM, S., LAMBERT, D.D., SHEE, S.R., SMITH, C.B.
& HAMILTON, R. 1999*b*. Re–Os and Sm–Nd
isotopic constraints on the sources of kimberlites
and melnoites, Earaheedy Basin, Western
Australia. *In*: GURNEY, J.J., GURNEY, J.L., PASCOE,
M.D. & RICHARDSON, S.H. (eds) *Proceedings of
the VIIth International Kimberlite Conference*,
Cape Town, South Africa, **1**, 280–290.

GRAHAM, S., LAMBERT, D.D., SHEE, S.R., SMITH, C.B.
& REEVES, S. 1999*c*. Re–Os isotopic evidence for
Archean lithospheric mantle beneath the
Kimberley Block, Western Australia. *Geology*, **27**,
431–434.

GRAINGER, C.J., GROVES, D.I. & COSTA, C.H.C. 2002.
*The epigenetic sediment-hosted Serra Pelada
Au–PGE deposit and its potential genetic associ-
ation with Fe-oxide Cu–Au mineralization within*

the Carajas Mineral Province, Amazon Craton, Brazil. Society of Economic Geologists Special Publication, **9**, 47–64.

GRIFFIN, W.L., ZHANG, A., O'REILLY, S.Y. & RYAN, C.G. 1998. Phanerozoic evolution of the lithosphere beneath the Sino-Korean craton. *In*: FLOWER, M.F.J., CHUNG, S.-L., LO, C.-H. & LEE, T.Y. (eds) *Mantle Dynamics and Plate Interactions in East Asia*. Washington DC. American Geophysical Union Geodynamics Series, **27**, 107–126.

GRIFFIN, W.L., O'REILLY, S.Y., ABE, N., AULBACH, S., DAVIES, R.M., PEARSON, N.J., DOYLE, B.J. & KIVI, K. 2003. The origin and evolution of Archean lithospheric mantle. *Precambrian Research*, **127**, 19–41.

GROSS, G.A. (1993) Industrial and genetic models for iron ore in iron-formations. *In*: KIRKHAM, R.V., SINCLAIR, W.D., THORPE, R.I. & DUKE, J.M. (eds) *Mineral Deposit Modeling*. Geological Association of Canada Special Paper, **40**, 151–170.

GROVES, D.I. 1993. The crustal continuum model for late-Archaean lode-gold deposits of the Yilgarn Block, Western Australia. *Mineralium Deposita*, **28**, 366–374.

GROVES, D.I. & VIELREICHER, N.M. 2001. The Phalabowra (Palabora) carbonatite-hosted magnetite–copper sulphide deposit, South Africa: An end-member of the iron oxide copper-gold-rare earth element group? *Mineralium Deposita*, **36**, 189–194.

GROVES, D.I., HO, S.E., ROCK, N.M.S., BARLEY, M.E. & MUGGERIDGE, M.Y. 1987. Archean cratons, diamond and platinum; evidence for coupled long-lived crust-mantle systems. *Geology*, **15**, 801–805.

GROVES, D.I., GOLDFARB, R.J., GEBRE-MARIAM, M., HAGEMANN, S.G. & ROBERT, F. 1998. Orogenic gold deposits – a proposed classification in the context of their crustal distribution and relationship to other gold deposit types. *Ore Geology Reviews*, **13**, 7–27.

GROVES, D.I., GOLDFARB, R.J., ROBERT, F. & HART, C.J.R. 2003. Gold deposits in metamorphic belts: overview of current understanding, outstanding problems, future research, and exploration significance. *Economic Geology*, **98**, 1–29.

GROVES, D.I., GRAINGER, C.J. & TALLARICO, F.H.B. 2004. Repeated alkaline granitoid magmatism and formation of Fe-oxide Cu–Au and related deposits in the giant Carajas Mineral Province of Brazil. *In*: *Dynamic Earth: Past, Present and Future*. 17th Australian Geological Convention, Hobart, Abstract Volume, **83**.

GROVES, D.I., CONDIE, K.C., GOLDFARB, R.J., HRONSKY, J.M.A. & VIELREICHER, R.M. 2005. Secular changes in global tectonic processes and their influence on the temporal distribution of gold-bearing mineral deposits. *Economic Geology*, **100**, 203–224.

GURNEY, J.J., GURNEY, J.L., PASCOE, M.D. & RICHARDSON, S.H. (eds). 1999. *Proceedings of the VIIth International Kimberlite Conference*, Cape Town, South Africa, 947pp.

GURVICH, Y.M. 1981. The carbonaceous manganiferous associations. *International Geological Review*, **23**, 584–590.

HAEUSSLER, P.J., BRADLEY, D., GOLDFARB, R.J., SNEE, L.W. & TAYLOR, C.D. 1995. Link between ridge subduction and gold mineralization in southern Alaska. *Geology*, **23**, 995–998.

HAMILTON, W.B. 1998. Archean magmatism and deformation were not products of plate tectonics. *Precambrian Research*, **91**, 143–179.

HAMLYN, P.R. & KEAYS, R.R. 1986. Sulphur saturation and second-stage melts: application to the Bushveld platinum metal deposits. *Economic Geology*, **81**, 1431–1445.

HANNINGTON, M.D., POULSEN, K.H., THOMPSON, J.F.H. & SILLITOE, R.H. 1999. Volcanogenic gold in the massive sulfide environment. *In*: BARRIE, T.C. & HANNINGTON, M.D. (eds) *Volcanic-Associated Massive Sulfide Deposits: Progresses and Examples in Modern and Ancient Settings*. Reviews in Economic Geology, **8**, 325–356.

HART, C.J.R., MAIR, J.L., GOLDFARB, R.J. & GROVES, D.I. 2004. Source and redox controls on metallogenic variations in intrusion-related ore systems, Tombstone–Tungsten Belt, Yukon Territory, Canada. *Transactions Royal Society Edinburgh*, **95**, 339–356.

HAWKESWORTH, C.J., LIGHTFOOT, P.C., FEDORENKO, V.A., BLAKE, S., NALDRETT, A.J., DOHERTY, W. & GORBACHEV, N.S. 1995. Magma differentiation and mineralisation in the Siberian continental flood basalts. *Lithos*, **34**, 61–88.

HAYNES, D.W. 2002. Giant iron oxide-copper-gold deposits: are they in distinctive geological settings? *In*: COOKE, D.R. & PONGRATZ, J. (eds) *Giant Ore Deposits: Characteristics, Genesis and Exploration*. University of Tasmania, Hobart, CODES Special Publication, **4**, 57–77.

HENLEY, R.W. & ADAMS, J. 1979. On the evolution of giant gold placers. *Transactions of the Institution of Mining and Metallurgy*, **88**, B41–B51.

HITZMAN, M.W. 2000. Iron oxide Cu–Au deposits: what, where, when, and why. *In*: PORTER, T.M. (ed.) *Hydrothermal Iron Oxide–Copper–Gold and Related Deposits: A Global Perspective*. Australian Mineral Foundation, Adelaide, vol. **1**, 9–25.

HITZMAN, M.W. 2001. Fe oxide–Cu–Au systems in the Lufilian orogen of southern Africa. *Geological Society of America, Abstracts with Program*, **33**, 1.

HITZMAN, M.W., ORESKES, N. & EINAUDI, M.T. 1992. Geological characteristics and tectonic setting of Proterozoic iron-oxide (Cu–U–Au–REE) deposits. *Precambrian Research*, **58**, 241–287.

HOFFMAN, P.E. 1988. The United Plates of America, the birth of a craton. *Annual Review of Earth and Planetary Sciences*, **16**, 543–604.

HOFFMAN, P.F. 1989. Speculations on Laurentia's first gigayear (2.0–1.0 Ga). *Geology*, **17**, 135–138.

HOFFMAN, P.F. 1999. The break-up of Rodinia, birth of Gondwana, true polar wander and the snowball Earth. *Journal of African Earth Sciences*, **28**, 17–33.

HOFFMAN, P.F. & SCHRAG, D.P. 2002. The snowball

Earth hypothesis: testing the limits of global change. *Terra Nova*, **14**, 129–155.

HOLLAND, H.D. 1994. Early Proterozoic atmosphere change. *In*: BENGTSON, S. (ed.) *Early Life on Earth*. Nobel Symposium 84, Columbia University Press, New York, 237–244.

HOLLAND, H.D. 1999. When did the Earth's atmosphere become oxic? A reply. *The Geochemistry News*, **100**, 857–858.

ILJINA, M. & HANSKI, E. 2002. Multimillion-ounce PGE deposits of the Portimo layered igneous complex, Finland. *9th International Platinum Symposium Abstracts*, 21–25 July 2002, Billings, Montana.

ISLEY, A.E. & ABBOTT, D.H. 1999. Plume-related mafic volcanism and the deposition of banded iron formation. *Journal of Geophysical Research*, **104**, 15 461–15 477.

JACKSON, M.J., SCOTT, D.L. & RAWLINGS, D.J. 2000. Stratigraphic framework for the Leichhardt and Calvert Superbasins: review and correlations of the pre-1700 Ma successions between Mt Isa and McArthur River. *Australian Journal of Earth Sciences*, **47**, 381–403.

JANSE, A.J.A. 1984. Kimberlites – where and when. *In*: GLOVER, J.E. & HARRIS, P.G. (eds) Kimberlite Occurrence and Origin. *The Geology Department and University Extension, The University of Western Australia Publication*, **8**, 19–61.

JAQUES, A.L., WEBB, A.W., FANNING, C.M., BLACK, L.P., PIDGEON, R.T., SMITH, C.B. & GREGORY, G.P. 1984. The age of the diamond-bearing pipes and associated leucite lamproites of the West Kimberley region, Western Australia. *BMR Journal of Australian Geology and Geophysics*, **9**, 1–7.

JENSEN, M.L. & BATEMAN, A.M. 1979. *Economic Mineral Deposits*. John Wiley and Sons, Toronto, 386–442.

KAMINSKY, F.V., SABLUKOV, S.M., SABLUKOVA, L.I., SHCHUKIN, V.S. & CANIL, D. 2002. Kimberlites from the Wawa area, Ontario. *Canadian Journal of Earth Sciences*, **39**, 1819–1838.

KEAYS, R.R. 1995. The role of komatiitic and picritic magmatism and S-saturation in the formation of ore deposits. *Lithos*, **34**, 1–18.

KERRICH, R., GOLDFARB, R.J., GROVES, D.I. & GARWIN, S. 2000. The geodynamics of world class gold deposits: characteristics, space-time distribution and origins. *In*: HAGEMANN, S.G. & BROWN, P.E. (eds) *Gold in 2000*. Reviews in Economic Geology, **13**, 501–551.

KESLER, S.E. 1997. Metallogenic evolution of convergent margins: selected ore deposit models. *Ore Geology Reviews*, **12**, 153–171.

KESLER, S.E. & REICH, M.H. in press. Precambrian Mississippi Valley-type deposits: relation to changes in composition of the hydrosphere and atmosphere. *In*: OHMOTO, H. & KESLER, S. (eds) *Ore Deposits as Indicators of Early Earth Atmosphere, Hydrosphere and Biosphere*. Geological Society of America Special Volume.

KESLER, S.E., APPOLD, M.S., MARTINI, A.M., WALKER, L.M., HUSTON, T.J. & KYLE, J.R., 1995. Na–Cl–Br systematics of mineralising brines in Mississippi Valley-type deposits. *Geology*, **23**, 641–644.

KIRK, J., RUIZ, J., CHESLEY, J., TITLEY, S. & WALSHE, J. 2001. A detrital model for the origin of gold and sulfides in the Witwatersrand basin based on Re–Os isotopes. *Geochimica et Cosmochimica Acta*, **65**, 2149–2159.

KIRK, J., RUIZ, J., CHESLEY, J., WALSHE, J. & ENGLAND, G. 2002. A major Archean gold and crust-forming event in the Kaapvaal Craton, South Africa. *Science*, **297**, 1856–1858.

KIRKHAM, R.V. & SINCLAIR, W.D. 1996. Porphyry copper, gold, molybdenum, tungsten, tin, silver. *In*: ECKSTRAND, O.R., SINCLAIR, W.D. & THORPE, R.I. (eds) *Geology of Canadian Mineral Deposits*. Geological Survey of Canada, Geology of Canada, **8**, 421–446.

KLEIN, C. & BEUKES, N.J. 1992. Time distribution, stratigraphy, sedimentologic setting, and geochemistry of Precambrian iron-formations. *In*: SCHOPF, J.W. & KLEIN, C. (eds) *The Proterozoic Biosphere*. Cambridge University Press, Cambridge, 139–146.

KLEMM, D.D. 2000. The formation of Palaeoproterozoic banded iron formations and their associated Fe and Mn deposits, with reference to Griqualand West deposits, South Africa. *Journal of African Earth Sciences*, **30**, 1–24.

KNIGHT, J., LOWE, J., JOY, S., CAMERON, J., MERRILLEES, J., NAG, S., SHAH, N., DUA, G. & JHALA, K. 2002. The Kheti copper belt, Rajasthan: iron oxide copper-gold terrane in the Proterozoic of NW India. *In*: PORTER, T.M. (ed.) *Hydrothermal Iron Oxide–Copper–Gold & Related Deposits: A Global Perspective*. PGC Publishing, Adelaide, **2**, 321–341.

KONSTANTINOV, M., DANKOVTSEV, R., SIMKIN, G. & CHERKASOV, S. 1999. Deep structure of the north Enisei gold district (Russia) and setting of ore deposits. *Geology of Ore Deposits*, **41**, 425–436.

KORNILOVA, V.P., SAFRONOV, A.F., ZAITSEV, A.I. & PHILIPPOV, N.D. 1999. Garnet–diamond association in lamprophyres of the Anabar Massif. *In*: GURNEY, J.J., GURNEY, J.L., PASCOE, M.D. & RICHARDSON, S.H. (eds) *Proceedings of the VIIth International Kimberlite Conference*, Cape Town, South Africa, **1**, 480–484.

KOSITCIN, N. & KRAPEZ, B. 2004. Relationship between detrital zircon age-spectra and the tectonic evolution of the Late Archaean Witwatersrand Basin, South Africa. *Precambrian Research*, **129**, 141–168.

KRUPP, R., OBERTHUER, T. & HIRDES, W. 1994. The early Precambrian atmosphere and hydrosphere: thermodynamic constraints from mineral deposits. *Economic Geology*, **89**, 1581–1598.

KUMP, L.R., KASTING, J.F. & BARLEY, M.E. 2001. Rise of atmospheric oxygen and the 'upside-down' Archean mantle. *Geochemistry, Geophysics, Geosystems*, **2**, Paper No. 2000GC000114.

LANDEFELD, L.A. 1988. The geology of the Mother Lode gold belt, Sierra Nevada Foothills metamorphic belt, California. *In*: GOODE, A.D.T. &

BOSMA, L.I. (eds) *Bicentennial Gold 88, Extended Abstracts*. Geological Society of Australia, Melbourne, 167–172.

LANG, J.R. & BAKER, T. 2001. Intrusion-related gold systems: the present level of understanding. *Mineralium Deposita*, **36**, 477–489.

LARGE, R.R. 1992. Australian volcanic-hosted massive sulfide deposits; features, styles, and genetic models. *Economic Geology*, **87**, 471–510.

LARGE, R.R., BULL, S.W., COOKE, D.R. & McGOLDRICK, P.J. 1998. A genetic model for the HYC deposit, Australia: based on regional sedimentology, geochemistry, and sulphide–sediment relationships. *Economic Geology*, **93**, 1345–1368.

LARGE, R.R., BULL, S., SELLEY, D., JIANWEN Yang, COOKE, D., GARVEN, G. & McGOLDRICK, P. 2002. Controls on the formation of giant stratiform sediment-hosted Zn–Pb–Ag deposits: with particular reference to the north Australian Proterozoic. *In*: COOKE, D.R. & PONGRATZ, J. (eds) *Giant Ore Deposits: Characteristics, Genesis and Exploration*. University of Tasmania, Hobart, CODES Special Publication, **4**, 107–150.

LARSON, R.L. 1991. Geological consequences of superplumes. *Geology*, **19**, 963–966.

LEACH, D.L., BRADLEY, D., LEWCHUK, M.T., SYMONS, D.T.A., de MARSILY, G. & BRANNON, J. 2001. Mississippi Valley-type lead–zinc deposits through geological time: implications from recent age-dating research. *Mineralium Deposita*, **36**, 711–740.

LECKIE, D.A., KJARSGAARD, B.A., BLOCH, J., McINTYRE, D., McNEIL, D., STASIUK, L.D. & HEAMAN, L.M. 1997. Emplacement and reworking of Cretaceous, diamond-bearing, crater facies kimberlite of central Saskatchewan, Canada. *Geological Society of America Bulletin*, **109**, 1000–1020.

LEISTEL, J.M., MARCOUX, E., THIEBLEMONT, D., QUESADA, C., SANCHEZ, A., ALMODOVAR, G.R., PASCUAL, E. & SAEZ, R. 1998. The volcanic-hosted massive sulphide deposits of the Iberian Pyrite Belt. *Mineralium Deposita*, **33**, 2–30.

LESHER, C.M. 1989. Komatiite-associated nickel sulfide deposits. *In*: WHITNEY, J.A. & NALDRETT, A.J. (eds) *Ore Deposition Associated With Magmas*. Reviews in Economic Geology, **4**, 45–101.

LEVINE, R. & WILBURN, D. 2003. *Russian PGM – Resources for 100+ years*. United States Geological Survey, Open-File Report **03-059**, 19pp.

LORENZ, V., ZIMANOWSKI, B., BUETTNER, R. & KURSZLAUKIS, S. 1999. Formation of kimberlite diatremes by explosive interaction of kimberlite magma with groundwater; field and experimental aspects. Extended Abstracts. *Proceedings of the VIIth International Kimberlite Conference*, Cape Town, South Africa, **1**, 971–973.

LOWMAN, J.P. & JARVIS, G.T. 1996. Continental collisions in wide aspect ratio and high Rayleigh number two-dimensional mantle convection models. *Journal of Geophysical Research*, **101**, 25 485–25 497.

LUDDEN, J.N. & HYNES, A. 2000. Lithoprobe

Abitibi–Grenville transect; two billion years of crust formation and recycling in the Precambrian Shield of Canada. *Canadian Journal of Earth Sciences*, **37**, 459–476.

LYDON, J.W. 1996. Sedimentary exhalative sulphides (SEDEX). *In*: ECKSTRAND, O.R., SINCLAIR, W.D. & THORPE, R.I. (eds) *Geology of Canadian Mineral Deposit Types*. Geological Survey of Canada, Geology of Canada, **8**, 130–152.

MACNAE, J. 1995. Applications of geophysics for the detection and exploration of kimberlites and lamproites. *Journal of Geochemical Exploration*, **53**, 213–243.

MARMONT, S. 1987. Ore deposit models – Unconformity-type uranium deposits. *Geoscience Canada*, **14**, 219–229.

MARSCHIK, R., CHIARADA, M. & FONTBOTÉ, L. 2003*a*. Implications of Pb isotope signatures of rocks and iron oxide Cu–Au ores in the Candelaria–Punta del Cobre district, Chile. *Mineralium Deposita*, **38**, 900–912.

MARSCHIK, R., FONTAGNIE, D., CHIARADA, M. & VOLDET, P. 2003*b*. Geochemical and Nd–Sr–Pb–O isotope characteristics of granitoids of the Early Cretaceous Copiapo plutonic complex (27°30'S), Chile. *Journal of South American Earth Sciences*, **16**, 381–398.

MARTIN, H. & MOYEN, J.F. 2002. Secular changes in tonalite–trondhjemite–granodiorite composition as markers of the progressive cooling of Earth. *Geology*, **30**, 319–322.

MATHUR, R., MARSCHIK, R., RUIZ, J., MUNIZAGA, F., LEVEILLE, R.A. & MARTIN, W. 2002. Age of mineralization of the Candelaria Fe oxide Cu–Au deposit and the origin of the Chilean iron belt, based on Re–Os isotopes. *Economic Geology*, **97**, 59–71.

MEERT, J.G. & POWELL, C. McA. 2001. Assembly and break-up of Rodinia: introduction to the special volume. *Precambrian Research*, **110**, 1–8.

MELON, G.B. 1962. Petrology of Upper Cretaceous oolitic Fe-rich rocks from northern Alberta. *Economic Geology*, **57**, 921–940.

MENZIES, A.H., CARLSON, R.W., SHIREY, S.B. & GURNEY, J.J. 1999. Re–Os systematics of Newlands peridotite xenoliths; implications for diamond and lithosphere formation. *In*: GURNEY, J.J., GURNEY, J.L., PASCOE, M.D. & RICHARDSON, S.H. (eds) *Proceedings of the VIIth International Kimberlite Conference*, Cape Town, South Africa, **2**, 566–573.

MEYER, C. 1981. Ore-forming processes in geologic history. *Economic Geology*, **75**, Anniversary Volume, 6–41.

MEYER, C. 1988. Ore deposits as guides to geologic history of the Earth. *Annual Review of Earth and Planetary Science*, **16**, 147–171.

MIKHAILOV, B.M. 2004. Hypergene metallogeny of the Urals. *Lithology and Mineral Resources*, **39**, 114–134.

MIKUCKI, E.J. 1998. Hydrothermal transport and depositional processes in Archean lode-gold systems: a review. *Ore Geology Reviews*, **13**, 307–321.

MILESI, J.P., LEDRU, P., MARCOUX, E., MOUGEOT, R., JOHAN, V., LEROUGE, C., SABATE, P., BAILLY, L., RESPAUT, J.P. & SKIPWITH, P. 2002. The Jacobina Paleoproterozoic gold-bearing conglomerates, Bahia, Brazil: a 'hydrothermal shear-reservoir' model. *Ore Geology Reviews*, **19**, 95–136.

MINTER, W.E.L. 1999. Irrefutable detrital origin of Witwatersrand gold and evidence of eolian signatures. *Economic Geology*, **94**, 665–670.

MINTER, W.E.L., GOEDHART, M.L., KNIGHT, J. & FRIMMEL, H.E. 1993. Morphology of Witwatersrand gold grains from the Basal Reef: Evidence for their detrital origin. *Economic Geology*, **88**, 237–248.

MITCHELL, A.H.G. & GARSON, M.S. 1981. *Mineral Deposits and Global Tectonic Settings*. Academic Press Geology Series, London.

MORALEV, V.M. & GLUKHOVSKY, M.Z. 2000. Diamond-bearing kimberlite fields of the Siberian Craton and the early Precambrian geodynamics. *Ore Geology Reviews*, **17**, 141–153.

MURPHY, J.B. & NANCE, R.D. 1992. Mountain belts and the supercontinent cycle. *Scientific American*, **266**, 84–91.

NALDRETT, A.J. 1997. Key factors in the genesis of Noril'sk, Sudbury, Jinchuan, Voisey's Bay and other world-class Ni–Cu–PGE deposits: implications for exploration. *Australian Journal of Earth Sciences*, **44**, 283–316.

NALDRETT, A.J. 1999. World-class Ni–Cu–PGE deposits: key factors in their genesis. *Mineralium Deposita*, **34**, 227–240.

NALDRETT, A.J. 2002. Requirements for forming giant Ni–Cu sulphide deposits. *In*: COOKE, D.R. & PONGRATZ, J. (eds) *Giant Ore Deposits: Characteristics, Genesis and Exploration*. University of Tasmania, Hobart, CODES Special Publication, **4**, 195–204.

NAVON, O. 1999. Diamond formation in the Earth's mantle. *In*: GURNEY, J.J., GURNEY, J.L., PASCOE, M.D. & RICHARDSON, S.H. (eds) *Proceedings of the VIIth International Kimberlite Conference*, Cape Town, South Africa, **2**, 584–604.

NEEDHAM, R.S. 1988. Geology of the Alligator Rivers uranium field, Northern Territory. Australian Bureau of Mineral Resources, *Geology and Geophysics Bulletin* **224**.

NISBET, B., COOKE, J., RICHARDS, M. & WILLIAMS, C. 2000. Exploration for iron oxide copper gold deposits in Zambia and Sweden: comparison with the Australian experience. *In*: PORTER, T.M. (ed.) *Hydrothermal Iron Oxide Copper–Gold & Related Deposits: A Global Perspective*. Australian Mineral Foundation, Adelaide, vol. **1**, 297–308.

O'DEA, M.G., LISTER, G.S., BETTS, P.G. & POUND, K.S. 1997. A shortened intraplate rift system in the Proterozoic Mt Isa terrain, NW Queensland, Australia. *Tectonics*, **16**, 425–441.

O'REILLY, S.Y. & GRIFFIN, W.L. 1996. 4-D lithospheric mapping: a review of the methodology with examples. *Tectonophysics*, **262**, 3–18.

OHMOTO, H. 1997. When did the Earth's atmosphere become oxic? *The Geochemistry News*, **93**, 13–13 & 26–27.

OHMOTO, H. 2003. Banded iron formations and the evolution of the atmosphere, hydrosphere, biosphere and lithosphere. *Transactions of the Institutions of Mining and Metallurgy*, **112**, B161–B162.

OHMOTO, H., WATANABE, Y., YAMAGUCHI, K.E., ONO, S., BAU, M., KAKEGAWA, T., NARAOKE, H., NEDOCHI, M. & LASAGA, A.C. 2001. The Archaean atmosphere, oceans, continents and life. *In*: CASSIDY, K.F., DUNPHY, J.M. & VAN KRANENDONK, M.J. (eds) *4th International Archaean Symposium – Extended Abstracts*, Australian Geological Survey Organization Record, 19–21.

PAN, Y. & XIE, Q. 2001. Extreme fractionation of platinum group elements in volcanogenic massive sulfide deposits. *Economic Geology*, **96**, 645–651.

PARSONS, B.A. 1982. Causes and consequences of the relationship between area and age of the ocean floor. *Journal of Geophysical Research*, **87**, 289–302.

PERELLO, J., COX, D., GARAMJAV, D., SANIDORJ, S., DIAKOV, S., SCHISSEL, D., MUNKHBAT, T.-O. & OYUN, G. 2001. Oyu Tolgoi, Mongolia: Siluro-Devonian porphyry Cu–Au–(Mo) and high-sulfidation Cu mineralization with a Cretaceous chalcocite blanket. *Economic Geology*, **96**, 1407–1428.

PESONEN, L.J., ELMING, S.Å., MERTANEN, S., PISAREVSKY, S., D'AGRELLA-FILHO, M.S., MEERT, J.G, SCHMIDT, P.W, ABRAHAMSEN, N. & BYLUNDET, G. 2003. Paleomagnetic configuration of continents during the Proterozoic. *Tectonophysics*, **375**, 289–324.

PHILLIPS, G.N. & MYERS, R.E. 1989. Witwatersrand gold fields. Part II. An origin for Witwatersrand gold during metamorphism and associated alteration. *Economic Geology Monograph*, **6**, 598–608.

PIGOIS, J.-P., GROVES, D.L., FLETCHER, I.R., MCNAUGHTON, N.J. & SNEE, L.W. 2003. Age constraints on Tarkwaian palaeoplacer and lode-gold formation in the Tarkwa–Damang district, SW Ghana. *Mineralium Deposita*, **38**, 695–714.

PIRAJNO, F. 2000. *Ore Deposits and Mantle Plumes*. Kluwer Academic Publishers, Dordrecht.

PLANT, J., SIMPSON, P.R., SMITH, B. & WINDLEY, B.F. 1999. Uranium ore deposits: products of the radioactive Earth. *In*: BURNS, P.C. & FINCH, R. (eds) *Uranium: Mineralogy, Geochemistry and the Environment*. Reviews in Mineralogy, **38**, 255–320.

POLLACK, H.N. 1986. Cratonization and thermal evolution of the mantle. *Earth and Planetary Science Letters*, **80**, 175–182.

POLLACK, H.N. 1997. Thermal characteristics of the Archaean. *In*: DE WIT, M.J. & ASHWAL, L.D. (eds) *Greenstone Belts*. Clarendon Press, Oxford, 223–232.

POUDJOM DJOMANI, Y.H., O'REILLY, S.Y., GRIFFIN, W.L. & MORGAN, P. 2001. The density structure of subcontinental lithosphere through time. *Earth and Planetary Science Letters*, **184**, 605–621.

POULSEN, K.H. & HANNINGTON, M.D. 1996.

Volcanic-associated massive sulphide gold. *In*: ECKSTRAND, O.R., SINCLAIR, W.D. & THORPE, R.I. (eds) *Geology of Canadian Mineral Deposit Types*. Geological Survey of Canada, Geology of Canada, **8**, 183–196.

PRETORIUS, D.A. 1981. Gold and uranium in quartz-pebble conglomerates. *Economic Geology*, **75**, Anniversary Volume, 117–138.

PROKIN, V.A. & BUSLAEV, F.P. 1999. Massive copper–zinc sulphide deposits in the Urals. *Ore Geology Reviews*, **14**, 1–69.

QIU, Y. & GROVES, D.I. 1999. Late Archean collision and delamination in the southwest Yilgarn craton: the driving force for Archean orogenic lode gold mineralization: *Economic Geology*, **94**, 115–122.

RASMUSSEN, B. 2000. Filamentous microfossils in a 3235-million-year-old volcanogenic massive sulphide deposit. *Nature*, **405**, 676–679.

RASMUSSEN, B. & BUICK, R. 1999. Redox state of the Archean atmosphere: Evidence from detrital minerals in *ca* 3250–2750 Ma sandstones from the Pilbara Craton, Australia. *Geology*, **27**, 115–118.

RASMUSSEN, B., FLETCHER, I.R. & MCNAUGHTON, N.J. 2001. Dating low-grade metamorphic events by SHRIMP U–Pb analysis of monazite in shales. *Geology*, **29**, 963–966.

RASMUSSEN, B., BENGTSON, S., FLETCHER, I.R. & MCNAUGHTON, N.J. 2002. Discoidal impressions and trace-like fossils more than 1200 million years old. *Science*, **296**, 1112–1115.

RELVAS, J.M.R.S., BARRIGA, F.J.A.S., PINTO, A., *ET AL.* 2002. *The Neves–Corvo Deposit, Iberian Pyrite Belt, Portugal: impacts and future, 25 years after the discovery*. Society of Economic Geologists Special Publications, **9**, 155–176.

RENNE, P.R., SWISHER, C.C., DEINO, A.L., KARNER, D.B., OWENS, T.L. & DEPAOLO, D.J. 1998. Intercalibration of standards, absolute ages and uncertainties in $^{40}Ar/^{39}Ar$ dating. *Chemical Geology*, **145**, 117–152.

REQUIA, K., STEIN, H., FONTBOTE, L. & CHIARADIA, M. 2003. Re–Os and Pb–Pb geochronology of the Archean Salobo iron oxide copper–gold deposit, Carajas mineral province, northern Brazil. *Mineralium Deposita*, **38**, 727–738.

RICHARDSON, S.H., GURNEY, J.J., ERLAK, A.J. & HARRIS, J.W. 1984. Origin of diamond in old enriched mantle. *Nature*, **310**, 198–202.

RICHARDSON, S.H., CHINN, I.L. & HARRIS, J.W. 1999. Age and origin of eclogitic diamonds from the Jwaneng Kimberlite, Botswana. *In*: GURNEY, J.J., GURNEY, J.L., PASCOE, M.D. & RICHARDSON, S.H. (eds) *Proceedings of the VIIth International Kimberlite Conference*, Cape Town, South Africa, **2**, 709–713.

RICHARDSON, S.H., SHIREY, S.B. & HARRIS, J.W. 2004. Episodic diamond genesis at Jwaneng, Botswana, and implications for Kaapvaal craton evolution. *Lithos*, **77**, 143–154.

ROGERS, J.J.W. 1996. A history of continents in the past three billion years. *Journal of Geology*, **104**, 91–107.

ROY, S. 1981. *Manganese Deposits*. Academic Press, London.

ROY, S. 1988. Manganese metallogenesis: a review. *Ore Geology Reviews*, **4**, 155–170.

ROY, S. 2000. Late Archean initiation of manganese metallogenesis: its significance and environmental controls. *Ore Geology Reviews*, **17**, 179–198.

RUSSELL, M.J. 2003. Origins and evolution of life: clues from ore deposits. *Transactions of the Institutions of Mining and Metallurgy*, **112**, B177–B178.

RUZICKA, V. 1996. Unconformity-associated uranium. *In*: ECKSTRAND, O.R., SINCLAIR, W.D. & THORPE, R.I. (eds) *Geology of Canadian Mineral Deposit Types*. Geological Survey of Canada, Geology of Canada, **8**, 197–210.

SAFONOV, Y.G. 1997. Hydrothermal gold deposits – distribution, geological/genetic types, and productivity of ore-forming systems. *Geology of Ore Deposits*, **39**, 20–32.

SAGE, R.P. 2000. *Kimberlites of the Attawapiskat area, James Bay Lowlands, northern Ontario*. Ontario Geological Survey, Open File Report **2000**, 341 pp.

SANDERS, T.S. 1999. *Mineralization of the Halls Creek Orogen, east Kimberley region, Western Australia*. Geological Survey of Western Australia, Report **44**.

SANGSTER, D.F. 1990. Mississippi Valley-type and SEDEX lead–zinc deposits: a comparative examination. *Transactions of the Institutions of Mining and Metallurgy*, **99**, B21–B42.

SANGSTER, D.F. 1996. Mississippi Valley-type lead–zinc. *In*: ECKSTRAND, O.R., SINCLAIR, W.D. & THORPE, R.I. (eds) *Geology of Canadian Mineral Deposit Types*. Geological Survey of Canada, Geology of Canada, **8**, 253–261.

SANO, Y., YOKOCHI, R., TERADA, K., CHAVES, M.L. & OZIMA, M. 2002. Ion microprobe Pb–Pb dating of carbonado, polycrystalline diamond. *Precambrian Research*, **113**, 155–168.

SAWKINS, F.J. 1984. *Metal Deposits in Relation to Plate Tectonics*. Minerals and Rocks, 17, Springer Verlag, Berlin.

SCHAEFER, M.O., GUTZMER, J. & BEUKES, N.J. 2001. Late Paleoproterozoic Mn-rich oncoids: Earliest evidence for microbially mediated Mn precipitation. *Geology*, **29**, 835–838.

SCHAERER, U., CORFU, F. & DEMAIFFE, D. 1997. U–Pb and Lu–Hf isotopes in baddeleyite and zircon megacrysts from the Mbuji-Mayi Kimberlite; constraints on the subcontinental mantle. *Chemical Geology*, **143**, 1–16.

SCHMITZ, M.D., BOWRING, S.A., DE WIT, M.J. & GARTZ, V. 2004. Subduction and terrane collision stabilize the western Kaapvaal craton tectosphere 2.9 billion years ago. *Earth and Planetary Science Letters*, **222**, 363–376.

SHANKER, R., NAG, S., GANGULY, A., ABSAR, A., RAWAT, B.P. & SINGH, G.S. 2001. Are Majhgawan–Hinota pipe rocks truly Group-I kimberlite? Proceedings of the Indian Academy of Sciences, *Earth and Planetary Sciences*, **110**, 63–76.

SHIREY, S.B., HARRIS, J.W., RICHARDSON, S.H., FOUCH, M., JAMES, D.E., CARTIGNY, P., DEINES, P. & VILJOEN, F. 2003. Regional patterns in the petrogenesis and age of inclusions in diamond, diamond composition, and the lithospheric seismic structure of Southern Africa. *Lithos*, **71**, 243–258.

SILLITOE, R.H. 1991. Intrusion-related gold deposits. *In*: FOSTER, R.P. (ed.), *Metallogeny and Exploration of Gold*. Blackie and Sons, Glasgow, 165–209.

SILLITOE, R.H. 1997. Characteristics and controls of the largest porphyry copper–gold and epithermal gold deposits in the circum-Pacific region. *Australian Journal of Earth Sciences*, **44**, 373–388.

SILLITOE, R.H. 2000. Gold-rich porphyry deposits: descriptive and genetic models and their role in exploration and discovery. *Reviews in Economic Geology*, **13**, 315–346.

SILLITOE, R.H. & THOMPSON, J.F.H. 1998. Intrusion-related vein gold deposits: types, tectono-magmatic settings, and difficulties of distinction from orogenic gold deposits. *Resource Geology*, **48**, 237–250.

SIMONSON, B.M. & HASSLER, S.W. 1997. Revised correlations in the Early Precambrian Hamersley basin based on a horizon of resedimented impact spherules. *Australian Journal of Earth Sciences*, **44**, 1–12.

SLEEP, N.H. 2003. Survival of Archean cratonal lithosphere. *Journal of Geophysical Research*, **108**, 8–1 to 8–25.

SLEEP, N.H. & WINDLEY, B.F. 1982. Archean plate tectonics: constraints and inferences. *Journal of Geology*, **90**, 363–379.

SMITH, P.E., SCHANDL, E.S. & YORK, D. 1993. Timing of metasomatic alteration of the Archean Kidd Creek massive sulfide deposit, Ontario, using $^{40}Ar/^{39}Ar$ laser dating of single crystals of fuchsite. *Economic Geology*, **88**, 1636–1642.

SOLOMON, M. & GROVES, D.I. 1994. *The Geology and Origin of Australia's Mineral Deposits*: Oxford Monographs in Geology and Geophysics, **24**. Oxford University Press, Oxford.

SOLOMON, M. & HEINRICH, C.A. 1992. Are high heat producing granites essential to the origin of giant lead–zinc deposits at McArthur River and Mount Isa, Australia. *Exploration Mining Geology*, **1**, 85–91.

SOLOMON, M. & QUESADA, C. 2003. Zn–Pb–Cu massive sulfide deposits: Brine-pool types occur in collisional orogens, black smoker types occur in backarc and/or arc basins. *Geology*, **31**, 1029–1032.

SOLOMON, M. & SUN, S-S. 1997. Earth's evolution and mineral resources, with particular emphasis on volcanic-hosted massive sulphide deposits and banded iron formations. AGSO *Journal of Australian Geology and Geophysics*, **17**, 33–48.

SONI, M.K., RAO, T.K. & JHA, D.K. 2001. Status of diamond exploration in Panna diamond belt, Madhya Pradesh. *Geological Survey of India, Special Publication Series*, **64**, 353–367.

SPETSIUS, Z.V., BELOUSSOVA, E.A., GRIFFIN, W.L.,

O'REILLY, S.Y. & PEARSON, N.J. 2002. Archean sulfide incluions in Paleozoic zircon megacrysts from the Mir Kimberlite, Yakutia; implications for the dating of diamonds. *Earth and Planetary Science Letters*, **199**, 111–126.

SPRAGUE, D. & POLLACK, H.N. 1980. Heat flow in the Mesozoic and Cenozopic. *Nature*, **285**, 393–395.

STANTON, R.L. 1972. *Ore Petrology*. McGraw-Hill, New York.

STEIN, M. & HOFFMANN, A.W. 1994. Mantle plumes and episodic crustal growth. *Nature*, **372**, 63–68.

STEIN, H.J., MARKEY, R.J., MORGAN, J.W., DU, A. & SUN, Y. 1997. Highly precise and accurate Re–Os ages for molybdenite from the East Qinling molybdenum belt, Shaanxi Province, China. *Economic Geology*, **92**, 827–835.

STOREY, M., MAHONEY, J.J., KROENKE, L.W. & SAUNDERS, A.D. 1991. Are oceanic plateaus sites of komatiite formation. *Geology*, **19**, 376–379.

STRIBRNY, B., WELLMER, F.-W., BURGATH, K.-P., OBERTHUER, T., TARKIAN, M. & PFEIFFER, T. 2000. Unconventional PGE occurrences and PGE mineralization in the Great Dyke: metallogenic and economic aspects. *Mineralium Deposita*, **25**, 260–281.

TALLARICO, F.H.B., FIGUEIREDO, B.R., GROVES, D.I., KOSITCIN, N., MCNAUGHTON, N.J., FLETCHER, I.R. & REGO, J.L. 2005. Geology and SHRIMP U–Pb geochronology of the Igarape Bahia deposit, Carajas Copper-Gold Belt, Brazil: an Archean (2.57 Ga) example of iron-oxide Cu–Au–(U–REE) mineralization. *Economic Geology*, **100**, 7–28.

TAYLOR, S.R. & MCLENNAN, S.M. 1985. *The Continental Crust: Its Composition and Evolution*. Blackwell, Oxford.

TAYLOR, D., DAHLSTRA, H.J., HARDING, A.E., BROADBENT, G.C. & BARLEY, M.E. 2001. Genesis of high-grade hematite orebodies of the Hamersley Province, Western Australia. *Economic Geology*, **96**, 837–873.

THOMPSON, J.F.H. & NEWBERRY, R.J. 2000. Gold deposits related to reduced granitic intrusions. *Reviews in Economic Geology*, **13**, 377–400.

THOMPSON, J.F.H., SILLITOE, R.H., BAKER, T., LANG, J.R. & MORTENSEN, J.K. 1999. Intrusion-related gold deposits associated with tungsten-tin provinces. *Mineralium Deposita*, **34**, 323–334.

TITLEY, S.R. 1993. Relationship of stratabound ores with tectonic cycles of the Phanerozoic and Proterozoic. *Precambrian Research*, **61**, 295–322.

TRENDALL, A.F. & MORRIS, R.C. (eds) 1983. *Iron Formations: Facts and Problems*. Elsevier, Amsterdam.

TRUBITSYN, V.P., MOONEY, W.D. & ABBOTT, D.H. 2003. Cold cratonic roots and thermal blankets: how continents affect mantle convection. *International Geology Reviews*, **45**, 479–496.

VEARNCOMBE, S., BARLEY, M.E., GROVES, D.I., MCNAUGHTON, N.J., MIKUCKI, E.J. & VEARNCOMBE, J.R. 1995. 3.26 Ga black smoker-type mineralization in the Strelley Belt, Pilbara Craton, Western Australia. *Journal of Geological Society of London*, **152**, 587–590.

VERMAAK, C.F. 1995. *The Platinum-Group Metals – A Global Perspective*. Mintek, Randburg, South Africa.

VIELREICHER, N.M., GROVES, D.I., FLETCHER, I.R., MCNAUGHTON, N.J. & RASMUSSEN, B. 2003. Hydrothermal monazite and xenotime geochronology: a new direction to precise dating of orogenic gold mineralization. *Society of Economic Geologists Newsletter*, **53**, 1&10–16.

VIETS, J.G., HOFSTRA, A.H. & EMSBO, P. 1996. Solute compositions of fluid inclusions in sphalerite from North America and European Mississippi Valley-type ore deposits: ore fluids derived from evaporated seawater. *In*: SANGSTER, D.F. (ed.) *Carbonate-Hosted Lead–Zinc Deposits*. Society of Economic Geologists Special Publication, **4**, 465–482.

VILJOEN, K.S., SMITH, C.B. & SHARP, Z.D. 1996. Stable and radiogenic isotope study of eclogite xenoliths from the Orapa kimberlite, Botswana. *Chemical Geology*, **131**, 235–255.

VLAAR, N.J. 2000. Continental emergence and growth on a cooling Earth. *Tectonophysics*, **322**, 191–202.

WHITE, N.C., LEAKE, M.J., MCCAUGHEY, S.N. & PARRIS, B.W. 1995. Epithermal gold deposits of the southwest Pacific. *Journal of Geochemical Exploration*, **54**, 87–136.

WILDE, A.R., MERNAGH, T.P., BLOOM, M.S. & HOFFMAN, C.F. 1989. Fluid inclusion evidence on the origin of some Australian unconformity-related uranium deposits. *Economic Geology*, **84**, 1627–1642.

WILLIAMS, P.J. & SKIRROW, R.G. 2000. Overview of iron oxide-copper-gold deposits in the Curnomona province and Cloncurry district (eastern Mount Isa block), Australia. *In*: PORTER, T.M. (ed.) *Hydrothermal Iron Oxide–Copper–Gold & Related Deposits: A Global Perspective*. Australian Mineral Foundation, Adelaide, vol. **1**, 105–122.

WYBORN, L.A.I., PAGE, R.W. & PARKER, A.J. (eds) 1987. *Geochemical and geochronological signatures in Australian Proterozoic igneous rocks*. Geological Society, London, Special Publications, **33**, 377–394.

WYMAN, D.A., KERRICH, R. & GROVES, D.I. 1999. Lode gold deposits and Archean mantle plume–island arc interaction, Abitibi subprovince, Canada. *Journal of Geology*, **107**, 715–725.

YAKUBCHUK, A.S. & EDWARDS, A.C. 1999. Auriferous Paleozoic accretionary terranes within the Mongol–Okholsk suture zone, Russian Far East. PACRIM'99, 10–13 October 1999, Bali, Indonesia, *The Australasian Institute of Mining and Metallogeny*, **4/99**, 347–358.

YAKUBCHUK, A., COLE, A., SELTMANN, R. & SHATOV, V. 2002. *Tectonic setting, characteristics, and regional exploration criteria for gold mineralization in the Altaid orogenic collage: the Tian Shan province as a key example*. Society of Economic Geologists Special Publication, **9**, 177–201.

ZEGERS, T.E., BARLEY, M.E., GROVES, D.I., MCNAUGHTON, N.J. & WHITE, W.J. 2002. Oldest gold: deformation and hydrothermal alteration in the Early Archean shear-zone hosted Bamboo Creek deposit, Pilbara, Western Australia. *Economic Geology*, **97**, 757–776.

ZHANG, H., SUN, M., LU, F., ZHOU, X., ZHOU, M., LIU, Y. & ZHANG, G. 2001. Geochemical significance of a garnet lherzolite from the Dahongshan Kimberlite, Yangtze Craton, southern China. *Geochemical Journal*, **35**, 315–331.

Pre-mineralization thermal evolution of the Palaeoproterozoic gold-rich Ashanti belt, Ghana

V. HARCOUËT[1,2], L. GUILLOU-FROTTIER[1], A. BONNEVILLE[2] &
J. L. FEYBESSE[1]

[1]*Bureau de Recherches Géologiques et Minières, Orléans, France*
(e-mail: v.harcouet@brgm.fr)
[2]*Institut de Physique du Globe, Paris, France*

Abstract: The region of the gold-rich Ashanti belt in southern Ghana was chosen as the subject for a detailed regional thermal modelling study. Geological studies, in addition to laboratory measurements of thermal properties and heat-production rates, allow us to constrain a finite-element thermal modelling. Scenarios integrating variations of the structure of the crust and various chronological settings were examined. We calculated the thermal regime before and after the thrust tectonism that affected the region during the Eburnean orogeny (2130–2095 Ma), just before ore deposit formation. This gives a new insight into the regional thermal state of the crust before the mineralizing events. To satisfy the thermobarometric observations, the most probable mantle heat flow must be 60 mW m^{-2}, which is at least three times greater than the present-day value. At shallow depths, our results also indicate anomalies of lateral heat flow reaching 25 mW m^{-2}, focused on the margins of each lithological unit, including the Ashanti belt. These anomalies are related to the distortion of the isotherms in the first few kilometres that can be explained mostly by lateral contrasts in thermal conductivity. Such anomalies could be of importance for the mineralizing events, as they would favour fluid circulation locally.

Over the past decade, the increasing volume of high-precision geochronological data has demonstrated that gold deposition was not a continuous process through geological time. The pattern for orogenic gold deposits in the Precambrian shows distinct peaks separated by several hundreds of millions of years (Goldfarb *et al.* 2001). Two main periods (at 2800–2500 and 2100–1800 Ma) correlate well with episodes of increased continental growth. Geochronological data on juvenile continental crust have been compiled and revised by Condie (2000), who suggested that the three major crustal formation episodes (estimated at 2700, 1900 and 1200 Ma) were related to 'superevents' in the mantle. For the Palaeoproterozoic peak (1900 Ma), a superplume event has been proposed (e.g. Condie *et al.* 2000). Similarly, the associated periods of orogenic gold formation are 'most commonly explained by major mantle overturning in the hotter early Earth, with associated plumes causing extreme heating at the base of the crust' (Goldfarb *et al.* 2001). Indeed, it is now generally agreed that global geodynamics played a key role in the formation of several classes of ore deposits (e.g. Barley *et al.* 1998; de Boorder *et al.* 1998; Isley & Abbott 1999; Pirajno 2001). If the occurrence of superevents in the mantle, the formation of super-

continents, and the genesis of world-class orogenic gold deposits (or 'superdeposits') are all related, then studies on the possible thermal conditions prevailing at the time of ore deposit formation could provide useful results.

Deep thermal processes beneath continents are well constrained below thermally stable areas for present times (e.g. Jaupart & Mareschal 1999; Lenardic *et al.* 2000). Heat flow studies in the Canadian and Fennoscandian shields (e.g. Kukkonen 1998; Rolandone *et al.* 2002) as well as petrological studies on mantle xenoliths (Kukkonen & Peltonen 1999; Russell & Kopylova 1999) have shown that mantle heat flow values at the base of the Precambrian crust range from 10 mW m^{-2} to 16 mW m^{-2}. However, these present estimates cannot be applied to Precambrian times, and the possibility of studying thermal regimes billions of years ago thus becomes a real challenge. A number of distinct arguments suggest that the thermal regime of the Archaean continental crust was not very different from that of the present-day crust (e.g. Richter 1985). If this were indeed the case, one would expect similar, low mantle heat flow values at the base of continents, unless some large-scale thermal perturbation, such as a superplume event, affected the base of the crust. The formation of world class deposits

From: McDonald, I., Boyce, A. J., Butler, I. B., Herrington, R. J. & Polya, D. A. (eds) 2005. *Mineral Deposits and Earth Evolution.* Geological Society, London, Special Publications, **248**, 103–118. 0305-8719/$15.00
© The Geological Society of London 2005.

probably requires anomalous thermal processes, located at shallow depth or at the crust–mantle interface. The use of various geological data, and in particular, thermobarometric data, should help to quantify thermal regimes of the past.

In this study, we start from the apparent relationship between Palaeoproterozoic orogenic gold deposits and the deep thermal processes prevailing at the time of their formation. Since numerous geological data have been acquired in mineralized areas, the modelling of past thermal regimes before and during mineralizing events may help to constrain thermal conditions at depth.

Among the major orogenic gold deposits, the Ashanti belt of Ghana possesses several features that are particularly well-suited to studying deep thermal processes 2000 Ma ago. First, the Ashanti belt of Ghana was formed during the 'Palaeoproterozoic event' of Condie (2000). Secondly, many thermobarometric data are available and can provide strong constraints to any thermal model. Moreover, numerous rock samples collected during several field campaigns in the area (Milesi *et al.* 1991, 1992; Feybesse & Milesi 1994) can be used to measure thermal properties of the main lithological units, thus reducing the number of unknown parameters in the models. Thirdly, the volcanic and sedimentary belts in Ghana are sufficiently parallel (over a distance reaching ten times the crustal thickness) to consider the area as a two-dimensional structure, thus justifying the 2D approach to the modelling.

This study presents the preliminary results of thermal modelling of the Ashanti and surrounding belts, before and during the collisional tectonics leading to the pre-mineralizing stage. After defining a geological model according to the available thermobarometric and geochronological data, various hypotheses on crustal structure, erosion rates and sediment thickness were tested, corresponding to different geological scenarios. As the model depends strongly on thermal properties, measurements of thermal conductivities and heat-production rates were performed. In all cases, thermobarometric data are the main constraints. As a consequence, our results argue for a suggested geological scheme, provide quantitative constraints on the deep thermal processes prevailing immediately before this particular Palaeoproterozoic mineralizing event, and suggest some shallow thermal processes that may lead to particular fluid–rock interactions. This latter consequence is the subject of an ongoing study.

Geological setting

Main stages of crustal evolution

The Palaeoproterozoic rocks covering a large parts of SW Ghana (Fig. 1) belong to the Birimian Supergroup and the Tarkwaian Group. The Birimian Supergroup includes sedimentary basins separated by subparallel NE-trending volcanic belts consisting of basalts and some interflow sediments. The Tarkwaian conglomerates overlie the Birimian rocks and were probably derived from the erosion of these rocks.

A detailed chronology for the emplacement of these different groups is proposed. We first consider a phase of magmatic accretion between 2250 and 2170 Ma, corresponding to a period of volcanoplutonism, leading to the emplacement of the greenstone belts constituting the so-called volcanic belts. The addition of extensive monzo-granitic plutonism, which occurred between 2160 and 2150 Ma (Opare-Addo *et al.* 1993*b*; Feybesse *et al.* 2000), led to the formation of a juvenile continental crust. After this crust-formation stage, basin opening and sedimentation took place within the Birim, Afema and Comoé basins between 2150 and 2100 Ma.

The emplacement of the Birimian rocks was followed by a period of tectonic accretion, the Eburnean orogeny, for which two main phases of deformation (D_1 and D_2) can be described (Fig. 2). The main deformation is associated with the initial D_1 thrust tectonism that took place between 2130 and 2105 Ma. This compressive deformation along a SE–NW direction results in crustal thickening by stacking of the different units. During this compressive stage, the Birimian rocks were deformed and subsequently uplifted and eroded. The resulting Tarkwaian detrital rocks were deposited between 2133 ± 4 and 2097 ± 2 Ma (Oberthür *et al.* 1998; Pigois *et al.* 2003). They were not affected by D_1 deformation and they were probably deposited mainly at the end of D_1 tectonism (Milesi *et al.* 1991; Milesi pers. comm.). Regional metamorphism associated with D_1 ranges from lower greenschist facies to lower amphibolite facies (Kleinschrot *et al.* 1993; John *et al.* 1999; Klemd *et al.* 2002; Pigois *et al.* 2003) and is dated at around 2100 Ma (Oberthür *et al.* 1998). A second tectonic phase, D_2, affecting both the Tarkwaian and Birimian rocks, began at 2095 Ma. The major epigenetic Birimian lode-gold event can be placed late in the Eburnean orogeny, after the peak of metamorphism was reached. The age of hydrothermal alteration has been determined by Oberthür *et al.* (1998), who

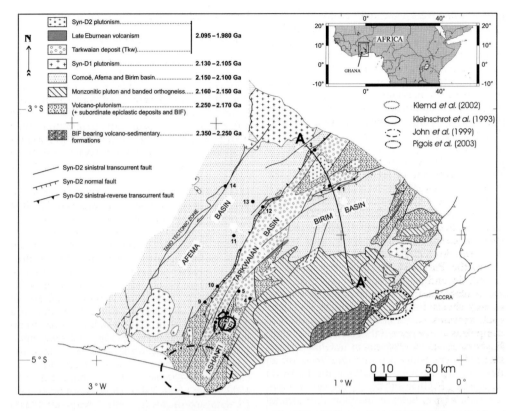

Fig. 1. Simplified geological map of south Ghana, West Africa. Cross section A–A′ is shown in Figure 3. Also shown are the locations of the major gold mines 1–14. Thermobarometric data used in this study are encircled (see Table 1).

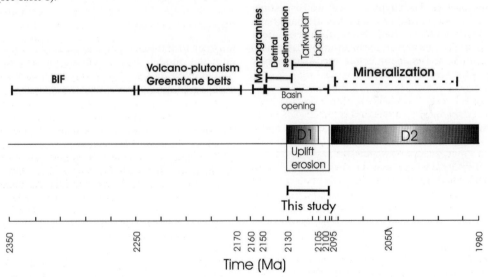

Fig. 2. Succession of the different lithological units (top) and the main tectonic phases (bottom). This simplified scheme allows the reconstruction of a geological scenario prior to the mineralizing event (after the D_1 phase). Time intervals are positioned according to geochronological data from several studies (see text for references).

give ages of 2092 ± 3 and 2086 ± 4 Ma. The most robust age for mineralization, determined at 2063 ± 9 Ma, was given by Pigois *et al.* (2003). Most of the gold deposits are concentrated along the western flank of the Ashanti belt. They are located in the vicinity of major fault zones, along NE-striking shear zones, at the contact between the Birimian sedimentary basins, Tarkwaian conglomerates and volcanic belts (Fig. 1).

The Ashanti belt

The Ashanti belt in south Ghana hosts the largest gold deposits known in West Africa. It contains potential resources of *c.* 5000 tons of gold (Milesi *et al.* 1991; Pigois *et al.* 2003). The mineralization is temporally related to the late stage of the Eburnean orogeny, which took place between 2130 and 1980 Ma and affected the supracrustal rocks. It has the characteristics of synorogenic, mesothermal gold deposits and occurs mainly along shear zones on the western margin of the belt. Figure 2 shows the main successive events that affected the Ashanti belt.

There is still no general agreement concerning the nature of the basement in Ghana. Some authors have proposed the presence of an Archaean basement, at least beneath the Birim basin and SE Ghana (Milesi pers. comm.), but the type of basement underlying the Ashanti belt is still controversial. The Ashanti belt may represent the boundary between a continental domain and the western units possibly being of oceanic type. Such hypotheses, as well as distinct scenarios for crustal evolution, could have been tested if there had been sufficient field constraints. However, in this study we assume a continental basement below the Ashanti belt and the western units.

Field constraints and thermal properties

Thermobarometric constraints

Pre-Eburnean. The pressure estimates from the migmatites yield pressures in excess of 5 kbar at temperatures between 600 and 1000 °C, corresponding to emplacement at a relatively deep crustal level (Opare-Addo *et al.* 1993*a*). The anatexis, indicated by the presence of deformed neosomes in these migmatites, began just before 2130 Ma (pre-D_1 stage) at a deep crustal level (*c.* 6 kbar). At this depth, the temperature conditions required for partial melting to occur fall between 650–800 °C, depending on water content. In the following, a value of 700 °C is adopted.

Syn-D_1 Eburnean tectonism. The minimum peak-metamorphism conditions reported for the Eburnean orogeny are 500–610 °C at *c.* 4.5–6 kbar (Table 1) in the southern Kibi. They were measured from samples of metapelites and amphibolites (Klemd *et al.* 2002). Pigois *et al.* (2003) suggest peak metamorphism conditions of 550 °C at minimum pressures of 5 kbar in the southern part of the Ashanti belt, which is consistent with previous estimates of 490–650 °C at 4–6 kbar (John *et al.* 1999; Klemd *et al.* 2002). These thermobarometric data are shown in Table 1 and were taken into account during construction of the geological model (section 4).

Thermal properties

In order to introduce petrophysical constraints into the model, measurements of thermal conductivity and heat-production rates were made on representative rock samples collected

Table 1. *Thermobarometric data (pressure and temperature estimates) for peak-metamorphism conditions at various locations in the Ashanti area. The spatial distribution of these estimates is shown on Figure 1*

Location	Rock type	Pressure (kbar)	Temperature (°C)	Reference
Nsuta Mine	Greenstone belts and metasediments	<5	500	Kleinschrot *et al.* (1993)
Southern Ashanti Belt	Amphibolites and metagranitoid rocks	5–6	500–650	John *et al.* (1999)
Southern Kibi–Winneba belt	Metapelites and amphibolites	4.5–6	500–610	Klemd *et al.* (2002)
Southern Ashanti belt and NW Sefwi belt	Birimian rocks	4–6	490–580	Klemd *et al.* (2002)
Damang	Argillite	<5	550	Pigois *et al.* (2003)

in Ghana (greenstones, monzogranites, sediments and Tarkwaian conglomerates) and Guinea (basement and sediments).

Thermal conductivity. For the thermal conductivity measurements, we used a QTM (Quick Thermal conductivity Meter) based on a modified hot wire method and commercialized by Showa Denko K.K. The measurement principle is the following (Tavman & Tavman 1999): a thin wire fixed to an insulating material of known thermal properties receives a constant electrical current. This process generates a constant heat (Q) per unit length and per unit time. According to Galson *et al.* (1987), specific experimental conditions yield measurements as precise as those obtained by the divided-bar apparatus method. These authors provide the experimental procedure required to use the QTM method. In our study, optimal conditions for the number of measurements on each sample were respected with at least six measurements per sample. In most cases, the size of the rock sample exceeded the minimum requirements ($20 \times 50 \times 70$ mm). With this procedure, the obtained reproducibility is $\pm 5\%$ and uncertainty remains lower than ± 10–20% (for dry samples).

By supplying a constant power to the heater element, a temperature rise (ΔT) of the wire is generated. The value of ΔT is then measured by a thermocouple as a function of time during a short heating interval. This temperature versus time record is used to calculate the thermal conductivity (k) according to the following equation (Carlsaw & Jaeger 1984):

$$k = F\frac{Q\ln(t_2/t_1)}{T_2 - T_1} - H, \qquad (1)$$

where T_1 and T_2 are temperatures at time t_1 and t_2 respectively, Q is the heat flow per unit time and per unit length of the heating wire, and F and H are specific constants of the probe, determined by calibration.

Measurements were made on 31 rock samples representative of the main lithological units (Table 2). The measured thermal conductivities for greenstones and monzogranites fall within the range of classical values for these rock types. The apparently high thermal conductivity of the Birimian sediments and Tarkwaian conglomerates is a result of metamorphism and burial, as compaction is the dominant controlling factor in the variation of thermal conductivity in the first few kilometres (e.g. Clauser & Huenges 1995). However, since the basins were still opening and erosion and conglomerate deposition were still occurring during D_1, we had to use a thermal conductivity value of sediments and conglomerates before metamorphism and burial. In accordance with the thermal conductivity values proposed by Jaupart & Provost (1985), an identical value of 1.8 W m^{-1} K^{-1} was taken for both lithologies during their deposition. After compaction due to burial, the measured thermal conductivity of the sediments was adopted, namely 2.9 W m^{-1} K^{-1} (see a similar case in Serban *et al.* 2001). For the basement and the mantle, mean values from the literature (Turcotte & Schubert 2002) were taken at 2.6 W m^{-1} K^{-1} and 3.3 W m^{-1} K^{-1}, respectively.

Heat-production rate. The concentration in heat-producing elements is also an important factor controlling the geotherm. Heat-production rates were therefore calculated for each unit, using the concentrations in the radiogenic elements U, Th and K (Table 3). U and Th concentrations were determined by ICP-MS, and K concentration by X-ray fluorescence spectrometry.

Since the emplacement of the Ashanti belt took place at around 2100 Ma, heat-production rates for this period had to be calculated using present-day concentrations of radiogenic elements. Mean production rates in the past can be calculated using the present concentrations in radiogenic elements with the following equation (Turcotte & Schubert 2002):

Table 2. *Thermal conductivity measurements of Birimian and Tarkwaian rock samples from the Ashanti belt and Guinea. For each sample, six measurements were made and thermal conductivity is the mean value of all the samples*

Rock type	Number of samples	Thermal conductivity (W m^{-1} K^{-1})	Standard deviation (W m^{-1} K^{-1})
Greenstones	7	2.93	0.74
Monzogranites	6	3.10	0.59
Sediments	12	2.89	0.90
Conglomerates (Tarkwaian)	6	3.12	0.62

Table 3. *Heat-production rates and their standard deviations (s.d.) for Birimian and Tarkwaian rock samples from the Ashanti belt, at present (measurements) and 2100 Ma ago (see equation (2))*

Rock type	Number of samples	U (ppm)	s.d.	K (%)	s.d.	Th (ppm)	s.d.	Heat-production rate µW m^{-3} present	s.d.	$t = 2100$ Ma	s.d.
Basement	48	0.98	1.03	2.00	1.29	8.47	9.60	1.05	0.90	1.71	1.28
Greenstones	12	0.81	0.64	0.60	0.42	1.76	1.49	0.40	0.28	0.69	0.45
Monzogranites	9	1.83	1.26	2.25	1.19	4.87	2.44	1.04	0.44	1.88	0.82
Sediments	25	1.71	1.38	1.96	1.44	4.32	2.17	0.77	0.43	1.38	0.75
Tarkwaian conglomerates	10	2.79	3.42	1.64	1.49	6.78	4.52	1.12	0.92	1.85	1.62

$$A(t) = \rho\{(9.3919 \ 10^{-5} \ e^{t\times0.155} \\ + 0.4399 \ 10^{-5} \ e^{t\times0.9846})[U_0] \\ + (2.64 \ 10^{-5} \ e^{t\times0.04951})[Th_0] \\ + (3.475 \ 10^{-5} \ e^{t\times0.5545}) \ [K_0]\} \quad (2)$$

where $A(t)$ is the heat-production rate at time t (Ga), for a material of density ρ.$[U_0]$, $[Th_0]$ and $[K_0]$ are the present concentrations of the different isotopes measured in the laboratory. Table 3 shows 104 present-day heat-production rates, measured on rock samples representative of the area and their corresponding values 2100 Ma ago.

Geological model for crustal evolution from the initial to the pre-D$_2$ stages

Structural constraints

The detailed structure and composition of the modelled area is described below. Figure 3 illustrates a geological cross-section inferred from more than 15 years of various field campaigns in the area. Figure 4 summarizes different stages of crustal evolution from the initial to the pre-D$_2$ stages. In order to build a realistic geological model, we first had to constrain the thickness and lateral extent of the different rock units.

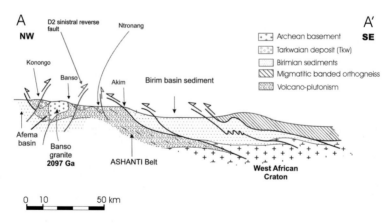

Fig. 3. Geological cross-section through A–A' (Fig. 1) as deduced from field campaigns. The NW direction of thrusting events is outlined.

Fig. 4. Model for the Ashanti and surrounding belts, between 2130 and 2105 Ma. The initial stage (**a**) accounts for thermobarometric data in the monzogranites. At 2105 Ma (**b**), the central unit has been overthrust by those from the east, implying the burial of sediments and metamorphism in the greenschist facies, while the Ashanti belt was recovered by sediments and barren conglomerates (Kawere). The third stage (**c**) is the result of erosion and deposition.

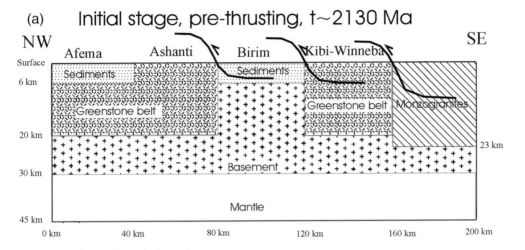

(a) **Initial stage, pre-thrusting, t~2130 Ma**

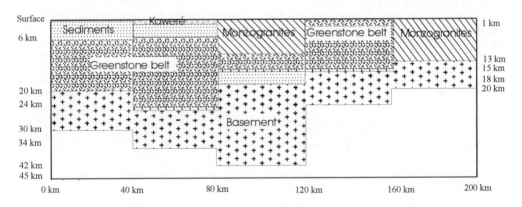

(b) **After instantaneous thrusting, t~2105 Ma**

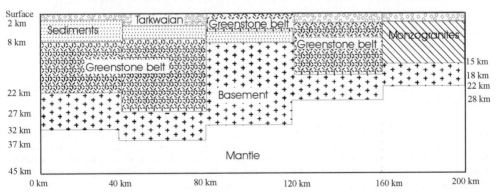

(c) **After deposition, erosion and uplift, t~2095 Ma**

Lateral extent of the rock units. In the following, lateral extent refers to a NW–SE direction, since a 2D model is considered. The volcanic belts are typically 15–40 km wide and 60–90 km apart (Taylor *et al.* 1992). From field observations and previous maps (Klemd *et al.* 2002; Pigois *et al.* 2003), the width of the Ashanti belt ranges from 15–50 km, and is surrounded to the west by the Afema basin (*c.* 70 km wide) and to the east by the Birim basin (*c.* 20–60 km wide), the Kibi belt (*c.* 40 km) and the monzogranite unit (*c.* 40 km wide). We chose a mean lateral extent of 40 km for all the different rock units, in order to simplify the calculations and to take into account a certain amount of crustal shortening that occurred during the Eburnean orogeny.

Thickness of the lithological units. The geophysical constraints on the region are poor and we did not have direct access to the data related to crustal thickness, for which a value of 30 km was assigned. Thermobarometric data were used to estimate the thickness of the different units. If partial melting started at the base of the monzogranites (at least 700 °C at *c.* 6 kbar) then these rocks were at least 23 km thick before thrusting. For our modelling approach, this partial melting condition is used solely to constrain temperature and depth at this given location.

Concerning the volcanic units, we chose a thickness of 20 km for the Ashanti and Kibi belts, in accordance with the peak pressure of 4–6 kbar that was recorded in these rock units. The minimum and maximum sediment thicknesses proposed for the sedimentary basins are 3 and 8 km respectively. Different values were tested and the results for 6 km are presented. Pigois *et al.* (2003) estimate the present minimum thickness of the sedimentary basins at 2 km (within the basin facies). This is the lower boundary for the thickness of the final sedimentary unit, after thrusting, erosion and the D_2 stage (which is beyond the scope of this study).

The thickness of the present-day Tarkwaian unit ranges from 1800–3000 m (see a detailed description in Pigois *et al.* 2003). This megasequence resulted from the stacking of different lithostratigraphic formations: the Kawere conglomerate (250–700 m), the Banket conglomerate (150–600 m), the Tarkwa formation (120–400 m) and the Huni formation (<1300 m). 1 km is taken for the Kawere, which was deposited just after the thrusting event; 3 km of conglomerates was then added to the pile, leading to a final value of 4 km for the Tarkwaian conglomerates.

Chronological constraints

The stage just before 2130 Ma is considered to represent the initial condition for the numerical modelling (Fig. 2). At that time, all the units except the Tarkwaian conglomerates were emplaced. From west to east, the 40 km-wide units are: the Afema basin, which covers the greenstone belts, the Ashanti volcanic belt, the Birim basin, the Kibi belt and the monzogranites.

The initial stage of tectonic accretion, D_1, began at 2130 Ma. It resulted in unit stacking that can be evaluated using the thermobarometric data at peak metamorphism conditions (Table 1). The pressure constraints of 4–6 kbar imply that the sediments of the Birim basin, the Ashanti greenstone belts and rocks of the Kibi belt were buried to a depth of 15–22.5 km, when a density of 2700 kg m^{-3} is used in the model. This value appears to be representative for most rocks of the area, except for fresh sediments (Table 4).

Deposition of the Kawere unit on the Ashanti belt was contemporaneous with D_1. It contains polygenic conglomerates deposited after 2132 ± 2 Ma (age of the youngest detrital zircon from the base of the series; Davis *et al.* 1994; Oberthür *et al.* 1998). Following D_1 tectonism, auriferous Tarkwaian conglomerates from the

Table 4. *Main lithological units with associated density values and thermal properties. Density measurements are from Turcotte & Schubert (2002) except ([1]) from Barritt & Kuma (1998). Thermal properties are from this study except ([2]) from Jaupart and Provost (1985)*

Rock type	Mean density (kg m^{-3})	Mean thermal conductivity (W m^{-1} K^{-1})	Mean heat-production rate (μW m^{-3})
Basement	2750	2.6	1.7
Greenstones	2870[1]	2.9	0.7
Sediments	2200 before D_1	1.8[2] before D_1	1.4
	2730[1] after D_1	2.9[2] after D_1	1.4
Monzogranites	2700	3.1	1.9
Tarkwaian conglomerates	2700[1]	1.8[2]	1.9

Banket series were deposited between 2104 ± 6 Ma (age of the youngest detrital zircon; Milesi *et al.* 1989) and 2097 ± 2 Ma (age of the Banso granitoid that is assumed to intrude the Tarkwaian deposits; Oberthür *et al.* 1998).

Constraints for erosion

The estimate of the erosion rate between the regional peak metamorphism (2100–2105 Ma) and the end of deposition of the Tarkwaian Banket series (2097 Ma) is constrained by the presence of mineralized pebbles of quartz in the conglomerate dated at 2105 Ma (Milesi *et al.* 1989). Orogenic gold deposits related to an earlier Eburnean compressional event represent one of the sources of gold for the auriferous conglomerates (Milesi *et al.* 1989; Klemd *et al.* 1993; Oberthür *et al.* 1997; Pigois *et al.* 2003). These pebbles result from the reworking of orogenic deposits that are estimated to have formed under the same conditions as the lode-gold deposits of the western margin of the Ashanti belt (Klemd *et al.* 1993). The estimated depth of formation for the deposits of the Ashanti area is *c.* 4–15 km (*c.* 1–4 kbar) (Oberthür *et al.* 1997; Schmidt Mumm *et al.* 1997; John *et al.* 1999; Yao *et al.* 2001). These data suggest the erosion of 4–15 km of strata within these 10 Ma. For modelling purposes, we adopted a value of 10 km, which is in accordance with the abundance of cataclastic and mylonitic pebbles (Milesi pers. comm.).

Thermal modelling

Equations, boundary conditions, parameters and variables

The general form of the heat conduction equation to account for heat transfer in a solid, involving both internal heat production and material advection in Cartesian coordinates, is:

$$\nabla(\mathbf{k}\nabla T) - \rho c \frac{\partial T}{\partial t} \\ - \rho_f c_f \mathbf{v}_f \nabla T \\ + A = 0 \quad (3)$$

where T is temperature in K, $\mathbf{k}$ the thermal conductivity tensor depending on direction and temperature (W m^{-1} K^{-1}), A the heat-production rate (W m^{-3}), ρ the density (kg m^{-3}), c the specific heat (J kg^{-1} K^{-1}), $\rho_f c_f$ the fluid heat capacity (J K^{-1} m^{-3}), and $\mathbf{v}_f$ the Darcy velocity field (m s^{-1}). In our present study, fluid advection is not taken into account, and $\mathbf{v}_f = 0$.

Moreover, $\mathbf{k}$ is considered to be isotropic, temperature independent and with the 2D approximation for the Ashanti belt, equation (3) reduces to:

$$k\left(\frac{\partial^2 T}{\partial x^2} + \frac{\partial^2 T}{\partial z^2}\right) - \rho c \frac{\partial T}{\partial t} + A = 0 \quad (4)$$

where x is the horizontal distance and z is depth.

Numerical modelling

The source code THERMIC, used to solve the 2D conductive and advective heat-transfer equation, was developed by Bonneville & Capolsini (1999). Based on a finite-element method it can incorporate realistic geometries, heterogeneous material properties and various boundary and initial conditions (e.g. Vasseur *et al.* 1993).

The original code was improved to take into account phenomena encountered in Earth sciences, such as erosion, deposition and instantaneous thrusting. Tests were performed to validate the modifications, and it turned out that our results are in excellent agreement with those of Nisbet & Fowler (1982*a*, *b*).

Description of the numerical model

The model is a rectangular grid 200 km long and 150 km thick. This represents the crust and the lithospheric mantle. This grid is divided into 9600 4-node isoparametric elements of 1 × 1 km, down to 47 km depth. We used the thickness deduced for each rock unit to build the initial model (Fig. 4a). Each 40 km-wide lithological unit corresponds to a specific material, whose thermal properties are defined by the thermal conductivity, the heat production rate and the heat capacity.

The solution of the thermal problem is a function of the imposed boundary conditions, and the thermal properties of the different rock types. The chosen boundary conditions involve a constant mantle heat flow q_m entering the base of the grid, and a constant surface temperature ($T = 0$ °C). Various values of the mantle heat flow are used in order to adjust thermobarometric data from the pre-Eburnean stage to the different stages of Eburnean tectonism.

Transient evolution

We chose to represent the tectonic evolution during D$_1$ stage using two different thrusting events (Fig. 4). First, the westward compression exerted by the Archaean craton caused thrusting in which part of the monzogranites (10 km) and half of the sediments of the Birim basin (3 km) were involved. The activation of two thrust faults located at the margin of these blocks led to an intermediate stage where the monzogranites and the sediments were over-thrust onto the Kibi greenstone belt and the Ashanti volcanic belt, respectively. The mean burial depth, estimated at 18 km (*c.* 5 kbar), at the base of the sedimentary Birim basins could only have been met if they were covered by 15 km of rocks. We thus assume that the unit overthrust onto the Birim basin a sequence comprising monzogranites (10 km) and volcanics from the Kibi belt (5 km).

Numerical calculations are simplified by keeping the Earth surface flat during the thrusting, which at the time scale of the model can be considered 'instantaneous'. The choice of instantaneous thrusting can be justified. The rate of burial does not influence the time lapse between the end of thrusting and maximum temperature. It plays only a minor role in determining the maximum temperature of rocks during the later unroofing stage (Ruppel & Hodges 1994). Moreover, these authors demonstrate that the maximum temperature depth is linked only slightly to the rate of burial or to the amount of syntectonic heating. As we are interested in the thermal state after D_1 tectonism, the instantaneous approach is acceptable.

The period of thrusting was followed by a stage of uplift and erosion, of which the Tark-waian conglomerates are the products. As a consequence of isostatic response to burial, the Birim sedimentary basin, the most overthrust unit, will be the most affected by erosion and uplift. Based on geological observations of Tark-waian samples, we assumed a rate of erosion of 1 km/Ma, so that 10 km were uplifted and eroded within the 10 Ma separating the end of instantaneous thrusting and the beginning of D_2 tectonism. In the model, erosion began immediately after instantaneous burial, with a rate of 1 km/Ma.

Results and discussion

Pre-thrusting

The thermobarometric conditions related to monzogranites, *c.* 700 °C at *c.* 6 kbar, appear to be reached only if the amount of heat at the base of this unit is sufficient. The pre-thrusting stage can be constrained using these data (Fig. 4a) by fitting temperature and pressure conditions with adequate model parameters. Those models

where the temperature at the base of the crust exceeded 900 °C were rejected.

Tests were initially performed using mean measured values of thermal conductivities and heat-production rates for the different litho-logical units (Table 4). Several experiments were performed to evaluate mantle heat flow values that would be consistent with the ther-mobarometric conditions. Using average thermal properties (Table 4), the minimum mantle heat flow that leads to approximately 700 °C at 22.5 km, is 60 mW m^{-2} (Fig. 5a). For lower mantle heat flow values, the 700 °C isotherm is located beneath the base of the unit whereas for higher heat flow values, tempera-tures at the base of the crust unit exceed 900 °C.

In Figure 5a, the thermal anomaly that appears at the base of the crust below the Birim basin is due entirely to the thermal properties of the rocks, the basal heat flow being constant at this depth. This figure also illustrates the effect of thermal refraction caused by a thermal conductivity contrast between the sediments (k = 1.8 W m^{-1} K^{-1}) and the surroundings (k = 2.9 W m^{-1} K^{-1}). Figure 5c shows lateral heat transfer before the thrusting event. The crustal scale anomalies are focused towards the bound-aries between the different rock units, empha-sizing the role of lateral conductivity contrasts in the heat transfer pattern. Lateral contrasts in heat-production rates also contribute to lateral heat transfer anomalies, but with a much lesser importance than conductivity contrasts.

Post-thrusting: Eburnean tectonism

The results need to fit the data recorded for regional peak metamorphism:

- in the greenstone belts, a temperature of 500–650 °C is required at 4–6 kbar;
- in the sediments of the Birim basin, green-schist facies is reached

The stage before thrusting was calculated with mean thermal conductivities and a mantle heat flow of 60 mW m^{-2}. Then, the different units were stacked instantaneously (Fig. 4b) and the resulting final thermal field, after thrusting, uplift and erosion, is shown in Figure 5b.

In order to validate the modelling, *P–T* paths from various rocks within the Ashanti belt were investigated. Figure 6 shows the paths for rocks situated in the Ashanti belt, in the middle of the unit at different initial depths (at the base of the greenstones, in the middle and in between these two levels). The rocks at an initial depth of 15 km were first buried to 18 km during the

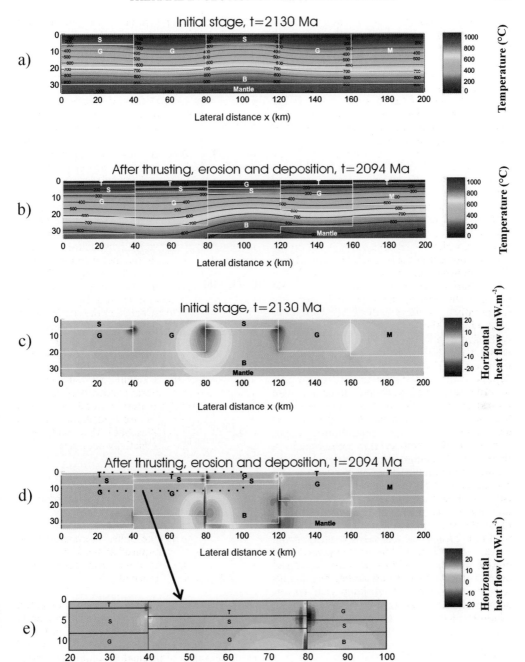

Fig. 5. Temperature field (**a**) before and (**b**) after the D_1 tectonic phase, as inferred from thermobarometric data and from a maximum temperature at the base of the crust of 900 °C. The undulated shape of the isotherms is due to the thermal properties of each lithological unit since mantle heat flow (60 mW m^{-2}) stays constant at the base of the model. This value allows the conditions for partial melting in the monzogranites to be reached. After the D_1 phase (**b**), the deposition of sediments and Tarkwaian conglomerates distorts the shallow isotherms and yields temperature gradients higher than 40 °C km^{-1} within the Ashanti belt. Lateral heat flow before the thrusting events (**c**) and after the entire D_1 phase (**d**). Positive values correspond to a westward heat flow. Ore deposits have formed after the D_1 phase and are mainly concentrated at the edges of the Ashanti belt, where shallow anomalous lateral heat flow is observed (see **e**).

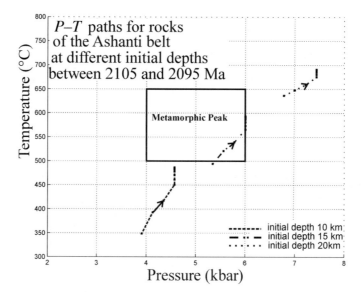

Fig. 6. Results of numerical experiments (*P–T* paths) compared with thermobarometric data. Each *P–T* path corresponds to the burial of greenstones of the Ashanti belt, after deposition of the Tarkwaian conglomerates. The greenstones buried to 19 km record *P–T* paths that are in agreement with the thermobarometric data.

instantaneous thrusting and then received deposition of the Tarkwaian conglomerates. They reach *P–T* conditions that are in good agreement with the amphibolite facies conditions recorded in the greenstones. A thickness of around 12 km of greenstones (equivalent to an initial burial of 15 km) would also enable the *P–T* domain defined by the peak metamorphism conditions to be reached, in accordance with the upper boundary estimates of Sylvester & Attoh (1992).

The *P–T* paths of sediments located at the base of the Birim basin were also investigated and are reported in Figure 7. The rock, originally located at the base of the basin (18 km depth), follows a retrograde *P–T* path as it travels towards the surface until it reaches its final location at 8 km depth. Computed values are in good agreement with thermobarometric data, which suggest that these rocks reached green-schist facies conditions. The retrograde *P–T* path is representative of the response to burial, with an increase in temperature during the first million years and then a decrease corresponding to the erosion phase. During the period of unroofing to shallower crustal levels, the time lag observed between the end of the compressional interval and the reaching of peak temperatures is in accordance with the theoretical prediction of England & Richardson (1977).

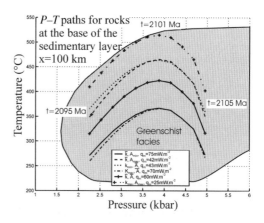

Fig. 7. Results of numerical experiments (*P–T* paths) for Birim sediments, compared with the greenschist facies domain. Each *P–T* path corresponds to the erosion phase after thrusting, but before the end of D_1 tectonism (i.e. from 2105 to 2095 Ma). The five curves reflect different values for thermal parameters: (k_{min} and k_{max}) = minimum and maximum values of thermal conductivity, (A_{min} and A_{max}) = minimum and maximum values of heat production rate, and ($\bar{k}, \bar{A}$) = mean values for thermal conductivity and heat production rate. The highest temperatures (peak metamorphism conditions) are reached at 2101 Ma, as suggested by Hirdes *et al.* (1996) and Oberthür *et al.* (1998).

Moreover, in all cases, the peak temperature is reached after 4 Ma of erosion, which corresponds to 2101 Ma ago. This correlates well with the timing of the regional peak metamorphism, dated between 2092 ± 3 and 2105 ± 3 Ma by Oberthür *et al.* (1998) and at 2100 ±3 Ma by Hirdes *et al.* (1996).

The lateral heat flow calculated 11 Ma after thrusting (corresponding to 2094 Ma ago) is illustrated in Fig. 5d. This shows the resulting shallow heat transfer mechanisms after D_1 tectonism and the associated erosion–deposition phase. Compared to the initial lateral heat flow (Fig. 5c), it shows an important phase of lateral heat transfer (20–25 mW m^{-2}) within the upper crust (maximum values are reached within the upper 5 km, Fig. 5d). These thermal anomalies are more important at each belt boundary, including the eastern and western boundaries of the Ashanti belt. As the thermal conductivity contrasts are the same as for the initial stage, the anomalies cannot be attributed to heat refraction effects alone. Indeed, it is the particular tectonic history of the Ashanti belt that could explain these shallow thermal anomalies: (a) before thrusting, isotherms were uplifted beneath the surrounding sedimentary basins (Afema, Birim) due to the insulating effect of fresh sediments; (b) thrusting of these anomalously hot sediments disturbed the shallow temperature field, yielding the observed lateral heat flow pattern (Fig. 5d); and (e) the deposition of insulating conglomerates enhanced this effect.

Sensitivity to thermal parameters

The thermal regime of the crust is controlled by boundary conditions (i.e. mantle heat flow in this study) and the thermal properties of representative rocks. In order to constrain our results on mantle heat flow values, the effects of variations (and uncertainty) in thermal conductivity and heat production rates were also investigated.

We tested the sensitivity of the modelling to extreme values of thermal conductivity. If the minimum values of thermal conductivity for all rock types are taken (mean values minus the standard deviation), then the temperature is higher than in previous cases because of the enhanced insulating effect. In this case, a value of 43 mW m^{-2} for the mantle heat flow would be sufficient to satisfy the thermobarometric conditions. On the contrary, if the highest values of thermal conductivity are considered, then a value of 70 mW m^{-2} for the basal heat flow is necessary to comply with thermobarometric conditions before and after thrusting.

Tests were also performed to constrain the influence of heat production rates. When maximum values of heat production rates are considered for all rock types, a value of 42 mW m^{-2} for mantle heat flow is necessary to fit the thermobarometric conditions. This value is much higher when minimum heat production rates are taken, since q_m must reach 75 mW m^{-2}. Increasing the heat production rate of the monzogranites to its maximum (2.7 µW m^{-3}) enables higher temperatures to be reached within this unit, without changing the temperature at the base of the crust dramatically.

When combining the effects of maximum heat production rates and minimum thermal conductivity, 25 mW m^{-2} is the minimum mantle heat flow value required to fit the thermobarometric data. This value is the lower valid boundary for basal heat flow. It represents the most extreme case, which is quite unlikely. Figure 7, which also shows the case for these parameters for heat production and thermal conductivity, demonstrates that the associated *P–T* paths fall in the uppermost limit of the greenschist facies domain. Tests using minimum heat production rates and maximum thermal conductivity present solutions that are not in accordance with the thermobarometric data, and were thus rejected.

The lateral heat flow value also depends on these thermal properties. Indeed, lateral thermal conductivity contrasts are the most determinant parameters for the presence of lateral heat flow anomalies. The contribution of lateral contrasts in heat production rate was estimated. First, by taking a thermal conductivity of 2.9 W m^{-1} K^{-1} for all the rock types, the effects of lateral thermal conductivity contrasts are cancelled. In this case, lateral heat flow values decrease dramatically compared to the standard case, and a maximum value of only 6 mW m^{-2} is observed, due to lateral contrasts in heat-production rates. Secondly, when uncertainties in heat production are considered, and contrasts between units are maximized (e.g. heat production rate minimum for the sediments and maximum for greenstone belts), lateral heat flow anomalies are modified by a factor of less than 25%.

Finally, the localization of the Tarkwaian conglomerates (here inferred from field observations) also influences the pattern of lateral heat flow anomalies. For example, if deposition only occurs within the surrounding units of the Birim basin, namely the belts of Ashanti and Kibi, then the anomalies will be higher than in the standard case and will be concentrated along the margins of these units.

Conclusion

The thermal evolution of southern Ghana from 2130 to 2095 Ma gives insight into the regional thermal state of the crust before the mineralizing events that led to the emplacement of the giant gold deposits of the Ashanti belt.

Thermal modelling of tectonic and surface processes with adequate thermal parameters has allowed us to reproduce the thermobarometric conditions. Amphibolite- to greenschist-facies conditions are encountered for the rocks of the volcanic belts and the Birimian sediments, which is in agreement with the regional peak metamorphism conditions indicated by previous studies (Kleinschrot *et al.* 1993; Hirdes *et al.* 1996; Oberthür *et al.* 1998; John *et al.* 1999; Klemd *et al.* 2002).

The obtained mantle heat flow values range from $25 \, \text{mW} \, \text{m}^{-2}$ to $75 \, \text{mW} \, \text{m}^{-2}$, with a most likely value of $60 \, \text{mW} \, \text{m}^{-2}$, which is at least three times greater than the present-day value. This relatively high value may be related to deep thermal perturbations. Indeed, the opening of sedimentary basins during the first stage of D_1 tectonism, not taken into account in the model, may be related to a regional increase in mantle heat flow. Nevertheless, when present-day mantle heat flow values are considered, a value of $60 \, \text{mW} \, \text{m}^{-2}$ does not represent a dramatic increase in subcontinental heat flow, and the 'mantle overturning' phenomenon (Condie 2000; Goldfarb *et al.* 2001) may not be so catastrophic. Indeed, even deep and large mantle plumes impinging at the base of the continental lithosphere does not necessarily induce sharp increases in mantle heat flow (Lenardic *et al.* 2000). If superevents in the mantle are related to anomalous deep processes such as superplumes, then the associated thermal signatures would not necessarily imply strong modifications in mantle heat flow values. One must also indicate that a temperature-dependent thermal conductivity (not tested in this study) would tend to lower the mantle heat flow values since the decreasing conductivities would enhance crustal temperatures. Finally, our results show that thermal equilibrium of the crust can be easily disturbed as soon as mantle heat flow increases by only a few tens of $\text{mW} \, \text{m}^{-2}$.

The results also indicate anomalies of lateral heat flow reaching $20–25 \, \text{mW} \, \text{m}^{-2}$, mainly focused on the western and eastern margins of the Ashanti belt and other belt boundaries. Such anomalies, where associated with fractures, may have been of importance for the mineralizing events, as they would have favoured fluid circulation locally. Indeed, field studies have emphasized the heterogeneous strain partitioning and the highly focused fluid flow at shallow depths, which could be triggered partly by such lateral heat transfer anomalies. The correlation between the location of these anomalies on both flanks of the Ashanti belt and the location of the major ore deposits is an encouraging result.

A number of first-order simplifications were taken in this study. On one hand, the choice of a depth-independent thermal conductivity may slightly overestimate the results since lateral contrasts may decrease with increasing temperature. Additional tests are required to quantify second-order effects, but the first-order physical mechanisms are expected to be sufficiently important to affect local fluid circulation.

Although our study does not take into account fluid circulation, which is critical in understanding the details of the mineralizing processes, it is a first and necessary step in the comprehension of the regional thermal evolution of the area. A second stage of modelling, investigating the mineralizing processes at a more local scale and integrating fluid circulation, is the objective of an ongoing project. It will be based on the results of this regional study.

V. H. was supported by a grant from BRGM (Bureau de Recherches Géologiques et Minières). The authors wish to thank sincerely Vincent Bouchot, Jean-Luc Lescuyer, Francis Lucazeau, and Jean-Pierre Milesi for their support and advice. We are grateful to the two reviewers, K. Gallagher and P. E. J. Pitfield, for their valuable comments and suggestions. We thank Rowena Stead for proofreading the final text and English editing. This is IPGP contribution #2919, and BRGM publication BRGM-CORP-03163.

References

BARLEY, M.E., KRAPEZ, B., GROVES, D.I. & KERRICH, R. 1998. The Late Archean bonanza : metallogenic and environmental consequences of the interaction between mantle plumes, lithospheric tectonics and global cyclicity. *Precambrian Research*, **91**, 65–90.

BARRITT, S.D. & KUMA, J.S. 1998. Constrained gravity models of the Ashanti belt, southwest Ghana. *Journal of African Earth Sciences*, **26**, 539–550.

BONNEVILLE, A. & CAPOLSINI, P. 1999. THERMIC: a 2-D finite-element tool to solve conductive and advective heat transfer problems in Earth sciences. *Computers and Geosciences*, **25**, 1137–1148.

CARSLAW, H.S. & JAEGER, J.C. 1984. *Conduction of heat in solids*. Clarendon Press, Oxford, 2nd edn, 510pp.

CLAUSER, C. & HUENGES, E. 1995. Thermal conductivity of rocks and minerals. *In*: AHRENS, T.J. (ed.)

Rock Physics and Phase Relations : A Handbook of Physical Constants. AGU, Washington, D.C., AGU Ref. Shelf, **3**, 105–126.

CONDIE, K.C. 2000. Episodic continental growth models: afterthoughts and extensions. *Tectonophysics*, **322**, 153–162.

CONDIE, K.C., DES MARAIS, D.J. & ABBOTT, D. 2000. Geologic evidence for a mantle superplume event at 1.9 Ga. *Geochemistry, Geophysics, and Geosystems*, **1**, Paper number 2000GC000095.

DAVIS, D.W., HIRDES, W., SCHALTEGGER, U. & NUNOO, E.A. 1994. U–Pb age constraints on deposition and provenance of Birimian and gold bearing Tarkwaian sediments in Ghana, West Africa. *Precambrian Research*, **67**, 89–1997.

DE BOORDER, H., SPAKMAN, W., WHITE, S.H. & WORTEL, M.J.R. 1998. Late Cenozoic mineralization, orogenic collapse and slab detachment in the European Alpine Belt. *Earth and Planetary Science Letters*, **164**, 569–575.

ENGLAND, P.C. & RICHARDSON, S.W. 1977. The influence of erosion upon mineral facies of rocks from different metamorphic environments. *Journal of the Geological Society, London*, **134**, 201–213.

FEYBESSE, J.-L. & MILESI, J.-P. 1994. The Archean/Proterozoic contact zone in West Africa: a mountain belt of decollement thrusting and folding on a continental margin related to 2.1 Ga convergence of Archaean cratons? *Precambrian Research*, **69**, 199–227.

FEYBESSE, J.-L., BILLA, M., MILESI, J.,-P., LEROUGE, C. & LE GOFF, E. 2000. Relationships between metamorphism-deformation-plutonism in the Archean-Paleoproterozoic contact zone of West Africa. 31st International Geological Congress, Rio, CD-ROM Abstract.

GALSON, D.A., WILSON, N.P., SCHÄRLI, U. & RYBACH, L. 1987. A comparison of the divided-bar and QTM methods of measuring thermal conductivity. *Geothermics*, **16**, 215–226.

GOLDFARB, R.J., GROVES, D.I. & GARDOLL, S. 2001. Orogenic gold and geologic time: a global synthesis. *Ore Geology Reviews*, **18**, 1–75.

HIRDES, W., DAVIS, D.W., LÜDTKE, G. & KONAN, G. 1996. Two generations of Birimian (Paleoproterozoic) volcanic belts in northeastern Côte d'Ivoire (West Africa): consequences for the 'Birimian controversy'. *Precambrian Research*, **80**, 173–191.

ISLEY, A.E. & ABBOTT, D.H. 1999. Plume-related mafic volcanism and the deposition of banded iron formation. *Journal of Geophysical Research*, **104**, 15 461–15 477.

JAUPART, C. & MARESCHAL, J.-C. 1999. The thermal structure and thickness of continental roots. *Lithos*, **48**, 93–114.

JAUPART, C. & PROVOST, A. 1985. Heat focussing, granite genesis and inverted metamorphic gradients in collision zones. *Earth and Planetary Science Letters*, **73**, 385–397.

JOHN, T., KLEMD, R., HIRDES, W. & LOH, G. 1999. The metamorphic evolution of the Paleoproterozoic (Birimian) volcanic Ashanti belt (Ghana, West Africa). *Precambrian Research*, **98**, 11–30.

KLEINSCHROT, D., KLEMD, R., BRÖCKER, M.,

OKRUSCH, M. & SCHMIDT, K. 1993. The Nsuta manganese deposit, Ghana: geological setting, ore-forming process and metamorphic evolution. *Zeitschrift des Angewandtes Geologie*, **39**, 48–50.

KLEMD, R., HIRDES, W., OLESCH, M. & OBERTHÜR, T. 1993. Fluid inclusions in quartz-pebbles of the gold-bearing Tarkwaian conglomerates of Ghana as guides to their provenance area. *Mineralium Deposita*, **28**, 334–343.

KLEMD, R., HÜNKEN, U. & OLESCH, M. 2002. Metamorphism of the country rocks hosting gold-sulfide-bearing quartz veins in the Paleoproterozoic southern Kibi-Winneba belt (SE-Ghana). *Journal of African Earth Sciences*, **35**, 199–211.

KUKKONEN, I.T. 1998. Temperature and heat flow density in a thick cratonic lithosphere: the sveka transect, Central Fennoscandian shield. *Journal of Geodynamics*, **26**, 111–136.

KUKKONEN, I.T. & PELTONEN, P. 1999. Xenolith-controlled geotherm for the central Fennoscandian Shield: implications for lithosphere–astenosphere relations. *Tectonophysics*, **304**, 301–315.

LENARDIC, A., GUILLOU-FROTTIER, L., MARESCHAL, J.-C., JAUPART, C., MORESI, L.-N. & KAULA, W.M. 2000. What the mantle sees: the effects of continents on mantle heat flow. *In*: RICHARDS, M. *ET AL.* (eds) *The history and dynamics of global plate motions.* AGU, Geophysical Monograph, **121**, 95–112.

MILESI, J.-P., FEYBESSE, J.L., LEDRU, P., *ET AL.* 1989. Mineralisations aurifères de l'Afrique de l'ouest, leurs relations avec l'évolution litho-structurale au Paleoproterozoïque inférieur. *Chroniques de la Recherche Minière*, **497**, 3–98.

MILESI, J.-P., LEDRU, P., ANKRAH, P., JOHAN, V., MARCOUX, E. & VINCHON, C. 1991. The metallogenic relationship between Birimian and tarkwaian gold deposits in Ghana. *Mineralium Deposita*, **26**, 228–238.

MILESI, J.-P., LEDRU, P., FEYBESSE, J.-L., DOMMANGET, A. & MARCOUX, E. 1992. Early Proterozoic ore deposits and tectonics of the Birimian orogenic belt, West Africa. *Precambrian Research*, **58**, 305–344.

NISBET, E.G. & FOWLER, C.M.R. 1982a. The thermal background to metamorphism – 1. Simple one-dimensional conductive models. *Geosciences Canada*, **9**, 161–164.

NISBET, E.G. & FOWLER, C.M.R. 1982b. The thermal background to metamorphism – II. Simple two-dimensional conductive models. *Geosciences Canada*, **9**, 208–214.

OBERTHÜR, T., WEISER, T., AMANOR, J.A. & CHRYSSOULIS, S.L. 1997. Mineralogical siting and distribution of gold in quartz veins and sulfide ores of the Ashanti mine and other deposits in the Ashanti belt of Ghana: genetic implications. *Mineralium Deposita*, **32**, 2–15.

OBERTHÜR, T., VETTER, U., DAVIS, D.W. & AMANOR, J.A. 1998. Age constraints on the gold mineralization and Paleoproterozoic crustal evolution in the Ashanti belt of southern Ghana. *Precambrian Research*, **89**, 129–143.

OPARE-ADDO, E., BROWING, P. & JOHN, B.E. 1993*a*. Pressure–temperature constraints on the evolution of an early Proterozoic plutonic suite in southern Ghana, West Africa. *Journal of African Earth Sciences*, **17**, 13–22.

OPARE-ADDO, E., JOHN, B.E. & BROWING, P. 1993*b*. Field and geochronologic (U–Pb) constraints on the age and generation of granitoids and migmatites in southern Ghana, AGU Spring meeting, Baltimore. *EOS Transactions Abstracts Supplement*, **74**, 301.

PIGOIS, J.P., GROVES, D.I., FLETCHER, I.R., MCNAUGHTON, N.J. & SNEE, L.W. 2003. Age constraints on Tarkwaian paleoplacer and lode-gold formation in the Tarkwa–Damang district, SW Ghana. *Mineralium Deposita*, **38**, 695–714.

PIRAJNO, F. 2001. *Ore deposits and mantle plumes*. Kluwer Academic Publishers, Dordrecht.

RICHTER, F.M. 1985. Models for the Archean thermal regime. *Earth and Planetary Science Letters*, **73**, 350–360.

ROLANDONE, F., JAUPART, C., MARESCHAL, J.-C., GARIÉPY, C., BIENFAIT, G., CARBONNE, C. & LAPOINTE, R. 2002. Surface heat flow, crustal temperatures and mantle heat flow in the Proterozoic Trans-Hudson Orogen, Canadian shield. *Journal of Geophysical Research*, **107**, B12, 2341.

RUPPEL, C. & HODGES, K.V. 1994. Pressure–temperature–time paths from two-dimensional thermal models: Prograde, retrograde, and inverted metamorphism. *Tectonics*, **13**, 17–44.

RUSSEL, J.K. & KOPYLOVA, M.G. 1999. A steady-state conductive geotherm for the north central Slave, Canada: inversion of petrological data from the Jericho kimberlite pipe. *Journal of Geophysical Research*, **104**, 7089–7101.

SCHMIDT MUMM, A., OBERTHÜR, T., VETTER, U. &

BLENKINSOP, T.G. 1997. High CO_2 content of fluid inclusions in gold mineralisations in the Ashanti belt, Ghana: a new category of ore forming fluids? *Mineralium Deposita*, **32**, 107–118.

SERBAN, D.Z., NIELSEN, S.B. & DEMETRESCU, C. 2001. Transylvanian heat flow in presence of topography, paleoclimate, and groundwater flow. *Tectonophysics*, **335**, 331–344.

SYLVESTER, P.J. & ATTOH, K. 1992. Lithostratigraphy and composition of 2.1 Ga greenstone belts of the West African craton and their bearing on crustal evolution and the Archean-Proterozoic boundary. *Journal of Geology*, **100**, 377–393.

TAVMAN, I.H. & TAVMAN, S. 1999. Measurement of thermal conductivity of dairy products. *Journal of Engineering*, **41**, 109–114.

TAYLOR, P.N., MOORBATH, S., LEUBE, A. & HIRDES, W. 1992. Early Proterozoic crustal evolution in the Birimian of Ghana: constraints from geochronology and isotope geochemistry. *Precambrian Research*, **56**, 97–111.

TAYLOR, S.R. & MCLENNAN, S.M. 1995. The geochemical evolution of the continental crust. *Reviews of Geophysics*, **33**, 241–265.

TURCOTTE, D. & SCHUBERT, G. 2002. *Geodynamics: Applications of Continuum Physics to Geological Problems*. 2nd edn. Cambridge Univrsity Press, New York.

VASSEUR, G., DEMONGODIN, L. & BONNEVILLE, A. 1993. Thermal effects arising from water circulation through thin inclined aquifers. *Geophysical Journal International*, **112**, 276–289.

YAO, Y., MURPHY, P.J. & ROBB, L.J. 2001. Fluid characteristics of granitoid-hosted gold deposits in the Birimian Terrane of Ghana: a fluid inclusion microthermometric and Raman spectroscopic study. *Economic Geology*, **96**, 1611–1643.

Geodynamic processes that control the global distribution of giant gold deposits

K. LEAHY[1], A. C. BARNICOAT[2], R. P. FOSTER[3], S. R. LAWRENCE[3] &
R. W. NAPIER[4]

[1]*Environmental Resources Management Ltd, Eaton House, Wallbrook Close, North
Hinksey Lane, Oxford, Oxon OX2 0QS, UK (e-mail: kevin.leahy@erm.com)*

[2]*RDR, School of Earth Sciences, University of Leeds LS2 9JT, UK (present address;
pmd*CRC, Geoscience Australia, PO Box 378, Canberra, ACT 2601, Australia)*

[3]*Exploration Consultants Ltd, Highlands Farm, Greys Road, Henley-on-Thames,
Oxon RG9 4PR, UK*

[4]*GFMS Mining & Exploration Consulting, Hedges House, 153–155 Regent Street,
London W1B 4JE, UK*

Abstract: This paper address the question of why giant gold deposits are so unevenly
spread over the continents, what processes control their distribution, and how more might
be found? Using the source–migration–trap paradigm, it is proposed that the regional
distribution of gold deposits is controlled by fluid access to gold sources on a regional scale,
and by large-scale migration mechanisms. Local distribution is controlled by migration and
trap processes, not discussed in this paper. Our current levels of understanding of gold
suggest a strong geodynamic control in the generation of enriched source rocks and the
fluids that may carry gold, particularly the influence of subduction and accretion during
orogeny. A new six-fold geodynamic classification system that emphasizes subduction and
accretion processes has been used here qualitatively to assess the potential for gold-bearing
source areas. The resulting classification is compared to the distribution of 181 known giant
gold deposits (those with more than 100 t contained gold). The results confirm the propo-
sition that the distribution of giant gold deposits is ultimately a function of the amount of
oceanic crust consumed during the orogenic episode that built that part of the crust. Of the
six geodynamic classes described, large ocean closure orogens were found to contain the
most gold, with nearly half of the world's gold held in known giant deposits. Implications
for understanding ore genesis, exploration for other giant deposits, and for other empirical
explanations of the distribution of gold are discussed further.

Giant gold deposits, defined in this paper as
those with a resource of more than 100 tonnes
of contained gold, are not spread evenly across
the globe and, intriguingly, are not evenly
distributed across the world's orogenic belts
(Fig. 1). The distribution of these deposits is
controlled by complex hydrothermal processes
that operate within specific geodynamic settings
that can be unravelled at a series of scales of
observation, from global to deposit-scale, by
consideration of mineralizing fluids in the
context of the source–migration–trap paradigm
and tectonic setting. In this paper gold-bearing
fluid sources and regional migration mechanisms are regarded as an integral part of a new
framework of crustal growth and orogeny.
Understanding these processes at the regional
scale sheds light on some interesting conun-
drums, such as why some orogenic belts are

endowed with giant gold deposits, whereas
others are entirely barren – a prime example
being why none are found in the Himalayas yet
the Altaids have a major gold endowment.

Source of gold

At the regional scale, gold prospectivity relates
predominantly to the presence of gold source
rocks in the crust or mantle, and to the availability of fluids with the capability of scavenging
and transporting gold.

Mantle sources

Subduction-related gold sourced from the
mantle has been demonstrated convincingly
(e.g. Sillitoe 1993). In this setting, fluids
mobilized from descending oceanic plate

From: McDonald, I., Boyce, A. J., Butler, I. B., Herrington, R. J. & Polya, D. A. (eds) 2005. *Mineral
Deposits and Earth Evolution.* Geological Society, London, Special Publications, **248**, 119–132. 0305-8719/$15.00
© The Geological Society of London 2005.

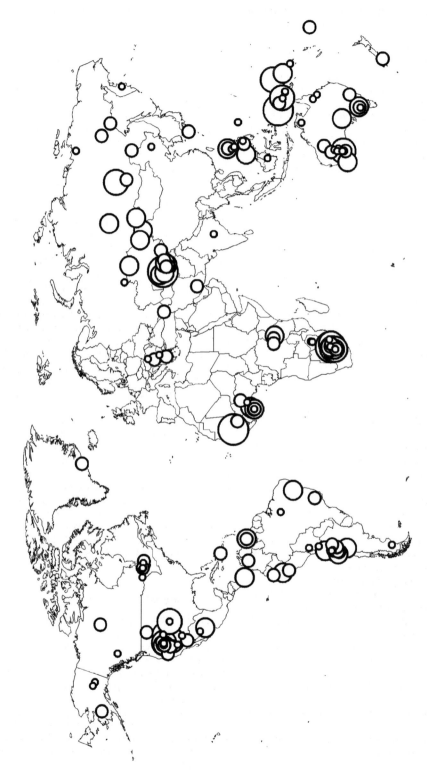

Fig. 1. Location of all known giant gold deposits. Symbol size relates to gold deposits size: smallest circles are deposits in the range of 100 t to 160 t, largest circles are deposits with >1500 t contained gold. Note that deposits with less than 100 t contained gold, and historical or exhausted mines are not shown.

hydrate the overlying lithospheric mantle of the continent and liberate gold into the magmatic–hydrothermal systems of the arc. A large volume of work, recently summarized by Blundell (2003), indicates that giant gold deposit formation is not a steady-state passive response to ongoing subduction, but rather a punctuated active response to the fluctuating stress regime in the overlying plate and to thermal variations in the lower plate. High heat-flow events in the subducting plate can be brought about by various mechanisms, including slab detachment, roll-back, and ridge subduction. These transient effects on the upper and lower plates are often linked dynamically, and can be caused by changes in plate configuration and, ultimately, by instabilities in mantle convection.

Gold-bearing fluids with a mantle source have only been demonstrated (so far) in an active margin context, as discussed above. Gold sources from non-subduction mantle environments, usually attributed to plumes and rifts, have yet to be proven, although these events are believed to be lead to higher heat flow, and hence indirect activation of gold sources in the crust may be possible.

Crustal sources

Given the considerable amount of continental crust available, it is important to recognize that potential source areas for gold are restricted to specific environments. Only crustal material with a predominantly basic bulk composition (i.e. ophiolites or other crustal mafic rocks, pelites and organic-rich sediments) contains enough gold as a trace element (Wedepohl 1978; Boyle 1979), together with sufficient quantities of mineral-hosted H_2O, to be considered a potential source on a regional scale. Significant devolatilization (and implied gold mobility) only starts to occur when pressure and temperature exceed those of sub-greenschist metamorphic facies (Cameron 1989; Groves et al. 2003; Jia et al. 2003). Furthermore, once a rock, even if basic in bulk composition, has exceeded amphibolite metamorphic facies, it is unlikely to contain sufficient H_2O to be capable of significant gold mobilization. These conditions rule out a great deal of the continental crust because:

- the crust is dominantly felsic in composition;
- much of the upper crust has been subjected to sub-greenschist metamorphic facies or lower grade; and

- much of the middle and lower crust has been subjected to metamorphism exceeding upper amphibolite facies.

The requirement of intermediate grades of metamorphism on a regional scale is met only during orogenesis, so the most favourable crustal gold-source areas are within orogenic belts (both inactive and active), in which rocks (or their protoliths) with hydrated 'bulk mafic' composition are dominant. Geodynamic terranes that may meet this requirement include: accretionary prisms, ophiolites (sensu lato), passive margin and foredeep sediment sequences and back-arc basins.

Hydrous fluids that can carry crustal gold are liberated most effectively during upper greenschist to amphibolite metamorphism and anatexis, as described for the Archaean Yilgarn (Kent et al. 1996; Weinberg et al. 2004), or the Mesozoic North China Craton (Yang et al. 2003), or in a high heat flow or other extraordinary, transient pressure–temperature regime (Kerrich 1999). These regimes are probably the same sort of punctuated active responses to plate reorganization described above for mantle sources above, and go some way to explaining the occurrence of accretionary (or 'slate') belt mesothermal giant gold deposits in active margin settings (Fig. 2) as discussed by Groves et al. (2003). It is possible that the orogenic trigger for mobilization of gold-bearing fluids is not related to oceanic subduction but could be brought about during continent–continent collision, particularly if this leads to delamination of the attached mantle lithosphere and associated thermal boost at the base of the crust (e.g. Yang et al. 2003). The fact that continent–continent collision can generate the required transient P–T events to lead to giant gold deposits is underpinned by the analysis of crustal setting and distribution of gold deposits presented in this paper.

Origins of gold-bearing fluids

Naturally-occurring fluids with the capability to carry significant amounts of gold fall into four broad groupings, all of which exhibit high sulphur contents but otherwise have varied physical and chemical parameters, as defined by field observations supported by studies using Geochemists Workbench™ and other published data (gold and associated elements generally in Heinrich & Eadington 1986; Gammons & Williams-Jones 1995; Huston 1998; Wood & Samson 1998; Heinrich et al. 1999; and for specific fluids as listed below):

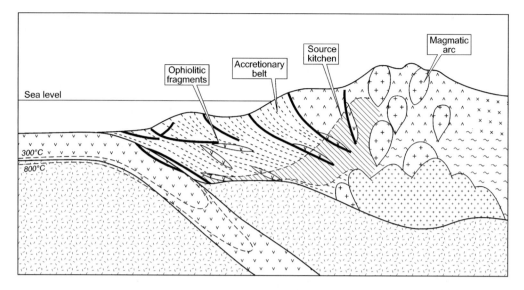

Fig. 2. Schematic cross-section of an accretionary subduction zone, with a shaded area indicating the 300–800 °C window typical of greenschist and lower amphibolite facies, where major devolatilization occurs, and from which gold may be mobilized.

Deep magma-dominated fluids (DMF); Davis & Bickle 1991; Hedenquist 1995; Thompson *et al.* 1999. These fluids are characterized by high temperature, high salinity, moderate fO_2 and weak to moderate acidity. They are predominantly magmatic and are associated with arc-intrusive hydrothermal–magmatic systems at depths from 2 km down to the base of the crust. Gold is derived from mantle sources in the subduction zone environment as discussed previously.

Shallow magma-dominated fluids (SMF); Hedenquist *et al.* 1998; Berger & Silberman 1985; Daliran 1999; Mehrabi *et al.* 1999; Richards 1995. These fluids are moderate to high temperature, moderate to high salinity, high fO_2 and weakly to moderately acidic. Gold source and fluid origin are similar to DMF, but the fluids are modified at high crustal levels by interaction with meteoric waters or other crustal fluids.

Multi-source fluids (MSF); Barnicoat *et al.* 1997; Bagby & Berger 1985; Colvine *et al.* 1988; Wood 1992. This diverse family of fluids encompasses a wide temperature range, low to moderate salinity, moderate fO_2 and are neutral to weakly acidic. They are rock-buffered fluids from a variety of crustal sources, mainly metamorphic, but also including sedimentary formation water and possible magmatic contributions.

Gold is derived predominantly from crustal sources, with possible mantle input in near-arc settings.

Basinal fluids (BF); Hoeve & Quirt 1986; Kirkham 1986; Wilde *et al.* 1989. These fluids are of low temperature and very high salinity, with the potential to carry gold, but have a limited capability to form giant gold deposits. However, they are very important for other giant deposit-types such as MVT and Athabasca-type uranium.

The discussion presented so far suggests that specific geodynamic processes of crustal growth, namely subduction and orogeny, are essential to the mobilization of fluid types that can form giant gold deposits. With this perspective, a new six-fold classification system has been constructed to subdivide regions of the Earth's continental crust firstly in terms of its orogenic constituents ('domains'), and subsequently in terms of the relative quantity and quality of fluid types each domain could generate.

Geodynamic classification

Plate tectonics has played a critical role in crustal growth throughout the Phanerozoic the Proterozoic, and possibly even as far back as the Late Archaean (e.g. de Wit & Hart 1993; Windley 1995; de Wit 1998). The recognizable temporal and spatial links between the accretionary process (as the main mechanism for crustal growth) and derivation of gold from both mantle (through subduction) and crustal sources (through metamorphism) thus forms the basis of the geodynamic classification and analysis presented in this paper.

Accretionary complexes can grow on the edges of continental blocks or within an oceanic setting as multiple arcs amalgamate, the key requirements being large oceans (e.g. Pacific-sized) and long time-periods (tens to hundreds of millions of years) that ensure the continual delivery of accretable material. This accreted material includes juvenile crust such as arcs, oceanic plateaux, and obducted oceanic crust, plus more diverse sedimentary terranes (fore-arc, intra-arc, back-arc and retro-arc basins, carbonate platforms, deep-sea fans), and micro-continental slivers and blocks spalled off other continents by rifting (Ben-Avraham *et al.* 1981; Howell 1989). The accretionary complexes are intruded by trenchward-advancing calcalkaline arcs that step, rather than creep, forward as large terranes become accreted (e.g. in the Central Asian Palaeozoic orogenic belt; Sengor & Natal'in 1996). Major crustal structures are dominated by transpression, especially along terrane boundaries, although pure compression and possible back-arc extension are also signifi-cant. As long as subduction continues around parts of the complex, the orogen may be termed 'open', or 'unconsolidated'. They become 'closed' or 'consolidated' by collision with another large continental mass which blocks the direct effects of subduction and deforms the whole complex. In this way the large ocean closure (LOC) orogenic type is formed (named 'Altaid' or 'Turkic' type by Sengor after the Central Asian Palaeozoic LOC).

Not all collisional orogenic belts comprise accreted terranes. Many of the world's best-studied orogenic belts have either none (e.g. Western Alps) or only one or two (e.g. Himalayas). This coincidence alone may explain why accretionary orogenic belts and complexes have not been satisfactorily understood in terms of their geodynamic evolution until workers started to synthesize the total framework of tectonics, magmatism, metamorphism and basin evolution for active complexes (starting with Coney *et al.* 1980), Phanerozoic complexes (e.g. Sengor & Natal'in 1996) and even Archaean accretionary complexes (de Wit *et al.* 1992).

There is a continuum between collisional orogens, that lack intervening accreted terranes, and the large ocean-closure orogens described above that contain vast tracts of accreted terranes. The key difference between these two

end-members is the width of oceanic plate that was consumed between the two colliding masses, because this controls both the amount of material that is available for accretion and the degree to which continental magmatic arcs can develop on the over-riding plate. To accommo-date this range we have classified consolidated orogenic domains into three types:

- large ocean closure (LOC) orogens, epitomized by the Altaids;
- moderate ocean closure (MOC) orogens, for example the Appalachians; and
- small ocean closure (SOC) orogens, for example the Western Alps.

Unconsolidated orogenic belts comprise those areas that are currently undergoing crustal growth by accretion and subduction. The most obvious of these are the large accretionary-arc complexes (such as Alaska) but they range through active continental margins to nascent island arcs in the Pacific. These are classified into three types:

- island arcs (IA), also known as intra-oceanic or juvenile arc, exemplified by the western Aleutian chain;
- continental arcs (CA), for example the Chilean Andes; and
- large accretionary-arc complexes (LAAC), for example the northern Rockies.

Schematic cross-sections illustrating the key features of known orogenic systems represent-ing the different domains are shown in Figure 3 (Pfiffner 1992); others drawn from crustal struc-tures discussed in Windley (1995) and Sengor & Natal'in (1996).

This classification is based on a series of 'critical features' and 'additional features' (Tables 1 & 2). Critical features describe defining elements of a domain that are essential to their characterization; additional features identify the commonly occurring features (e.g. terrane types) that are not defining but contrib-ute to the definition process. Terranes within these domains are subdivided into accreted types (or exotic terranes; micro-continental blocks, island arcs, oceanic fragments and sedi-mentary terranes), and local types (basement massif, accretionary wedges, passive margins, extinct continental arcs, and foreland, back-arc and intermontane orogenic basins)

The consolidated domains are the end-products of continent–continent collision events that have caused subduction to cease. Key discriminators between the different

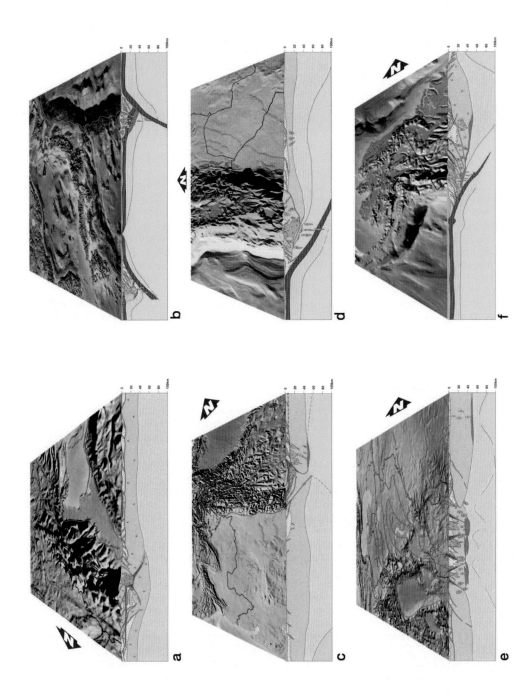

Table 1. *Definition of the consolidated domain classification*

Domain type	Critical definition	Additional features	Example
Small ocean closure orogen	Linear orogenic belts with aspect ratios of >4–1. No continental arc, or highly discontinuous and poorly developed if present, no back-arc terrane.	Local accretionary wedge and foreland basin terranes, basement massif, passive margin sequence, few or no accreted terranes (those present do not touch one another). They have experienced less than a few hundred kilometres of oceanic plate subduction, usually over less than 50 Ma.	Damara belt, Western Alps; Figure 3a
Moderate ocean closure orogen	Linear orogenic belts with aspect ratios of >4–1. Extinct continental arc ranging from slightly discontinuous to very well developed.	Various local terranes (especially forelands and accretionary wedges), basement massif, passive margin sequence, few to many accreted terranes (may be adjacent), post orogenic granites. They have experienced a few hundred to a few thousand kilometres of oceanic plate subduction, usually over less than 100 Ma.	Appalachians, Himalayas, Figure 3c
Large ocean closure orogen	Ovoid or irregular orogenic belts with aspect ratios of <3–1. Includes many accreted and local terranes of variable type, but especially extinct continental arcs. These large areas are the consolidated equivalents of LAAC.	Common post-orogenic granites, basement massifs rare, especially in the interior. Because of their size, more than two continents may be required to encircle the orogen to define the domain margins. They have experienced more than a few thousand kilometres of oceanic plate subduction, usually over many hundred of millions of years.	Pan-African North Africa, Yilgarn Block, Altaids, Figure 3e

consolidated types are the identification of continental arcs formed prior to collision, the number of entrained terranes, and the aspect ratio (length–width) of the orogenic tract (Table 1).

The unconsolidated domains (Table 2) are those with active or recently active margins. The oldest allowable age for this classification depends on preservation. Inactive arcs are defined here as those in which the upper 1–2 km of arc crust have not yet been removed by erosion. The oldest examples are probably of Miocene age.

To be able to classify the entire continental area of the world, a further non-orogenic type of Domain is required to describe areas of undeformed sedimentary or volcanic cover (e.g. intra-cratonic sag basins and flood basalt provinces) that completely obscure the underlying crust.

Fig. 3. Schematic crustal sections of the type-examples of the six orogenic domains, surface area tiles have variable scales. (**a**) Small ocean closure (SOC), view of the Western Alps, with Italy to the right. (**b**) Island arc (IA) view of the Isu–Bonin arc, Japan to the north. (**c**) Moderate ocean closure (MOC), view of the Himalayas, northern India and Pakistan to the left. (**d**) Continental arc (CA), view of the Andean arc in northern Chile. (**e**) Large ocean closure (LOC), view of the Altaids, Tarim basin on the left. In the schematic sections lithospheric mantle is yellow, asthenospheric mantle is green, oceanic crust is dark grey, continental crust is pink, basins are yellow with fine stipple, mafic underplating is purple, orogenic and post-orogenic granites are red, and magmatic arc is in blue. (**f**) Large accretionary-arc complexes (LAAC), for example the northern Rockies.

Table 2. *Definition of the unconsolidated domain classification*

Domain type	Critical definition	Additional features	Example
Island arc	Subduction-related volcano-plutonic belt on oceanic basement, active or recently active.	Various local terranes (especially back-arc and accretionary wedge), plus 1 other accreted terrane. If further terranes have been added it is then considered an LAAC.	Aleutian chain, Izu-Bonin Arc; Figure 3b
Continental arc	Continental or AAC basement, active or recent arc, with less than 3 accreted terranes.	If on a continental basement; various local terranes (especially retro-forelands and extinct arcs), overprinted passive margin sequence, plus 2 other accreted terranes (3 or more will constitute an AAC). If on an AAC basement, only the actual active chain and any adjacent local accretionary wedge terrane is delineated as part of the domain, all the other associated terranes are included in the AAC domain that the continental arc is overprinting.	AAC basement – eastern Alaska arc, continental basement – Peruvian Andes; Figure 3d
Large accretionary arc complex	3 or more accreted terranes, with adjacent or overprinting active continental arc. May be either intra-oceanic e.g. Borneo, or peri-continental e.g. Alaska.	Local-type terranes (especially extinct arc belts and retro-foreland), overprinted passive margin sequence (peri-continental only). Possibly hundreds of mappable accreted terranes.	Borneo or Alaska; Figure 3f

The seven-fold domain classification is summarized in Table 3.

Gold prospectivity within domains

Potential fluid source regions are not evenly distributed throughout the crust, and may occur within specific settings within each domain type, as illustrated on crustal sections through representative orogens (Fig. 4). The overall potential for each fluid type in a given domain is a function of the presence and number of all potential fluid source regions within the crustal section. For example SMF potential in SOC orogens is nil because magmatic arcs are, by definition, absent or very poorly developed, whereas in IA orogens the potential is very high. The fluid potentials are summarized in Table 4.

The implication of fluid habitat variation according to domain-type is that gold source areas are primarily related to subduction and to the presence of 'bulk basic' accretionary

Table 3. *General features of each domain classification*

Domain	Accretionary	Collisional	Terranes	Arcs
Large ocean closure	yes	yes	very many	many
Moderate ocean closure	possibly	yes	several	yes
Small ocean closure	no	yes	few	no
Large accretionary arc complex	yes	no	very many	yes
Continental arc	possibly	no	several	yes
Island arc	possibly	no	few	yes
Non-orogenic (e.g. sag basin)	no	no	no	no

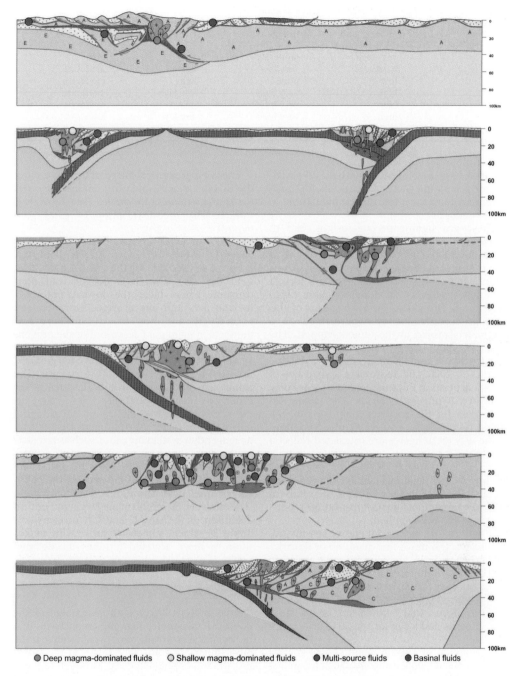

● Deep magma-dominated fluids   ○ Shallow magma-dominated fluids   ● Multi-source fluids   ● Basinal fluids

Fig. 4. Crustal sections of the type-examples of the six orogenic domains, showing schematic potential fluid habitats in circles.

terranes. Thus the domains with the greatest potential are LOC, LAAC and CA orogens, as these have had greatest exposure to subduction and arc activity, and contain the largest volumes of 'bulk basic' accreted material, often including vast swaths of 'slate belt' rocks. By the same criteria SOC orogens have the lowest potential, lacking arcs and any accreted

Table 4. *Summary of fluid-type potential in each domain*

Domain	DMF Potential	SMF Potential	MSF Potential	BF Potential
LOC	very high	moderate	very high	high
MOC	moderate	very low	high	moderate
SOC	low	none	high	moderate
LAAC	high	very high	very high	moderate
CA	high	very high	high	moderate
IA	high	very high	very low	low
Non-orogenic	none	none	none	high

terranes. Non-orogenic domains have no potential for generating giant gold deposits, although potential does exist for the accumulation of placer deposits.

These are significant conclusions, and to check whether any of the high-potential source areas have actually contributed to the generation of giant gold deposits a spatial analysis was undertaken of the distribution of the largest gold deposits against the domains within which they occur. The global review in terms of this domain classification was performed as a commercial project by the authors, supported by Brian Windley. The gold deposit database was constructed from commercially available sources in 2000, and represents a snapshot of known deposits at this time, with the focus being solely on hypogene deposits. The 'contained gold' data for the deposits comprise publicly-stated resource plus reserve figures. Variations in the way these figures are calculated reflect a number of factors:

- errors and inconsistencies in the reporting standards, and in what specific rules of government or financial institution reporting have been applied;
- political or commercial bias or sensitivities (can act to move the values up or down);
- accounts only for deposits as declared in 2000, with no reference back to previously exploited deposits, even if they had produced more than 100 t gold (we assume, therefore, that mined-out giant gold deposits had a similar spatial distribution); and
- definition of a 'giant gold deposit' at 100 t contained gold is arbitrary and at the judgement of the authors, rather than any international standard.

A total of 181 deposits were used in the spatial analysis: their names, contained gold (with the caveats described above) and host domain is presented in Table 5 below.

The comparison of tonnage of gold in each of the domain types (Table 6) demonstrates that the source potential we predict (see Table 4) is reflected by the distribution of giant gold deposits.

Those domains predicted to have the greatest potential were the LOC, LAAC and CA orogens, and accordingly these contain the most gold deposits. The SOC orogen and non-orogenic areas have the lowest potential, lacking arcs and any accreted terranes, and accordingly are seen to contain very low numbers of giant gold deposits. For the orogens we predicted to have high potential for gold it was expected that LOC would have the greatest, followed by LAAC and then CA orogens. The actual spatial distribution indicates clearly the order was LOC, CA and then LAAC, suggesting either an over-estimation of the potential fluid habitats in the LAAC orogen (see Table 4), or some other systematic cause, such as artefacts in the data, e.g. under-exploration of LAAC domains relative to CA, due to either the strategy of exploration companies, or the political or logistical situation of the orogens. Alternatively, the problem may lie in our critical definition of the LAAC and CA orogens (see Table 2) which, somewhat arbitrarily, makes three or more accreted terranes the defining point.

Discussion

Spatial analysis linked with assessment of the potential fluid-source regions demonstrates that the presence of an appropriate gold source is the most important controlling factor in the global distribution of gold deposits. Although regional to local-scale geodynamic variations, such as upper-plate stresses and lower-plate thermal perturbations, are recognized as important in the hydrothermal mobilization and transportation (and hence localization) of the gold, ultimately there must be a source for the gold and for the fluids that mobilize the element. This

Table 5. *Giant gold deposits of the world in 2000 (see text for limitations)*

Deposit Name	Au (t)	Host domain	Deposit name	Au (t)	Host domain
Abosso	204	LOC	Gaby	188	CA
Agua Rica	173	CA	Geita	311	MOC
Ajo	174	CA	Getchell	498	CA
Alumbrera	523	CA	Giant – Lolor	225	LOC
Angostura	342	CA	Globe & Phoenix	120	LOC
Angren Region	270	LOC	Gold Quarry	251	CA
Ankerite – Aurnor – Delni	143	LOC	Gold Ridge	154	IA
Antamok	350	IA	Goldstrike	659	CA
Ashanti	710	LOC	Grasburg	2127	CA
Atlas-Lutopan	135	IA	Grass Valley	323	CA
Bakyrchik	412	LOC	Gunung Pongkor	102	IA
Ballarat – Buninyong	161	LOC	H J Joel	205	LOC
Barnat – Canadia – East	165	LOC	Harmony	125	LOC
Batangas	129	IA	Hemlo	597	LOC
Battle Mountain	180	CA	Hidden Valley	100	CA
Batau Hijau	500	LAAC	High Desert JV	124	CA
Bendigo	684	LOC	Highland Valley	600	CA
Bingham Canyon	1219	CA	Hill 50	227	LOC
Blyvooruitzicht	727	LOC	Hishikari	260	CA
Boddington	486	LOC	Homestake	1200	CA
Bralorne	120	LAAC	Jerritt Canyon	215	CA
Brisas Del Cuyuni	174	LOC	Jundee	210	MOC
Bronzewing	158	LOC	Kalgoorlie Consolidated	1282	LOC
Bulyanhulu	390	LOC	Kal'makyr	450	LOC
Cadia Hill	222	LOC	Kamchatka	155	LAAC
Cam & Motor	146	LOC	Kanowna Belle	155	LOC
Campbell – Red Lake	435	LOC	Kassandra Gold	196	LAAC
Candelaria	100	CA	Kelian	113	LAAC
Carlin	135	CA	Kemess	127	LAAC
Castlemaine – Chewton	180	LOC	Kerr Addison	324	LOC
Cerro Casale	676	CA	Kidston	139	LOC
Cerro Leste	148	LOC	Kokpatass	613	LOC
Cerro Vanguardia	100	CA	Kori Kollo	161	CA
Chelopech	114	LAAC	Kumtor	343	LOC
City Deep/Crown Tlngs	331	LOC	Kyuchus	132	LAAC
Comstock	260	CA	Las Cristinas	428	LOC
Con – Rycon – Negus	165	LOC	La Coipa	120	CA
Cortez/Pipeline	276	CA	La Escondida	422	CA
Cripple Creek	750	CA	Lamaque-Sigma	258	LOC
Dalneye	875	LOC	Lihir Island	1692	IA
Darasun	304	LOC	Lobo-Marte	139	CA
Daugiztau	540	LOC	Lone Tree	133	CA
Dizon	130	IA	Macraes Flat	206	CA
Dome	143	LOC	Majdanpek	300	LAAC
Donlin Creek	170	LAAC	Malanjkand	158	MOC
Doyon	129	LOC	Masbate	155	IA
Driefontein Consolidated	1326	LOC	Mcdonald	218	LAAC
El Indio	293	CA	Meikle	147	CA
El Sauzal	112	CA	Mesquite	147	CA
Elandsrand	577	LOC	Metates	382	CA
Elnichny	192	LAAC	Midas	313	CA
Emperor	270	IA	Minas Conga	354	CA
Evander	1331	LOC	Mokrsko	100	SOC
Famatina	178	CA	Morila	242	LOC
Far Southeast	295	IA	Morro Velho	278	MOC
Fort Knox	155	LAAC	Mt Charlotte	103	LOC
Frazenda Braziliero	600	LOC	Mt Kare	114	CA
Free State Consolidated	1683	LOC	Muruntau	4500	LOC
Frieda River	302	CA	Myutenbai	618	LOC

Table 5. (*Continued*)

Deposit Name	Au (t)	Host domain	Deposit name	Au (t)	Host domain
Nezhdaninskoe	235	LAAC	Sons of Gwalia	105	LOC
Norseman	150	LOC	South Pipeline	324	CA
Obuasi	423	LOC	St Helena	228	LOC
Ok Tedi	321	CA	St Ives	113	LOC
Olimpiada	700	LOC	Sukhoi Log	1035	LOC
Olympic Dam	391	MOC	Sunrise Dam	162	LOC
Original Goldstrike	771	CA	Svetlinskoye	147	LOC
Oryx	848	LOC	Syama	276	LOC
Pajingo	150	LOC	Target	446	LOC
Pamour-Hahnor	145	LOC	Tarkwa	417	LOC
Panguna	529	IA	Teberebie	143	LOC
Pascua	395	CA	Telfer	321	LOC
Pavlik	303	LAAC	Timmins Division	111	LOC
Petaquila	672	CA	Tuanjigou	120	LOC
Pierina	203	CA	Twangiza	222	MOC
Pikes Peak	134	CA	Twin Creeks	342	CA
Pogo	161	LAAC	Unisel	101	LOC
Poplar	308	LOC	Upper Kumar	300	LOC
Porgera	980	CA	Vaal River	1129	LOC
Prestea	219	LOC	Vasilkovskoye	414	LOC
Pueblo Viejo	338	LAAC	Veladero	182	CA
Randfontein	400	LOC	Wafi	155	CA
Refugio	263	CA	West Witwatersrand	158	LOC
Rosia Montana	145	LAAC	Western Areas	4280	LOC
Round Mountain	201	CA	Western Deeps	726	LOC
Sadiola Hill	2494	LOC	Williams	138	LOC
Santo Tomas	233	IA	Wilson Creek	156	CA
Sar Cheshmeh	324	CA	Yamfo-Sefwi	135	LOC
Sheba – Fairview	126	LOC	Yanacocha	633	CA
Sipalay	301	IA	Yimuyn Manjerr	101	LOC
Skaergaard	166	SOC	Zarmitan	200	LOC
Skouries	235	LAAC			

Table 6. *Distribution of gold in giant deposits by tonnage and by number of deposits*

Domain	% of world's total by tonnage of gold	% of world's total by number of deposits
LOC	57.4%	46.4%
CA	28.7%	31.5%
LAAC	6.1%	10.5%
IA	5.3%	7.2%
MOC	2.1%	3.3%
SOC	0.4%	1.1%
Non-orogenic	0.0%	0.0%

The data used in this analysis are partly derived from commercial sources, but consist of the locations of 181 deposits, with a total contained gold of 73 286 t. The Witwatersrand Basin accounts for about 20% of world's tonnage and strongly skews this data. It formed in a foreland basin of the Kalahari LOC.

is clearly demonstrated by the notably small number of large gold deposits that have been found in SOC orogens. This is because SOC orogens by definition lack well-developed arcs and accretionary terranes. This conclusion also provides the mechanism that explains the observations about orogenic geometry (i.e. 'rotund'

and 'skinny') being in some way related to gold endowment (Goldfarb *et al.* 2001).

The gold-transporting capabilities of the fluid types alone are not likely to define the potential to generate giant gold deposits; they merely define a potential for generation of any size of deposit. In the case of crustal sources, exceptional volumes of fluid and access to major structures are key factors in forming the giant deposits; in magmatic systems it is the highly focused nature of fluid (and magma), but interactions of unusually enhanced physical (e.g. temperature or pressure) and chemical (e.g. salinity) characteristics also play key roles. Clearly, these mineralizing systems are complex, but the source–migration–trap framework can be applied to successively finer-scale analysis. Further subdivision of domains on the basis of their arc and accreted components (terranes) is an essential exercise to understanding fully the controls on the genesis of giant gold deposits.

Conclusions

In an effort to understand the global distribution of giant gold deposits the geodynamic setting of both potential gold source areas and the fluids that can mobilize the gold were analysed. The analysis confirms the premise that the global distribution of giant gold deposits is controlled by the occurrence of gold in a specific range of lithologies and by the availability of fluids that can actually mobilize the metal. Gold source areas are generated during normal crustal growth processes, but in many cases unusual (non-steady-state) conditions are required to mobilize both fluids and the metal. Our evaluation suggests a simple paradigm for gold mineralization: the more long-lived the subduction, (i) the more the arc develops, and (ii) the more voluminous the accretionary belt, and hence (iii) the greater the potential for the non-steady-state events that lead to giant gold deposits. These findings explain the processes that drive the distribution of gold deposits within the crust, and are in contrast to the empirical approach of time- and space-controlled 'golden epochs' adopted by some workers. We conclude that apparent epochs of gold mineralization are either spurious (artefacts introduced by focusing on known deposit groups or districts) or coincident (e.g. with major periods of crustal growth), or both. Either way, consideration of 'golden epochs' does not contribute to the understanding of the processes that form gold deposits, and is an example of empirical observation rather than scientific thought.

It is our conviction that a fundamental understanding of the spatial distribution and geodynamic context of gold- and fluid-sources is an essential foundation for fine-scale analysis of orogenic components and their timing to produce a better-resolved gold prospectivity model. This will lead to improved exploration and target-generation strategies at the continent, country and district-scale, and a predictive insight of the process that may have formed giant gold deposits in the study area.

The authors extend their thanks to the following organizations for financial support in the production of this paper: Exploration Consultants Ltd, Environmental Resources Management Ltd, GFMS Mining Exploration Consulting and pmd*CRC at Geoscience Australia.

References

BAGBY, W.C. & BERGER, B.R. 1985. Geologic characteristics of sediment-hosted, disseminated precious-metal deposits in the western United States. *In*: BERGER, B.R. & BETHKE, P.M. (eds) *Reviews in Economic Geology*, **2**, 169–202.

BARNICOAT, A., HENDERSON, I., KNIPE, R., ET AL. 1997. Hydrothermal gold mineralization in the Witwatersrand basin. *Nature*, **386**, 820–824.

BERGER, B.R. & SILBERMAN, M.L. 1985. Relationships of trace-metal patterns to geology in hot-spring type precious-metal deposits. *Reviews in Economic Geology*, **2**, 233–247.

BEN-AVRAHAM, Z., NUR, A., JONES, D. & COX, A. 1981. Continental accretion: from oceanic plateaus to allochthonous terranes. *Science*, **213**, 47–54.

BLUNDELL, D. 2003. Tectonic processes conducive to magmatic-hydrothermal mineralisation. Applied Earth Sciences Transactions of the Institute of Mining and Metallurgy B, August 2003, **112**, B107–109.

BOYLE, R.W. 1979. *The Geochemistry of Gold and its Deposits*. Geological Survey of Canada Bulletin 280, 584p.

CAMERON, E.M. 1989. Scouring of gold from the lower crust. *Geology*, **17**, 26–29.

COLVINE, A.C., FYON, J.A., HEATHER, K.B., MARMONT, S., SMITH, P.M. & TROOP, D.G. 1988. *Archean lode gold deposits in Ontario*. Ontario Geological Survey Miscellaneous Paper 139, p. 136.

CONEY, P., JONES, D. & MONGER, J.W. 1980. Cordilleran suspect terranes. *Nature*, **288**, 329–333.

DALIRAN, F., WALTHER, J. & STÜBEN, D. 1999. Sediment-hosted disseminated gold mineralisation in the North Takab geothermal field, NW Iran. *In*: STANLEY, C.J., RANKIN, A.H. ET AL. (eds) *Mineral deposits: process to processing*, **2**, 837–840.

DAVIES, J. & BICKLE, M. 1991. A physical model for the volume and composition of melt produced by hydrous fluxing above subduction zones. *Philosophical Transactions of the Royal Society of London*, **A335**, 355–364.

DE WIT, M.J. 1998. On Archean granites, greenstones, cratons and tectonics: does the evidence demand a verdict? *Precambrian Research*, **91**, 181–226.

DE WIT, M.J. & HART, R.A. 1993. Earth's earliest continental lithosphere, hydrothermal flux and crustal recycling. *Lithos*, **30**, 309–335.

DE WIT, M.J., ROERING, C., HART, R.J., *ET AL.* 1992. Formation of an Archaean continent. *Nature*, **357**, 553–562.

GAMMONS, C.J. & WILLIAMS-JONES, A.E. 1995. Hydrothermal geochemistry of electrum: thermodynamic constraints. *Economic Geology*, **90**, 420–432.

GROVES, D., GOLDFARB, R., ROBERT, F. & HART, C. 2003. Gold deposits in metamorphic belts: Current understanding, outstanding problems, future research and exploration significance. *Economic Geology*, **98**, 1–30.

GOLDFARB, R.J., GROVES, D.I. & GARDOLL, S. 2001. Rotund versus skinny orogens: Well-nourished or malnourished gold? *Geology*, **29**, 539–542.

HEDENQUIST, J.W. 1995. *The ascent of magmatic fluids: discharge versus mineralisation.* Mineralogical Association of Canada Short Course Series 23, 263–289.

HEDENQUIST, J., ARRIBAS, A. & REYNOLDS, T.J. 1998. Evolution of intrusion-centered hydrothermal systems: Far Southeast-Lepanto porphyry and epithermal Cu–Au deposits, Philippines. *Economic Geology*, **93**, 373–404.

HEINRICH, C.A. & EADINGTON, P.J. 1986. Thermodynamic predictions of the hydrothermal chemistry of arsenic, and their significance for the paragenetic sequence of some cassiterite-arsenopyrite-base metal sulfide deposits. *Economic Geology*, **81**, 511–529.

HEINRICH, C.A., GÜNTHER, D., AUDÉTAT, A., ULRICH, T. & FRISCHKNECHT, R. 1999. Metal fractionation between magmatic brine and vapor, determined by microanalysis of fluid inclusions. *Geology*, **27**, 755–758.

HOEVE, J. & QUIRT, D. 1986. *A common diagenetic-hydrothermal origin for unconformity-type uranium and stratiform copper deposits?* Geological Association of Canada Special Paper **36**, 151–172.

HOWELL, D.G. 1989. *Tectonics of suspect terranes: mountain building and continental growth.* Chapman and Hall, London.

HUSTON, D.L. 1998. The hydrothermal environment. *AGSO Journal of Australian Geology and Geophysics*, **17**, 15–30.

JIA, Y., KERRICH, R. & GOLDFARB, R. 2003. Origin of the ore-forming fluid for orogenic gold quartz vein systems in the western North American Cordillera: Constraints from $\Delta^{15}N$, ΔD, $\Delta^{18}O$, and Se/S studies. *Economic Geology*, **98**, 109–124.

KENT, A., CASSIDY, K. & FANNING, C. 1996. Archean gold mineralisation synchronous with the final stages of cratonisation, Yilgarn Craton, Western Australia. *Geology*, **24**, 879–882.

KERRICH, R. 1999. Nature's gold factory. *Science*, **284**, 2101–2102.

KIRKHAM, R.V., 1986. *Distribution, settings and genesis of sediment-hosted stratiform copper deposits.* Geological Association of Canada Special Paper, **36**, 3–38.

MEHRABI, B., YARDLEY, B.W.D. & CANN, J.R. 1999. Sediment-hosted disseminated gold mineralisation at Zarshuran, NW Iran. *Mineralium Deposita*, **34**, 673–696.

PFIFFNER, O.A. 1992. Alpine Orogeny. *In*: BLUNDELL, D., FREEMAN, R. & MUELLER, S. (eds) *A Continent Revealed – The European Geotraverse.* Cambridge University Press, Cambridge, 180–190.

RICHARDS, J.P. 1995. *Alkalic-type epithermal gold deposits – a review.* Mineralogical Association of Canada Short Course Series **23**, 367–400.

SENGOR, A. & NATAL'IN, B. 1996. Paleotectonics of Asia: fragments of a synthesis. *In*: YIN, A. & HARRISON, T. (eds) *The tectonic evolution of Asia.* Cambridge University Press, Cambridge, 486–640.

SILLITOE, R. 1993. Epithermal models: genetic types, geometric controls and shallow features. *In*: KIRKHAM, R.V., SINCLAIR, W.D., THORPE, R.I. & DUKE, J.M. (eds) *Ore Deposits Modeling.* Geological Association of Canada, Special Volume **40**, 403–417.

THOMPSON, J.F.H., SILLITOE, R.H., BAKER, T., LANG, J.R. & MORTENSON, J.K. 1999. Intrusion-related gold deposits associated with tungsten-tin provinces. *Mineralium Deposita*, **34**, 323–344.

WEDEPOHL, K.H. (ed.) 1978. *Handbook of Geochemistry.* Springer-Verlag, Berlin.

WEINBERG, R.F., HODKIEWICZ, P.F. & GROVES, D.I. 2004. What controls gold distribution in Archean terranes? *Geology*, **32**, 545–548.

WILDE, A., BLOOM, M. & WALL, V. 1989. Transport and deposition of gold, uranium, and platinum-group elements in unconformity-related uranium deposits. *In*: KEAYS, R., RAMSAY, W. & GROVES, D. (eds) *The Geology of Gold Deposits – The Perspective in 1988.* Economic Geology Monograph **6**, 637–650.

WINDLEY, B.F. 1995. *The Evolving Continents.* Wiley, Chichester. 3rd edn.

WOOD, S.A. 1992. Experimental determination of the solubility of $WO_3(s)$ and the thermodynamic properties of $H_2WO_4(aq)$ in the range 300–600°C at 1 kbar: calculation of scheelite solubility. *Geochimica et Cosmochimica Acta*, **56**, 1827–1836.

WOOD, S.A. & SAMSON, I.M. 1998. Solubility of ore minerals and complexation of ore metals in hydrothermal solutions. *Reviews in Economic Geology*, **10**, 33–80.

YANG, J.H., WU, F.Y. & WILDE, S.A. 2003. A review of the geodynamic setting of large-scale Late Mesozoic gold mineralization in the North China Craton: an association with lithospheric thinning. *Ore Geology Reviews*, **23**, 125–152.

Terrane and basement discrimination in northern Britain using sulphur isotopes and mineralogy of ore deposits

D. LOWRY[1], A. J. BOYCE[2], A. E. FALLICK[2], W. E. STEPHENS[3] & N. V. GRASSINEAU[1]

[1]*Department of Geology, Royal Holloway University of London, Egham Hill, Egham, Surrey TW20 0EX, UK (e-mail: d.lowry@gl.rhul.ac.uk)*
[2]*Isotope Geology Unit, Scottish Universities Environmental Research Centre, East Kilbride, Glasgow G75 0QF, UK*
[3]*The School of Geography and Geosciences, University of St Andrews, Fife KY16 9AL, UK*

Abstract: This study of four well characterized and adjacent terranes in Northern Britain outlines the sulphur isotope variations, assesses the overall importance of crustal and mantle sulphur, and presents a model that can be applied to terrane distinction throughout the North Atlantic Caledonides. The characteristics of metal components within the mineralization provide additional information that can be related to the nature of underlying basement and events from the onset of sedimentation to the cessation of mineralization within stratigraphically linked packages of rock.

The $\delta^{34}S$ data show that the dominant crustal units in each terrane, whether upper crustal sediments or cratonic basement, provide the main alternative sulphur source to the mantle and act also as the main contaminant of subcrustal melts. The $\delta^{34}S$ values of granitoid-related mineralization are either within the subcrustal melt-range of $-3‰$ to $+3‰$ or deviate toward the values of major crustal units in the terrane, i.e. toward ^{34}S depletion in the Southern Uplands and toward ^{34}S enrichment in the Lakesman and Grampian terranes. More complex mineralization in the Northern Highland terrane is linked to the presence of thick North Atlantic craton beneath upper crustal metasediments. Across the region the vein systems beyond the influence of magmatic components represent homogenized sulphur, metals and fluids from local upper crustal units. The sulphur isotope data and style of mineralization for the British terranes are compared with terranes of similar age along strike in Eastern Canada revealing notable correlations.

The Caledonian hinterland of northern Britain comprises mainly Late Proterozoic metasediments and metavolcanics (locally termed Moinian and Dalradian Supergroups) and Lower Palaeozoic (Cambrian to Devonian) sediments, volcanics and intrusives that have been metamorphosed and/or amalgamated during the Caledonian Orogeny. The orogeny resulted from the complex closure of the Iapetus Ocean mostly over the period 470 Ma to 390 Ma. The closure lineament, the Iapetus Suture zone, extends to Newfoundland and mainland Canada in the west and eastern Greenland in the north (Fig. 1). On the basis mainly of key faunal, structural, radiometric and geophysical parameters (e.g. McKerrow & Cocks 1976; Bamford *et al.* 1977; Anderton *et al.* 1979), these Caledonian rocks have been divided into a number of discrete terranes. Six main terranes have been identified, separated by major fault systems (Hutton 1987). Our work relates to four terranes in particular which are best known as (from north–south) the Northern Highland, Grampian, Southern Upland and Lakesman terranes (Fig. 2).

A large dataset in excess of 600 $\delta^{34}S$ analyses now exists for sulphides across these four Caledonian terranes from sulphide disseminations in Caledonian intrusions and their related mineralizations (e.g. Fletcher *et al.* 1989, 1997; Laouar *et al.* 1990; Lowry *et al.* 1995, 1997), base-metal vein, stratiform and stratabound deposits (Willan & Coleman 1983; Pattrick & Russell 1989; Anderson *et al.* 1989; Hall *et al.* 1991; Lowry *et al.* 1991a; Scott *et al.* 1991; Steed & Morris 1997), and pyrite in metasediments (e.g. Hall *et al.* 1987, 1988, 1994a, b). Our analysis of this dataset highlights distinctive isotopic variations, which not only provides further evidence for the delineation of the putative terrane boundaries, but also suggests deep crustal and possibly subcrustal variations across the region.

From: McDonald, I., Boyce, A. J., Butler, I. B., Herrington, R. J. & Polya, D. A. (eds) 2005. *Mineral Deposits and Earth Evolution.* Geological Society, London, Special Publications, **248**, 133–151. 0305-8719/$15.00
© The Geological Society of London 2005.

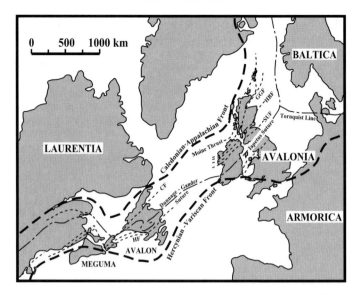

Fig. 1. Pre-Atlantic configuration of the Caledonide orogen and adjacent regions, outlining major terrane boundaries, faults and sutures between the Caledonian–Appalachian and Hercynian–Variscan fronts, based on a three plate model. Information derived from Powell *et al.* (1988; Fig. 1), Stillman (1988; Fig. 1) and Barr *et al.* (1998; Fig. 1). The pre-collisional continents are labelled in upper case with white background, and the proposed suture traces in lower case. The Avalon and Meguma terranes of Canada are marked in upper case. GGF, Great Glen Fault; HBF, Highland Boundary Fault; SUF, Southern Uplands Fault; CF, Cabot Fault; HF, Hermitage Bay Fault.

There are also distinct variations in the tenor of $\delta^{34}S$ across the Iapetus Suture zone, in lithologies of similar age, implying a significant difference in the history of diagenetic sulphide on either side of the palaeo-ocean (Fig. 3).

Although sulphur data provide information mostly about sub-crustal and sedimentation processes, the metals present in the deposits provide additional information about basement rocks contributing to the ore-forming events.

Geology of northern Britain

The small crustal section which represents our study area (Fig. 1) is unique in that it records multiple episodes of crustal growth, basin development and orogenic events dating back beyond 2 Ga, which culminated in tectonic collision related to the closure of the Iapetus Ocean (bringing together Laurentia and Pan-Africa) during the Palaeozoic period. As a result of the complex collision, many terranes are clearly unrelated in terms of their geological histories (Bluck 1990). Below, we describe briefly the main elements of each of the four terranes under discussion.

The dominant psammitic Moinian metasediments of the Northern Highland terrane are intruded by pre-, syn- and post-tectonic

granitoids, mineralization being largely restricted to the post-tectonic ultramafic to monzogranitic complexes. The mineralization consists mostly of $Cu – Pb – Zn \pm Au – Ag – Bi$ veins with fluorite and barite as important gangue at Grudie, Loch Shin and Ratagain (Lowry 1991), the main exception being a large base metal sulphide and barite vein deposit which was emplaced in a dilation zone in country rocks around the northern margin of the Strontian Igneous Complex. The Northern Highland terrane is separated from the high grade Archaean–Early Proterozoic gneisses of the North Atlantic craton and the overlying unmetamorphosed Precambrian and Cambrian sediments by the Moine Thrust zone (Fig. 2).

The Grampian Terrane is separated from the Northern Highland terrane by the NNE-trending Great Glen stike–slip fault, although there are sedimentological similarities in the dominantly amphibolite facies Late Proterozoic (1200–800 Ma) Moinian schists and psammites to the NW with the psammites of the lowermost Grampian Group of the Dalradian Supergroup to the SE. The overlying Dalradian groups consist of a wide range of metasediments and metabasalts (Fig. 2) dating mostly from the 800–500 Ma period (Oliver 2002). The terrane is intruded by plutons that are pre-, syn- and

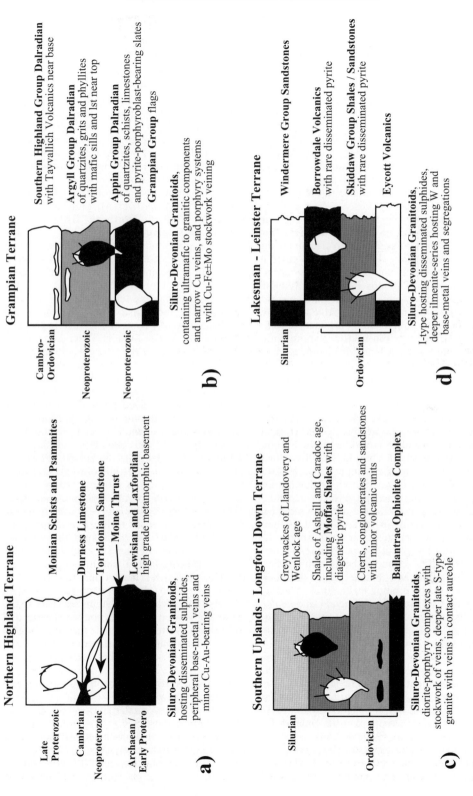

Fig. 2. Schematic vertical sections through the four terranes that were studied in detail. (**a**) Northern Highland terrane and Foreland; (**b**) Grampian terrane; (**c**) Southern Uplands–Longford Down terrane; and (**d**) Lakesman–Leinster terrane.

post-tectonic and ranging in composition from ultramafic to high-silica granites. The Dalradian metamorphic sequence is host to diagenetic sulphides, mostly in fine-grained units such as the Ballachullish Slates (Hall *et al.* 1987), Easdale Slates (Hall *et al.* 1988) and Ardrishaig Phyllites (Hall *et al.* 1994*a*), base-metal vein and stratabound sulphides such as Tyndrum (Scott *et al.* 1991), and stratiform barite deposits (e.g. Aberfeldy, Willan & Coleman 1983; Hall *et al.* 1991). The granitoids are host to disseminated sulphides and porphyry Cu–Mo deposits that are most abundant in the south and west of the terrane (e.g. Tomnadashan and the Kilmelford region, Lowry *et al.* 1995).

The Precambrian–Cambrian Grampian terrane is separated from the Devono-Carboniferous Midland Valley terrane by the Highland Boundary Fault. This in turn is separated from the Southern Uplands terrane by the Southern Uplands Fault. The predominant fault direction is SSW–NNE in the Grampian terrane. South of the Midland Valley a series of subparallel faults trending WSW–ENE divide the major terrane units (Fig. 4). The Southern Uplands terrane is made up of two similar and juxtaposed Siluro-Ordovician units. The base of this sequence in the north is the Ballantrae Ophiolite Complex consisting mainly of dykes, pillow lavas and deep-water sediments (Fig. 2). Overlying Ordovician sediments are mostly fine-grained, anoxic, and intercalated with the products of the gradual cessation of volcanic activity. Black shales contain diagenetic sulphides (Anderson *et al.* 1989). Overlying Silurian sediments are dominantly fine-grained greywackes becoming purer sandstones toward the top. The sequence is intruded by post-tectonic Siluro-Devonian diorite–granite complexes, minor porphyry Cu systems and Devonian S-type granites (Lowry *et al.* 1997). The sediments contain major base metal sulphide and barite veins, the most important of these being at Leadhills (Pattrick & Russell 1987) and Wanlockhead (Anderson *et al.* 1989). Peripheral to the granites there are numerous base metal, copper and arsenic-bearing veins (Lowry *et al.* 1997).

South of the postulated Iapetus suture zone is the Cambro-Silurian Lakesman-Leinster volcano-sedimentary terrane (Figs 2, 4), the southern margin of which lies beneath younger cover in England but is bounded by the metamorphic Rosslare Complex in SE Ireland. The terrane in northern England consists of two Ordovician volcanic groups, the Eycott and Borrowdale, separated by the Ordovician Skiddaw Group of mainly fine-grained

sediments. Overlying the Borrowdale Group is the Windermere Group of Silurian fine sandstones. The Ordovician strata are host to numerous copper and base-metal sulphide veins, with the finer-grained units of the Skiddaw Group containing some occurrences of diagenetic sulphide (Lowry *et al.* 1991*a*). The Ordovician groups are intruded by mostly post-tectonic Siluro-Devonian S- and I-type granites, the former hosting tungsten and base-metal mineral veins, the latter containing disseminations of sulphide and hosting molybdenite-bearing veins and segregations.

During a study of the genesis of Late Caledonian (Siluro-Devonian) granitoid-related mineralization in Northern Britain which focused on sulphur isotopes (Lowry 1991) and an earlier pilot study of the $\delta^{34}S$ of disseminated sulphides in Caledonian granitoids (Laouar *et al.* 1990), it became clear that $\delta^{34}S$ characteristics were not only useful in assessing the source of sulphur and temperatures of mineralization in individual deposits, but also contrasted greatly from one terrane to the next (Lowry *et al.* 1991*b*; Table 1). These data provide further evidence for the positions of the terrane boundaries discussed earlier but also indicate deep crustal and even possibly subcrustal variations across the region (Lowry *et al.* 1992). In order to avoid involvement in the complexities of sulphur isotopes and different granitoid classifications (e.g. S- and I-type: Coleman 1977, 1979; magnetite- and ilmenite-series: Sasaki & Ishihara 1979: Ishihara & Sasaki 1989), the granitoids will be referred to as: subcrustal; I-type (representing an igneous crustal source such as the North Atlantic Craton or subducted slab); or S-type (sedimentary- or metasedimentary-dominated crustal melts). In addition to the granitoid-hosted mineralization the data for bacterial and diagenetic sulphides plus sediment-hosted vein sulphur enlarges the database for each terrane (Fig. 3).

The granitoids provide important $\delta^{34}S$ data (Table 1) because in many terranes the sediment-hosted sulphur is not widespread, and because the various granitoid types in the region provide information on different subcrustal and crustal sulphur sources. These include:

a) subcrustal-sourced gabbros, diorites and porphyry systems;
b) tonalites and monzodiorites contaminated by lower crust;
c) lower crust and subducted slab-sourced I-type monzogranites; and
d) upper crustal-sourced S-type granites which vary in mineralogy and chemistry

Table 1. *Mean δ³⁴S data for granitoids within four Caledonide terranes of northern Britain*

Granitoid	Sulphur species	Mean δ^{34}S	Grouping
Northern Highland terrane			
1 Grudie	sulphides	−10.9‰ (15)	□
	pyrite	−1.6‰ (5)	○
	sulphates	+8.4‰ (8)	
2 Helmsdale	pyrite	−2.6‰ (1)	○
3 Loch Shin	sulphides	−8.3‰ (12)	□
	pyrite	−4.1‰ (5)	○
	sulphates	+10.2‰ (2)	
4 Ratagain	sulphides	−3.9‰ (26)	○
	pyrite	−2.2‰ (14)	
	sulphates	+14.9‰ (6)	
5 Strontian − outer	sulphides	−1.7‰ (6)	○
− inner	sulphides	+7.0‰ (3)	■
Grampian terrane + Irish Dalradian			
6 Aberdeen	pyrite	+8.3‰ (1)	■
7 Ardsheal	sulphides	+2.6‰ (12)	●
	pyrite	+3.8‰ (7)	
8 Arrochar	sulphides	+3.0‰ (18)	●
9 Ballachulish	sulphides	+2.5‰ (12)	●
10 Comrie	sulphides	+4.5‰ (5)	●
11 Etive	sulphides	+2.1‰ (2)	●
12 Galway	pyrite	+3.6‰ (3)	●
13 Garbh Achadh	pyrite	−0.4‰ (5)	⊕
14 Glen Doll	sulphides	+1.4‰ (3)	●
15 Inish	pyrite	+0.6‰ (3)	⊕
16 Kilmelford − porphyry	pyrite	+0.6‰ (24)	⊕
− plutonics	pyrite	+2.8‰ (10)	●
17 Lagalochan	sulphides	+0.3‰ (33)	⊕
18 Oughterard	sulphides	+10.1‰ (7)	■
19 Souter Head	sulphides	+2.5‰ (5)	●
20 Tomnadashan − porphyry		+0.5‰ (28)	⊕
− plutonics		+2.0‰ (4)	●
Southern Uplands terrane			
21 Black Stockarton Moor		+0.2‰ (17)	⊕
22 Cairngarroch Bay		−2.2‰ (8)	○
23 Fleet − margin	sulphides	−1.5‰ (3)	○
24 Fleet − inner	sulphides	−8.6‰ (4)	□
25 Hare Hill	sulphides	−4.2‰	○
Lakesman terrane + Leinster			
26 Dhoon	sulphides	+6.9‰ (2)	■
27 Foxdale	sulphides	+6.4‰ (2)	■
28 Leinster	sulphides	+5.7‰ (47)	■
	sphalerite	+6.6‰ (15)	
	sulphates	+22.0‰ (5)	
29 Shap	sulphides	+1.1‰ (12)	●
30 Skiddaw	sulphides	+8.0‰ (22)	■
31 Weardale	sulphides	+13.8‰ (12)	■

Data derived from Williams & Kennan (1983); Caulfield *et al.* (1987); Laouar *et al.* (1990); Lowry (1991); Lowry *et al.* (1991a, 1995, 1997). The key to the groupings is shown in Fig. 4.

depending on the arkosic or argillaceous nature of the source metasediments.

The granitoids often develop to batholithic proportions, notably the Lake District and Grampian batholiths. In regions of low grade metamorphism such as the Lake District, intrusion of the batholith set up circulation cells and the ascent of fluids as the overlying sediments were dewatered. The resultant mineral veins

therefore have $\delta^{34}S$ representing homogenization of the sedimentary and volcanosedimentary sulphur within that terrane and this feature
is especially notable in the Southern Uplands
and Lakesman terranes (Lowry *et al.* 1991*a*).
The Grampian and Northern Highland terranes
were metamorphosed to a minimum of greenschist facies prior to intrusion of Late
Caledonian granitoids. Granitoid-related
mineral deposits in this region that contain
significant external fluid and sulphur inputs
from the host rocks tend to be located in the
zones of lower metamorphic grade. Notable
mineral veins of intrusion age are rare in the
higher-grade host metasediments due to their
prior dewatering.

Sulphur isotopes

Sulphur isotopes have been used over the last
forty years to understand ore genesis, the source
and processes affecting igneous sulphur, and the
environment of deposition of sediments. This
study of four well characterized and adjacent
terranes in Northern Britain, outlines the variations, assesses the overall importance of crustal
and mantle sulphur, and presents a sulphur
isotope terrane model which further strengthens
existing terrane distinctions within the North
Atlantic Caledonides.

Analytical techniques for $\delta^{34}S$ analysis

The granitoid-related sulphides and sulphates
and most of the sedimentary sulphur reported
here were separated using standard heavy
liquid, magnetic and hand picking techniques.
Where necessary, XRD analyses were
performed to check the purity of the phases.
SO_2 was produced from sulphides and sulphates
at SUERC (Scottish Universities Environmental Research Centre) following the methods
of Robinson & Kusakabe (1975) and Coleman
& Moore (1978) respectively, and analysed on
either Isospec 64 or VG Micromass 602 mass
spectrometers. Standard corrections were
applied to the raw data (e.g. Craig 1957). The
results are reported in standard notation ($\delta^{34}S$)
as per mil (‰) deviations relative to Cañon
Diablo Troilite (CDT). Analytical uncertainty
(1σ) based on repeat analysis of internal
standards and selected samples is better than
$\pm0.2‰$. Data from other studies in the literature
and used here followed similar techniques.

New sulphur isotope analyses of the
Lewisian/Laxfordian basement were carried out
at RHUL (Royal Holloway, University of
London) using the continuous flow technique

outlined in Grassineau *et al.* (2001). Analytical
uncertainty based on repeat analyses of standards is better than $\pm0.1‰$ (1σ), with all samples
replicating to better than $\pm0.15‰$.

Results

Northern Highland terrane

The Northern Highland Terrane contains inliers
of cratonic basement directly beneath the
Moinian succession, so the $\delta^{34}S$ characteristics
of sulphides within the foreland terrane are
also discussed here. The main characteristics
(Table 1, Figs 2–4) are:

1   New analyses of pyrite segregations from
the Lewisian/Laxfordian basement,
sampled in roadcuts between Ullapool and
Scourie and analysed at RHUL, give values
ranging from $-1.4‰$ in a mafic band up to
$+5.5‰$ in granitic gneiss ($+1.7 \pm 1.7‰$, $n =$
13), a similar range seen in subcrustal intrusives of the Grampian terrane. Lower
Proterozoic metasedimentary units of the
Loch Maree Supergroup in the cratonic
sequence of the foreland are host to both
disseminated and stratiform sulphides.
Boyce *et al.* (1988) report values of $-1.0 \pm$
$0.3‰$ for stratiform sulphides hosted by
hornblende and quartz–mica schists, and
values of $-3.2‰$ to $-2.9‰$ for hanging wall
Flowerdale schists. A value of $-2.9‰$ has
been recorded for sulphide in the overlying
Torridonian sediments from the Isle of
Rum (Hulbert *et al.* 1992) and one pyrite
sample hosted by the Cambrian Durness
Limestone records $-21‰$ (Lowry 1991). It
is therefore considered that the foreland
terrane (with the exception of the Durness
Limestone outlier) and its extension
beneath the Northern Highland terrane is
a source of slightly ^{34}S-enriched sulphur in
the range $-3‰$ to $+5‰$.
2   Moinian host rocks have so far yielded only
two pyrite analyses from Morar Division
psammites with values of $+3.4‰$ and $+4.6‰$
(Lowry 1991).
3   Disseminated sulphides in the granitoids
(appinites, diorites, granodiorites and
monzogranites) at Strontian, Ratagain and
Grudie (Laouar *et al.* 1990; Lowry 1991)
have $\delta^{34}S$ values in the range $-4.5‰$ to $0‰$
and are considered to represent predominantly an igneous source, but with varying
degrees of contamination of subcrustal
magmas by the thick cratonic sequence (see
later metals section).

4 Disseminated sulphides from the young inner biotite granite at Strontian have $\delta^{34}S$ values of +5.6‰ to +8.5‰, either representing large scale addition of crust to the magma which formed the outer appinites and granodiorite (Laouar et al. 1990) or suggesting that the granite is the result of melting of upper crustal Moinian metasediments/Lewisian basement (enriched in ^{34}S), intrusion being permitted during movement on the Great Glen Fault (Hutton 1988).

5 Early vein mineralization (normally pyrite) in monzogranites at Ratagain, Grudie, Loch Shin and Helmsdale has $\delta^{34}S$ values of −4.5‰ to 0‰, like the disseminated sulphides in the granitoids, and representing a magmatic sulphur source (Lowry 1991).

6 Later in the vein parageneses at Grudie, Loch Shin and Ratagain the near neutral pH vein fluids underwent significant relative increases in f_{O_2} and significant ^{34}S fractionation between H_2S (precipitating as sulphides) and SO_4^{2-} (precipitating as sulphates) components of the magmatic fluids. This resulted in sulphide $\delta^{34}S$ values decreasing beyond −5‰ and reaching −21‰ for the lowest temperature ($\approx$200 °C) galena at Grudie. Barite co-precipitating with late galena has $\delta^{34}S$ of +7‰ to +11‰. Only at Ratagain, where there was an input of ^{34}S-enriched sulphur from the host Moinian metasediments after a shearing event, is there a second sulphur source. This caused the $\delta^{34}S_{H_2S}$ of the fluid to shift from c. −13‰ prior to shearing to c. −6.5‰ for late galena and celestite. At Grudie and Loch Shin the $\delta^{34}S_{H_2S}$ of the precipitating fluid decreased to −18‰. At all three of these mineral deposits the calculated $\delta^{34}S_{\Sigma S\,(H_2S+SO_4)}$ is close to −1‰ for the main part of the vein parageneses, similar to the early vein pyrite and disseminated pyrite values. The sulphur isotope variations have been modelled for these deposits and are related to the unique way that these generally sulphur-poor mineralizing systems developed from sulphides to sulphates to oxides as the sulphur reservoir was used up and meteoric water entered the mineralizing fluids along Orcadian Basin bounding faults (Lowry 1991).

7 Pb–Zn–Fe–Ba mineralization at the Strontian Mine, outcropping at the Northern Margin of the Strontian granodiorite is much younger and unrelated to the intrusives (Gallagher 1963). Most of the sulphides have values between +3‰ and +12‰ and an upper crustal Moinian or Lewisian basement sulphur source is considered most likely.

From the available data on the Northern Highland terrane only the inner granite of the Strontian Complex and the mineralization at Strontian Mine can be related to a dominant upper crustal sulphur source (Figs 3 & 4). The main body of granitoids and the mineralization that they host represent subcrustal sulphur with varying degrees of contamination by components from the Late Archaean/Early Proterozoic cratonic sequence, the monzogranites possibly representing melts with a cratonic component. This basement has given the granitoids specific characteristics and resulted in the near neutral pH, fluorine-rich fluids and the anomalous fluid and sulphur isotopic evolution found in the mineral veins.

Grampian terrane

The metasediments of the Grampian terrane can be subdivided into three main sections:

1 The oldest Central Highland division that is comparable both in age and lithological features with the Moinian succession of the adjacent Northern Highland terrane.
2 The overlying Grampian Group (the oldest Dalradian) that has many lithological characteristics similar to Moinian rocks.
3 The overlying and strongly variable lithologies of the Dalradian Supergroup.

Sulphur isotope data are not available for either of the first two sequences; this can be attributed partly to their sulphur-deficient arkosic and psammitic nature, and partly to the lack of sampling. Mineralization associated with granites within these sequences is also extremely rare (Laouar 1987). There is a large dataset for the remainder of the Dalradian metasedimentary/metavolcanic sequence and the granites hosted by this sequence.

Disseminations in the Dalradian metamorphic sequence. The $\delta^{34}S$ characteristics of diagenetic/metamorphic disseminations are discussed from the base upward (Table 1, Figs 2–4).

1 Ballachulish Slates of the lower Appin Group contain pyrrhotite and pyrite with $\delta^{34}S$ values ranging from +12.8‰ to +16.3‰ (Hall et al. 1987; Lowry 1991). The suggested environment of deposition is an

anoxic basin (Anderton 1985). The over-lying Appin Quartzites contain pyrite with $\delta^{34}S$ of +12.2‰ to +13.6‰.

2   The middle Dalradian Argyll Group shows the widest $\delta^{34}S$ variations and generally represents the widening, deepening and eventually failed rifting of the Dalradian marginal basin. Near the base of this sequence, exceptional values of +42‰ have been recorded for pyrite in the Bonahaven dolomite (Hall *et al.* 1994*b*), although it is unlikely that the Dalradian seawater ever became so enriched in ^{34}S. Pyrite and pyrrhotite have $\delta^{34}S$ values of +12‰ to +22‰ in the Easdale Slates of the west coast (Hall *et al.* 1988) and in the Ben Eagach Schist which is at a similar strati-graphic level further east (Scott *et al.* 1987, 1991). After a period of basin filling (e.g. Anderton *et al.* 1979), the main period of basin opening and failed rifting commenced with the shelf muds, now represented by the Ardrishaig/Craignish phyllites. These have values of –15‰ to –1‰ (mean ≈–10‰) (Willan & Coleman 1983; Hall *et al.* 1994*a*). Values then become enriched in ^{34}S again, through the Ben Lawers Schist (–4‰ to +4‰, Scott *et al.* 1987) then the Crinan Grit (+1‰ to +8‰; Willan & Coleman 1983; Lowry 1991) and Ben Challum Quartzite units (+8‰ to +15‰, Scott *et al.* 1987).

3   The Southern Highland Group of the Dalradian includes the Tayvallich pillow lavas and associated dyke and sill complex which have $\delta^{34}S$ of 0 ± 2‰ (Lowry 1991). Intercalated with these are the Tayvallich shelf edge and turbiditic limestones. These and the Loch Tay limestone and overlying Ben Ledi Grits further east have $\delta^{34}S$ values between –2‰ and +4‰ (Willan & Coleman 1983; Lowry 1991).

Undifferentiated Dalradian metasediments (limestones and schists) of the Shetland Islands give $\delta^{34}S$ of –8 to +2‰ (Maynard *et al.* 1997), but the presence of volcanic-related sediments suggest that these represent part of the upper Dalradian at or above the level of the Ardr-ishaig phyllites.

Within the Dalradian sequence as a whole, only the Ardrishaig/Craignish phyllite units have values typical of open-system bacterial reduction (fractionation of 35‰–50‰ from seawater sulphate values). The other units are dominated by fractionations relative to seawater of 10‰–30‰, representing closed- or semi-closed-system bacterial reduction.

Synsedimentary mineralization horizons in the Middle Dalradian. The most important examples of this type are the Aberfeldy barite mineralization occurring within the Ben Eagach Schist formation, and the smaller Ba–Pb–Zn Loch Lyon horizon. Sulphide $\delta^{34}S$ values range from +16‰ to +26‰ and barite values from +27‰ to +40‰ (Willan & Coleman 1983; Hall *et al.* 1987; Scott *et al.* 1991), the sources being sediments and hydrothermal sulphur for the sulphides and seawater sulphate for the barite.

Vein mineralization hosted by the Middle Dal-radian. Devonian (380–400 Ma) Au-bearing veins at Tyndrum and Cononish have $\delta^{34}S$ values of –2‰ to +14‰ and are thought to have a sulphur and fluid input from Late Caledonian granitoids (Pattrick *et al.* 1988; Curtis *et al.* 1993; Lowry *et al.* 1995). Lower Carboniferous Pb–Zn–Ba veins at Tyndrum and Castletown have $\delta^{34}S$ values for sulphides of +3‰ to +13‰, the barite having values of +13‰ to +17‰ (Pattrick *et al.* 1983, 1988). The sources are thought to be the Dalradian sequence for the sulphides and a continental groundwater source for the sulphates (Pattrick & Russell 1989).

Intrusive-hosted mineralization. The mineral-ization of this type can be divided into five categories based on intrusive type and $\delta^{34}S$ data.

1   Sulphides related to the Ordovician Huntly-Knock gabbroic intrusions of Aberdeenshire have $\delta^{34}S$ values ranging from –2‰ to +6‰, the values above +3‰ representing contamination with Dalradian sedimentary sulphur (Fletcher *et al.* 1989). Further to the east the Arthrath mafic intrusion has also incorporated crustal sulphur. The intrusive-hosted sulphides have $\delta^{34}S$ of –3 to 0‰ and surrounding hornfelsed metasediments –12 to –7‰ (Fletcher *et al.* 1997), which is unusual for the Dalradian but similar to the Ardrishaig phyllite unit (Hall *et al.* 1994*a*).

2   Appinite pipes from the Ardsheal Penin-sula, Cruachan Cruinn, Garabal Hill and Arrochar also have values in the range –2‰ to +6‰. Values heavier than +3‰ are represented by strong enrichment in disseminated sulphides at the margins of the Ardsheal pipes, indicating the incorpo-ration of local sedimentary sulphur (Lowry 1991; Lowry *et al.* 1995).

3   Late Caledonian (435–395 Ma) porphyry systems at Lagalochan, Kilmelford, Garbh Achadh and Tomnadashan have $\delta^{34}S$ values restricted to the range –3‰ to +3‰ (mean

+0.4‰, $n = 136$) with all but four analyses in the range –2‰ to +2‰. Again the source of the sulphur is thought to be subcrustal magmatic, but modification from the sulphur source value has been minimal due to rapid ascent and cooling and pyrite precipitation above 400 °C (Lowry et al. 1995).

4   Late Caledonian plutonic systems containing disseminated sulphides and late stage quartz–sulphide veins with pyrite–chalcopyrite ± molybdenite have $\delta^{34}S$ values of –1‰ to +5‰, with rare values indicative of crustal inputs up to +7‰ (mean +2.6‰, $n = 68$) (Laouar et al. 1990; Lowry 1991; Lowry et al. 1995). Despite the general small enrichment in ^{34}S, these granitoids of dioritic to monzogranitic affinity (e.g. Ballachulish, Etive, Arrochar, Kilmelford, Glen Doll) are thought to contain solely subcrustal sulphur and that oxidation and fractionation of the magmatic sulphur during slow cooling are the main causes of the c. 2‰ enrichment relative to the porphyry systems (Lowry et al. 1995).

5   The older granites (Ordovician) which are thought to have a metasedimentary crustal magma source have values of +8.3‰ for the Aberdeen granite and up to +12‰ for Dalradian-hosted granites of western Ireland (Laouar et al. 1990).

The Unst ophiolite in the Shetland Islands contains sulphides having $\delta^{34}S$ of –1.6 to +6.3‰, with the exception of mineralization associated with pegmatites and plagiogranites at Mu Ness. The ophiolite amalgamated with the Dalradian at approximately 480 Ma. Although possibly representing an exotic terrane, the values suggest incorporation of Dalradian sulphur in serpentinites and talc close to the basal thrust, with earlier seawater alteration in the higher levels of the ophiolite (Maynard et al. 1997).

Overall the Grampian terrane is a region enriched in ^{34}S (Table 1, Figs 3 & 4). The influence of the metasedimentary hosts on the $\delta^{34}S$ of the intruding subcrustal granites as a whole is low and only those granites with a large sedimentary component are strongly enriched in ^{34}S (>+7‰).

Southern Upland terrane

The main sulphur isotopic characteristics of this terrane (Table 1, Figs 2–4) are:

1   Ordovician diagenetic pyrite in the Moffat Shale Unit with $\delta^{34}S$ of –17‰ to 0‰ (mean

–8.4‰), representing open system reduction of Ordovician seawater sulphate (+28‰) to H_2S and then pyrite (Anderson et al. 1989). Values of –16 to –15‰ are recorded for similar black shale units in the Clontibret area of NE Ireland (Steed & Morris 1997).

2   Subcrustal or I-type magmatic sulphur with $\delta^{34}S$ of –1‰ to +3‰ with values of –5‰ to +2‰ for veins associated with these granites e.g. porphyry systems at Black Stockarton Moor and Cairngarroch Bay (mean –0.6‰, Lowry 1991; Lowry et al. 1997). An antimony-rich vein in the Hare Hill granodiorite with a mean of –4.2‰ (Caulfield et al. 1987; Naden & Caulfield 1989) represents a variable input of sulphur from the intruded sediments, as do antimony-rich veins at Clontibret and Glendinning where vein stibnite has $\delta^{34}S$ means of –3.4 and –3.9‰, respectively (Duller et al. 1997), despite there being no visible evidence of intrusives at either of these deposits.

3   S-type magmatic sulphides in the central regions of the Fleet biotite–muscovite granite which formed within the 2 m wide Orchars Vein have $\delta^{34}S$ values of –12‰ to –4‰ (mean –8.6‰, Lowry et al. 1997).

4   Lower Carboniferous Pb–Zn-rich vein systems hosted by the Lower Palaeozoic sediments at Wanlockhead and Leadhills (and additionally the Salterstown mineralization in NE Ireland) which have $\delta^{34}S$ of –10‰ to –3‰ (mean –7.1‰) (Anderson et al. 1989; Pattrick & Russell 1989).

The Southern Uplands terrane and its continuation into Longford Down, Ireland, therefore represents a ^{34}S-depleted terrane (Fig. 3), a characteristic inherited by central parts of the Fleet granite.

Lakesman terrane

The main sulphur isotopic characteristics of this terrane (Lowry et al. 1991a; Table 1, Figs 2–4) are:

1   Ordovician Skiddaw Group (and similarly the Ribband group of SE Ireland), dominantly argillaceous sediments, which accumulated in a basin that was closed to, or only periodically recharged with, Ordovician seawater sulphate (+28‰, Claypool et al. 1980), resulting in diagenetic pyrite $\delta^{34}S$ values of +11‰ to +28‰ (mean +18.5‰) formed by closed system bacterial reduction of SO_4 (Lowry et al. 1991a).

2 I-type or subcrustal magmatic sulphur (Shap granite) having $\delta^{34}S$ values of $-1‰$ to $+3‰$ (mean $+1.3‰$), like those of the Southern Uplands Terrane (Laouar *et al.* 1990; Lowry *et al.* 1991*a*).

3 Sulphur in mineral deposits hosted by S-type or heavily crustally contaminated granites with $\delta^{34}S$ of $+5‰$ to $+10‰$ (mean $+7.6‰$), and thus largely reflecting the ^{34}S-enrichment of the host sequence (Solomon *et al.* 1971; Williams & Kennan 1983; Gallacher 1989; Lowry 1991). This includes granites at Weardale, Skiddaw, Foxdale and Leinster (SE Ireland).

4 Devonian and Carboniferous veins hosted by the Ordovician volcanosedimentary sequence with $\delta^{34}S$ of $+13‰$ to $+23‰$ (mean $+18.8‰$) representing partially homogenized sulphur from a Skiddaw Group (or similar) source (Pattrick & Russell 1989; Lowry *et al.* 1991*a*).

The Lakesman terrane and its continuation into Leinster, SE Ireland, therefore represents a ^{34}S-enriched sequence, a characteristic which is, at least in some part, inherited by most of the intruding granites within the terrane (Fig. 4). The intruding granites also acted as the heat source that initiated dewatering and vein formation in the overlying portions of the volcanosedimentary sequence (Lowry *et al.* 1991*a*).

Discussion of sulphur isotope data

Sulphur isotope analysis of widespread mineralization in Northern Britain, hosted by sedimentary, volcanic, intrusive and metamorphic rocks, reveals a wide range of $\delta^{34}S$ values from $-22‰$ to $+42‰$ (Fig. 3). The data can be grouped and many distinctions made.

Firstly there is great contrast between the Lower Palaeozoic Lakesman–Leinster and the adjacent Southern Uplands–Longford Down terranes, which represent sources of ^{34}S-enriched and ^{34}S-depleted sulphur respectively. These values can be related to different modes of bacterial seawater sulphur reduction either side of the Iapetus Suture (Lowry *et al.* 1991*a*). The Dalradian and Moinian successions of the Grampian and Northern Highland terranes respectively, are both dominated by ^{34}S-enriched metasedimentary sequences. The causes of variations in sedimentary sulphide isotopes over the Neoproterozoic to Silurian period in the Caledonides are referred to by Lowry *et al.* (2003) and will be discussed in more detail elsewhere as they are contentious and

may be related to ocean anoxia, oxygenation of the atmosphere, climate change or the commonly cited open- or closed-system reduction.

Vein and stratiform mineralization that did not form under direct influence of Late Caledonian intrusive rocks show $\delta^{34}S$ signatures representative of homogenized sulphur sources within these upper crustal units. Where the granitoid-related mineralization falls out of the range which could represent wholly subcrustal sulphur ($-3‰$ to $+3‰$, Ohmoto 1986), the deviations from these values are normally toward the $\delta^{34}S$ values typical of the upper crustal host rocks and so could therefore represent either crustal contamination at any time from magma ascent to circulation of the mineralizing fluid, or they could solely represent upper crustal melts. The deviations are toward ^{34}S-enrichment in the Lakesman and Grampian terranes and ^{34}S-depletion in the Southern Uplands terrane.

The story is slightly different for much of the Northern Highland terrane. This is due to the presence of thick cratonic basement beneath what is in some parts of the terrane a relatively thin veneer of Moinian upper crustal metasediments. These systems have low sulphur contents and have higher and relatively increasing f_{O_2} and pH resulting in primary sulphides with a slight depletion in ^{34}S. These represent near stability boundary conditions for H_2S and small changes in fluid chemistry drive the system toward sulphate precipitation. This cratonic basement input is emphasized by the presence of very high Ba, inherited zircons older than 1500 Ma in host granitoids (Pidgeon & Aftalion 1978) and general similarities with oxidized Archaean magmas. Gold-bearing oxidized hydrothermal fluids emanating from Archaean plutons, such as those in Ontario and Western Australia, contain sulphides as light as $-17‰$ and magmatic sulphate between $+6$ and $+14‰$ (Cameron & Hattori 1985, 1987; Hattori & Cameron 1986). Outside of Archaean and fringing terranes, barite and celestite are rare as magmatic sulphates (Hattori 1989) and the Northern Highlands may be the first Palaeozoic example of such mineralization. All these features point to a basement control on the metals and sulphur behaviour in the granitoids of this terrane.

Notable continental basement is absent beneath most, if not all, of the Lakesman and Southern Uplands terranes (Bamford *et al.* 1977). The large number of relatively primitive subcrustal magmas in the south and SE part of the Grampian terrane and the presence of

δ^{34}S (‰)

Scale: −25 −20 −15 −10 −5 0 +5 +10 +15 +20 +25 +30 +35 +40 +45

Terrane	Host Lithology	Mineralization Style	N
NORTHERN HIGHLAND TERRANE AND FORELAND	Appinite / diorite / granodiorite	Disseminated py-po	13
	Strontian biotite granite	Disseminations	4
	Monzonite-related	Vein Pb-Zn-Fe-Cu-Ba	46 - sulphide / 16 - sulphate
	Lewisian / Laxfordian	Disseminations +	13
	Loch Maree supracrustals	Stratiform + disseminations	
	Moinian metasediments	Disseminations + veinlets	2
	Durness Limestone	Disseminated py	1
GRAMPIAN TERRANE	Ultramafic / mafic intrusions	Disseminated and matrix	51
	Unst ophiolite (exotic?)	Whole rock +	53
	Porphyry-related	Cu±Sb±As veins +	136
	Magnetite-series granitoids	Cu-Fe±Mo veins + dissem.	68
	S-type granites (incl. Connemara)	Disseminated	10
	Dalradian metamorphic sequence	Vein, stratiform and stratabound	161 - sulphide / 33 - sulphate
	Dalradian metasediments	Disseminated (diagenetic + metamorphic)	90
SOUTHERN UPLANDS TERRANE	Porphyry-related	Cu-Fe veinlets + dissem.	25
	Fleet granite (S-type)	Veins	7
	Greywacke, shale	Au-Sb-As veins +	36
	The Knipe granodiorite	Sb vein	
	Greywacke, shale	Pb-Zn veins (e.g. Leadhills)	27
	Moffat shales, Clontibret shales	Disseminated	12
LAKESMAN TERRANE	Shap granite	Disseminations + veinlets	11
	Ilmenite-series gts (Weardale, Skiddaw, Foxdale, Leinster I+III)	Disseminations + veins	45
	Lower Palaeozoic-sediments and volcanics	i) Cu-Fe-As veins / ii) Pb-Zn veins	33 / 11
	Skiddaw Group sediments	Disseminated	10

Fig. 3. Summary bar charts of δ^{34}S for Palaeozoic and older mineralization in Caledonide terranes of northern Britain. Note the striking variation between the Southern Uplands terrane and the Lakesman terrane (updated after Lowry 1991). py, pyrite; po, pyrrhotite; gts, granites.

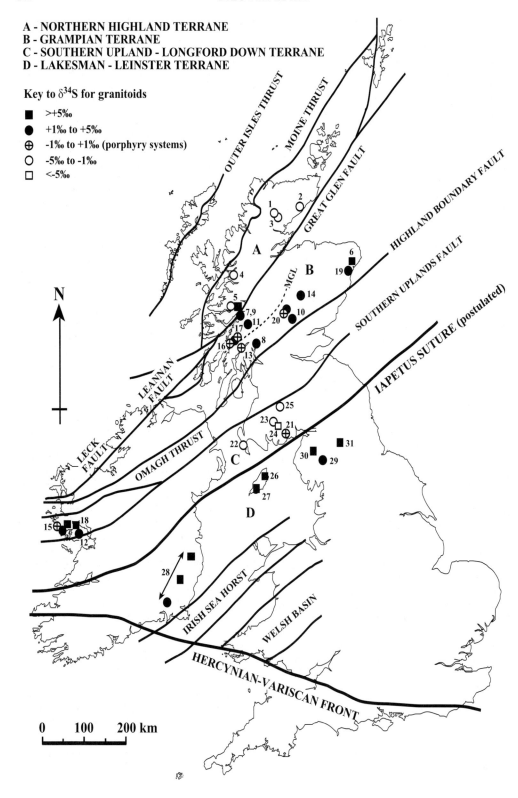

A - NORTHERN HIGHLAND TERRANE
B - GRAMPIAN TERRANE
C - SOUTHERN UPLAND - LONGFORD DOWN TERRANE
D - LAKESMAN - LEINSTER TERRANE

Key to $\delta^{34}S$ for granitoids

■   >+5‰
●   +1‰ to +5‰
⊕   -1‰ to +1‰ (porphyry systems)
○   -5‰ to -1‰
□   <-5‰

N

OUTER ISLES THRUST

MOINE THRUST

GREAT GLEN FAULT

HIGHLAND BOUNDARY FAULT

SOUTHERN UPLANDS FAULT

IAPETUS SUTURE (postulated)

A

B

MGL

C

D

LEANNAN FAULT

LECK FAULT

OMAGH THRUST

IRISH SEA HORST

WELSH BASIN

HERCYNIAN-VARISCAN FRONT

0   100   200 km

porphyry Cu deposits, together with the presence of the Dalradian rift volcanics in the centre of the terrane, suggests that basement was thin to absent SE of the Mid Grampian Line (which roughly correlates with the NW margin of the rift sequence; Fig. 4). $\delta^{34}S$ data for granitoids and host rocks NW of this line in the Grampian terrane are restricted to the area of Dalradian rocks in western Argyll to the north of the Kilmelford/Lagalochan intrusive centres. This region is represented by intrusive complexes such as Ballachulish and Etive that have central monzogranites and Mo-deposits. These complexes are thought to contain cratonic components (e.g. Clayburn et al. 1983; Stephens & Halliday 1984). The $\delta^{34}S$ values though are no different from other parts of the Grampian Terrane, as the Lewisian basement still has $\delta^{34}S$ close to 0‰ (average +2‰). A detailed investigation of sulphur isotopes in granitoids and mineralization hosted by the Central Highland Division and Grampian Group metasediments of the Grampian Terrane may give further information about the inputs from a lower crustal sulphur reservoir.

In a similar way, detailed sulphur isotope investigations may assist in resolving poorly-constrained terrane boundary locations and also, have the potential to constrain the position of the Iapetus Suture as it extends westwards through Ireland (probably directly beneath the Navan orebody; see Anderson et al. 1989) and on into Canada (Fig. 1). $\delta^{34}S$ data for Lower Palaeozoic sequences in North Wales (Bottrell & Spiro 1988; Miller et al. 1992), Shropshire (Pattrick & Bowell 1991) and Cornwall (Jackson et al. 1989) suggest that the technique could also delineate more terranes in the British Isles (see following sections).

In addition to its importance in terrane discrimination the $\delta^{34}S$ dataset can also be added to other chemical, isotopic and mineralogical datasets to aid the understanding of the sources and genesis of intrusive magmas and mineralizing fluids. The $\delta^{34}S$ data for Late Caledonian granitoid-related mineralization in Northern Britain has been combined with metal, fluid and mineralogical characteristics to provide additional evidence for different basement components and/or formational environments.

Constraints from metals, fluids and styles of mineralization

Russell (1985) and Plant et al. (1997) have linked the distribution patterns of certain metals to enrichments in basement and crustal sequences. Barium, enriched in the Lewisian/Laxfordian basement, is strongly enriched in the overlying Moinian psammites derived from this basement. Barium is a key component of the granitoid-related mineral deposits hosted by the Moinian rocks. Molybdenum is thought to be scavenged from the basement by ascending melts and is predominant in granitoid-related deposits NW of the Mid Grampian line, which marks the edge of the cratonic basement (Halliday 1984), also confirmed by very negative ϵNd values for the intrusives (< -5, Halliday et al. 1985). These are often accompanied by minor gold and copper showings where mafic assemblages were incorporated into the monzogranitic fluid/melt systems.

In terms of metals, granites and basement involvement, the Dalradian terrane SE of the Mid Grampian line has much in common with the Southern Uplands terrane (Stephens & Halliday 1984). In the intrusive-related deposits, the high level magmas are porphyrytic rather than monzogranitic, Cu minerals dominate over Mo and phases bearing As and Sb also occur (tetrahedrite–tennantite, stibnite) reflecting in part a thicker upper crustal sediment-dominated sequence over oceanic crust. Arsenopyrite is also present in sediment-hosted vein systems in the vicinity of granitoids. In general the levels of As, B and Cu are very low in metasediments north of the Mid-Grampian line (Plant et al. 1997), hence so are levels of arsenopyrite, tourmaline and chalcopyrite in granitoid-related mineralization.

In the Lakesman terrane, W and As are key indicators of a sediment-sourced component in granite-related mineralization, with Cu and As in Devonian vein mineralization related to high temperature deep fluid circulation through the Ordovician slates and volcanics at the time of granite emplacement (Lowry et al. 1991a). Additionally some granitoids contain molybdenite (Shap, south Leinster). There are also shifts in mineralization styles further south through the North Wales copper belt (porphyry and sediment-hosted veins) before Sn becomes the key indicator metal for mineralization associated with the S-type Cornubian batholith.

Fig. 4. Sulphur isotope characteristics of granitoid-hosted mineralization in Caledonian terranes of northern Britain derived from Table 1. The locality numbers correspond to those given in Table 1. MGL is the Mid-Grampian Line (after Halliday 1984).

The metals present in the granitoid-related mineral deposits are therefore useful for distinguishing magmatic from external components, but are of less use in terrane distinctions unless there are differing basement and magma source components across the terrane boundaries. For example, metals such as As, Pb and Zn are characteristic of mineralization in both the Southern Uplands and Lakesman terranes and represent a source in the sedimentary hosts, whereas $\delta^{34}S$ characteristics distinguish between the two terranes. The metal assemblage may indicate the likely presence of cratonic components in the granitoid-related mineralizing systems, the presence of significant Mo, Bi and Ag-bearing minerals in the absence of notable As-bearing minerals tending to be typical of the cratonic component as seen in the Grampian terrane.

The characterization of fluids in the mineral deposits provides much information about ore forming events and paragenesis. This can distinguish between magmatic, sedimentary, and surface water inputs but provides little evidence for use in terrane discrimination. One aspect that can augment the S-isotope data is the identification of surficial fluids in the mineralizing systems; the identification of basinal brines or meteoric water above or within a terrane at the time of mineralization possibly providing diagnostic information. Most notable of these signatures are those from Carboniferous seawater in the Lakesman terrane (Lowry et al. 1991) and Devonian modified meteoric water in the Northern Highlands terrane (Lowry 1991), but unless these events pre-date terrane amalgamation they will not be diagnostic.

Sulphur and metal correlations along strike to the Appalachians of Eastern Canada

Russell (1985) considered that Scotland was part of a metallogenic subprovince also including East Greenland, the northern part of Ireland and the eastern fringes of North America. The correlation has already been demonstrated into Ireland for sulphur isotopes, intrusive styles and metals present, but also exists into Newfoundland, Nova Scotia and New Brunswick. Correlations are shown between the terranes of Great Britain and Ireland and eastern Canada in Table 2. The correlations are clear despite the terranes being generally more fragmented (like NW Ireland) and sometimes less distinct than in Britain.

The equivalents of the Moinian and

Dalradian rocks, the Piedmont, Fleur de Lys and Blair River zones have a paucity of published $\delta^{34}S$ values to date but are host to polymetallic veins bearing Cu and Au. The Notre Dame and Exploits subzones of the Dunnage zone contain porphyry Cu and ophiolitic Cu and are equivalents to the Southern Uplands of Scotland. Both contain sediments and vein systems with sulphides depleted in ^{34}S (down to –17‰) (Shelton & Rye 1982; Anderson et al. 1989) and represent open-system bacterial reduction.

South of the proposed Iapetus suture in Canada, the Gander/Aspy zone is the equivalent of the Lakesman/Leinster terrane with granitoid-related W – Mo ± Sb mineralization and Cu – As – Sb – Pb – Zn-bearing veins. The host sediments and mineral veins contain sulphur enriched in ^{34}S from +7 to +28‰ (Lusk & Crockett 1969; Goodfellow & Peter 1996; Lowry et al. 1991a) due to formation in small basins in irregular island arc settings. South of these is a narrow zone of higher grade metamorphism, perhaps the root zone of a late Proterozoic arc system, known as the Bras d'Or terrane in Canada, the Rosslare Complex at the SE tip of Ireland and extending to NW Wales and Anglesey. This terrane is characterized by granitoid–porphyry Cu mineralization dominated by a magmatic sulphur component. The Avalon/Mira terrane of southern Cape Breton Island, southern Newfoundland and central Wales are not clearly distinct in terms of styles of mineralization. Cu veins predominate, particularly in relation to volcanic massifs and there is some Mo and Sn related to minor granites, but thick sequences of Late Precambrian to Silurian sediments dominate. $\delta^{34}S$ of sulphides have a wide range from –21 to +52‰ in sediments and from –24 to +22‰ in sediment-hosted vein mineralization (Bottrell & Spiro 1988; Bottrell & Morton 1992; Miller et al. 1992; Strauss et al. 1992; Haggerty & Bottrell 1997).

South of Avalon is the Variscan front, behind which are the Meguma and Cornubian S-type batholiths and major tin deposits. In Cornubia the batholith is intruding Middle Devonian–Lower Carboniferous sediments. The pre-Variscan metasedimentary sulphur is significantly depleted in ^{34}S (–34 to –26‰; Clayton & Spiro 2000) and veins hosted by these contain sulphides with $\delta^{34}S$ of –15 to +2‰ (Jackson et al. 1989; Clayton & Spiro 2000), reflecting mixing between sedimentary and igneous sulphur sources. In Nova Scotia the Variscan front tracks further north and the South Mountain batholith is emplaced into the Cambro-Ordovician Meguma Group (Avalonian) shales and greywackes that have a

Table 2. *Comparison of $\delta^{34}S$ and granitoid-related mineralization styles between the British terranes and those along strike in Eastern Canada*

Canadian terranes	Mineralization styles	Sulphur isotopes	British/Irish terranes	Mineralization styles	Sulphur isotopes
Appalachian Front			Caledonian Front		
Humber			Lewisian Foreland	Minor stratiform Cu – Zn	–3 to +6 ‰
Blair River Complex			Moinian	Granitoid Pb – Zn – Ba – Fl	+3 to +4 ‰
Piedmont/Fleur de Lys	Polymetallic veins Cu – Fe – Zn – Pb – Au – Ag		Dalradian	Polymetallic veins Cu – Fe – Zn – Pb – Au – Ag	–15 to +50 ‰
				Porphyry Cu – Mo – Au	
Dunnage a) Notre Dame b) Exploits	Ophiolite Cu Porphyry Cu	–17 to +8 ‰	Southern Uplands a) N b) S	Granitoid Cu – Sb – As Porphyry Cu	–17 to 0 ‰
Iapetus Suture					
Gander/Aspy	Granitoid W–Mo ± Sn Zn – Cu – Pb – Sb veins	+7 to +26 ‰	Lakesman/Leinster	Granitoid W – Mo ± Sn Cu – As – Pb – Zn veins	+11 to +28 ‰
Bras d'Or	Granitoid-porphyry-Cu		Anglesey/Rosslare	Porphyry-Cu	
Avalon/Mira	Mo ± Sn	–20 to +53 ‰	Avalon (incl. Welsh Basin)	Few granitoids, Cu veins	–19 to +21 ‰
Hercynian/Variscan Front					
Meguma (Ordovician lithologies)	Sn ± W	–4 to +26 ‰	Cornubia (Devonian lithologies)	Sn province	–13 to +2‰

Appalachian mineral localities derived from Chatterjee & Strong (1985). All sulphur isotope values reported as $\delta^{34}S_{CDT}$.

dominant positive tenor (–4 to +26‰; Poulsen *et al.* 1991). In this instance the metals and granitoids define the Variscan activity, whereas the sulphur in the sediments records evidence of the Lower Palaeozoic terrane.

Conclusions

A large database is now available for δ^{34}S and associated isotopic, geochemical and mineralogical characteristics of disseminated porphyry-style and vein mineralization associated with Caledonian granitoids in Northern Britain. This can be used in conjunction with data available for host lithologies to provide an alternative route for distinguishing terranes in this, and possibly other regions, particularly those that contain no fossil evidence. The dominant crustal units in each terrane, whether they are upper crustal sediments, volcanics or metamorphics, or thick cratonic basement, provide the main alternative source of melts other than the mantle or subducted oceanic crust and also act as the main contaminant for subcrustal melts. The δ^{34}S data shows that in the four terranes under discussion the values for granitoid-related mineralization are either typical of subcrustal melts (–3‰ to +3‰) or deviate toward the values of the major crustal units in the terrane. The deviations are toward ^{34}S depletion in the Northern Highland terrane (cratonic influence) and Southern Uplands terrane (Lower Palaeozoic greywackes) and ^{34}S enrichment in the Grampian terrane (metamorphosed marginal basin sediments and volcanics) and Lakesman terrane (arc basin sediments). In all four terranes the vein systems outside the influence of magmatic sulphur have δ^{34}S values representing homogenized sulphur from local upper crustal sediments.

The δ^{34}S tenors are not only useful in distinguishing major terrane units and boundaries but may also provide information about the tectonic and climatic environments and seawater chemistry at the time of sediment formation (Lowry *et al.* 2003). Although these distinctions are marked clearly in Northern Britain by major fault-bounded terranes that have distinct metamorphic, tectonic and faunal characteristics, they also provide a basis for additional clarification of the dissected North Atlantic Caledonide terranes from Canada, through Ireland and Scotland to Greenland and Svalbard. Granitoid types and metal assemblages in the mineralization also confirm these discriminations.

DL wishes to thank St Andrews University for the provision of a research studentship during which much of this research was carried out. SUERC is supported by NERC and the Scottish Universities. AJB is funded by NERC support of the Isotope Community Support facility at SUERC. Clive Rice and Gus Gunn are thanked for their constructive reviews of the manuscript.

References

ANDERSON, I.K., ANDREW, C.J., ASHTON, J.H., BOYCE, A.J., CAULFIELD, J.B.D., FALLICK, A.E. & RUSSELL, M.J. 1989. Preliminary sulphur isotope data of diagenetic and vein sulphides in the Lower Palaeozoic strata of Ireland and southern Scotland: implications for Zn + Pb + Ba mineralization. *Journal of the Geological Society, London*, **146**, 715–720.

ANDERTON, R. 1985. Sedimentation and tectonics in the Scottish Dalradian. *Scottish Journal of Geology*, **21**, 513–545.

ANDERTON, R., BRIDGES, P.H., LEEDER, M.R. & SELLWOOD, B.W. 1979. *A Dynamic Stratigraphy of the British Isles*. George Allen & Unwin, London, 301pp.

BAMFORD, D., NUNN, K., PRODEHL, C. & JACOBS, B. 1977. LISPB-III. Upper crustal structure of Northern Britain. *Journal of the Geological Society, London*, **133**, 481–488.

BARR, S.M., RAESIDE, R.P. & WHITE, C.E. 1998. Geological correlations between Cape Breton Island and Newfoundland, northern Appalachian orogen. *Canadian Journal of Earth Sciences*, **35**, 1252–1270.

BLUCK, B.J. 1990. Terrane provenance and amalgamation – examples from the Caledonides. *Philosophical Transactions of the Royal Society of London, Series A*, **331**, 599–609.

BOTTRELL, S.H. & MORTON, M.D.B. 1992. A reinterpretation of the genesis of the Cae Coch pyrite deposit, North Wales. *Journal of the Geological Society, London,* **149**, 581–584.

BOTTRELL, S.H. & SPIRO, B. 1988. A stable isotope study of black shale-hosted mineralization in the Dolgellau Gold Belt, North Wales. *Journal of the Geological Society, London,* **145**, 941–949.

BOYCE, A.J., FALLICK, A.E. & RICE, C.M. 1988. A sulphur isotope study of the Lower Proterozoic stratiform sulphides of Gairloch, NW Scotland (Abstract). *In: Oil and Ore.* Mineral Deposits Studies Group A.G.M., Royal Holloway and Bedford New College.

CAMERON, E.M. & HATTORI, K. 1985. The Hemlo gold deposit, Ontario: A geochemical and isotopic study. *Geochimica et Cosmochimica Acta*, **29**, 2041–2050.

CAMERON, E.M. & HATTORI, K. 1987. Archaean gold mineralization and oxidized hydrothermal fluids. *Economic Geology*, **82**, 1177–1191.

CAULFIELD, J.B.D., BOYCE, A.J. & FALLICK, A.E. 1987. A reconnaissance sulphur isotope study of mineralization in the Southern Uplands of Scotland (Abstract). *In: Oil and Ore.* Mineral Deposits Studies Group AGM, Royal Holloway and Bedford New College.

CHATTERJEE, A.K. & STRONG, D.F. 1985. Review of some chemical and mineralogical characteristics of granitoid rocks hosting Sn, W, U, Mo deposits in Newfoundland and Nova Scotia. *In: High heat production (HHP) granites, hydrothermal circulation and ore genesis.* Institution of Mining and Metallurgy, London, 489–516.

CLAYBURN, J.A.P., HARMON, R.S., PANKHURST, R.J. & BROWN, J.F. 1983. Sr, O, and Pb isotope evidence for origin and evolution of Etive Igneous Complex, Scotland. *Nature*, **303**, 492–497.

CLAYPOOL, G.E., HOLSER, W.T., KAPLAN, I.R., SAKAI, H. & ZAK, I. 1980. The age curves of sulphur and oxygen isotopes in marine sulphate and their mutual interpretation. *Chemical Geology*, **28**, 199–260.

CLAYTON, R.E. & SPIRO, B. 2000. Sulphur, carbon and oxygen isotope studies of early Variscan mineralization and Pb–Sb vein deposits in the Cornubian orefield: implications for the scale of fluid movements during Variscan deformation. *Mineralium Deposita*, **35**, 315–331.

COLEMAN, M.L. 1977. Sulphur isotopes in petrology. *Journal of the Geological Society, London*, **133**, 593–608.

COLEMAN, M.L. 1979. Isotopic analysis of trace sulphur from some S- and I-type granites: heredity or environment? *In:* ATHERTON, M.P. & TARNEY, J. (eds) *Origin of Granite Batholiths – Geochemical Evidence.* Shiva, Nantwich, 129–133.

COLEMAN, M.L. & MOORE, M.P. 1978. Direct reduction of sulphates to sulphur dioxide for isotopic analysis. *Analytical Chemistry*, **50**, 1594–1595.

CRAIG, H. 1957. Isotope standards for carbon and oxygen and correction factors for mass-spectrometric analysis of carbon dioxide. *Geochimica et Cosmochimica Acta*, **12**, 133–149.

CURTIS, S.F., PATTRICK, R.A.D., JENKIN, G.R.T., FALLICK, A.E., BOYCE, A.J. & TREAGUS, J.E. 1993. Fluid inclusion and stable isotope study of fault-related mineralization in the Tyndrum area, Scotland. *Transactions of the Institution of Mining and Metallurgy, Section B*, **102**, B39–47.

DULLER, P.R., GALLAGHER, M.J., HALL, A.J. & RUSSELL, M.J. 1997. Glendinning deposit – an example of turbidite-hosted arsenic-antimony-gold mineralization in the Southern Uplands, Scotland. *Transactions of the Institution of Mining and Metallurgy, Section B*, **106**, 119–134.

FLETCHER, T.A., BOYCE, A.J. & FALLICK, A.E. 1989. A sulphur isotope study of Ni–Cu mineralization in the Huntly-Knock Caledonian mafic and ultramafic intrusions of NE Scotland. *Journal of the Geological Society, London*, **146**, 675–684.

FLETCHER, T.A., BOYCE, A.J., FALLICK, A.E., RICE, C.M. & KAY, R.L.F. 1997. Geology and stable isotope study of Arthrath mafic intrusion and Ni–Cu mineralization, northeast Scotland. *Transactions of the Institution of Mining and Metallurgy, Section B*, **106**, 169–178.

GALLACHER, V. 1989. Geological and isotope studies of microtonalite-hosted W–Sn mineralization in SE Ireland. *Mineralium Deposita*, **24**, 19–28.

GALLAGHER, M.J. 1963. Lamprophyre dykes from Argyll. *Mineralogical Magazine*, **27**, 415–430.

GOODFELLOW, W.D. & PETER, J.M. 1996. Sulphur isotope composition of the Brunswick No. 12 massive sulphide deposit, Bathurst Mining Camp, New Brunswick: implications for ambient environment, sulphur source and ore genesis. *Canadian Journal of Earth Sciences*, **33**, 231–251.

GRASSINEAU, N.V., MATTEY, D.P. & LOWRY, D. 2001. Rapid sulfur isotopic analysis of sulfide and sulfate minerals by continuous flow – isotope ratio mass spectrometry (CF-IRMS). *Analytical Chemistry*, **73**, 220–225.

HAGGERTY, R. & BOTTRELL, S.H. 1997. The genesis of the Llanrwst and Llanfair veinfields, North Wales: evidence from fluid inclusions and stable isotopes. *Geological Magazine*, **134**, 249–260.

HALL, A.J., BOYCE, A.J. & FALLICK, A.E. 1987. Iron sulphides in metasediments: isotopic support for a retrogressive pyrrhotite to pyrite reaction. *Chemical Geology (Isotope Geoscience Section)*, **65**, 305–310.

HALL, A.J., BOYCE, A.J. & FALLICK, A.J. 1988. A sulphur isotope study of iron sulphides in the Late Precambrian Dalradian Easdale Slate Formation, Argyll, Scotland. *Mineralogical Magazine*, **52**, 483–490.

HALL, A.J., BOYCE, A.J., FALLICK, A.E. & HAMILTON, P.J. 1991. Isotopic evidence of the depositional environment of Late Proterozoic stratiform barite mineralization, Aberfeldy, Scotland. *Chemical Geology*, **87**, 99–114.

HALL, A.J., BOYCE, A.J. & FALLICK, A.J. 1994a. A sulphur isotope study of iron sulphides in the Late Precambrian Dalradian Ardrishaig Phyllite Formation, Knapdale, Scotland. *Scottish Journal of Geology*, **30**, 63–71.

HALL, A.J., MCCONVILLE, P., BOYCE, A.J. & FALLICK, A.E. 1994b. Sulphides with high δ^{34}S from the Late Precambrian Bonahaven Dolomite, Argyll, Scotland. *Mineralogical Magazine*, **58**, 486–490.

HALLIDAY, A.N. 1984. Coupled Sm–Nd and U–Pb systematics in Late Caledonian granites and basement under northern Britain. *Nature*, **207**, 229–233.

HALLIDAY, A.N., STEPHENS, W.E., HUNTER, R.H., MENZIES, M.A., DICKIN, A.P. & HAMILTON, P.J. 1985. Isotopic and chemical constraints on the building of the deep scottish lithosphere. *Scottish Journal of Geology*, **21**, 465–491.

HATTORI, K. 1989. Barite-celestine intergrowths in Archaean plutons: The product of oxidizing hydrothermal activity related to alkaline intrusions. *American Mineralogist*, **74**, 1270–1277.

HATTORI, K. & CAMERON, E.M. 1986. Archaean magmatic sulphate. *Nature*, **319**, 45–47.

HULBERT, L.J. ET AL. 1992. Metallogenesis and geochemical evolution of cyclic unit 1, Lower eastern layered series, Rhum. *Mineral deposit modelling in relation to crustal reservoirs of the ore-forming elements. Extended abstracts with programmes*, Keyworth, 4pp.

HUTTON, D.H.W. 1987. Strike-slip terranes and a

model for the evolution of the British and Irish Caledonides. *Geological Magazine*, **124**, 405–425.

HUTTON, D.H.W. 1988. Granite emplacement mechanisms and tectonic controls: inferences from deformation studies. *Transactions of the Royal Society, Edinburgh*, **79**, 245–255.

ISHIHARA, S. & SASAKI, A. 1989. Sulfur isotopic ratios of the magnetite-series and ilmenite-series granitoids of the Sierra Nevada batholith – a reconnaissance study. *Geology*, **17**, 788–791.

JACKSON, N.J., WILLIS-RICHARDS, J., MANNING, D.A.C. & SAMS, M.S. 1989. Evolution of the Cornubian orefield, southwest England, 2: Mineral deposits and ore-forming processes. *Economic Geology*, **84**, 1101–1133.

LAOUAR, R. 1987. *A sulphur isotope study of the Caledonian Granites of Britain and Ireland*. Unpublished MSc thesis, University of Glasgow.

LAOUAR, R., BOYCE, A.J., FALLICK, A.E. & LEAKE, B.E. 1990. A sulphur isotope study on selected Caledonian granites of Britain and Ireland. *Geological Journal*, **25**, 359–369.

LOWRY, D. 1991. *The genesis of Late Caledonian granitoid-related mineralization in Northern Britain*. Unpublished PhD thesis, University of St Andrews.

LOWRY, D., BOYCE, A.J., PATTRICK, R.A.D., FALLICK, A.E. & STANLEY, C.J. 1991a. A sulphur isotopic investigation of the potential sulphur sources for Lower Palaeozoic-hosted vein mineralization in the English Lake District. *Journal of the Geological Society, London*, **148**, 993–1004.

LOWRY, D., BOYCE, A.J., FALLICK, A.E. & STEPHENS, W.E. 1991b. Granitoid-related sulphide mineralization: implications for magma sources and terrane boundaries in the Caledonian of Northern Britain. *Frontiers in Isotope Geosciences, Abstracts with Programmes*, Keyworth, *Terra Abstracts supplement to Terra Nova* **3**.

LOWRY, D., BOYCE, A.J., FALLICK, A.E. & STEPHENS, W.E. 1992. Terrane characterisation in the British Isles using sulphur isotopes. *Mineral deposit modelling in relation to crustal reservoirs of the ore-forming elements, Extended abstracts with programmes*, Keyworth, 4pp.

LOWRY, D., BOYCE, A.J., FALLICK, A.E. & STEPHENS, W.E. 1995. Genesis of porphyry-style and plutonic-hosted mineralization related to metaluminous granitoids in the Grampian Terrane, Scotland. *Transactions of the Royal Society, Edinburgh*, **78**, 221–237.

LOWRY, D., BOYCE, A.J., FALLICK, A.E. & STEPHENS, W.E. 1997. Sources of sulphur, metals and fluids in granitoid-related mineralization of the Southern Uplands Terrane. *Transactions of the Institution of Mining and Metallurgy, Section B*, **106**, 157–168.

LOWRY, D., BOYCE, A.J., FALLICK, A.E., STEPHENS, W.E., HALL, A.J. & GRASSINEAU, N.V. 2003. Sulphur isotope signatures of Neoproterozoic and Palaeozoic terranes of Northern Britain – environments of formation. *Applied Earth Science, Transactions of the Institution of Mining and Metallurgy B*, **112**, 122–124.

LUSK, J. & CROCKETT, J.H. 1969. Sulfur isotope fractionation in co-existing sulfides from the Heath Steele B-1 orebody, New Brunswick, Canada. *Economic Geology*, **64**, 147–155.

MAYNARD, J., PRICHARD, H.M., IXER, R.A., LORD, R.A., WRIGHT, I.P., PILLINGER, C.T., MCCONVILLE, P., BOYCE, A.J. & FALLICK, A.E. 1997. Sulphur isotope study of Ni–Fe–Cu mineralisation in the Shetland ophiolite. *Transactions of the Institution of Mining and Metallurgy, Section B*, **106**, 215–226.

MCKERROW, W.S. & COCKS, L.R.M. 1976. Progressive faunal migration across the Iapetus Ocean. *Nature*, **263**, 304–306.

MILLER, O., BOYCE, A.J., FALLICK, A.E. & RICE, C.M. 1992. A stable isotope and fluid inclusion study of mineralization associated with the Coed-y-Brenin porphyry copper system, N. Wales. *Mineral Deposits Studies Group, Abstracts with Programmes*, AGM, University of Aberdeen.

NADEN, J. & CAULFIELD, J.B.D. 1989. Fluid inclusion and isotopic studies of gold mineralization in the Southern Uplands of Scotland. *Transactions of the Institution of Mining and Metallurgy, Section B*, **98**, 46–58.

OHMOTO, H. 1986. Stable isotope geochemistry of ore deposits. Ch.14. *In*: VALLEY, J.W., TAYLOR, H.P. Jnr & O'NEIL, J.R. (eds) *Stable isotopes in high temperature geological processes*. Mineralogical Society of America, Reviews. *In*: *Mineralogy*, **16**, 491–559.

OLIVER, G.J.H. 2002. Chronology and terrane assembly, new and old controversies. *In*: TREWIN, N.H. (ed.) *The Geology of Scotland*. The Geological Society, London, 201–211.

PATTRICK, R.A.D. & RUSSELL, M.J. 1989. Sulphur isotopic investigation of Lower Carboniferous vein deposits of the British Isles. *Mineralium Deposita*, **24**, 148–153.

PATTRICK, R.A.D. & BOWELL, R.J. 1991. The genesis of the West Shropshire Orefield: evidence from fluid inclusions, sphalerite chemistry, and sulphur isotopic ratios. *Geological Journal*, **26**, 101–115.

PATTRICK, R.A.D., COLEMAN, M.L. & RUSSELL, M.J. 1983. Sulphur isotopic investigation of vein Pb–Zn mineralization at Tyndrum, Scotland. *Mineralium Deposita*, **18**, 477–485.

PATTRICK, R.A.D., BOYCE, A.J. & MACINTYRE, R.M. 1988. Gold–silver mineralization at Tyndrum, Scotland. *Mineralogy and Petrology*, **38**, 61–76.

PIDGEON, R.T. & AFTALION, M. 1978. Cogenetic and inherited zircon U–Pb systems in Granites: Palaeozoic granites of Scotland and England. *In*: BOWES, D.R. & LEAKE, B.E. (eds) *Crustal evolution in northwestern Britain and adjacent regions*. Geological Journal Special Issue **10**, 183–248.

PLANT, J.A., STONE, P., FLIGHT, D.M.A., GREEN, P.M. & SIMPSON, P.R. 1997. Geochemistry of the british Caledonides: the setting for metallogeny. *Transactions of the Institution of Mining and Metallurgy, Section B*, **106**, 67–78.

POULSEN, S.R., KUBILIUS, W.P. & OHMOTO, H. 1991. Geochemical behavior of sulfur in granitoids during intrusion of the South Mountain

Batholith, Nova Scotia, Canada. *Geochimica et Cosmochimica Acta*, **55**, 3809–3830.

POWELL, T.B., ANDERSEN, T.B., DRAKE, A.A. JR., HALL, L. & KEPPIE, J.D. 1988. The age and distribution of basement rocks in the Caledonide orogen of the North Atlantic. *In*: HARRIS, A.L. & FETTES, D.J. (eds) *The Caledonide–Appalachian Orogen*, Geological Society of London, Special Publications, **38**, 63–74.

ROBINSON, B.W. & KUSAKABE, M. 1975. Quantitative preparation of SO$_2$ for ^{34}S/^{32}S analysis from sulphides by combustion with cuprous oxide. *Analytical Chemistry*, **47**, 1179–1181.

RUSSELL, M.J. 1985. The evolution of the Scottish mineral sub-province. *Scottish Journal of Geology*, **21**, 513–545.

SASAKI, A. & ISHIHARA, S. 1979. Sulfur isotopic composition of the magnetite-series and ilmenite-series granitoids in Japan. *Contributions to Mineralogy and Petrology*, **68**, 107–115.

SCOTT, R.A., PATTRICK, R.A.D. & POLYA, D.A. 1987. *Sulphur isotopic and related studies on Dalradian stratabound mineralization in the Tyndrum region, Scotland*. British Geological Survey Stable Isotope Report **130**, 40pp.

SCOTT, R.A., PATTRICK, R.A.D. & POLYA, D.A. 1991. Origin of sulphur in metamorphosed stratabound mineralization from the Argyll Group Dalradian of Scotland. *Transactions of the Royal Society, Edinburgh*, **82**, 91–98.

SHELTON, K.L. & RYE, D.M. 1982. Sulfur isotopic compositions of ores from Mines Gaspe, Quebec: An example of sulfate-sulfide isotopic disequilibria in ore-forming fluids with applications to other porphyry-type deposits. *Economic Geology*, **77**, 1688–1709.

SOLOMON, M., RAFTER, T.A. & DUNHAM, K.C. 1971. Sulphur and oxygen isotope studies in the North Pennines in relation to ore genesis. *Transactions of the Institution of Mining and Metallurgy, Section B*, **80**, 259–275.

STEED, G.M. & MORRIS, J.H. 1997. Isotopic evidence for the origins of a Caledonian gold–arsenopyrite–pyrite deposit at Clontibret, Ireland. *Transactions of the Institution of Mining and Metallurgy, Section B*, **106**, 109–118.

STEPHENS, W.E. & HALLIDAY, A.N. 1984. Geochemical contrasts between late Caledonian granitoid plutons of northern, central and southern Scotland. *Transactions of the Royal Society, Edinburgh*, **75**, 259–273.

STILLMAN, C.J. 1988. Ordovician to Silurian volcanism in the Appalachian–Caledonian orogen. *In*: HARRIS, A.L. & FETTES, D.J. (eds) *The Caledonide–Appalachian Orogen*. Geological Society of London, Special Publications, **38**, 275–290.

STRAUSS, H., BENGTSON, S., MYROW, P.M. & VIDAL, G. 1992. Stable isotope geochemistry and palynology of the late Precambrian to Early Cambrian sequence in Newfoundland. *Canadian Journal of Earth Sciences*, **29**, 1662–1673.

WILLAN, R.C.R. & COLEMAN, M.L. 1983. Sulfur isotope study of the Aberfeldy barite, zinc, lead deposit and minor sulfide mineralization in the Dalradian Metamorphic Terrain, Scotland. *Economic Geology*, **78**, 1619–1656.

WILLIAMS, F.M. & KENNAN, P.S. 1983. Stable isotope studies of sulphide mineralization on the Leinster granite margin and some observations on its relationship to coticule and tourmalinite rocks in the aureole. *Mineralium Deposita*, **18**, 399–410.

A reassessment of the tectonic zonation of the Uralides: implications for metallogeny

R. J. HERRINGTON[1], V. N. PUCHKOV[2] & A. S. YAKUBCHUK[1,3]

[1]Centre for Russian and Central Asian Mineral Studies, Natural History Museum, Cromwell Road, London SW7 5BD, UK

[2]Institute of Geology, Ufimian Scientific Centre, Russian Academy of Sciences, K.Marx Str., 16/2, Ufa-centre, 450000, Russia

[3]Geological Institute, Russian Academy of Sciences, Pyzhevsky Pereulok 7, Moscow 109017, Russia

Abstract: A review of the structural zonation of the 'oceanic' Urals shows that only its westernmost Sakamara, Tagil and Magnitogorsk zones reveal the presence of thrust structures, whereas in the East Uralian megazone and Trans-Uralian zone, the classic zonation rather reflects late- or post-collisional granitic welding and strike–slip displacement of the orogen for 100–300 km. This sinistral strike–slip displacement is responsible for the lens-shaped structure of the individual zones in the Urals.

Metallogenically, these orogen-parallel faults and the eastern boundary of the East European craton control the distribution of the orogenic Au deposits. Restoration of the individual zones into their pre-strike–slip fault positions suggests that the Urals contains only two magmatic arcs, one in the west and one in the east. The western Tagil–Magnitogorsk immature arc hosts a variety of chromite, Alaska-type PGE and major VMS deposits. The eastern Valerianovka arc effectively stitches together the Kazakh–Tien Shan structures and is host to important copper–gold and giant iron(–copper) skarn deposits.

The geodynamic evolution of the Urals can be observed with the generation of the immature Tagil–Magnitogorsk magmatic arc in the Late Ordovician. Metallogenic zoning of the VMS deposits supports the petrological data that the arc developed due to eastward subduction inside the oceanic back-arc basin that existed in the rear of the Kazakhstan–Tien Shan arcs. In the late Palaeozoic, these arcs collided with each other and were together thrust onto the East European craton. Syncollisional granitoid intrusions welded the magmatic arcs, which were soon displaced into presently observed fragments along the post-collisional orogen-parallel strike–slip faults.

The geographic region of the Urals is an important source of Russia's metals. Around 8 million tonnes of iron ore were produced in 1998 feeding important steelworks in the region. Copper and zinc production in the Urals is also important, largely sourced from a range of diverse VMS deposits, which collectively contain around 15% of known world reserves. Gold has been mined in the Urals since the eighteenth century and the region contains a number of large orogenic gold deposits including the c. 17 Moz Berezovskoe deposit. Nickel has been produced in the Urals from lateritic silicate bodies developed on exhumed ultramafic parts of Palaeozoic ocean crust. The Uralian region is a potentially important Palaeozoic porphyry copper province. Platinum placers of the Urals were the premier source of the metal until discovery of Bushveld with some 15 Moz produced, some from hardrock deposits, which are currently under re-evaluation. Lastly, the Urals is a bauxite-producing region from Devonian age laterites, and the presence of alluvial diamonds is noted.

Tectonically, the Urals Mountains are dominated by units belonging to the Uralides, which is a Palaeozoic orogen at the eastern margin of the East European craton, and west of the Altaid orogenic collage (Sengor et al. 1993). The Uralides and Altaids together are also known as the Ural–Mongolian or Central Asian fold belt, as identified by Muratov (Muratov 1974; Zonenshain et al. 1990; Mossakovskiy et al. 1993). Much has been recently published about Uralide tectonics (Berzin et al. 1996; Brown et al. 1998; Puchkov 1997, 2000, 2002, 2003; Friberg et al. 2002), its high-pressure metamorphic rocks (Matte et al. 1993; Lennykh et al. 1995; Leech & Ernst 2000; Leech & Stockli 2000) and petrology of magmatic arc rocks (Spadea et al. 1998; Seravkin et al. 1994), which serve to illuminate features of the structural and geodynamic

From: McDonald, I., Boyce, A. J., Butler, I. B., Herrington, R. J. & Polya, D. A. (eds) 2005. *Mineral Deposits and Earth Evolution*. Geological Society, London, Special Publications, **248**, 153–166. 0305-8719/$15.00
© The Geological Society of London 2005.

evolution of the orogen, but largely without relation to its mineral resources. Recent studies highlight the relationship between gold mineralization of the Urals and its late- and post-collisional strike–slip tectonics (Sazonov *et al.* 2001). The formation of VMS and porphyry deposits has been clearly linked to the pre-collisional magmatic evolution of the Uralides (Herrington *et al.* 2002*a*). Such metallogenic studies have highlighted some of the problems of Uralide tectonics and in this paper we intend to constrain the traditional tectonic zonation and tectonic evolution of the eastern Urals using mineral deposits.

Tectonic zonation of the Urals

Numerous studies of the Urals tectonics and metallogeny have identified a western and eastern slope (Ivanov *et al.* 1975; Puchkov 1997, 2000). The western slope is universally recognized as a deformed Palaeozoic continental passive margin (Puchkov 2002), whereas different plate tectonic models have been proposed in the past for its eastern, 'oceanic', slope, consisting of welded Palaeozoic magmatic arcs, micro-continents and sutured oceanic basins (Zonenshain & Matveenkov, 1984; Zonenshain *et al.* 1990; Puchkov 1993, 1997, 2000; Sengor *et al.* 1993; Matte *et al.* 1993; Brown *et al.* 1998; Friberg *et al.* 2002). The number of magmatic arcs defined in these interpretations varies between two and four, clearly highlighting the need to better define these features.

A simplified structural subdivision of the Urals is proposed and shown in Figure 1. The eastern slope is fully exposed only in the southern Urals (Zonenshain & Matveenkov 1984, Zonenshain *et al.* 1990; Puchkov 1993, 2000). In this part, from the west to the east, are the Sakmara, Magnitogorsk, East Urals, Trans-Uralian, and Valerianovka zones. The Magnitogorsk and Valerianovka zones consist mostly of magmatic arc rocks.

Airborne magnetic data (Fig. 2) define the continuity of the Magnitogorsk and Valerianovka magmatic arcs beyond the exposed portion of the Urals. The Magnitogorsk zone starts in the southern Urals, north of the Ust–Yurt plateau. The Tagil zone in the Middle and Polar Urals reveals the same magnetic pattern, as well as parts of the East Urals megazone. In the Polar Urals, both of them turn to the southeast under Mesozoic–Cenozoic sediments of the West Siberian basin towards the Zharma–Saur zone in eastern Kazakhstan.

The Valerianovka arc is sometimes considered as an additional arc accreted to the Uralides (Puchkov 2000). However, the magnetic data suggest it may be traced to the Beltau–Kurama arc in the Tien Shan, thus indicating its affinity to the Kazakh–Tien Shan structures. Magnetic data cannot be used for certain identification of the concealed Carboniferous suture east of the Valerianovka arc. Geological data are also insufficient to determine the position of a possible Carboniferous suture in the poorly exposed part of the southern Urals. This means the dip of the Valerianovka subduction zone is still a matter of a serious debate (Puchkov 2000), but nevertheless, both Valerianovka and Beltau-Kurama arcs host consanguineous porphyry-Cu deposits, which is additional support for their close affinity.

Sakmara and Tagil zones

The Sakmara zone lies to the west of the Main Urals fault zone (MUFZ; Fig. 1), along with other exotic fragments. These fragments comprise thrust sheets of Ordovician to Devonian ophiolites, which overlie deformed and in some cases also partly allochthonous passive margin sequences of the East European craton (EEC). These allochthons are rooted in the MUFZ, which contains fragments of rocks of the same ages and affinities and as such is interpreted as a main suture of the Urals Paleo-Ocean (Zonenshain *et al.* 1990; Puchkov 1993, 2000). A similar situation is seen in the northern part of the southern Urals and in the middle Urals, where both the Kraka and Bardym–Nyazepetrovsk allochthons contain some ophiolitic and island-arc complexes.

In the northern Urals, the most westerly zone of the 'eastern slope' is formed by the Tagil zone of ophiolitic to arc rocks that contain major Fe-Ti, Cr, PGE and VMS deposits. The Tagil zone forms a distant structural, lithological and metallogenic analogue of the upper thrust sheets of the Sakmara zone, but it constitutes a much larger, thick-skinned feature, up to 20 km thick. This is contrasting with the partly eroded, thin-skinned accretionary allochthons preserved in the Sakmara zone (Figs 1 & 2). In addition, seismic and structural data permit interpretation of the Tagil zone as an entirely allochthonous structure (Friberg *et al.* 2002).

The Tagil zone consists of ophiolites at the base (Savelieva *et al.* 2002), followed by Silurian island-arc rocks, Lower Devonian clastic sediments and Middle to Upper Devonian limestones and andesitic volcanics (Yazeva & Bochkarev 1996). On this basis, it is interpreted as an intra-oceanic arc that was active from the

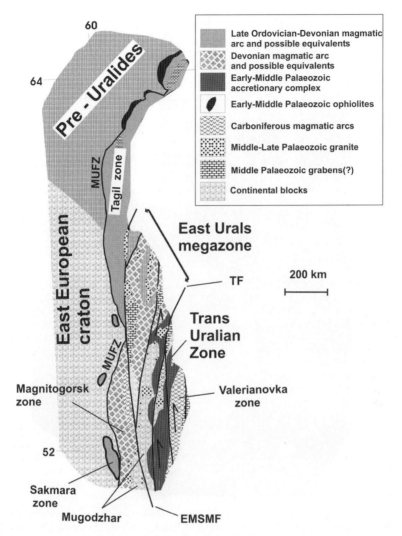

Fig. 1. Proposed schematic map of tectonic domains of the Urals (modified from various sources). MUFZ, Main Urals fault zone; TF, Troitsk Fault; EMSMF, East Mugodzhar-Serov-Mauk fault.

Late Ordovician to Middle Devonian. A specific and very prominent feature of the Tagil zone is the platinum-bearing belt of mafic–ultramafic plutons, which will be discussed later.

Magnitogorsk zone

In the southern Urals, the westward thrust bivergent Magnitogorsk zone occurs to the east of the MUFZ. It consists of Devonian oceanic magmatic arc rocks with superimposed Early Carboniferous rift-related volcanics, sediments and intrusives. According to both seismic and interpreted geological data, the zone has been

thrust onto the EEC, but probably still retains its root and is terminated in the east by west-dipping faults. To the south, the Magnitogorsk zone abuts the East Mugodzhar Precambrian massif (microcontinent) and eastwards these two structures are truncated by a north–south trending fault zone. Based on the structural analysis of the geological maps, Yakubchuk (2001) proposed that this fault can be traced northward to the MUFZ at the eastern edge of the East European craton, between the Tagil and Magnitogorsk zones. This fault can be then traced northward to join the Serov–Mauk fault that forms the eastern boundary of the Tagil

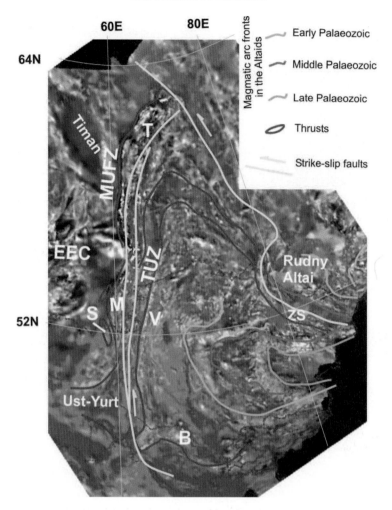

Fig. 2. Magnetic map and proposed correlations (map sourced from National Geophysical Data Center, 1996). T, Tagil arc; M, Magnitogorsk arc; V, Valerianovka arc; B, Beltau-Kurama arc; ZS, Zharma-Saur arc; EEC, East European craton; S, Sakmara allochthon; MUFZ, Main Urals fault zone.

zone and this should be recognised as a single structure to be called the East Magnitogorsk–Serov–Mauk fault (EMSMF) (Brown *et al.* 2002). Seismic studies (Brown *et al.* 1998; Friberg *et al.* 2002) show that this fault system dips westward (Fig. 3a), evidence taken by previous authors to support the bivergent thrust style of the boundaries to the Magnitogorsk zone (Brown *et al.* 1998).

East Uralian megazone

To the east of the EMSMF lies the East Uralian megazone (EUZ). Traditionally, this has been recognized as 'a granite axis of the Urals' due to

the presence of the numerous Carboniferous and Permian granitoid intrusives forming a north–south-oriented chain (Ivanov *et al.* 1975; Fershtater *et al.* 1997; Puchkov 2000). The pre-granitic units of the EUZ demonstrate that it is actually a composite structure, consisting of NE-striking subzones (Necheukhin 2000), which are en echelon and oblique to the north–south orientation of the main trend of the tradition-ally interpreted EUZ, following the linear faults, such as the Serov–Mauk fault (Fig. 1). From the north to the south, these include the Salda and Murzinka–Aduy Precambrian meta-morphic blocks, the Petrokamensk and Alapaevsk zones with Silurian to Middle

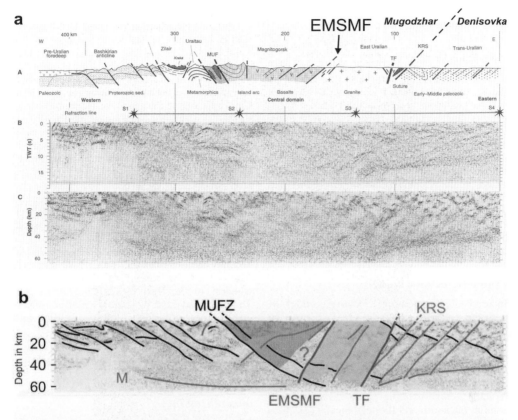

Fig. 3. (**a**) Above: URSEIS section (after Echtler *et al.* 1996). EMSMF, East Mugodzhar-Serov-Mauk fault; MUF, Main Uralian Fault Zone; KRS, Kartaly Reflection Zone (**b**) Below: URSEIS reinterpretation (this study). M, Moho.

Devonian magmatic arc rocks as well as melange and dismembered ophiolites in the Alapaevsk and Rezh massifs, which are thrust onto the Precambrian blocks.

These units are very similar to the Magnitogorsk and Tagil zones (Yazeva & Bochkarev 1995; Puchkov 2000) and if so, they may be offset northward for 300 km along the EMSMF. This means that this structure can be interpreted as a west-dipping strike–slip fault, not a simple thrust fault. Within the traditional EUZ, there are several subparallel faults with sinistral displacement for up to 100–200 km, mostly identified by offset of the metamorphic rocks in the Mugodzhar massif (Fig. 1).

Trans-Uralian zone

East of the Mugodzhar massif in the extreme east of the southern Urals lies the Trans-Uralian zone (TUZ). The TUZ consists of various accretionary complex rocks, including Late Precambrian and Cambrian continental fragments, Ordovician rift and ophiolite rocks, Silurian calc-alkaline rocks, Devonian shallow-water and continental sandstones along with deep-water olistostromic cherty facies and some basalts (Puchkov 2000). Geological maps suggest that this zone may be thrust westward onto the Mugodzhar massif. In contrast to this, the published URSEIS-95 geological cross-sections (Berzin *et al.* 1996; Echtler *et al.* 1996) show that the west-dipping Kartaly fault is the main boundary between the Mugodzhar and Trans-Uralian zone (Fig. 3a). Nevertheless, the same section clearly shows east-dipping reflectors at the upper crust level, which would correspond very well to the westward thrusts, interpreted from the geological maps. The identification of the thrusts is also supported by the presence of the serpentinite massifs in the Mugodzhar block (Shevchenko, Dzhetygara, etc), which can be considered as fragments of eroded allochthon (Puchkov 2000).

Valerianovka zone

The most easterly zone in the southern Urals is the Valerianovka zone. It consists of Devonian to Carboniferous intermediate volcanic, terrigenous and carbonate rocks, overlying older formations of the Trans-Uralian zone (Puchkov 2000). As it was mentioned above, the zone can be traced into the Beltau–Kurama continental arc in Uzbekistan, and this represents the active margin of the Kazakh continent during the Late Palaeozoic (Fig. 2). In this case, the complete Trans-Uralian zone, which also hosts Carboniferous intermediate volcanics, may be interpreted as an accretionary complex that accumulated in front of the Valerianovka arc.

Proposed tectonic model

Instead of the complex existing model of multiple collided magmatic arcs, we propose that there are only two magmatic arcs in the Urals. The west is marked by the rather immature arc, whose western portions form frontal allochthons (Sakmara, Tagil and others) and whose eastern portions may form a rooted structure, corresponding to the Magnitogorsk zone in the south and the tectonically complex, Petrokamensk–Alapaevsk zone farther north. To the east, the separate Valerianovka continental arc belongs to the Kazakh–Tien Shan structures. These two arc systems collided during the late Palaeozoic and were then significantly offset along the north–south trending faults, which can be interpreted as late- or post-collisional sinistral strike-slip faults (Figs 1 & 2).

In sectional view, the relationships between these units can be understood using the published seismic traverses (Brown *et al.* 1998; Friberg *et al.* 2002). The interpretations of the URSEIS traverse can be reconsidered in the view of above-described structural relationships. Indeed, the seismic data allow a different interpretation of the southern Urals (Fig. 3b). The published URSEIS interpretations (Brown *et al.* 1998) for the western zones of the Urals, including the Magnitogorsk zone, remain valid, but the East Uralian and Trans-Uralian zones can be viewed differently as most of the west-dipping faults on the published cross-sections are likely to represent late orogenic features. The clear definition of east-dipping reflectors at the eastern part of the cross-section allows us to interpret the features in terms of the westward thrusting of the Trans-Uralian zone on top of the

East Uralian megazone. In this case the lower crustal part underneath the Trans-Uralian zone should be reinterpreted as a fragment of the East Uralian (East Mugodzhar) zone. This means that the entire orogen can be viewed as a west-vergent pre-Late Carboniferous structure, which was then cross-cut by the late (Late Carboniferous to Permian) west-dipping strike–slip faults with a thrust component. The dips of the faults in this poorly exposed area are very difficult to identify and steeply dipping structures are difficult to interpret from 2D seismics. However, the western dip of at least one of the faults underlined by serpentinitic melange and probably belonging to the Kartaly reflection zone (KRS) of the URSEIS profile, has been tested by a series of boreholes (Kosarev *et al.* 2001).

This review of the structural zonation of the 'oceanic' Urals shows that only its westernmost zones reveal the presence of thrust structures, whereas in the East Uralian megazone and Trans-Uralian zone, the classic zonation reflects late- or post-collisional granitic welding and strike–slip deformation of the orogen.

Implications for metallogeny of the Urals

Oceanic-arc-related base metal mineralization

Sakmara and Main Urals Fault zones. In the Sakmara zone, Early Silurian volcanics in the Mednogorsk district show strong affinities with volcanics formed in a primitive arc setting, where extensive ophiolites of the Kempirsai massif indicate that it developed in a rift, which evolved to a full oceanic basin (Savelieva *et al.* 1997). It has been suggested that the PGE-enriched chromites of Kempirsai formed in a supra-subduction zone setting of an oceanic arc at around 400–370 Ma (Melcher *et al.* 1999). In the Mednogorsk VMS district, the Silurian volcanic sequences show a transition from tholeiite to calc-alkaline nature and a degree of alkaline enrichment, atypical for a mid-oceanic ridge setting. Very low REE and Zr/Y ratios of between 1 and 2 are indicative of a primitive arc setting (Herrington *et al.* 2002a). Small, ultramafic-hosted massive sulphide deposits in the Main Urals fault zone are identified as having a forearc setting (Nimis *et al.* 2003; Gannoun *et al.* 2003). Upper Silurian and Lower Devonian sedimentary sequences in the MUFZ are amagmatic along the MUFZ zone, which serves to demonstrate that the southern Magnitogorsk arc was evolving quite separately from the Tagil arc at this time.

Tagil zone. This zone contains fragments of dismembered ophiolites, located close to the MUFZ (Savelieva *et al.* 2002). In the Tagil zone, the ophiolite fragments are found along with zoned mafic–ultramafic complexes termed 'platinum-bearing', or Ural–Alaska-type complexes (Zoloev *et al.* 2001). The platinum-bearing ultramafics, chromites, some sulphides and rare gabbros are likely to be feeders to the calc-alkaline lavas, which are part of the proper arc higher in the sequence (Friberg 2000). The more than 5 km-thick volcano-sedimentary package of the Tagil arc progresses west to east, chronologically younging to the centre, from Late Ordovician or Early Silurian bimodal basalt and rhyolite with cherts (hosting signifi-cant Cu–Zn VMS deposits) and carbonates through a sequence of Silurian tuffs and lavas with basalt, andesite–basalt to the more calc-alkaline arc-like andesites (with polymetallic VMS deposits) referred to as being contem-poraneous with the platinum-belt intrusives. The massifs of the platinum-bearing belt contain also vanadium-titanomagnetite deposits at Kachkanar and magnetite-apatite copper with PGE ores at Volkovo.

Narkisova *et al.* (1999) note a more 'continen-tal' contribution to the magma source as the EEC margin approached the MUFZ, confirm-ing subduction polarity. Upwards in section through the Tagil zone, there is also an evolution through subalkaline to calc-alkaline volcanism and intrusive activity through the Lower and Middle Devonian (Yazeva & Bochkarev 1993) and Frasnian, comparable to the age of the Magnitogorsk zone (see below). The eastern part of the Tagil zone hosts Fe-skarn and Cu–Fe-skarn deposits related to Lower Devonian subalkaline (syenite, diorite, granodiorite) and Middle-Devonian calc-alkaline intrusions (Yazeva & Bochkarev 1993).

Unlike the Magnitogorsk arc, there is no direct evidence of collision between the Tagil zone and the EEC in the Devonian (or earlier), and the turbidites sourced from the arc sequences in the east only reached the eastern margin of EEC in Early Carboniferous times (Puchkov 1997, 2002), possibly indicating a diachronous arc–continent collision from south to north.

Magnitogorsk zone. Some features of the Magnitogorsk zone are still somewhat enig-matic. Silurian magmatic arc rocks are found in the adjacent East Uralian and Trans-Uralian zones, to the east, and in the Sakmara zone, to the west (Yazeva & Bochkarev 1995). The Magnitogorsk zone mostly consists of Devonian

arc-related volcanism, whereas the Upper Silurian and Lower Devonian (pre-Emsian) rocks immediately in the flanks and under them are deep-water sedimentary, and reveal no magmatism at all. Therefore, the Magnitogorsk arc did not simply inherit the Silurian arc, but there may have been some intra-arc spreading. This is supported by the presence of back-arc ophiolites in the western Mugodzhars, behind the Magnitogorsk arc to the south (Zaykov *et al.* 1996).

The formation of highly significant economic VMS deposits in the Magnitogorsk zone took place during two stages of the arc evolution in the Early to Middle Devonian times. It is restricted to the first accretionary arc stage and to tholeiitic and calc-alkaline stages of intra-arc volcanism (e.g. Herrington *et al.* 2002*a*). Small to medium-sized low-grade Cu-pyritic to Cu–Zn deposits are associated with basaltic to andesitic host rocks, whereas higher grade Cu–Zn and polymetallic deposits are associated with more evolved andesitic to felsic host rocks of the upper parts of the succession.

Parts of these deposits are hosted by half-graben structures, pointing to an intra-arc exten-sional setting of VMS formation. Medium to large-sized low-grade Cu–Zn deposits of the Karamalytash stage are confined to bimodal suites that occur in the upper part of the succes-sion. The distribution of the VMS deposits and the trace element patterns of the felsic host rocks point to their formation during intra-arc rifting. Small polymetallic VMS deposits were generated in calc-alkaline successions at the top of the eastern parts of the Karamalytash complex. These features are supportive of an eastward dipping subduction zone beneath the Magnitogorsk arc.

Conondont biostratigraphy and isotopic age determinations record very fast formation of the VMS deposits, during 1 to 2 Ma for both the Upper Emsian and the Upper Eifelian volcanic arc successions (Maslov *et al.* 1993). Hence, the formation of economic VMS deposits in the Magnitgorsk Zone was associated with only brief periods of intra-arc and back-arc extension that lasted during approximately 15–20 Ma of oceanic arc development.

Minor porphyry copper deposits and possible high-level epithermal gold–sulphide mineraliza-tion at Bereznyakovskoe (Lehmann *et al.* 1999) are found in Upper Devonian parts of the volcanic arc sequences (Gusev *et al.* 2000). These systems have probably recorded changes in the nature of subduction along the Magnito-gorsk arc through time. Post arc–continent collision rifting during the Carboniferous led to the generation of A-type gabbro–granite complexes to which the skarn magnetite and titanium–iron mineralization in the Magnitogorsk region is associated (Fershtater *et al.* 1997; Herrington *et al.* 2002*b*).

Continental arc related porphyry copper and magnetite 'skarn' mineralization

Significant porphyry prospects are known in the Valerianovka arc (Koroteev *et al.* 1997). These subduction-related volcanics formed prior to the final closure of the oceanic basin between the Magnitogorsk and Valerianovka arcs. The suture of this basin could correspond to the position of the Trans-Uralian zone, though no Carboniferous ophiolites or other proof of relics of a Carboniferous oceanic basin have been found there. The porphyries show a spatial relationship to the distribution of important copper–magnetite skarn deposits of Benkala and Varvarinskoe and enormous hydrothermal magnetite 'skarn' deposits, such as Sokolovsko–Sarbaiskoe, which may have formed in a setting somewhat similar to the central Andes (Sillitoe 2003).

The hydrothermal and skarn magnetite bodies in the Valerianovka arc are associated with the dominantly mafic, largely subaerial calc-alkaline volcanism intruded by comagmatic stocks. Volcanism, sedimentation and mineralization all seem to be controlled by north to NNE-trending structures. Much of the mafic volcanic sequence shows hematitization, evidence of early oxidation of the lava–tuff packages. Base metals are common in the deposits, often appearing late in the paragenetic sequence, with some bodies having almost economic copper grades (0.6% Cu) and significant precious metals. On closer inspection, the classification of the magnetite deposits as simple contact skarns is not so evident and many of these deposits have poorly defined relationships with intrusive rocks, more related to zones of regional scapolite alteration in common with the Proterozoic magnetite camps such as Kiruna (Herrington *et al.* 2002*b*). Less than 20% of the magnetite bodies actually form in contact with igneous bodies. The largest deposit in the Turgai district, Kachar, lies some 18 km laterally and probably more than 2 km vertically (based on geophysical modelling) from the contact zone of any prospective intrusive body.

Post-collision orogenic gold mineralization

The distribution of the bulk of Urals gold deposits are controlled by structures such as shear zones, faults, and their intersections (Sazonov *et al.* 2001). Throughout the Urals, structurally-controlled gold mineralization relates to regional deformation and hydrothermal fluid–rock interaction in the Late Carboniferous to Early Triassic, which has been documented by Sazonov *et al.* (2001). These events overprint and remobilize gold contents of earlier deposits, but more significantly gold-bearing quartz vein lodes, which effectively encompass the majority of the larger gold deposits in the Urals, were also formed in structural–chemical traps.

The Mindyak deposit is hosted by tectonic melange of the MUFZ, and it is a typical late-orogenic lode gold deposit. Others occur within and along the margins of Early to Middle Carboniferous granitoids and, although local orthomagmatic relationships have been described (Bortnikov *et al.* 2001), most observations favour a strong structural control on the lodes and no direct genetic association with adjacent granitic intrusion (Kisters *et al.* 1999). In the two largest gold deposits, Kochkar and Berezovskoe, gold mineralization was controlled by structural and combined structural–chemical traps in dilational jogs in both shear and contact zones. Locally, gold mineralization may constitute NW-trending clusters that formed during a change from orthogonal to transpressional compression in relation to strike–slip faults (Fig. 4).

The inferred ages of the lode gold deposits suggest they are mostly coeval with the generally undeformed Permian granites although a spatial relationship with gold mineralization has not been observed. The granites were formed in an extensional regime following Ural-wide magma generation, caused by a thermal flux related to a thickening of the crust and formation of a crustal root resulting from continent–continent collision (Puchkov 2000).

We suggest that the widespread distribution of this late thermal event, during changes in stress regime, and the involvement of magma-generating mantle processes were responsible for regional delamination in the lithosphere and concomitant upwelling of the asthenosphere. These processes may have occurred mainly prior to the formation of the prominent crustal root of the Uralides, which formed during transpressional convergence.

Sazonov *et al.* (2001) propose that a major indenter of the EEC is responsible for a singular, transcrustal, sidewall ramp in the subsurface of the middle Urals. This indenter could be reassembled into a pre-strike–slip position by juxtaposing the Salda block and the

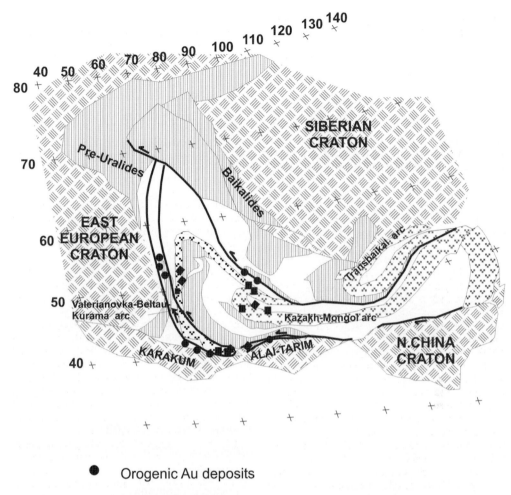

●   Orogenic Au deposits

◆   Skarn deposits

■   Cu-Mo-(Au) porphyry and Au epithermal deposits

Fig. 4. Strike–slip faults of the Urals and major orogenic gold deposits (based on Sazonov *et al.* 2001).

EEC promontory in the middle Urals, but the Precambrian age of the Salda block requires further investigation as a considerable part of its stratigraphy has been shown, from conodonts, to be Palaeozoic (Puchkov 2000). The isotope ages are mostly Palaeozoic and are suggestive of island-arc-related metamorphism (Petrov & Friberg 1999; Friberg *et al.* 2000).

The indenter may extend into the mantle, forming the principal conduit zone for mineralizing fluids of the orogenic gold deposits. The world-class Kochkar and Berezovskoe gold

deposits would occur symmetrically adjacent to this proposed conduit (*see* Sazonov *et al.* 2001).

Geodynamic and metallogenic evolution of the Uralides

The previous discussion leads to some re-evaluation of the geodynamic evolution of the Uralides. Figure 5 summarizes this and its highly simplified relationship to key styles of mineralization.

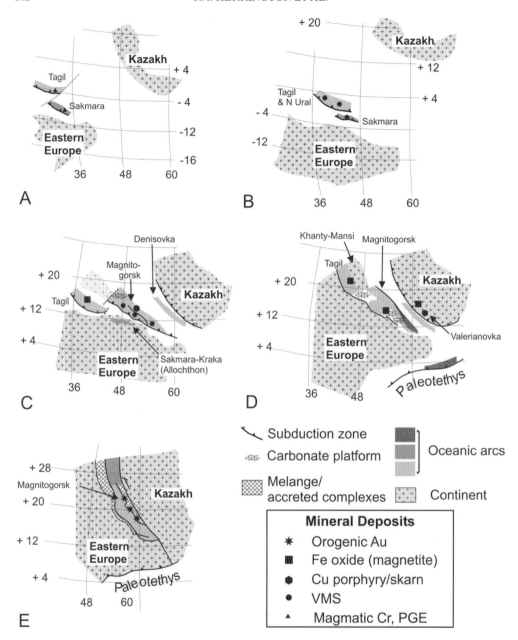

Fig. 5. Palinspastic reconstruction of the Urals with major mineral deposits located (based on Filippova *et al.* 2001). A to E represent highly simplified time slices: A, 430 Ma; B, 390 Ma; C, 360 Ma; D, 330 Ma; E, 280 Ma.

During the Ordovician, oceanic crust, whose fragments are now found in the Urals ophiolites, developed a new Paleo-Uralian ocean (Zonenshain & Matveenkov 1984; Puchkov 2000) between the East European craton and Kazakh–Tien Shan cratonic slivers. The latter formed the basement of some Palaeozoic magmatic arcs, such as the Kipchak arc of Sengor *et al.* (1993).

In Late Ordovician to Silurian times, there is

the first evidence of arc volcanism in the Sakmara zone and further north in the Tagil zone. In the Sakmara zone at Kempirsai, harzburgites and lherzolites are host to world-class chromite deposits with clear evidence of subduction-derived fluids upgrading the chromite bodies. In the Mednogorsk district, VMS deposits formed in the Lower Silurian (Llandoverian). These events may correlate with the beginning of the island arc magmatism. No evidence exists to confirm the polarity of this arc during its initiation, but this arc separated the formerly single oceanic back-arc basin into two basins, whose sutures are now found in the Main Urals fault zone in the west and Trans-Uralian zone in the east. The Tagil–Magnitogorsk arc, therefore, started to form behind the contemporaneous magmatic arcs of Kazakhstan–Tien Shan (Fig. 5a).

In the Early Devonian (Fig. 5b), arc volcanism commenced in the Magnitogorsk zone with a major episode of VMS formation. Arc volcanism also episodically continued in the Tagil zone and East Uralian megazone (Friberg et al. 2002), relics here hosting VMS deposits like Safyanovka north of Ekaterinburg (Koroteev et al. 1997). Therefore, the Tagil–Magnitogorsk arc system was clearly active along all its strike. The presence of boninites in the west of the Magnitogorsk zone and zonation of mineralized volcanic sequences is evidence of the east-dipping subduction polarity under this arc.

In general, the Magnitogorsk arc evolution records an eastward dipping subduction of the oceanic crust attached to the passive margin of the East European craton during the Emsian to Eifelian followed by subduction of the distal continental margin that probably entered the subduction zone during the Givetian (Brown et al. 2001). The temporal and spatial retreat of arc complexes could therefore simply record changes in the subduction angle from the west through time and not younging due to suggested westward subduction on the eastern flank of the arc (Zonenshain et al. 1990; Sengor et al. 1993). During the Early to Middle Devonian, continental arc magmatism is also recorded in the Valerianovka zone at the western flank of the Kazakh–Tien Shan structures. The accretionary complexes of the Trans-Uralian zone could be affiliated with this arc (Fig. 5c).

During the Late Devonian, arc–continent collision occurred in the Magnitogorsk zone with descent of the leading edge of the East European craton into the subduction zone and its subsequent exhumation. Due to the entrance of less dense, and, hence, buoyant continental

crust in the subduction zone during the Ulutau stage, the subduction angle shallowed up to promote the renewed formation of calc-alkaline complexes at eastern parts of the Karamalytash complex and during the subsequent Ulutau stage. The arrival of thicker continental margin crust at the convergent plate margin terminated the oceanic arc evolution in the west and induced a jump of the subduction process, with new subduction, initiated in the east of the TUZ to produce the Valerianovka arc.

The Magnitogorsk arc became anchored between the East European craton and the East Mugodzhar Precambrian massifs in the Southern Urals. The origin of the East Mugodzhar massif to the east of the Magnitogorsk zone is not quite clear. The explanations vary from an exposed basement of the East European craton and totally allochthonous (rootless) nature of the Magnitogorsk zone (Perfiliev 1979) to a passively 'floating' and later amalgamated massif (Puchkov 1993, 1997, 2000). Based on the structural relationships of the Mugodzhar massif with the surrounding structures and its termination in the north, we propose that it was detached from the Ust–Yurt massif in the SE of the East European craton and was translated northward along the strike of the Urals during westward intrusion and oroclinal bending of the Kazakh–Tien Shan arcs (Fig. 5c).

In the Early to Middle Carboniferous, the final stages of the arc–continent collision and closure of the Main Urals basin occurred with rift-related magmatism along the axis of the Magnitogorsk zone, which may relate to the possible slab detachment beneath. Simultaneously, the new episode of the Andean-style subduction-related volcanism with significant porphyry copper mineralization started in the Valerianovka zone that indicates that oceanic crust of the Trans-Uralian basin began to subduct at the western flank of the Kazakh–Tien Shan structures (Fig. 5d).

In Middle Carboniferous times, final locking of the Trans-Uralian basin and its thrusting onto the Mugodzhar massif with possible eastward backthrusting of the Magnitogorsk zone took place. The continental masses of the East European craton and Kazakh–Tien Shan arcs amalgamated, and associated north–south-trending and superimposed NW-trending strike–slip transpressive deformation was accompanied by major orogenic gold mineralization in the Urals (Fig. 5e). Location of the latter strike–slip structures was largely controlled by the presence of a lobe of the East European craton (Fig. 4), which clearly indents the Middle Urals (Sazonov *et al.* 2001). In the Late Carboniferous, marine sedimentation was completely terminated in the Urals, with exception of the foredeep in the Middle to North Urals.

The orogeny *sensu stricto* commenced at this time and continued through the Permian. Orogenesis resulted in mountain growth and their erosion with formation of flysch and molasse in the foredeep. Thrusts formed in the foreland, and Main Granitic axis in the hinterland. In the Trans-Uralian zone was the contemporaneous east-vergent thrusting and strike-slip faulting with creation of a bivergent orogen structure.

This was a culmination of the major orogenic event in the Uralides. The orogen collapsed at the Permian–Triassic transition, when central Eurasia was affected by the Siberian superplume (Dobretsov & Vernikovsky 2001; Nikishin *et al.* 2002). In the Urals, it is represented by graben structures that host basalts and coal deposits.

Conclusions

Using structural, petrological, and geophysical data we propose a tectonic and geodynamic interpretation for the evolution of the 'oceanic' sector of the Urals. This is additionally constrained by the metallogenic data.

The Urals orogen consists of Precambrian microcontinents, such as Mugodzhar and Salda, and Middle Palaeozoic magmatic arcs exposed in the Magnitogorsk, Tagil and East Uralian zones, always with ophiolites at the base of their sequences. The magmatic arc rocks in these zones are very similar and host similar VMS deposits. However, they presently form apparently separate tectonic units. Our study identifies that they are separated by major north–south trending strike–slip faults. In plan view, these faults offset for 100–300 km the Precambrian slivers, ophiolite sutures and magmatic arcs into presently observed fragments.

The sinistral strike–slip pattern controls the distribution of the clusters of orogenic gold deposits. Recognition of these faults simplifies the pre-strike–slip structure of the Urals. As a result, we recognize that the 'oceanic' part of the orogen, from the west to the east, contains the following relic structures: the Main Urals fault zone as a suture of the Main Urals oceanic (back-arc) basin; Tagil–Magnitogorsk intraoceanic arc; Trans-Uralian oceanic (back-arc) basin; and Valerianovka arc of the Kazakh–Tien Shan structures.

The Tagil–Magnitogorsk arc system (assemblage) is thrust onto the passive margin of the East European craton in the Late Devonian to Early Carboniferous, but its Magnitogorsk zone remained a rooted structure during northward translation of the Mugodzhar massif and subsequent collision of the arc with the East European craton.

The subsequent collision of the Tagil–Magnitogorsk and Valerianovka arcs produced the westward-thrust Trans-Uralian zone and affected the Uralides with emplacement of granitoid intrusions.

These two arc systems were later fragmented significantly by the trans-regional strike–slip faults extending from the Tien Shan to the Polar Urals. These faults developed in the Late Palaeozoic during oroclinal bending, westward intrusion and clockwise rotation of the external Kazakh–Tien Shan arcs relative to the East European craton.

The nucleus of this paper developed at a GEODE Urals workshop hosted in Amlwch, North Wales and the often animated discussions with colleagues Victor Zaykov, Bernd Buschmann, Peter Jonas, Valeriy Maslennikov, Igor Seravkin, Vasily Prokin, Robin Armstrong and Nicola Holland were highly appreciated. Chris Stanley and Andy Fleet of the NHM are thanked for early comments. This is an ESF GEODE publication and the authors thank the ESF for financial support to attend the workshop. RJH and VP acknowledge support from EU INCO2 Project MinUrals. This is a contribution to IGCP-473.

References

BERZIN, R., ONCKEN, O., KNAPP, J.H., PEREZ-ESTAUN, A., HISMATULIN, T., YUNUSOV, N. & LIPILIN, A. 1996. Orogenic evolution of the Ural Mountains; results from an integrated seismic experiment. *Science*, **274**, 220–221.

BORTNIKOV, N.S., VIKENTYEVA, O.V., FALLICK, A. & SAZONOV, V.N. 2001. REE distribution and O and D isotope data of minerals from quartz-carbonate wall rocks at the giant Berezovsk mesothermal lode gold deposit, Urals, Russia: Evidence for a contribution from a magmatic fluid. *In: Mineral deposits at the beginning of the 21st century*. Proceedings Joint 6th Biennial SGA-SEG Meeting, August 2001. Krakow, Poland. Swets & Zeitlinger, Lisse, 707–710.

BROWN, D., JUHLIN, C., ALVAREZ-MARRON, A., PEREZ-ESTAUN, A. & OLIANSKI, A. 1998. Crustal-scale structure and evolution of an arc-continent collision zone in the southern Urals, Russia. *Tectonics*, **17**, 158–171.

BROWN, D., ALVAREZ-MARRON, J., PEREZ-ESTAUN, A., PUCHKOV, V., GOROZHANINA, Y. & AYARZA, P. 2001. Structure and evolution of the Manitogorsk forearc basin: Identifying into upper crustal processes during arc-continent collision in the southern Urals. *Tectonics*, **20**, 364–375.

BROWN, D., JUHLIN, C. & PUCHKOV, V.N. 2002. Introduction. *In*: BROWN, D., JUHLIN, C. & PUCHKOV, V. (eds) *Mountain Building in the Uralides: Pangea to the Present*. AGU Geophysical Monograph **132**, 1–9.

DOBRETSOV, N.L. & VERNIKOVSKY, V.A. 2001. Mantle plumes and their geologic manifestations. *International Geology Review*, **43**, 771–787.

ECHTLER, H.P., STILLER, M., STEINHOFF, F., ET AL. 1996. Preserved collisional crustal architecture of the southern Urals – Vibroseis CMP – profiling. *Science*, **274**, 224–226.

FERSHTATER, G.B., MONTERO, P., BORODINA, N.S., PUSHKAREV, E.V., SMIRNOV, V., ZIN'KOVA, E. & BEA, F. 1997. Uralian magmatism: an overview. *Tectonophysics*, **276**, 87–102.

FILIPPOVA, I.B., BUSH, V.A. & DIDENKO, A.N. 2001. Middle Paleozoic subduction belts: the leading factor in the formation of the Central Asian fold-and-thrust belt. *Russian Journal of Earth Sciences*, **3**, 405–426.

FRIBERG, M. 2000. *Tectonics of the Middle Urals*. PhD Thesis, Uppsala University.

FRIBERG, M., LARIONOV, A., PETROV, G.A. & GEE, D.G. 2000. Paleozoic amphibolite-granulite facies magmatic complexes in the hinterland of the Uralide Orogen. *International Journal Earth Science*, **89**, 21–39.

FRIBERG, M., JUHLIN, C., BECKHOLMEN, M., PETROV, G.A. & GREEN, A.G. 2002. Palaeozoic tectonic evolution of the Middle Urals in the light of the ESRU seismic experiments. *Journal of the Geological Society of London*, **159**, 295–306.

GANNOUN, A., TESSALINA, S., BOURDON, B., ORGEVAL, J.J., BIRCK, J.L. & ALLEGRE, C.J. 2003. Re–Os isotopic constraints on the genesis and evolution of the Dergamish and Ivanovka Cu (Co, Au) massive sulphide deposits, South Urals, Russia. *Chemical Geology*, **196**, 193–207.

GUSEV, G.S., GUSHCHIN, A.V. ET AL. 2000. Geology and metallogeny of island arcs. *In*: MEZHELOVSKY, N.V. ET AL. (eds) *Geodynamics and Metallogeny: Theory and Implications for Applied Geology*. Ministry of Natural Resources of the RF & GEOKART Ltd, Moscow, 213–295.

HERRINGTON, R.J., ARMSTRONG, R.N., ZAYKOV, V.V., MASLENNIKOV, V.V., TESSALINA, S.G., ORGEVAL, J-J. & TAYLOR, R.N. 2002a. *In*: BROWN, D., JUHLIN, C. & PUCHKOV, V. (eds) *Mountain Building in the Uralides: Pangea to the Present*. AGU Geophysical Monograph **132**, 155–182.

HERRINGTON, R., SMITH, M., MASLENNIKOV, V., BELOGUB, E. & ARMSTRONG, R. 2002b. A short review of Paleozoic hydrothermal magnetite iron-oxide deposits of the south and central Urals and their geological setting. *In*: PORTER, T.M. (ed.) *Hydrothermal Iron Oxide Copper–Gold and Related Deposits: A Global Perspective*. Vol. 2, PGC Publishing, South Australia, 243–253.

IVANOV, S.N., PERFILYEV, A.S., YEFIMOV, A.A., SMIRNOV, G.A., NECHEUKHIN, V.M. & FERSH-

TATER, G.B. 1975. Fundamental features in the structure and evolution of the Urals. *American Journal of Science*, 107–130.

KISTERS, A.F.M., MEYER, F.M. & SERAVKIN, I.B. ET AL. 1999. The geological setting of lode-gold deposits in the central southern Ural: a review. *Geologische Rundschau*, **87**, 603–616.

KOROTEEV, V.A., BOORDER, H. de, NECHEUKHIN, V.M. & SAZONOV, V.N. 1997. Geodynamic setting of the mineral deposits of the Urals. *Tectonophysics*, **276**, 291–300.

LEECH, M.L. & ERNST, W.G. 2000. Petrotectonic evolution of the high- to ultrahigh-pressure Maksyutov Complex, Karayanova area, South Ural Mountains: structural and oxygen isotope constraints. *Lithos*, **52**, 235–252.

LEECH, M.L. & STOCKLI, D.F. 2000. The late exhumation history of the ultrahigh-pressure Maksyutov Complex, South Ural Mountains, from new apatite fission track data. *Tectonics*, **19**, 153–167.

LEHMANN, B.J., HEINHORST, J., HEIN, U., NEUMANN, M., WEISSER, J.D. & FEDOSEYEV, V. 1999. The Bereznyakovskoye gold trend, southern Urals, Russia. *Mineralium Deposita*, **34**, 241–249.

LENNYKH, V.I., VALISER, P.M., BEANE, R., LEACH, M. & ERNST, W.G. 1995. Petrotectonics of the Maksyutov ultrahigh-pressure/high-pressure metamorphic complex, South Ural Mountains, Russia, AGU 1995 fall meeting, *Eos, Transactions, American Geophysical Union*, **76**, 668.

MASLOV, V.A., CHERKASOV, V.L., TISCHCHENKO, V.T., SMIRNOVA, I.A., ARTYUSHKOVA, O.V. & PAVLOV, V.V. 1993. [*On the stratigraphy and correlation of the Middle Paleozoic complexes of the main copper-pyritic areas of the Southern Urals*.] Ufa Science Centre, Ufa, 217 [in Russian].

MATTE, P.H., MALUSKI, H., CABY, R., NICOLAS, A., KEPEZHINSKAS, P. & SOBOLEV, S. 1993. Geodynamic model and ^{39}Ar/^{40}Ar dating for the generation and emplacement of the high pressure (HP) metamorphic rocks in SW Urals. *Comptes Rendus Académies des Sciences, Series II*, **317**, 1667–1674.

MELCHER, F., GRUM, W., THALHAMMER, T.V. & THALHAMMER, O.A.R. 1999. The giant chromite deposits at Kempirsai, Urals: constraints from trace element (PGE, REE) and isotope data. *Mineralium Deposita*, **34**, 250–272.

MOSSAKOVSKIY, A.A., RUZHENTSEV, S.V., SAMYGIN, S.G. & KHERASKOVA, T.N. 1993. [Central Asian fold belt: Geodynamic evolution and history of formation] *Geotektonika*, **6**, 3–33 [in Russian].

MURATOV, M.V. 1974. [Ural–Mongolian fold belt.] *In*: MURATOV, M.V. (ed.) [*Tectonics of the Ural–Mongolian fold belt*.] Nauka, Moscow, 5–11 [in Russian].

NARKISOVA, V.V., SAZONOVA, L.V. & NOSOVA, A.A. 1999. [Magmatic bodies within the sequence of the Uralian ultradeep borehole.] *In*: KOROTEEV, V.A. (ed.) [*Paleosubduction zones: Tectonics, Magmatism, Metamorphism and Sedimentology*.] Urals Branch, RAS, Ekaterinburg, 104–106 [in Russian].

NATIONAL GEOPHYSICAL DATA CENTER 1996.

Magnetic anomaly data of the former Soviet Union, Boulder CO, CD-ROM.

NECHEUKHIN, V.M. 2000. [Sedimentary basins of Urals and adjacent regions: structural features and minerageny.] *In*: KOROTEEV, A.V. (ed.) [Proceedings of the 4th regional Urals lithological conference] Rossiyskaya Akademiya Nauk, Ural'skoye Otdeleniye. Institut Geologii i Geokhimii, Yekaterinburg, Russian Federation, 88–97 [in Russian].

NIKISHIN, A.M., ZIEGLER, P.A., ABBOTT, D., BRUNET, M.F. & CLOETINGH, S. 2002. Permo-Triassic intraplate magmatism and rifting in Eurasia; implications for mantle plumes and mantle dynamics. *Tectonophysics*, **351**, 3–39.

NIMIS, P., OMENETTO, P., TESALINA, S.G., ZAYKOV, V.V., TARTAROTTI, P. & ORGEVAL, J.-J. 2003. Peculiarities of some mafic–ultramafic-hosted massive sulfide deposits from southern Urals. A likely forearc occurrence. *In*: ELIOPOULOS, D.G., ALLEN, C. *ET AL.* (eds) *Mineral Exploration and Sustainable Development*. Millpress, Rotterdam, 627–630.

PERFILIEV, A. 1979. *Ophiolite Belt of the Urals*. Geological Society of America Map. Chart Series, 9–12.

PETROV, G.A. & FRIBERG, M. 1999. The Salda metamorphic complex (Middle Urals, Russia) – a root of a Middle Paleozoic island arc. *Journal of Conference Abstracts*, EUG-10, **4**, 81.

PROKIN, V.A. & POLTAVETS, YU.A. 1996. [Geodynamic formation conditions of endogenic copper and iron deposits in the Urals.] *Yearbook – 1995*. Inst. Geol. Geochem., Yekaterinburg, 161–165 [in Russian].

PUCHKOV, V.N. 1993. [Paleo-oceanic structures of the Urals.] *Geotektonika*, **3**, 18–33 [in Russian].

PUCHKOV, V.N. 1997. Structure and geodynamics of the Uralian orogen. *In*: BURG, J-P. & FORD, M. (eds) *Orogeny Through Time*. Geological Society, London, Special Publications, **121**, 201–236.

PUCHKOV, V.N. 2000. [*Paleogeodynamics of the Central and Southern Urals*.] Dauriya, Ufa, 145 pp. [in Russian].

PUCHKOV, V. 2002. Paleozoic evolution of the East European continental margin involved in the Urals. *In*: BROWN, D., JUHLIN C. & PUCHKOV,V. (eds) *Mountain Building in the Uralides: Pangea to the Present*. AGU Geophysical Monograph, **132**, 9–31.

PUCHKOV, V.N. 2003. Uralides and Timanides: their structural relationship and position in the geologic history of the Ural-Mongolian fold belt. *Russian Geology and Geophysics*, **44**, 28–39.

SAVELIEVA, G.N., SHARASKIN, A.Y., SAVLIEV, A.A., SPADEA, P. & GAGGERO, L. 1997. Ophiolites of the southern Uralides adjacent to the East European continental margin. *Tectonophysics*, **276**, 117–138.

SAVELIEVA, G.N., SHARASKIN, A.YA., SAVELIEV, A.A., SPADEA, P., PERTSEV, A.N. & BABARINA, I.I. 2002. Ophiolites and zoned mafic-ultramafic massifs of the Urals: A comparative analysis and some tectonic implications. *In*: BROWN, D., JUHLIN, C.

& PUCHKOV, V. (eds) *Mountain Building in the Uralides: Pangea to the Present*. AGU Geophysical Monograph, **132**, 135–153.

SAZONOV, V.N., VAN HERK, A.H. & BOORDER, H. DE 2001. Spatial and temporal distribution of gold deposits in the Urals. *Economic Geology*, **96**, 683–701.

SENGOR, A.M.C., NATAL'IN, B.A. & BURTMAN, V.S. 1993. Evolution of the Altaid tectonic collage and Paleozoic crustal growth in Eurasia. *Nature*, **364**, 299–307.

SERAVKIN, I.B., ZNAMENSKY, S.E., KOSAREV, A.M., RYKUS, M.V., SALIKHOV, D.N., SNACHEV, V.I. & MOSEICHUK, W.M. 1994. *Vulkanogennaya metallogeniya Yuzhnogo Urala* [*Volcanic Metallogenesis of the Southern Urals*]. Nauka, Moscow, 152 pp. [in Russian].

SILLITOE, R.H. 2003. Iron oxide–copper–gold deposits: an Andean view. *Mineralium Deposita*, **38**, 787–812.

SPADEA, P., KABANOVA, L.YA. & SCARROW, J.H. 1998. Petrology, geochemistry and geodynamic significance of mid-Devonian boninitic rocks from the Baimak-Buribai area (Magnitogorsk Zone, Southern Urals). *Ofioliti*, **23**, 17–36.

YAKUBCHUK, A.S. 2001. New opportunities in deciphering the tectonic zonation of the Urals: A metallogenic implication. GEODE Workshop Abstract Volume, European Science Foundation, p. 9.

YAZEVA, R.G. & BOCHKAREV, V.V. 1993. [Post-collisional Devonian magmatism of Northern Urals.] *Geotektonika*, **4**, 56–65 [in Russian].

YAZEVA, R.G. & BOCHKAREV, V.V. 1995. [Urals' Silurian island arc: structure, development, geodynamics.] *Geotectonika*, **6**, 32–44 [in Russian].

YAZEVA, V.V. & BOCHKAREV, R.G. 1996. The magmatic formation of the Variscides of the Urals as indicators of a collision between island arc and a continent. *Geotectonics*, **35**, 62–71.

ZAYKOV, V.V., MASLENNIKOV, V.V., ZAYKOVA, E.V. & HERRINGTON, R.J. 1996. Hydrothermal activity and segmentation in the Magnitogorsk – West Mugodjarian zone on the margins of the Urals paleo-ocean, Tectonic, Magmatic, Hydrothermal and Biological Segmentation of the Mid Ocean Ridges. *In*: MACLEOD, C.J., TYLER, P. & WALKER, C.L. (eds) *Tectonic, Magmatic, Hydrothermal & Biological Segmentation at Mid Ocean Ridges*. Geological Society, London, Special Publications, **118**, 199–210.

ZOLOEV, K.K., VOLCHENKO, YU.A., KOROTEEV, V.A., MALAKHOV, I.A., MARDIROSYAN, A.N. & KHRYPOV, V.N. 2001. [*The Platinum-Metal Ores in the Geological Complexes of the Urals*.] Ekaterinburg, 199pp. [in Russian].

ZONENSHAIN, L. & MATVEENKOV, V. (eds) 1984. [*History of development of the Uralian Palaeo-ocean*.] Institute of Oceanology, USSR Academy of Sciences, Moscow, 164 pp. [in Russian].

ZONENSHAIN, L.P., KUZMIN, M.I. & NATAPOV, L.M. 1990. *Geology of the USSR: A plate tectonic synthesis*. American Geophysical Union Geodynamics Series Monograph **21**, 242 pp.

The terrestrial record of stable sulphur isotopes: a review of the implications for evolution of Earth's sulphur cycle

JAMES FARQUHAR & BOSWELL A. WING

Earth System Science Interdisciplinary Center and Department of Geology, University of Maryland, College Park, Maryland 20742 (e-mail: jfarquha@essic.umd.edu)

Abstract: The observation of anomalous (non mass-dependent) sulphur isotope compositions in Archaean and early Proterozoic rocks but not in rocks younger than approximately 2 Ga has been interpreted to reflect fundamental change in the terrestrial sulphur cycle, in atmospheric chemistry, and in atmospheric oxygen content. Similar non mass-dependent sulphur isotope compositions in present-day samples (atmospheric aerosols and ice-core horizons containing remnants of stratosphere-piercing volcanic eruptions) are interpreted to carry information about modern atmospheric chemistry and transport. The interpretation of these observations hinges on our understanding of the processes that produce non mass-dependent sulphur isotope compositions, the processes that transport and transfer the isotopic signals throughout the sulphur cycle, and the processes that act to preserve or erase these isotopic signals once they are established. The growing dataset and hypotheses related to non mass-dependent sulphur are evaluated, emphasizing that which remains to be learned about the evolution of the record, the compositions of key reservoirs, and the transfer of the signal from the atmosphere to the surface and ultimately to the deep Earth.

This review developed from a lecture given at the 2003 Fermor Flagship Meeting in Cardiff, Wales. It examines the geological record of the minor sulphur isotopes (^{33}S and ^{36}S). The distinguishing feature of this record is the presence of anomalous sulphur isotope compositions of Archaean and early Proterozoic sulphides and sulphates that cannot be explained by typical geological and biological processes. Our goal is to present a summary of minor isotope data, to examine hypotheses that attempt to account for the data, and to discuss the significance of these hypotheses with regard to the evolution of early Earth's surface environment.

The anomalous sulphur isotope signatures found in a variety of Archaean and Proterozoic rocks by Farquhar *et al.* (2000*a*) were anticipated by Tom Hoering's discovery of the signature in pyrite and barite from a single sample of the high-grade gneissic Sargur Group of India (Hoering 1989). The anomalous signature in many rocks older than *c.* 2.45 Ga was similar to signatures observed in the products of photochemical experiments, and led to the suggestion that an atmospheric influence on Earth's early sulphur cycle was preserved in the sulphur isotope record. The preservation of an atmospheric signal implies that the persistent redox recycling of surface sulphur by biological and weathering processes was not as vigorous early in Earth's history as it is today. An abrupt diminution of the signature occurred in rocks younger than *c.* 2.45 Ga, which was interpreted

as an invigoration of the redox recycling of sulphur and as an indication of the global rise of atmospheric oxygen (Farquhar *et al.* 2000*a*).

Preliminary experiments demonstrated that photochemical reactions involving sulphur dioxide produce isotopic characteristics that are similar to the isotopic signal in the Archaean rock record (Farquhar *et al.* 2000*b*, 2001). The wavelength dependence of these reactions led to the suggestion that they will operate significantly only in low oxygen atmospheres. Accordingly, the presence of isotopic effects in the rock record could be used to place quantitative upper limits on the amount of oxygen in the Archaean atmosphere. Kasting (2001) and Pavlov & Kasting (2002) examined atmospheric networks for sulphur chemistry and suggested that, in low-oxygen atmospheres, elemental sulphur aerosols are stabilized. In concert with the deposition of sulphate aerosols, this enables at least two pathways for transfer of sulphur to Earth's surface. Increased transfer efficiency of the atmospheric isotopic signal, therefore, places another constraint on oxygen levels in the atmosphere of early Earth.

Recent studies have provided confirmation of the original observations and expanded the interpretations of Farquhar *et al.* (2000*a*, 2001), Kasting (2001), and Pavlov & Kasting (2002). These studies demonstrate that anomalous sulphur isotope variations at scales from metres (Hu *et al.* 2003; Ono *et al.* 2003) to a single thin section (Mojzsis *et al.* 2003) can capture a

From: MCDONALD, I., BOYCE, A. J., BUTLER, I. B., HERRINGTON, R. J. & POLYA, D. A. (eds) 2005. *Mineral Deposits and Earth Evolution.* Geological Society, London, Special Publications, **248**, 167–177. 0305-8719/$15.00
© The Geological Society of London 2005.

significant portion of the entire range of observations for the Proterozoic and Archaean rock record. Significant stratigraphic control indicates a complex sulphur cycle dominated by at least two sources of sulphur during the Archaean (Ono *et al.* 2003) whereas a diminished sulphur isotope anomaly in radiometrically-dated early Proterozoic samples suggests that the dominant atmospheric influence on early Earth's sulphur cycle ended between 2.322 and 2.45 Ga (Bekker *et al.* 2004).

This review starts with a discussion of the sulphur isotope notation and a description of the experimental studies that form the foundation for many of the hypotheses in this study. These experiments are used to formulate the working model that was used as a basis for interpreting the evolution of Earth's sulphur cycle. This is followed by a description of the sulphur isotope data from the geological record and an interpretation of the variation in the abundance of minor sulphur isotopes (^{33}S, ^{36}S) with time, and of their covariation with each other and with ^{34}S abundance. Finally, there is a discussion of the implications of the isotope record for the evolution of the sulphur cycle.

Mass-dependent and non mass-dependent isotopic fractionations

Isotopic fractionations occur because of small differences in the chemical and physical behaviour of isotopically-substituted species. Most fractionation processes can be thought of as mass-dependent because the drive for fractionation is primarily due to small differences in the masses of the isotopes. Almost all geologically relevant fractionations (equilibrium, kinetic, and biological) are mass-dependent and lead to characteristic fractionation arrays with a slope that is proportional to the relative difference between the isotopes being fractionated (see Urey 1947; Hulston & Thode 1965; Matsuhisa *et al.* 1978; Mook 2000; Young *et al.* 2002). For example Figure 1 illustrates one of the characteristic arrays that is produced by mass-dependent fractionation of sulphur isotopes. Figure 1 plots $\delta^{33}S$ [= $((^{33}S/^{32}S)_{sample}/(^{33}S/^{32}S)_{reference} - 1) \times 1000$] versus $\delta^{34}S$ [= $((^{34}S/^{32}S)_{sample}/(^{34}S/^{32}S)_{reference} - 1) \times 1000$], and which fall on an array $\delta^{33}S \sim 0.5 \, \delta^{34}S$. A second group of isotopic fractionation processes has been called non mass-dependent because they do not follow the traditional mass-dependent fractionation relationships, and it is inferred that factors other than mass contribute to these isotopic fractionations. This designation of non

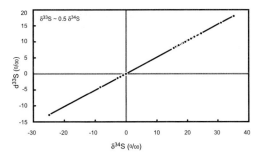

Fig. 1. Plot of $\delta^{33}S$ versus $\delta^{34}S$ for terrestrial sulphide and sulphate (data from Farquhar *et al.* 2002; Heyman *et al.* 1998). The array of $\delta^{33}S \sim 0.5 \, \delta^{34}S$ can best be described by text Equation 1 with a λ value of 0.515. The array reflects a collection of mass-dependent isotopic fractionation processes and the relative mass differences of 1 and 2 atomic mass units that are associated with the $^{33}S/^{32}S$ and $^{34}S/^{32}S$.

mass-dependent does not imply that mass terms are absent from the physics and chemistry that control these fractionation processes.

Both mass-dependent and non mass-dependent effects can be described using an exponential term (λ) to capture the mass-dependent variation within an array and another term (k) to describe the relationship between mass dependent arrays (Miller 2002). Using ^{33}S as an example, this leads to

$$(1 + \delta^{33}S/1000) = (1 + \delta^{34}S/1000)^{33\lambda}(1 + {}^{33}k). \quad (1)$$

Values for λ are not constant for all processes, but the variability for λ values is small. Typical values of λ for $\delta^{33}S$ and $\delta^{34}S$ range from 0.514 to 0.516 for equilibrium exchange (Hulston & Thode 1965; Farquhar & Wing 2003), and from 0.510 to 0.516 for biological fractionation processes (Farquhar *et al.* 2003). In this study we use 0.515 for $^{33}\lambda$ and 1.90 for $^{36}\lambda$. The quantities $\Delta^{33}S$ and $\Delta^{36}S$ are given as $1000^{33}k$ and $1000^{36}k$, respectively. $\Delta^{33}S$ and $\Delta^{36}S$ are often used to describe non mass-dependent isotopic fractionation effects.

The theoretical treatment for mass-dependent isotope effects are well understood; however, the theoretical understanding of non mass-dependent fractionation effects is incomplete, but treatments of non mass-dependent fractionation effects have been made for specific cases (see Hathorn & Marcus 1999, 2000; Zmolek *et al.* 1999; Bhattacharya *et al.* 2000; Gao & Marcus 2001, 2002). The present understanding of non mass-dependent isotope effects places these effects for the most part with gas-phase reactions. The only exception to this is a class of isotope effects called magnetic isotope effects that are limited to fractionations involving odd-mass isotopes (see Turro 1983 for a discussion). Since the $\Delta^{36}S$ involves only even-mass sulphur isotopes, it is not susceptible to disturbance by magnetic isotope effects and can be used to test whether an observed non mass-dependent isotopic fractionation is consistent with a magnetic isotope effect or a non mass-dependent isotope effect associated with a gas-phase reaction.

Laboratory model for the origin of non mass-dependent sulphur isotope fractionations

A number of studies have investigated the sulphur isotope fractionations produced when sulphur-bearing gases are irradiated by ultraviolet and visible radiation (e.g. Colman *et al.* 1996; Zmolek *et al.* 1999; Farquhar *et al.* 2000*b*, 2001). These studies have demonstrated that it is possible to produce large non mass-dependent sulphur isotope effects involving all sulphur isotopes in gas-phase reactions.

A detailed description of the isotopic effects of experimental photolysis of sulphur dioxide is reported in Farquhar *et al.* (2000*b*, 2001), and these results are summarized here. Photolysis under a continuum light source that emitted at >220 nm yielded sulphate that was enriched in ^{33}S, ^{34}S, and ^{36}S relative to residual SO_2. The isotopic effects were non mass-dependent and followed trends of $\delta^{33}S$ *c.* 0.7 $\delta^{34}S$, $\delta^{36}S$ *c.* 1.8 $\delta^{34}S$, and $\Delta^{36}S/\Delta^{33}S$ *c.* –1.0. Experiments conducted with a low pressure mercury resonance lamp that emitted two lines at 184.9 nm and 253.7 nm yielded sulphate that was enriched in ^{33}S and ^{36}S but depleted in ^{34}S relative to residual SO_2 and elemental sulphur that was enriched in ^{33}S but depleted in ^{34}S and ^{36}S relative to residual SO_2. The isotopic effects followed trends of $\delta^{33}S$ *c.* –1.1 $\delta^{34}S$, $\delta^{36}S$ *c.* 0.6 $\delta^{34}S$, and $\Delta^{36}S/\Delta^{33}S$ *c.* 1.0. Experiments with 193 nm light from an excimer laser yielded sulphate that was enriched in ^{33}S, ^{34}S, and ^{36}S relative to residual SO_2 and elemental sulphur that was enriched in ^{33}S but depleted in ^{34}S and ^{36}S relative to residual SO_2. The isotopic effects followed trends of $\delta^{33}S$ *c.* –0.6 $\delta^{34}S$, $\delta^{36}S$ *c.* 1.5 $\delta^{34}S$, and $\Delta^{36}S/\Delta^{33}S$ *c.* –0.9. Experiments with 248 nm light from an excimer laser with ArF fill gas yielded sulphate that was enriched in ^{33}S, ^{34}S, and ^{36}S relative to residual SO_2. The isotopic effects followed trends of $\delta^{33}S$ *c.* 0.6 $\delta^{34}S$, $\delta^{36}S$ *c.* 1.5 $\delta^{34}S$, and $\Delta^{36}S/\Delta^{33}S$ *c.* –4.0. These results provide a laboratory-based model to explain the origin of anomalous sulphur isotopic signatures in the rock record. Discrimination diagrams of Δ–δ and Δ–Δ allow the most likely experimental candidates for the measured anomalies to be determined (Fig. 2a, b). In these

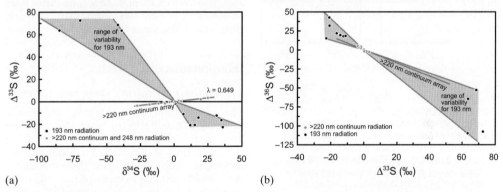

(a)　　　　　　　　　　　　　　　(b)

Fig. 2. Plots of (**a**) $\Delta^{33}S$ versus $\delta^{34}S$ (**b**) and $\Delta^{36}S$ versus $\Delta^{33}S$ for closed cell experiments with sulphur dioxide to illustrate the dependence of isotopic fractionation effects associated with sulphur dioxide chemistry that are produced by changes in incident radiation. (**a**) Plot illustrates the different fractionation behaviour of sulphur dioxide as expressed by $\Delta^{33}S$ and $\delta^{34}S$ when it is exposed to ultraviolet radiation at wavelengths between 193 nm and >220 nm. Filled black circles are for data from experiments undertaken with 193 nm radiation (argon fluoride excimer laser). Filled grey circles are for data produced by experiments undertaken with a high pressure xenon lamp producing radiation at wavelengths >220 nm, and an Excimer laser with krypton fluoride fill gas that produces radiation at 248 nm (data from Farquhar *et al.* 2000*b*, 2001). (**b**) Plot illustrates the different fractionation behaviour of sulphur dioxide as expressed by $\Delta^{36}S$ and $\Delta^{33}S$ when it is exposed to ultraviolet radiation at wavelengths between 193 nm and >220 nm. The grey field represents the range of variability for the 193 nm experiments and the grey line labelled >220 nm continuum array is fit to the data from experiments with the xenon arc lamp. These fields are the same fields that are plotted as reference fields on Figure 4b.

figures we emphasize results from the 193 nm and >220 nm experiments because they bracket the observations from the geological record, and they are thought to be relevant in primitive Earth atmospheres.

Observations: the nature of the Δ^{33}S record

An updated diagram of Δ^{33}S versus time is presented in Figure 3. These data can be used to divide the geological record of Δ^{33}S into three stages. Stage I extends from 2.45 Ga to the earliest Archaean and is characterized by large magnitude non-zero Δ^{33}S values. Stage II extends from *c.* 2 Ga to 2.45 Ga and is characterized by smaller magnitude Δ^{33}S variation ($< \pm 0.5$‰), and Stage III extends from present to *c.* 2 Ga and is characterized by mass-dependent sulphur isotope fractionations. The use of Δ^{33}S to subdivide Earth's sulphur cycle into different stages is complementary to the use of

δ^{34}S as a criterion for describing temporal changes in the sulphur cycle.

Additional information about the sulphur isotopic effect can be obtained by comparing data for Δ^{36}S with data for Δ^{33}S. Farquhar *et al.* (2000*a*) observed a negative correlation between Δ^{33}S and Δ^{36}S for Stage I samples (Fig. 4b) that they interpreted to indicate that the same process was responsible for both the Δ^{33}S and the Δ^{36}S variations. The Δ^{36}S/Δ^{33}S correlation is not significantly affected by mass-dependent fractionation processes, and therefore, is characteristic of the non mass-dependent reactions that produced the signature in the geological record. It also forms the basis for interpreting the non mass-dependent signature in the Stage I record as a gas-phase signature rather than the product of a magnetic isotope effect. To date the difficulty associated with measurement of Δ^{36}S has left unresolved the issue of whether the Δ^{36}S/Δ^{33}S of the Stage II samples is the same as that for the Δ^{36}S/Δ^{33}S of the Stage I samples. Recent data for sulphate recovered from ice cores has been shown to have non-zero Δ^{36}S and Δ^{33}S, but a Δ^{36}S/Δ^{33}S that is different from that observed for Stage I samples (Savarino *et al.* 2003). The origin of the isotopic anomaly in this ice core sulphate has been attributed to stratospheric chemistry involving sulphur dioxide that was injected into the stratosphere by stratosphere-piercing plinian volcanic eruptions. Small, but measureable non-zero Δ^{36}S and Δ^{33}S have also been observed in present-day atmospheric aerosols (Romero & Thiemans 2003).

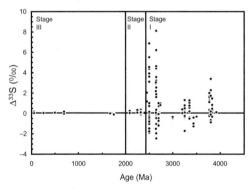

Fig. 3. Plot of Δ^{33}S versus time using published data (Farquhar *et al.* 2000*a*; Hu *et al.* 2003; Mojzsis *et al.* 2003; Ono *et al.* 2003; Bekker *et al.* 2004). The record is divided into three stages (I, II, & III). Stage I is characterized by large positive and negative Δ^{33}S (greater than +/– 0.5 ‰) and includes samples 2.45 Ga and older. Stage II is characterized by a smaller range of Δ^{33}S values, but still preserves evidence for non mass-dependent fractionation processes in rock samples. Stage II begins after 2.45 Ga, but the end of Stage II is not well established. It is presently set at *c.* 2 Ga. Stage III is characterized by isotopic fractionations without obvious contributions from non mass-dependent isotopic effects. The onset of Stage III occurs some time after *c.* 2 Ga and continues until present. The transition between Stage I and Stage II is attributed to the initial rise of atmospheric oxygen. The transition from Stage II to Stage III may reflect the establishment of a stratospheric ozone shield, or alternatively, the importance of biology and weathering in the sulphur cycle.

Interpretations

Implications for Earth's early atmosphere

The observation of a negative correlation between Δ^{33}S and Δ^{36}S for Stage I samples was used to rule out one class of isotope effects (magnetic isotope effects) and to argue that the anomalous sulphur signature observed in the Archaean record has an atmospheric origin. The similarity between arrays produced by experiments with sulphur dioxide and the Stage I sulphur isotope data for Δ^{36}S versus Δ^{33}S (Fig. 4b) and Δ^{33}S versus δ^{34}S (Fig. 4a) provides our best candidate for the atmospheric source reaction. Farquhar *et al.* (2001) observed variable Δ^{36}S/Δ^{33}S for experiments at different wavelengths, and among the different products of some experiments at single wavelengths. These were interpreted to indicate the operation of multiple non mass-dependent isotopic fractionation steps and this possibility should be

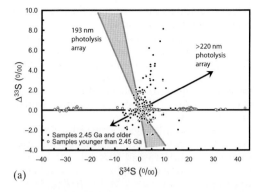

(a)

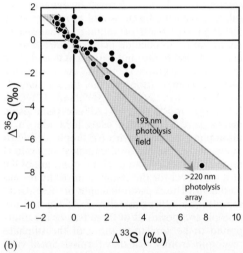

(b)

Fig. 4. Plots of (a) $\Delta^{33}S$ versus $\delta^{34}S$ (b) and $\Delta^{36}S$ versus $\Delta^{33}S$ for published data for sulphide and sulphate from the terrestrial rock record (Farquhar *et al.* 2000a; Hu *et al.* 2003; Mojzsis *et al.* 2003; Ono *et al.* 2003; Bekker *et al.* 2004) and for data collected at UMCP. Filled black circles are data for samples 2.45 Ga and older. Open grey circles are data for samples younger than 2.45 Ga. The plots illustrate the nature of the isotopic variability for sulphur in Stage I samples and Stage II and II samples. The plots also illustrate the similarities between data for Stage I samples and the arrays produced by closed cell photochemical experiments involving sulphur dioxide illustrated in Figure 2a and 2b. (a) ($\Delta^{33}S$ versus $\delta^{34}S$)Plot also includes vectors representing the sense of arrays produced by photolysis of sulphur dioxide with 193 nm and >220 nm radiation that is illustrated in Figure 2a. (b) ($\Delta^{36}S$ versus $\Delta^{33}S$) Plot illustrates the relationship between $\Delta^{36}S$ and $\Delta^{33}S$ for data older than 2.45 Ga. The fields on this plot for the 193 nm photolysis and for >220 nm photolysis are the fields produced by experiments with sulphur dioxide illustrated in Figure 2b. The scattering of the data to higher $\Delta^{36}S$ values above the origin is attributed to trace contaminants that contributed an interference to the 131 ion beam ($^{36}SF_5^+$) in the mass spectrometer for some of the samples analysed at UCSD by Farquhar *et al.* (2000a).

considered when evaluating the Stage I and Stage II data. The present $\Delta^{36}S/\Delta^{33}S$ data are not sufficient to evaluate this possibility, but work in the foreseeable future will allow these possibilities to be investigated.

Implications for Earth's pre 2.45 Ga surface environments

The variability of sulphur isotope compositions in the Archaean rock record is consistent with the hypothesis that deposition of atmospheric sulphur compounds with non-zero $\Delta^{33}S$ and $\Delta^{36}S$ strongly influenced the Archaean sulphur cycle. Published data for $\Delta^{33}S$, grouped by rock type, are presented in Figure 5. Analyses of shale, carbonate, and a variety of metamorphosed supracrustal rocks from Isua, Greenland have yielded $\Delta^{33}S$ values that are both positive and negative; however, analyses of barite, and sulphides from volcanogenic massive sulphide (VMS) deposits and banded iron formations (BIF) have yielded predominantly negative $\Delta^{33}S$. The negative $\Delta^{33}S$ values of barite and sulphides from VMS deposits and BIFs have been interpreted to indicate that the $\Delta^{33}S$ of surface sulphate is negative. The positive $\Delta^{33}S$ values of pyrite from many shale samples and from other samples has been interpreted to indicate that neutral or reduced sulphur species with positive $\Delta^{33}S$ were transferred from the atmosphere to the surface. The transfer of this sulphur isotope signature to pyrite may reflect biological activity or it may reflect abiological oxidation of FeS by S to form pyrite (e.g. Farquhar *et al.* 2001; Ono *et al.* 2003).

The variability of sulphur isotope compositions also indicates that surface cycling of sulphur during the Archaean was not sufficiently vigorous to homogenize $\Delta^{33}S$. If oxidation and reduction reactions allowed sulphur to be freely interconverted between sulphate and sulphide during the Archaean, the $\Delta^{33}S$ variations would not be preserved. Evidence for reduction of sulphate to pyrite is present and includes the negative $\Delta^{33}S$ of some shale samples, VMS sulphides, pyrites from the Carawine dolomite, sulphides from Isua, and pyrite from BIF. Strong evidence for active oxidation of sulphide to sulphate during the Archaean is lacking, and this may reflect the limited dataset for sulphate (barite) or it may reflect the nature of the sulphate reservoir as a predominantly dissolved reservoir that was well mixed, and less prone to large swings of $\Delta^{33}S$. The observation that barite samples from different localities have analytically resolvable

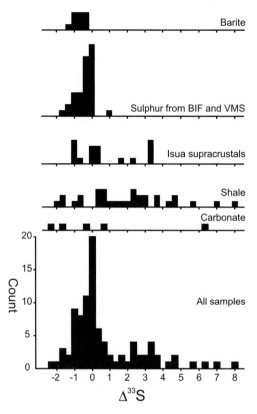

Fig. 5. Histograms of $\Delta^{33}S$ for samples older than 2.45 Ga. The plot illustrates the negative $\Delta^{33}S$ of barite and sulphide from VMS and BIF, the range of $\Delta^{33}S$ values extending to positive values for samples of shale and for the Isua supracrustals, and the range of values for pyrites from carbonate. Samples with negative $\Delta^{33}S$ are thought to have affinities with water-soluble (oceanic and other) sulphate, while samples exhibiting positive $\Delta^{33}S$ are thought to derive sulphur from another source that is ^{33}S enriched. This second source is thought to be an elemental sulphur of atmospheric origin that is reduced by reactions such as $FeS + S = FeS_2$.

negative $\Delta^{33}S$ values suggests that the $\Delta^{33}S$ of surface sulphate reservoirs was influenced either by variable contributions from atmospheric sources or by the addition of positive $\Delta^{33}S$ from sulphide.

Implications for Earth's pre 2.45 Ga oceanic sulphate reservoirs

A bootstrap technique was used to evaluate the mean of the published $\Delta^{33}S$ data. The mean for hand samples (e.g. not including measurements by SIMS) is 0.89‰ ± 0.21‰ (s). Since all

evidence points to a near zero $\Delta^{33}S$ value for juvenile sulphur, this positive $\Delta^{33}S$ for measured samples indicates a bias in our present sample set that underrepresents a negative $\Delta^{33}S$ Archaean sulphur reservoir. The mean for Isua supracrustals has also been evaluated using these techniques. It yields a similar value of 0.83‰ ± 0.40‰ (s) and is consistent with a 'missing' negative $\Delta^{33}S$ reservoir.

Three of the lithologies analysed so far have yielded negative $\Delta^{33}S$ values: barite, and sulphides from VMS and BIF. All of these may have inherited their sulphur from a sulphate reservoir with negative $\Delta^{33}S$. The BIF and VMS sulphides would have formed by reduction of sulphate, and the barite would have formed by direct precipitation of sulphate. Analyses of Archaean barite include massive hydrothermal barite, bladed barite, and barite sands, and all of these have negative $\Delta^{33}S$. The mean sulphur isotopic composition of the barite data is −0.73‰ ± 0.10‰ (s) for $\Delta^{33}S$ and 4.10‰ ± 0.25‰ (s) for $\delta^{34}S$ (Fig. 6). (All uncertainties on means were determined using a Monte Carlo resampling technique with 1000 replicates. Estimates of uncertainty on the means were determined using the resampled data and are given at the 1s level for the population of the resampled data.) The depositional ages of these barite samples are approximately 3.3–3.5 Ga and the isotopic compositions of these barites are interpreted to be similar to those of the sulphate reservoirs from which they formed. Small variations for $\Delta^{33}S$ of barite from different localities suggest that the $\Delta^{33}S$ of the sulphate reservoirs was not constant.

Other indicators of an Archaean oceanic sulphate reservoir with negative $\Delta^{33}S$ include the observations of pyrite in the Carawine dolomite, pyrites with negative $\Delta^{33}S$ that have been reported for shale samples from the McRae formation (Ono *et al.* 2003), and pyrites from the Dales Gorge member BIF of the Hammersley group (Mojzsis *et al.* 2003). All of these have been attributed to bacterial reduction of oceanic sulphate with negative $\Delta^{33}S$. Three pyrite analyses from the Carawine dolomite by Ono *et al.* (2003) are particularly interesting when considered in the context of Archaean oceanic sulphate. The $\delta^{34}S$ values of these three samples are 4.6‰, 9.7‰, and 16.0‰, and the $\Delta^{33}S$ values are −2.46‰, −0.45‰, and −1.50‰, respectively. Ono *et al.* (2003) suggest that these pyrites reflect nearly complete reduction of sulphate and indicate that the $\delta^{34}S$ of oceanic sulphate was >10‰ at the time of the Carawine dolomite formation. This implies the presence and influence of sulphate reducers on

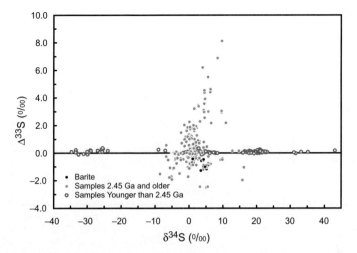

Fig. 6. Plot of $\Delta^{33}S$ versus $\delta^{34}S$ for published data for sulphide and sulphate from the terrestrial rock record (Farquhar *et al.* 2000a; Hu *et al.* 2003; Mojzsis *et al.* 2003; Ono *et al.* 2003; Bekker *et al.* 2004). The plot illustrates the negative $\Delta^{33}S$ and small range of $\delta^{34}S$ for barite samples. These data form the basis for the hypothesis that the water-soluble (oceanic) sulphate reservoir had negative $\Delta^{33}S$ during the first half of Earth's history. Filled black circles are data for barite samples 2.45 Ga and older. Open grey circles are data for samples younger than 2.45 Ga. Filled grey circles are for rock types other than barite that are older than 2.45 Ga as presented in Figure 4a.

Carawine environments and possibly high enough sulphate concentrations so that their actions would influence the isotopic composition of this inferred oceanic sulphate reservoir by *c.* 2.6 Ga.

An operating hypothesis for the Archaean oceans includes an ocean with low sulphate concentrations that is maintained by the balance between sulphate sources that are predominantly atmospheric and sulphate sinks that include both biological and abiological sulphate reduction. The concentration of sulphate is inferred to be significantly lower than at present because atmospheric sources were not large, and the isotopic composition of oceanic sulphate is inferred to have negative $\Delta^{33}S$ values. The concentration and isotopic composition of sulphate is not inferred to be constant in space or time.

Implications for recycling of sulphur to the pre 2.45 Ga mantle

Farquhar *et al.* (2002) reported SIMS analyses of $\Delta^{33}S$ and $\delta^{34}S$ for sulphide inclusions from diamonds mined at the Orapa kimberlite pipe, Botswana, Africa (Fig. 7). Prior studies of sulphide inclusions from this kimberlite pipe have demonstrated a wide range of $\delta^{34}S$ values

(Chaussidon *et al.* 1987; Eldridge *et al.* 1991) and also the existence of two distinct populations of inclusions with Re–Os ages of 2.9 Ga and 1.0 Ga (Shirey *et al.* 2001, 2002). For the minor sulphur isotope study, sulphide inclusions were liberated from twelve diamonds and analysed by multi-collector secondary ion mass spectrometry. The sulphide inclusions that were studied have low Ni contents (2.5 wt% mean), and high Cu and Co contents (0.55 wt% and 0.27 wt%, respectively) suggesting an eclogite-type affinity. One of the diamonds also contained minute silicate inclusions of e-type garnet and clinopyroxene.

Three of the diamonds yielded populations of sulphide inclusions with $\Delta^{33}S$ values that were statistically resolvable at the three-sigma level or higher from a $\Delta^{33}S$ value of 0‰. A fourth was resolvable at the two-sigma level. The $\Delta^{33}S$ and $\delta^{34}S$ values of these inclusions overlie the field of data for 2.45 Ga samples and have been interpreted to indicate that the sulphur in these sulphide inclusions reflected recycling of sulphur from an Archaean surface reservoir to the diamond source area. Farquhar *et al.* (2002) did not favour the interpretation that the variable $\Delta^{33}S$ reflected a primordial signature of mantle heterogeneity. Although some minor meteoritic components have been shown to have non-zero $\Delta^{33}S$, numerous studies of bulk extracts of sulphide and sulphate from

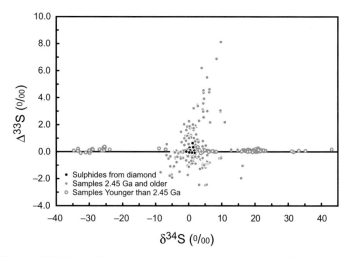

Fig. 7. Plot of $\Delta^{33}S$ versus $\delta^{34}S$ for published data for sulphide and sulphate from the terrestrial rock record (Farquhar *et al.* 2000*a*; Hu *et al.* 2003; Mojzsis *et al.* 2003; Ono *et al.* 2003; Bekker *et al.* 2004). The plot illustrates the variable $\Delta^{33}S$ for sulphide inclusions from diamond that support recycling of sulphur to the diamond source area and has implications for surface–mantle sulphur exchange and tectonic recycling processes early in Earth history. Filled black circles are data for sulphide inclusions from diamonds. The diamonds are from the Orapa Kimberlite pipe, Botswana, and data are from Farquhar *et al.* (2002). Open grey circles are data for samples younger than 2.45 Ga. Filled grey circles are for rock types other than barite that are older than 2.45 Ga as presented in Figure 4a.

meteorites indicate a very homgenous $\Delta^{33}S$ of between –0.01 and 0.04‰ for $\Delta^{33}S$ (cf. Gao & Thiemens 1991, 1993*a*, *b*). Their preferred hypothesis extends the scale of the Archaean sulphur cycle to also include recycling of sulphur from the surface into the mantle and raises the possibility that sulphur with anomalous $\Delta^{33}S$ may have been recycled to other mantle reservoirs.

The record from 2.45 Ga to 2.0 Ga

The presence of a damped but measurable non mass-dependent sulphur isotope signal in samples that are younger than 2.45 Ga but older than 2.0 Ga implies a change in Earth's sulphur cycle. This time interval is also coincident with the larger range of observed $\delta^{34}S$ values (cf. Canfield 2001) that has been attributed to an increase in the concentration of sulphate in the oceans. The amplification of the $\delta^{34}S$ and the dampening $\Delta^{33}S$ signal can be attributed to the transition from a sulphur cycle like the one described above where the atmosphere exerts a significant influence on the isotopic composition and oxidation states of sulphur in surface environments, to a sulphur cycle where biology and oxidative weathering play a significant role. The $\delta^{34}S$ of sulphate and sulphide since *c.* 2.45 Ga reflect biological reduction in

environments with higher sulphate concentrations that were maintained by more significant inputs from oxidative weathering and biological activity (e.g. Canfield *et al.* 2000; Canfield 2001; Habicht *et al.* 2002; Strauss 2003).

Pavlov & Kasting (2002) argued that efficient transfer of more than one form of non mass-dependent sulphur from the atmosphere to the surface was a necessary for the establishment of a large $\Delta^{33}S$ signal in the rock record. Their models of atmospheric chemistry implied that the efficiency of this transfer depends directly to the oxidation state of the atmosphere. In model atmospheres with less than *c.* 10^{-5} of the present level of atmospheric oxygen (PAL), elemental sulphur aerosols (S_8) are stabilized and the transfer of sulphur by aerosols from the atmosphere to the surface occurs both as oxidized (H_2SO_4) and neutral (S_8) species. At higher oxygen levels, the production of S_8 ceases, and most sulphur is transferred in oxidized forms. The dramatic step in the range of observed $\Delta^{33}S$ in the sedimentary record after 2.45 Ga has been interpreted to reflect the rise of atmospheric oxygen to levels above 10^{-5} PAL (Pavlov & Kasting 2002; Bekker *et al.* 2004).

Calculations of SO_2 photolysis rates have been evaluated in the context of present-day constraints for the atmospheric lifetime of SO_2 to argue that photolysis of SO_2 will be important in atmospheres with less than $c.$ $10^{-1.5}$ PAL of atmospheric oxygen (Farquhar *et al.* 2001). If the $\Delta^{33}S$ signal derives from chemistry associated with SO_2 photolysis, the $\Delta^{33}S$ signal would be present in atmospheres with oxygen levels intermediate between 10^{-5} and $c.$ $10^{-1.5}$ PAL. The damped $\Delta^{33}S$ signal observed in samples younger than 2.45 Ga may reflect stabilization of atmospheric oxygen to levels between 10^{-5} and $c.$ $10^{-1.5}$ of PAL. According to this hypothesis, the disappearance of the $\Delta^{33}S$ signal would reflect the rise of oxygen levels above $c.$ $10^{-1.5}$ PAL. It is not known whether atmospheric oxygen levels rose in singular, unidirectional steps, or whether oscillations occurred during the rise of atmospheric oxygen. The present data do not allow us to resolve the possibility that the large $\Delta^{33}S$ signal did not disappear at some time before 2.45 Ga, nor does our present data allow us to examine the possibility that atmospheric oxygen levels oscillated between intermediate levels during the period immediately after 2.45 Ga. Current age constraints indicate that the step from a large non-zero $\Delta^{33}S$ to a damped signal occurs between 2.45 Ga and 2.322 Ga (Bekker *et al.* 2004).

Oceanic sulphate from 2.45 Ga to 2.0 Ga

Little is known about the evolution of $\Delta^{33}S$ of oceanic sulphate. Measurements of $\Delta^{33}S$ in 2.0–2.45 Ga sediments have yielded mostly positive values. The onset of oxidative weathering during this interval could have caused a significant influx of sulphate with positive $\Delta^{33}S$ weathered from exposed Archaean shales into the oceans. However, oxidative weathering of igneous rocks would have contributed sulphate with $\Delta^{33}S$ $c.$ 0‰ and the combined action of these two new sources for sulphate could have caused the $\Delta^{33}S$ of oceanic sulphate to shift from negative to positive values. The answer to this question may not be obtained with $\Delta^{33}S$ data alone. Recently, Savarino *et al.* (2003) reported significant non-zero $\Delta^{33}S$ for sulphate recovered from ice cores. The sulphate-rich horizons were deposited following stratosphere-piercing plinian volcanic eruptions, and the non-zero $\Delta^{33}S$ values were attributed to photochemical reactions in the stratosphere. The $\Delta^{36}S/\Delta^{33}S$ of the Savarino *et al.* (2003) data was distinct from that observed for Archaean samples and raises the possibility that $\Delta^{36}S/\Delta^{33}S$ might provide a means of further understanding the significance of the Stage II sulphur isotope record.

The post 2.0 Ga record

Data from sedimentary rocks studied to date indicates that Phanerozoic and late Proterozoic (Neoproterozoic) rocks do not possess a $\Delta^{33}S$ that can be attributed to non mass-dependent chemical processes. There is very little data for Mesoproterozoic samples and it is possible that Stage III does not begin until significantly later than 2.0 Ga.

The reason for the change from Stage II to Stage III remains unknown. It may reflect the time when biological reduction and oxidative weathering overwhelmed the atmospheric contributions to the Earth's surface sulphur cycle. It may also reflect a fundamental change in the atmospheric sulphur chemistry and its influence on the sulphur cycle as a whole. If the isotopic signal in Stage II derives from chemistry that requires ultraviolet radiation at wavelengths shorter than that allowed by atmospheric ozone, the change from Stage II to Stage III may reflect a change in the efficiency or altitude of Earth's ozone shield. Stratospheric ozone shields ultraviolet radiation and affects tropospheric chemistry, specifically chemistry induced by UV radiation at wavelengths shorter than 300 nm. The efficiency of the ozone shield for blocking UV radiation changes with atmospheric oxygen content (Kasting & Donahue 1980; Levine 1985). Farquhar *et al.* (2001) argued that oxygen levels higher than $c.$ $10^{-1.5}$ present atmospheric level (PAL) produced enough of an ozone shield to prevent sulphur dioxide from undergoing photolytic destruction at a rate higher than that of other chemical and physical sinks. The distinction between Stage II and Stage III may reflect the rise of oxygen above $c.$ $10^{-1.5}$ PAL and the formation of a stratospheric ozone shield, but this hypothesis remains unproven, and alternative interpretations that attribute the transition from Stage II to Stage III to increased biological activity and oxidative weathering remain viable.

Summary

The record of $\Delta^{33}S$, $\Delta^{36}S$, and $\delta^{34}S$ has been used to subdivide Earth's history into stages. The first stage (I) is characterized by a small range of $\delta^{34}S$ variations, a large range of $\Delta^{33}S$ variations, and $\Delta^{36}S/\Delta^{33}S$ around −1. A second stage (II) that begins after 2.45 Ga is characterized by a larger range of $\delta^{34}S$ variations and a smaller range of $\Delta^{33}S$ variations.

The significance of transition from Stage I to Stage II has been interpreted to reflect the rise of atmospheric oxygen. The assignment of the signal to an atmospheric source hinges on the observation of $\Delta^{36}S$ variations associated with the $\Delta^{33}S$ isotopic signal of Stage I because they point to a gas-phase rather than a magnetic isotope effect. The best present candidates for the source reaction of the isotope signature can produce variations for $\Delta^{33}S$, $\Delta^{36}S$, and $\delta^{34}S$, but they do not provide a clear cut match to the data for the Stage I record. Identification of new candidate reactions and refinement of our understanding of the present candidates remains a top priority. The transition from Stage I to Stage II also marks a transition from a Stage I sulphur cycle where the isotopic signatures and the oxidation of sulphide to sulphate are strongly influenced by atmospheric reactions to a Stage II sulphur cycle that still carries a small signature of non mass-dependent atmospheric reactions, but bears the imprint and isotopic signatures of oxidation and reduction reactions associated with weathering and biological activity.

There is also a suggestion of a third stage (III) for Earth's sulphur cycle that does not carry the signature of non mass-dependent atmospheric reactions. Recent studies with aerosol sulphate and ice-core sulphate, that have been attributed to stratosphere-piercing volcanic eruptions, suggest that a component of this atmospheric chemistry is still part of the sulphur cycle, but it remains to be demonstrated that this damped signature is altogether absent from the geological record, and also whether this is the same chemistry that produced the Stage I and Stage II records.

We gratefully acknowledge the assistance of David Johnston and Lisa Tuit. This work was supported by grants from the NSF-EAR, NASA-Exobiology and Astrobiology, and the ACS-PRF.

References

BEKKER A., HOLLAND, H.D., WANG, P.L., RUMBLE, D., STEIN, H.J., HANNAH, J.L., COETZEE, L.L. & BEUKES, N.J. 2004. Dating the rise of atmospheric oxygen. *Nature*, **427**, 117–120.

BHATTACHARYA, S.K., SAVARINO, J. & THIEMENS, M.H. 2000. A new class of oxygen isotope fractionation in photodissociation of carbon dioxide: Potential implications for atmospheres of mars and earth. *Geophysical Research Letters*, **27**, 1459–1462.

CANFIELD, D.E. 2001. Biogeochemistry of sulfur isotopes. *In*: Stable isotope geochemistry. *Reviews in Mineralogy and Geochemistry*, **43**, 607–636.

CANFIELD, D.E., HABICHT, K.S. & THAMDRUP, B. 2000. The Archaean sulphur cycle and the early history of atmospheric oxygen. *Science*, **288**, 658–661.

CHAUSSIDON, M., ALBAREDE, F. & SHEPPARD, S.M.F. 1987. Sulfur isotope heterogeneity in the mantle from ion microprobe measurements of sulfide inclusions in diamonds. *Nature*, **330**, 242–244.

COLMAN, J.J., XU, X.P., THIEMENS, M.H. & TROGLER, W.C. 1996. Photopolymerization and mass-independent sulfur isotope fractionations in carbon disulfide. *Science*, **273**, 774–776.

ELDRIDGE, C.S., COMPSTON, W., WILLIAMS, I.S., HARRIS, J.W. & BRISTOW, J.W. 1991. Isotope evidence for the involvement of recycled sediments in diamond formation. *Nature*, **353**, 649–653.

FARQUHAR, J. & WING, B.A. 2003. Multiple sulfur isotopes and the evolution of the atmosphere. *Earth and Planetary Science Letters*, **213**, 1–13.

FARQUHAR, J., BAO, H.M. & THIEMENS, M. 2000*a*. Atmospheric influence of earth's earliest sulfur cycle. *Science*, **289**, 756–758.

FARQUHAR, J., SAVARINO, J., JACKSON, T.L. & THIEMENS, M.H. 2000*b*. Evidence of atmospheric sulphur in the martian regolith from sulphur isotopes in meteorites. *Nature*, **404**, 50–52.

FARQUHAR, J., SAVARINO, J., AIRIEAU, S. & THIEMENS, M.H. 2001. Observation of wavelength-sensitive mass-independent sulfur isotope effects during SO$_2$ photolysis: Implications for the early atmosphere. *Journal of Geophysical Research – Planets*, **106**, 32 829–32 839.

FARQUHAR, J., WING, B.A., MCKEEGAN, K.D., HARRIS, J.W., CARTIGNY, P. & THIEMENS, M.H. 2002. Mass-independent sulfur of inclusions in diamond and sulfur recycling on early earth. *Science*, **298**, 2369–2372.

FARQUHAR, J., JOHNSTON, D.T., WING, B.A., HABICHT, K.S., CANFIELD, D.E., AIRIEAU, S.A. & THIEMENS, M.H. 2003. Multiple sulfur isotopic interpretations of biosynthetic pathways: Implications for biological signatures in the sulphur isotope record. *Geobiology*, **1**, 27–36.

GAO, X. & THIEMENS, M.H. 1991. Systematic study of sulfur isotopic composition in iron meteorites and the occurrence of excess ^{33}S and ^{36}S. *Geochimica et Cosmochimica Acta*, **55**, 2671–2679.

GAO, X. & THIEMENS, M.H. 1993*a*. Isotopic composition and concentration of sulfur in carbonaceous chondrites. *Geochimica et Cosmochimica Acta*, **57**, 3159–3169.

GAO, X. & THIEMENS, M.H. 1993*b*. Variations of the isotopic composition of sulfur in enstatite and ordinary chondrites. *Geochimica et Cosmochimica Acta*, **57**, 3171–3176.

GAO, Y.Q. & MARCUS, R.A. 2001. Strange and unconventional isotope effects in ozone formation. *Science*, **293**, 259–263.

GAO, Y.Q. & MARCUS, R.A. 2002. On the theory of the strange and unconventional isotopic effects in ozone formation. *Journal of Chemical Physics*, **116**, 137–154.

HABICHT, K.S., GADE, M., THAMDRUP, B., BERG, P. & CANFIELD, D.E. 2002. Calibration of sulfate levels in the Archaean ocean. *Science*, **298**, 2372–2374.

HATHORN, B.C. & MARCUS, R.A. 1999. An intramolecular theory of the mass-independent isotope effect for ozone. I. *Journal of Chemical Physics*, **111**, 4087–4100.

HATHORN, B.C. & MARCUS, R.A. 2000. An intra-molecular theory of the mass-independent isotope effect for ozone. II. Numerical implementation at low pressures using a loose transition state. *Journal of Chemical Physics*, **113**, 9497–9509.

HEYMANN, D., YANCEY, T.E., WOLBACH, W.S., THIEMENS, M.H., JOHNSON, E.A., ROACH, D. & MOECKER, S. 1998. Geochemical markers of the Cretaceous-Tertiary boundary event at Brazos River, Texas, USA. *Geochimica et Cosmochimica Acta*, **62**, 173–181.

HOERING, T.C. 1989. The isotopic composition of bedded barites from the Archaean of southern India. *Journal of the Geological Society of India*, **34**, 461–466.

HU, G.X., RUMBLE, D. & WANG, P.L. 2003. An ultraviolet laser microprobe for the in situ analysis of multisulfur isotopes and its use in measuring Archaean sulfur isotope mass-independent anomalies. *Geochimica et Cosmochimica Acta*, **67**, 3101–3118.

HULSTON, J.R. & THODE, H.G. 1965. Variations in the S^{33}, S^{34}, and S^{36} contents of meteorites and their relation to chemical and nuclear effects. *Journal of Geophysical Research*, **70**, 3475–3484.

KASTING, J.F. 2001. Earth history – the rise of atmospheric oxygen. *Science*, **293**, 819–820.

KASTING, J.F. & DONAHUE, T.M. 1980. The evolution of atmospheric ozone. *Journal of Geophysical Research*, **85**, 3255–3263.

LEVINE, J.S. 1985. *The photochemistry of atmospheres: Earth, the other planets, and comets.* Orlando, Flora: Academic Press. xxiv, 518 pp.

MATSUHISA, Y., GOLDSMITH, J.R. & CLAYTON, R.N. 1978. Mechanisms of hydrothermal crystallization of quartz at 250°C and 15 kilobar. *Geochimica et Cosmochimica Acta*, **42**, 173–182.

MILLER, M.F. 2002. Isotopic fractionation and the quantification of O-17 anomalies in the oxygen three-isotope system: An appraisal and geochemical significance. *Geochimica et Cosmochimica Acta*, **66**, 1881–1889.

MOJZSIS, S.J., COATH, C.D., GREENWOOD, J.P., MCKEEGAN, K.D. & HARRISON, T.M. 2003. Mass-independent isotope effects in Archean (2.5 to 3.8 Ga) sedimentary sulfides determined by ion microprobe multicollection. *Geochimica et Cosmochimica Acta*, **67**, 1635–1658.

MOOK, W.G. 2000. *Environmental isotopes in the hydrological cycle principles and applications, v i: Introduction – theory, methods, review.* Paris: UNESCO/IAEA.

ONO, S., EIGENBRODE, J.L., PAVLOV, A.A., KHARECHA, P., RUMBLE, D., KASTING, J.F. & FREEMAN, K.H. 2003. New insights into Archean sulphur cycle from mass-independent sulfur isotope records from the Hamersley basin, Australia. *Earth and Planetary Science Letters*, **213**, 15–30.

PAVLOV, A.A. & KASTING, J.F. 2002. Mass-independent fractionation of sulfur isotopes in Archaean sediments: Strong evidence for an anoxic Archaean atmosphere. *Astrobiology*, **2**, 27–41.

ROMERO, A.B. & THIEMENS, M.H. 2003. Mass-independent sulphur isotopic compositions in present-day sulfate aerosols. *Journal of Geophysical Research – Atmospheres*, **108**, (D16), 4524; doi: 10.1029/2003JD003660.

SAVARINO, J., ROMERO, A., COLE-DAI, J., BEKKI, S. & THIEMENS, M.H. 2003. UV induced mass-independent sulphur isotope fractionation in stratospheric volcanic sulfate. *Geophysical Research Letters*, **30**, (21) 2131; doi: 10.1029/2003GL018134.

SHIREY, S.B., CARLSON, R.W., RICHARDSON, S.H., MENZIES, A., GURNEY, J.J., PEARSON, D.G., HARRIS, J.W. & WIECHERT, U. 2001. Archaean emplacement of eclogitic components into the lithospheric mantle during formation of the Kaapvaal craton. *Geophysical Research Letters*, **28**, 2509–2512.

SHIREY, S.B., HARRIS, J.W., RICHARDSON, S.H., FOUCH, M.J., JAMES, D.E., CARTIGNY, P., DEINES, P. & VILJOEN, F. 2002. Diamond genesis, seismic structure, and evolution of the Kaapvaal–Zimbabwe craton. *Science*, **297**, 1683–1686.

STRAUSS, H. 2003. Sulphur isotopes and the early Archaean sulphur cycle. *Precambrian Research*, **126**, 349–361.

TURRO, N.J. 1983. Influence of nuclear spin on chemical reactions: Magnetic field and magnetic isotope effects (a review). *Proceedings of the National Academy of Sciences of USA*, **80**, 609–621.

UREY, H.C. 1947. The thermodynamic properties of isotopic substances. *Journal of the Chemical Society*, 562–581.

YOUNG, E.D., GALY, A. & NAGAHARA, H. 2002. Kinetic and equilibrium mass-dependent isotope fractionation laws in nature and their geochemical and cosmochemical significance. *Geochimica et Cosmochimica Acta*, **66**, 1095–1104.

ZMOLEK, P., XU, X.P., JACKSON, T., THIEMENS, M.H. & TROGLER, W.C. 1999. Large mass independent sulfur isotope fractionations during the photopolymerization of $(CS_2)–^{12}C$ and $(CS_2)–^{13}C$. *Journal of Physical Chemistry a*, **103**, 2477–2480.

Reactive iron enrichment in sediments deposited beneath euxinic bottom waters: constraints on supply by shelf recycling

R. RAISWELL[1] & T. F. ANDERSON[2]

[1]*School of Earth Sciences, University of Leeds, Leeds LS2 9JT, UK*
(e-mail: r.raiswell@earth.leeds.ac.uk)
[2]*Department of Geology, University of Illinois at Urbana-Champaign, 1301 W. Green St,*
Urbana, IL 61801, USA

Abstract: Modern and ancient euxinic sediments are often enriched in iron that is highly reactive towards dissolved sulphide, compared to continental margin and deep-sea sediments. It is proposed that iron enrichment results from the mobilization of dissolved iron from anoxic porewaters into overlying seawater, followed by transport into deep-basin environments, precipitation as iron sulphides, and deposition into sediments. A diagenetic model shows that diffusive iron fluxes are controlled mainly by porewater dissolved iron concentrations, the thickness of the surface oxygenated layer of sediment and to a lesser extent by pH and temperature. Under typical diagenetic conditions (pH < 8, porewater $Fe^{2+} = 10^{-6}$ g cm^{-3}) iron can diffuse from the porewaters in continental marginal sediments to the oxygenated overlying seawater at fluxes of 100–1000 µg cm^{-2} a^{-1}. The addition of reactive iron to deep-basin sediments is determined by the magnitude of this diffusive flux, the export efficiency (ε) of recycled iron from the shelf, the ratio of source area (S) to basin sink area (B) and the trapping of reactive iron in the deep basin. Values of ε are poorly constrained but modern enclosed or semi-enclosed sedimentary basins show a wide variation in S/B ratios (0.25–13) where the shelf source area is defined as sediments at less than 200 m water depth. Diffusive fluxes in the range 100–1000 µg cm^{-2} a^{-1} are able to produce the observed reactive iron enrichments in the Black Sea, the Cariaco Basin and the Gotland Deep for values of $\varepsilon \times S/B$ from 0.1–5. Transported reactive iron can be trapped physically and/or chemically in deep basins. Physical trapping is controlled by basin geometry and chemical capture by the presence of euxinic bottom water. The S/B ratios in modern basins may not be representative of those in ancient euxinic/semi-euxinic sediments but preliminary data suggest that $\varepsilon \times S/B$ in ancient euxinic sediments has a similar range as in modern euxinic sediments.

Recent studies of the Black Sea and the Cariaco Basin (Canfield *et al.* 1996; Raiswell & Canfield 1998; Wijsman *et al.* 2001; Wilkin & Arthur 2001; Lyons *et al.* 2003) have shown that their deep-basin sediments, deposited under euxinic bottom waters, are enriched in iron that is highly reactive towards sulphide compared to oxic continental margin and deep-sea sediments. Similar highly reactive iron enrichments are also found in ancient well-laminated black shales with ages ranging from Jurassic to Proterozoic (Schieber 1995; Raiswell & Canfield 1998; Raiswell *et al.* 2001; Poulton & Raiswell 2002; Shen *et al.* 2002, 2003; Werne *et al.* 2002). It has been suggested that highly reactive iron enrichment results from the mobilization of iron from basin margin sediments, followed by lateral transport from the basin margin to the deep basin (e.g. Canfield *et al.* 1996; Lyons 1997; Raiswell & Canfield 1998; Wijsman *et al.* 2001). Wijsman *et al.* (2001) also demonstrated that the extent of highly reactive iron enrichment in

Black Sea deep-basin sediments was within the range of iron release from oxic and dysoxic continental shelf sediments, as estimated from a model of diagenetic iron recycling and *in situ* measurements of iron fluxes. It was proposed that highly reactive iron is first dissolved from iron oxides during anoxic diagenesis of the shelf sediments, and then released as dissolved Fe^{2+} to the overlying seawater. A significant fraction of the dissolved iron is re-oxidized and redeposited into the marginal sediments, but some escapes and is transferred to the deep basin, where the dissolved iron is precipitated as iron sulphide and deposited.

Mass balance calculations for the Black Sea (Anderson & Raiswell 2004) also suggest that the recycling of highly reactive iron from the continental margin makes an important contribution (40 ± 20%) to the iron enrichment found in the deep-basin sediments. Anderson & Raiswell (2004) suggested that the remaining enrichment was due to enhanced reactivity of

From: McDonald, I., Boyce, A. J., Butler, I. B., Herrington, R. J. & Polya, D. A. (eds) 2005. *Mineral Deposits and Earth Evolution.* Geological Society, London, Special Publications, **248**, 179–194. 0305-8719/$15.00
© The Geological Society of London 2005.

lithogenous iron but were unable to demonstrate quantitatively the mechanism(s) of enrichment, i.e. chemocline extraction, microbial enhancement of iron silicate reactivity or fractionation of the riverine sediment flux. Wijsman *et al.* (2001) suggest that iron enrichment by mobilization from shelf sediments probably occurs in other basins and, consistent with this, Landing & Bruland (1987) and Saager *et al.* (1989), have reported the shelf mobilization and lateral transport of iron in the Pacific and Indian Oceans. These observations indicate that a detailed study of intrabasinal mobilization and transport would be valuable. It is the purpose of the paper to use a simple model of diagenetic iron mobilization and recycling to examine the factors which determine the extent of iron mobilization from basin margin sediments and enrichment in deep-basin sediments.

Recognition of iron enrichment

Anderson & Raiswell (2004) discussed two different approaches by which reactive iron enrichment may be identified, as is summarized here briefly. The first approach (Raiswell & Canfield 1998; Raiswell *et al.* 2001) uses the ratio of highly reactive iron (Fe_{HR}) to total iron (FeT), and the second the ratio of FeT to Al (Werne *et al.* 2002; Lyons *et al.* 2003).

The first approach is based on the formation of pyrite in anoxic marine sediments, where dissolved sulphide produced by microbial sulphate reduction reacts with iron-bearing minerals (Berner 1970, 1984). Anoxic conditions can be established either within the water column or within the sediment. In these environments, sulphate-reducing bacteria oxidize organic matter to generate H_2S (Berner 1970, 1984), a fraction of which reacts with detrital iron minerals to form pyrite. The initial products are usually iron sulphide minerals, which are transformed to pyrite by reaction with dissolved sulphide (Rickard 1997; Rickard & Luther 1997) and/or polysulphides (Goldhaber & Kaplan 1974; Rickard 1975; Luther 1991; Schoonen & Barnes 1991).

Modern sediment studies (Canfield 1989; Canfield & Raiswell 1991; Canfield *et al.* 1992; Poulton *et al.* 2004) have shown that iron oxides (ferrihydrite, goethite, hematite and lepidocrocite) react rapidly with dissolved sulphide. However, only relatively small concentrations of silicate iron react even over timescales of millions of years (Canfield *et al.* 1992; Raiswell & Canfield 1996). In general (except in deep-sea sediments) the amounts of iron oxide buried into the zone of sulphate reduction are exceeded by the amounts of H_2S that are produced microbially. Thus, in many basin margin sediments, the concentrations of pyrite are limited by concentrations of iron oxide. The concentration of iron that is highly reactive towards dissolved sulphide (Fe_{HR}) is measured as the sum of the iron that is extracted by dithionite (FeD, iron present mainly as oxides with small concentrations of silicates; Raiswell *et al.* 1994) plus that which has already reacted with sulphide (FeP, iron present as pyrite or other sulphides). Use of the Fe_{HR}/FeT ratio corrects for the dilution effects of biogenous sediment and represents the maximum proportion of lithogenous iron that can be pyritized in such sediments.

The second approach for the recognition of highly reactive iron enrichments in euxinic sediments is through changes in the FeT/Al ratio (Werne *et al.* 2002; Lyons *et al.* 2003). This approach assumes that values of FeT/Al higher than base-line values (i.e. the FeT/Al ratio in lithogenous sediments deposited in shelf environments) are due solely to the addition of Fe_{HR}. Use of the FeT/Al ratio also eliminates the effects of dilution by biogenous sediment, but small changes in the Fe_{HR} fraction may be difficult to identify within a large FeT content.

This paper uses elevated Fe_{HR}/FeT ratios to identify highly reactive iron enrichment. The base-line for recognizing enrichment was suggested by Raiswell & Canfield (1998) who found that an Fe_{HR}/FeT ratio of 0.4 represented the maximum value for a geographically widespread compilation of oxic continental margin and deep-sea sediments; the average Fe_{HR}/FeT ratio for those sediments is 0.26 (± 0.09). However, this approach may fail to detect iron enrichments imposed on local detrital sediment fluxes that have an Fe_{HR}/FeT ratio below the threshold ($Fe_{HR}/FeT = 0.4$) and in principle it is better to identify reactive iron enrichments by comparison with shelf sediments within the same basin. Werne *et al.* (2002) and Lyons *et al.* (2003) use this approach for the detection of reactive iron enrichments via FeT/Al ratios. This approach assumes that there is relatively little fractionation of iron within the basin, and thus the composition of lithogenous components deposited in shelf and deep basin sites are identical (but see Wijsman *et al.* 2001; Anderson & Raiswell 2004).

Approach

A schematic model of intrabasinal iron mobilization and transport from margin to deep-basin sediments is shown in Figure 1. Iron

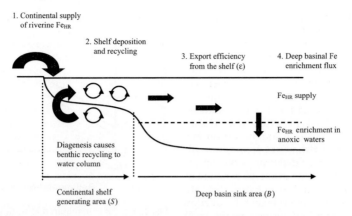

Fig. 1. Schematic diagram of the processes involved in enriching iron in deep-basin sediments.

enrichment requires a diagenetically mobilized source, a transportation mechanism and a deep-basin sink (Wijsman *et al.* 2001). First, iron must be reduced in the margin sediments and then released to the overlying seawater. This process requires margin sediments that contain sufficient organic C to become anoxic and undergo microbial reduction of iron oxides to produce dissolved Fe^{2+} usually by direct reduction (e.g. Canfield 1989):

$$CH_2O + 4FeOOH + 8H^+ \rightarrow$$
$$CO_2 + 4Fe^{2+} + 7H_2O.$$

Iron reduction is followed closely by microbial sulphate reduction in most continental margin sediments (e.g. Canfield & Raiswell 1991):

$$2CH_2O + SO_4^{2-} \rightarrow 2HCO_3^- + H_2S,$$

and the dissolved sulphide precipitates the iron reduced by microbial reduction. The rapid reaction between dissolved sulphide and dissolved Fe^{2+} (plus solid iron oxides) buffers dissolved sulphide at low concentrations until the reactive iron oxides are exhausted. The porewaters are then dominated by dissolved sulphide. Thus a depth profile through typical continental margin sediments shows a layer of iron-rich porewaters overlying a layer of dissolved sulphide-rich porewaters (Canfield & Raiswell 1991). Dissolved iron can diffuse from the upper layer into the overlying seawater. Here the dissolved iron may be oxidized and redeposited into the sediments, but dissolved iron that escapes oxidation may become available for transport to the deep basin. Anderson & Raiswell (2004) suggest that the iron oxides formed by the oxidation of Fe^{2+} after release to

seawater may not all be redeposited into the shelf sediments and may be transported to the deep basin sediments. There is independent evidence for this process in the Black Sea, where Murray *et al.* (1995) have shown that the oxidation of sulphide at the oxic/anoxic interface requires the transport of Mn and iron oxides from the basin margins to the deep basin.

Secondly, the iron must be transported to the deep basin. In the Black Sea currents at the depths of the suboxic zone and the sulphide interface have lateral velocities of about 0.5 cm s^{-1} (Buessler *et al.* 1991) and are able to transport dissolved and particulate iron to the deep basin (Kempe *et al.* 1990; Wijsman *et al.* 2001). Thirdly, the iron must be precipitated by reaction with dissolved sulphide in the water column and then deposited into the deep-basin sediments. In the deep basin of the Black Sea, eddy diffusion (Brewer & Spencer 1974; Lewis & Landing 1991) is an effective mechanism for mixing the upper dissolved iron-rich water with the lower sulphide-rich water, thus allowing the precipitation of iron sulphides in the water column and their deposition into sediments. Sediment trap data (Honjo *et al.* 1987; Muramoto *et al.* 1991; Canfield *et al.* 1996), pyrite S isotope composition (Calvert *et al.* 1996; Lyons 1997) and pyrite framboid size distributions (Wilkin & Arthur 2001) confirm that iron sulphides are precipitating in the Black Sea water column. Consistent with this, the water column is supersaturated with iron sulphides at depths below approximately 120–180 m (Landing & Lewis 1991). The Black Sea microlaminated Unit I sediments deposited under euxinic conditions are remarkably uniform and laminae can be traced for distances of more than 1000 km over the deep basin floor (Lyons 1991;

Lyons & Berner 1992; Arthur *et al.* 1994). Thus iron released from the margin must be dispersed evenly over the entire area of the deep basin.

Iron enrichments generated by recycling from the shelf clearly require that transport mechanisms similar to those observed in the Black Sea are a typical feature of other modern and ancient euxinic depositional environments. Studies of other modern basins may clarify whether such mechanisms are common but this may be difficult to substantiate in ancient sediments. Nonetheless provided transport mechanisms are available then the extent of enrichment in the deep basin is determined by: (1) the rate of iron release from the margin; (2) the fraction of the released iron which escapes the shelf; (3) the ratio between the margin source area and the area of the deep-basin sink; (4) the extent to which recycled iron is trapped in the deep-basin sink; and (5) the relative significance of the recycled iron compared to the background clastic (lithogenous) supply.

A useful starting point is to identify the extent of iron enrichment associated with the Fe_{HR}/FeT ratios observed in modern and ancient euxinic sediments and then estimate the magnitude of Fe_{HR} flux which would need to be supplied by recycling to produce such enrichments. Anderson & Raiswell (2004) show that Fe_{HR}/FeT ratios in the Black Sea average 0.70 ± 0.19, and similar values have been found by Poulton & Raiswell (2002) for ancient euxinic sediments. We shall assume that these Fe_{HR}/FeT ratios are obtained solely by the addition of recycled iron to a background lithogenous flux, J(lith.) (in units of g cm^{-2} a^{-1}) with $FeT_{lith.}$ = 4.0% (Raiswell & Canfield 1998) and an $(Fe_{HR}/FeT)_{lith.}$ ratio of 0.26 (see above). The flux of additional Fe_{HR}, defined as J(aHR), required to yield a particular value of Fe_{HR}/FeT is:

$$J(aHR) = FeT_{lith.} \times J(lith.) \times [Fe_{HR}/FeT - (Fe_{HR}/FeT)_{lith.}]. \quad (1)$$

Figure 2 shows the benthic flux of recycled iron which must be added to a background sedimentation rate of lithogenous material to produce sediments with Fe_{HR}/FeT ratios of 0.5, 0.7 and 0.9. Note that J(lith.) represents the lithogenous flux and must be corrected for the presence of biogenous, essentially Fe_{HR}-free components such as calcareous and siliceous material. The mass of recycled iron is deposited as oxides or sulphides and represents $\leq$ 5% of the lithogenous flux at any values of J(lith.) or Fe_{HR}/FeT. To illustrate the information contained in Figure 2, let us calculate the necessary J(aHR) required to yield an Fe_{HR}/FeT ratio of 0.7 in a typical

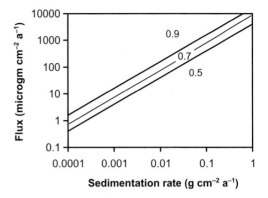

Fig. 2. Flux of iron required to be delivered to the deep basin to enrich Fe_{HR}/FeT ratios to 0.5, 0.7 and 0.9 at varying sedimentation rates.

lithogenous sediment deposited at the rate of 0.01 cm a^{-1}. Assuming a grain density of 2.5 g cm^{-3} and a porosity of 90%, this sedimentation rate is equivalent to a flux of 0.0025 g cm^{-2} a^{-1}. Substituting into Equation (1) yields J(aHR) = 0.04 × 0.0025 g cm^{-2} a^{-1} × (0.70 − 0.26) = 44 µg cm^{-2} a^{-1}.

Quantifying iron release from the basin-margin sediments by diagenetic recycling

A diagenetic model for diffusive iron recycling

A general one-dimensional diagenetic equation for a solute in porewater can be written to describe the effects of diffusion, advection, adsorption and reaction on variations in solute concentration with time (Berner 1980; Boudreau 1996). A simplified version of this equation has been used by Boudreau & Scott (1978) to describe the flux of dissolved Mn^{2+} from a reduced porewater through an oxygenated layer of sediment to the sediment–water interface. This equation:

$$D_s\frac{d^2C}{dx^2} - k_1(C - C_s) = 0 \quad (2)$$

can be used to estimate the analogous diffusive flux of Fe^{2+}, where x is the depth (cm) below the sediment–water interface, D_s (cm^2 s^{-1}) is the diffusion coefficient corrected for tortuosity effects and C is the porewater concentration of Fe^{2+} (g cm^{-3}). It is assumed that: (i) steady state diagenesis occurs; (ii) there are no changes in porosity and D_s with depth; (iii) the effects of advection are small relative to diffusion and reaction; and (iv) Fe^{2+} is removed only by oxidation which occurs at a rate given by $k_1(C - C_s)$, where C_s is the saturation concentration of Fe^{2+} (g cm^{-3}) with respect to $Fe(OH)_3$ and k_1 is the first order rate constant (s^{-1}) for Fe^{2+} oxidation. This model (Fig. 3) envisages that transport processes maintain a negligible concentration of Fe^{2+} above the sediment–water interface, below which there exists a thin zone of oxygenated sediment where solid phase iron oxides are present. There is a gradient in Fe^{2+} through this zone from the sediment–water interface to the base of the oxygenated layer. Porewaters below the oxygenated layer are reducing and maintain a uniform Fe^{2+} concentration with time. Boudreau & Scott (1978) show that the diffusive flux of Fe^{2+} in g cm^{-2} s^{-1} is then given by:

$$J(Fe^{2+}) = \frac{\varphi(D_s k_1)^{0.5} C_P}{\sinh\left[(k_1 / D_s)^{0.5} L\right]}, \qquad (3)$$

where φ is the porosity (assumed constant at 0.85), C_P is the porewater concentration of Fe^{2+} (g cm^{-3}) and L (cm) is the thickness of the oxygenated layer. The infinite dilution diffusion coefficient (D^o) for Fe^{2+} is derived from Boudreau (1996) as D^o (cm^2 s^{-1}) = (3.31 + 0.15 T °C) 10^{-6}. The corrections for tortuosity are derived from Ullman & Aller (1982) who give $D_s = D^o/\varphi F$, where φ is the porosity and the formation factor F is approximated as $1/\varphi^m$ and $m = 2.5$ to 3.0 for muddy sediments. Simplifying thus approximately produces:

$$D_s = D^o \varphi^{1.7}$$
and hence $\quad D_s = \varphi^{1.7} (3.31 + 0.15\ T°C)\ 10^{-6}.$

The oxidation of reduced iron can occur chemically or be catalysed biologically. Chemical oxidation may be retarded by the presence of humics and other natural organic species (Theis & Singer 1974) but is very rapid and is believed to predominate over biological oxidation (Nealson 1997; Santschi et al. 1990), which

probably only occurs under acid conditions (Kristensen 2000). The chemical oxidation kinetics of Fe^{2+} in seawater have been studied by many workers (e.g. Stumm & Morgan 1980; Davison & Seed 1983; Roekens & van Grieken 1983; Millero et al. 1987) and there is general agreement that the rate law can be expressed as:

$$d[FeII]/dt = -k[FeII]\,[O_2]\,[OH^-]^2. \qquad (4)$$

Millero et al. (1987) have derived the following expression for k (mol^{-3} kg^3 min^{-1}):

$$\log k = 21.56 - 1545/T - 3.29 I^{0.5} + 1.52 I.$$

Where T is temperature in kelvin and I is ionic strength. Table 1 shows values of log k from 0–20 °C, assuming $I = 0.723$ for seawater. Equation (4) can be modified to derive an apparent first order rate constant k_1, where:

$$d[FeII]/dt = -k_1[FeII]$$
and $\quad k_1 = k[O_2]\,[OH^-]^2. \qquad (5)$

Values of $k_1(s^{-1})$ are given in Table 1 for the appropriate oxygen concentrations (Benson & Krause 1984) to achieve saturation at the specified temperature, and for varying pH (using the apparent K_w for seawater from Millero 2001). The results in Table 1 show that it is necessary to define values of temperature, pH and bottom water O_2 concentrations in order to make reasonable estimates of k_1 for shelf sediments. Furthermore Equation (3) shows that fluxes of Fe^{2+} from shelf sediments also depend on porewater Fe^{2+} concentrations (C_P) and the thickness of the oxygenated layer (L) in addition to k_1. Some reasonable limits are derived for these variables below.

Surface waters in the present oceans range from approximately 2 °C in the polar seas to approximately 28 °C in the equatorial regions (Chester 2000). The models used here have a slightly lower range (0–20 °C) in order to account for dealing with cooler bottom waters at depths down to 200 m (see below). The observed pH of porewaters in modern shelf sediments undergoing anoxic diagenesis is usually between 7 and 8 (e.g. Boudreau & Canfield 1988). This relatively small range in pH is a consequence of the combined effects of iron reduction and sulphate reduction, and has been modelled by Ben-Yaakov (1973), Gardner (1973) and Boudreau & Canfield (1993). These models are consistent with field data in indicating that pH varies only between approximately 7 and 8, with the lower values occurring where greater proportions of sulphide accumulate in

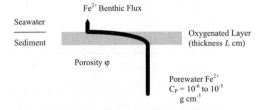

Fig. 3. Schematic model for the diffusive flux of diagenetically recycled iron from sediments to the overlying seawater.

Table 1. *Variation in the apparent first-order rate constant (k$_1$) for Fe^{2+} oxidation in seawater with temperature and pH*

Temp (°C)	pK_w	log k	Sat. O$_2$ (μM kg^{-1})	k$_1$ (sec^{-1}) at varying pH		
				pH 7	pH 7.5	pH 8
0	14.3	14.2	348	2.3×10^{-6}	2.3×10^{-5}	2.3×10^{-4}
5	14.1	14.3	307	6.3×10^{-6}	6.3×10^{-5}	6.3×10^{-4}
10	13.8	14.4	275	2.8×10^{-5}	2.8×10^{-4}	2.8×10^{-3}
20	13.4	14.6	226	2.4×10^{-4}	2.4×10^{-3}	2.4×10^{-2}

Values of k$_1$ assume saturation with respect to the atmospheric oxygen. Data for pK_w from Millero (2001), log k from Millero *et al.* (1987) and oxygen saturation concentrations from Benson & Krause (1984).

the porewaters (as opposed to being precipitated as iron sulphide). A pH range of 7–8 is used here to accommodate the observed and model ranges of pH. A literature survey (Fig. 4) of porewater Fe^{2+} concentrations (averaged over depths of 10–30 cm below the surface oxidized layer) in anoxic shelf sediments shows that values of 10^{-6} to 10^{-5} g cm^{-3} are common, and this range is used in the present models. Relatively high porewater dissolved iron concentrations are found in suboxic sediments where sulphate reduction is suppressed. For example, values as high as 10^{-4} g cm^{-3} are observed, in depositional environments such as the Amazon inner shelf (Aller *et al.* 1986) where physical reworking regenerates iron oxides and allows iron reduction (rather than sulphate

reduction) to dominate sediment diagenesis. Aller *et al.* (1986) point out that reworking on this scale promotes the export of colloidal iron oxides.

Lower values of k$_1$ produce slower rates of oxidation (in turn enhancing escape through the oxygenated layer) and the variations in k$_1$ with temperature and pH are shown in Figure 5a. A temperature change of 10 °C produces a change of about one order of magnitude in k$_1$, as does a change of 0.5 pH. A decrease in bottom water

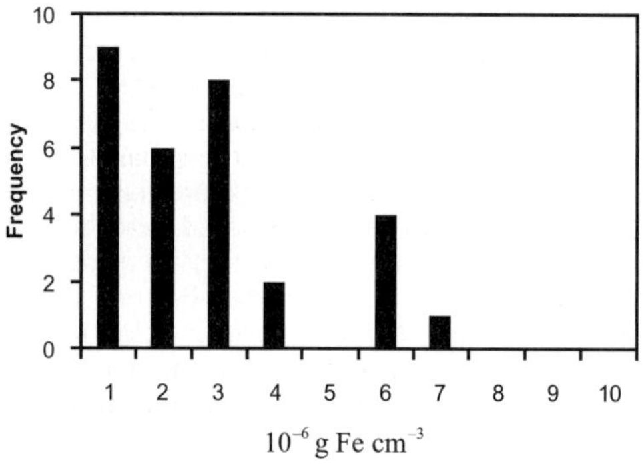

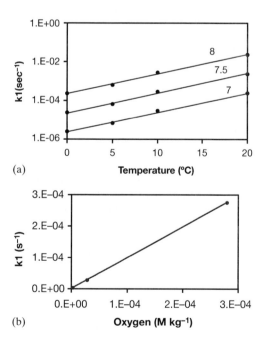

Fig. 4. Mean porewater dissolved iron concentrations (averaged over depths of 10–30 cm below the surface oxygenated layer). Data from Aller (1977), Elderfield *et al.* (1981), Trefry & Presley (1982), Chanton (1985), Sorenson & Jorgensen (1987), Canfield (1989), Canfield *et al.* (1993), Thamdrup *et al.* (1994), Thamdrup & Canfield (1996), Haese *et al.* (2000), Deflandre *et al.* (2002), Krom *et al.* (2002).

Fig. 5. Variations in the first order rate constant (k$_1$) for iron oxidation (**a**) as a function of temperature at pH 7, 7.5 and 8, (**b**) with dissolved oxygen concentrations.

O_2 concentrations from saturation produces a substantial decrease in k_1 (Fig. 5b), with concentrations at 1% saturation producing a decrease in k_1 of two orders of magnitude. Thus variations in temperature, pH and bottom water O_2 concentrations are clearly capable of producing substantial variations in k_1, and thus also in the flux of recycled iron from the sediment. Equation (3) also shows that variations in C_P produce equivalent variations in the flux of recycled iron from the sediment.

Model results: validation, implications and constraints

The estimated diffusive fluxes of recycled iron from Equation (3) can be validated to a limited extent by comparison with more sophisticated diffusion–advection–reaction models (e.g. Wang & Van Cappellen 1996; Wijsman et al. 2002; Berg et al. 2003). Such models take account of the rates of all the main carbon mineralization reactions (aerobic respiration, denitrification, manganese oxide reduction, iron oxide reduction, sulphate reduction and methanogenesis) together with the effects of oxidation on porewater and solid phase species (by O_2, iron and manganese oxides) and the precipitation of iron sulphides. In general, these models show that there is extensive redox recycling of iron in the upper few centimetres of sediment, in contrast to the single oxidation pathway modelled here. The model of Wijsman (2001) and Wijsman et al. (2002) predicts that the Black Sea shelf sediments generate a diffusive iron flux of approximately 7 nmol cm^{-2} day^{-1} (equivalent to 140 μg cm^{-2} a^{-1}) for bottom waters saturated with O_2 (Wijsman et al. 2001). Wijsman (2001) gives data for one shelf station (pH = 7.2, C_P = 3×10^{-6} g cm^{-3}, ϕ = 0.88, T = 5.8 °C and L = 0.42 cm) that can be used in Equation (3) to derive a flux of 425 μg cm^{-2} a^{-1}. Some discrepancies of a similar magnitude (up to a factor of 3) occur in comparisons with calculated diffusive fluxes from other sites (McManus et al. 1997; Elrod et al. 2004). Thus the simple model presented here should be regarded principally as identifying the main variables that affect diffusive recycling of iron and the flux estimates should only be used semi-quantitatively, as here in examining iron recycling in a basinal context.

Friedl et al. (1998) and Friedrich et al. (2002) also report benthic flux measurements for the shelf sediments of the Black Sea which range from 0.5–184 nmol cm^{-2} day^{-1} and average 46 nmol cm^{-2} day^{-1} (excluding one unusually high value of 1633 nmol cm^{-2} day^{-1}). The average is equivalent to 920 μg cm^{-2} a^{-1}. However the benthic flux measurements may be elevated because the benthic chambers became depleted in O_2 which significantly decreases the rate of re-oxidation in the surface layer and thus enhances the benthic flux (see above). McManus et al. (1997) and Elrod et al. (2004) also report benthic flux measurements of iron from sites on the California continental margin. The data show two significant features. First, shallow-water bioirrigated sites (99 m depth) with well-oxygenated bottom waters (101–185 μM kg^{-1}) have benthic flux measurements in the range 2–22 μg cm^{-2} a^{-1} that are on average 75 times higher than predicted by diffusion flux estimates based on porewater dissolved iron concentrations (Elrod et al. 2004). Ferro et al. (2003) have observed experimentally that bioirrigation maintains porewater iron at low levels and thus results in low estimates of diffusive fluxes. Together these data suggest that bioirrigation decreases porewater dissolved iron by transporting iron to the overlying waters, producing low diffusive fluxes that are augmented by bioirrigation fluxes.

Secondly, Elrod et al. (2004) show that deeper-water sites with oxygen-depleted bottom waters (< 100 μM kg^{-1}) have benthic fluxes in the range 0.2–37 μg cm^{-2} a^{-1} that are roughly in agreement with estimated diffusive fluxes. However, exceptions occur for the highest benthic flux measurements which are significantly smaller (by a factor of > 10) than the estimated diffusive fluxes. Elrod et al. (2004) suggest that dissolved iron was re-oxidized in the sediment rather than being transported into the benthic chamber. It is also possible that iron was released but re-oxidized and precipitated from seawater in the benthic chamber without being collected and measured. In the absence of the chamber such iron might have been transported away from the site and some fraction exported from the shelf (see later). Overall the benthic flux measurements show that iron fluxes can be significant but are clearly influenced by processes that are poorly-understood. These are further reasons to regard the present model as semi-quantitative (see earlier).

The model partially accommodates the effects of bioirrigation and bioturbation through their effects on C_P and L. High rates of iron reduction require the rapid redox cycling of iron which is aided by bioturbation and bio-irrigation (Canfield et al. 1993; Thamdrup 2000; Ferro et al. 2003). These processes transport iron oxides down into the reduced zone by particle mixing and transport reduced iron upwards by particle mixing, diffusion and irrigation. The overall effect is to maintain porewater dissolved iron at low levels (Ferro *et al.* 2003) and thus our use of mean observed porewater values accommodates the effect of bioturbation and bioirrigation on the diffusive benthic fluxes but does not account for any additional fluxes resulting from bioirrigation.

Figure 6 shows the model estimates of recycled iron fluxes as a function of the thickness of the oxygenated layer ($L = 0.1$–10 cm).

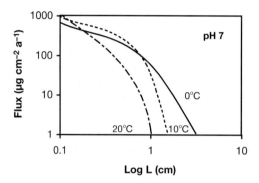

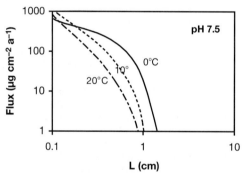

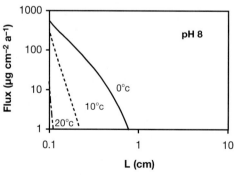

Fig. 6. Variations in the diffusive flux of diagenetically recycled iron with oxygenated zone thickness (L cm) for $Fe^{2+} = 10^{-6}$ g cm^{-3}, temperatures of 0, 10 and 20 °C, at pH 7, 7.5 and 8.

Smaller values of L approach the thickness of the benthic boundary layer (400–1200 µm in coastal marine sediments; Kristensen 2000) and would thus be unrealistic. The three graphs in Figure 6 have been derived for pH values of 8.0, 7.5 and 7.0, assuming bottom waters saturated with O_2 at the designated temperatures ($T = 0$, 10 and 20 °C) of each plot. All three graphs use a value of $C_P = 10^{-6}$ g cm^{-3} (see Fig. 2). Equation 3 shows that the diffusive flux varies directly with C_P and using $C_P = 10^{-5}$ g cm^{-3} simply produces an order of magnitude increase in the diffusive flux. Figure 6 shows that the diffusive flux of recycled iron increases sharply for decreasing values of L but approaches a maximum as L becomes relatively small. The graphs for pH 7 and 7.5 show similar maximum fluxes of 100–1000 µg cm^{-2} a^{-1} are reached for $L < 1$ cm at any temperatures from 0–20 °C. However the graph for pH 8 only approaches maximum values of 1000 µg cm^{-2} a^{-1} for $L < 0.1$ cm and $T = 0$ to 10 °C. Diffusive fluxes of recycled iron are clearly very sensitive to variations in L for any given C_P.

Diffusive iron fluxes are not directly affected by bottom-water oxygenation concentrations. Figure 7 shows that diffusive fluxes of recycled iron remain uniform as bottom-water oxygen concentrations decrease from saturation. However, bottom-water oxygen levels are likely to affect the thickness of the oxygenated layer (L). Jorgensen & Boudreau (2001) show that values of L decrease with increasing rates of organic C mineralization (and thus vary seasonally in many shelf sediments). Hence lower bottom-water oxygen levels will also tend to decrease L, for any given mineralization rate. Conversely, bioturbation and bioirrigation act to increase the volume of oxic sediment by a factor of 1–3 (Kristensen 2000) and thus increase L (which may be of irregular thickness). Nonetheless, the thickness of the oxygenated zone is often only a few millimetres in shelf sediments (Jorgensen & Boudreau 2001). Figure 6 suggests that maximum diffusive iron fluxes from shelf sediments with fully oxygenated bottom waters will typically be in the range 100–1000 µg cm^{-2} a^{-1} for porewater C_P values of 10^{-6} g cm^{-3}. Note, however, that steady-state conditions require that the addition of reactive iron by sedimentation at least equals the loss by mobilization to overlying seawater. The mean reactive iron content of continental margin sediments is 0.83 ± 0.21% (Raiswell & Canfield 1998). Sediments with this composition that deposit at rates of 0.012 ± 0.003 and 0.12 ± 0.03 g cm^{-2} a^{-1} are potentially capable of supplying reactive iron fluxes of 100–1000 µg cm^{-2} a^{-1}

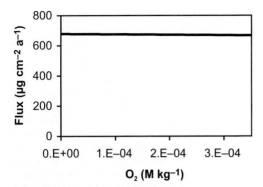

Fig. 7. Variations in the diffusive flux of diagenetically recycled iron with bottom-water oxygen concentrations for $L = 0.1$ cm, $Fe^{2+} = 10^{-6}$ g cm^{-3}, pH 7.5 and a temperature of 0 °C.

respectively, provided all the deposited reactive iron is recycled.

Factors influencing the flux of Fe_{HR} to deep basins

Evidence from modern sediments

In addition to the gross flux of iron from shelf sediments derived above, the flux of highly reactive iron to deep-basin sediments is dependent on the escape efficiency of iron from the shelf source region and the relative sizes of the source generating (S) and basin sink (B) areas (and by their spatial interrelationships). The escape efficiency, ε, can be defined as:

$$\varepsilon = J(nHR)/J(gHR), \qquad (6)$$

where $J(gHR)$ is the gross flux of Fe_{HR} from source sediments and $J(nHR)$ is the net flux of Fe_{HR} from source sediments to the deep basin. The net flux from the source is determined by the Fe_{HR} flux into deep-basin sediments (given by $J(aHR)$ in Equation (1)), and the source (shelf): basin area ratio, S/B:

$$J(aHR) = (S/B) \times J(nHR). \qquad (7)$$

Combining Equations (6) and (7) gives:

$$J(aHR) = \varepsilon \times (S/B) \times J(gHR). \qquad (8)$$

The evaluation of iron fluxes in the Black Sea by Wijsman et al. (2001) and Anderson & Raiswell (2004) permits the estimate of the escape efficiency for that basin. The diagenetic model of Wijsman et al. (2001) yields a gross

flux of recycled iron [$J(gHR)$] from the Black Sea shelf sediments over the range 7–22 nmol cm^{-2} day^{-1} (equivalent to 140–440 µg cm^{-2} a^{-1}) with the higher values occurring in oxygen-depleted bottom waters. Based on the composition of one deep-basin core, Wijsman et al. (2001) estimated an additional highly reactive iron flux to the deep basin [$J(aHR)$] of 0.82–2.75 nmol cm^{-2} day^{-1} (16–35 µg cm^{-2} a^{-1}). Given that the relative areas of the shelf (S) and deep basin (B) are 27% and 73%, respectively, then by rearranging Equation (7) the net flux of Fe_{HR} from the shelf is $J(nHR) = J(aHR) / (S/B) = 44$–150 µg cm^{-2} a^{-1}. The corresponding escape efficiency is approximately 30%. Anderson & Raiswell (2004) used different assumptions in the Wijsman et al. (2001) diagenetic model to derive a gross flux of recycled iron from the shelf of 140 ± 120 µg cm^{-2} a^{-1}. Based on their evaluation of Fe_{HR} enrichment in three deep-basin sediment cores, Anderson & Raiswell (2004) estimated that the flux of recycled iron to the deep basin is 38 ± 28 µg cm^{-2} a^{-1}. Using the same S/B ratio, the required net flux from the shelf is 120 ± 100 µg cm^{-2} a^{-1}, corresponding to an escape efficiency of 80–90%. We can take a third approach to estimate the gross flux of recycled iron from the Black Sea shelf. The average dissolved oxygen concentration at the margin sites reported by Wijsman et al. (2001) is 250 ± 23 µM (quoted by Anderson & Raiswell 2004). Using this value the diagenetic model of Wijsman et al. (2001) gives an average gross flux of recycled iron of 220 ± 40 µg cm^{-2} a^{-1}. Assuming the basinal flux of Anderson & Raiswell (2004) the corresponding escape efficiency is about 50%, intermediate between the two previous estimates. These differing estimates of ε indicate that $\varepsilon \times S/B$ for the Black Sea ranges from 0.1 to 0.3.

The spatial relationships between the source-generating and basin-sink areas also influence the extent of reactive iron enrichment in deep-basin sediments. Anderson & Raiswell (2004) define the source area of the Black Sea basin as being those sediments at depths of < 200 m. These sediments are anoxic below the sediment–water interface and have the potential to recycle iron diagenetically because their overlying waters are non-sulphidic. The top 95 m of the Black Sea water column are oxic or suboxic (Murray et al. 1989; Jorgensen et al. 1991) and below this depth there is approximately a further 100 m depth of waters that contain dissolved iron but which are undersaturated with respect to iron sulphides (Brewer & Spencer 1974; Lewis & Landing 1991; Landing

& Lewis 1991). However, deeper waters are saturated with FeS and high dissolved sulphide concentrations maintain low concentrations of dissolved iron. Porewaters in sediments beneath these euxinic waters have little source potential. Thus, the Black Sea basin source area is constrained by oceanographic variables that determine the depths below which euxinicity occurs. This is also likely to be true for ancient black shales.

The capacity of the source area to mobilize iron may also vary between different types of sediment and with the degree of oxygenation of the bottom waters (see earlier). The Black Sea sediments at < 200 m depth (Muller & Stoffers 1974; Wijsman 2001) comprise a mixture of sand, silt and clay sediment with variable proportions of carbonate (< 50%) and organic C (< 5%). In some areas these sediments support suboxic diagenesis with elevated concentrations of porewater dissolved iron (Lyons et al. 1993). As described earlier, the fluxes of recycled iron from the shelf sediments have been measured directly using benthic chambers (Friedl et al. 1998; Friedrich et al. 2002) and indirectly via a diagenetic model (Wijsman et al. 2001). Both methods show considerable variation between localities across the shelf. Accurate basinal-scale estimates of recycled iron clearly require that measurements by either method are made at a range of shelf

locations that encompasses the main sediment types and depositional environments in order to obtain an integrated value for the benthic flux over the entire source area.

Anderson & Raiswell (2004) conclude that iron enrichments in the Black Sea are readily discernible partly because of a favourable (i.e. high) ratio of the source area to the deep basinal sink area. However, Lyons et al. (2003) have shown that euxinic sediments on the margins of the Cariaco Basin show little (or no) iron enrichments in contrast to the marked enrichments in the euxinic central basin. This is attributed to rapid sedimentation at the margins (up to 0.8 cm a^{-1}) where lithogenous supply dilutes the iron enrichment in contrast to the more slowly deposited sediments (approximately 0.05 cm a^{-1}) in the central basin. Consistent with this, Table 2 shows the ratio of the source shelf area (S defined as sediments at < 200 m depth which is typically the limit of surface water mixing) to the remaining deeper area (B) for a range of modern enclosed and semi-enclosed basins. The S/B ratios in Table 2 do not imply that these areas are capable of exhibiting source–sink characteristics nor that iron enrichments occur in their deep-basinal areas. However, they demonstrate that the Black Sea does not have an unusually favourable S/B ratio, and that other enclosed and semi-enclosed basins have comparable or larger S/B

Table 2. *Ratios of shelf area (< 200 m depth) to deep (> 200 m) basin area in modern basins (from Dietrich et al. 1980 except as indicated*)*

Enclosed/semi-enclosed basins	Basin area ($\times 10^6$ km^2)	S/B
Gulf of California	0.15	0.47/0.53 = 0.89
Baltic Sea	0.39	0.99/0.01 = 99
Caspian Sea*	0.41	0.70/0.30 = 2.3
Black Sea*	0.42	0.27/0.73 = 0.37
Red Sea	0.45	0.42/0.58 = 0.72
Sea of Japan	1.01	0.23/0.77 = 0.30
East China Sea	1.20	0.81/0.19 = 4.3
Hudson Bay*	1.23	0.93/0.07 = 13
Gulf of Mexico*	1.29	0.48/0.52 = 0.92
Sea of Okhotsch	1.39	0.27/0.73 = 0.37
Bering Sea	2.26	0.46/0.54 = 0.85
Marginal basins		
S. Australian Basin*	1.53	0.18/0.82 = 0.22
Argentine Basin*	4.40	0.19/0.81 = 0.23
World oceans	362	0.075/0.925 = 0.075

ratios. The data also reflect present tectonic, sea-level and oceanographic conditions and are not necessarily representative of S/B ratios in enclosed or semi-enclosed basins in the geological record. The effective S/B ratio for any basin will be determined primarily by the depth at which euxinic bottom waters occur, which in turn limits the area of the remaining regions of the basin that have source potential.

The Baltic Sea is a comparable semi-enclosed basin to the Black Sea, although the deepest areas (the Gotland Deep and the Bornholm Deep) are not persistently anoxic/euxinic. The Bornholm Deep has oxygen-depleted bottom waters (0.1–1.0 ml O_2 l^{-1}) for periods of up to a few years, but the Gotland Deep is euxinic below 150 m depth for periods of at least 10 years (Boesen & Postma 1988). Recent studies of sediments of the Gotland Deep have been concerned mainly with their manganese enrichment (e.g. Burke & Kemp 2002) and analytical data for Fe_{HR} and FeT on the same samples are rare. Belmans et al. (1993) have analysed two Gotland Deep sediments which have a mean FeP/FeT = 0.7 (which approximates to Fe_{HR}/FeT; see Poulton & Raiswell 2002). Boesen & Postma (1988) give a mean sedimentation rate of 0.02 cm a^{-1} for the Gotland Deep but rates in the deepest part of the basin vary greatly. Assuming that the mean sedimentation rate is equivalent to a mean sediment flux of 0.005 g cm^{-2} a^{-1}, then the $J(aHR)$ required to produce the observed Fe_{HR}/FeT ratio is obtained from Equation (1): $J(aHR)$ = 0.04 × 0.005 g cm^{-2} a^{-1} × (0.70 – 0.26) = 88 µg cm^{-2} a^{-1}. This flux could be derived from a diffusive flux of recycled iron of 100–1,000 µg cm^{-2} a^{-1} from shelf sediments with oxygenated bottom waters by any combination of ε and the S/B ratio such that $\varepsilon \times S/B$ = 0.1 to 1. A detailed study of the Baltic shelf sediments that produced an integrated measure of the benthic flux of recycled iron, and a comparison with the Fe_{HR}/FeT ratio in the deep euxinic sediments would provide valuable data for ε and the S/B ratio for this basin.

The Cariaco Basin occupies a depression approximately 175 km long and 50 km wide located on the continental shelf north of Venezuela, and thus provides a useful contrast to the semi-enclosed basins of the Black Sea and the Baltic Sea. Water exchange with the Caribbean Sea is restricted by a series of sills with maximum depths of 120–140 m. Euxinic bottom waters exist from approximately 300 m depth to maximum depths of approximately 1400 m (Lyons et al. 2003). Sediment geochemical data in Lyons et al. (2003) allow a mean

Fe_{HR}/FeT = 0.48 to be calculated for a core deposited at a depth of 900 m and at a sedimentation rate of 0.45 cm a^{-1}. This sedimentation rate includes the deposition of approximately 30% calcareous and up to 30% opaline material (Raiswell & Canfield 1998; Lyons et al. 2003). Thus the deposition rate of lithogenous material is approximately 0.2 cm a^{-1}. Assuming a lithogenous sediment flux of 0.05 g cm^{-2} a^{-1}, then the $J(aHR)$ required to produce the observed Fe_{HR}/FeT ratio of 0.48 is $J(aHR)$ = 0.04 × 0.05 g cm^{-2} a^{-1} × (0.48 – 0.26) = 440 µg cm^{-2} a^{-1}. This flux could be derived from a benthic recycled iron flux of 100–1000 µg cm^{-2} a^{-1} from shelf sediments with oxygenated bottom waters by any combination of ε and the S/B ratio such that $\varepsilon \times S/B$ = 0.5–5 (a higher range than in the semi-enclosed basins of the Black Sea and the Baltic Sea). A more detailed study of the Cariaco Basin is required to identify the potential source area and produce an integrated measure of the benthic recycled iron flux over that area. The resulting estimates of ε and the S/B ratio for this basin would provide a better comparison with the semi-enclosed basins of the Black Sea and the Baltic Sea.

Evidence from ancient sediments

Many ancient black shales accumulated in the deeper parts of epicontinental seas (Wignall 1994) rather than in enclosed or semi-enclosed basins such as those in Table 2. The spatial relationships between the source area and the basinal sink area may be complex in these circumstances. Shallow seas which have become euxinic over substantial areas may have relatively low S/B ratios, especially where a narrow, essentially linear, shelf source area borders on a deeper euxinic basin. Present-day marginal basins may provide an indication of the S/B ratios of such ancient black shales that are iron-enriched. Table 2 shows the S/B ratios for the South Australian and Argentine Basins, which are well-defined areas of deep water adjacent to the continental shelf. The source areas are defined as being the coastal sediments deposited at depths of < 200 m that are laterally adjacent to the basin and, on this basis, these basins have S/B ratios of approximately 0.2. For comparison the S/B ratio of the world oceans is 0.08 (see Table 2).

The Toarcian (early Jurassic) period was an interval of widespread black shale deposition (notably the Jet Rock and the Posidoniaschiefer; Jenkyns 1988) in an epicontinental sea that covered much of Europe. These two sediments show different behaviour in respect of their highly reactive iron contents (Raiswell & Berner 1985; Poulton & Raiswell 2002), which are more enriched in the Jet Rock (where highly reactive iron is also correlated with organic C) than in the Posidoniaschiefer. Poulton & Raiswell (2002) give a mean $Fe_{HR}/FeT = 0.66$ for the Jet Rock, which was deposited at a rate of 0.007 cm a^{-1} (McArthur *et al.* 2000), equivalent to approximately 0.07 cm a^{-1} of uncompacted sediment. Assuming a lithogenous sediment flux of 0.02 g cm^{-2} a^{-1}, then the $J(aHR)$ required to produce the mean Fe_{HR}/FeT ratio of 0.66 is $J(aHR) = 0.04 \times 0.02$ g cm^{-2} a$^{-1} \times [0.66 - 0.26]$ $= 320$ µg cm^{-2} a^{-1}. This flux could be derived from a benthic recycled iron flux of $100–1,000$ µg cm^{-2} a^{-1} from shelf sediments deposited beneath oxygenated bottom waters by any combination of ε and the S/B ratio such that $\varepsilon \times S/B = 0.3–3$. Estimates of ε varied from $0.3–0.8$ (see earlier) which then suggests that S/B for the Jet Rock ranged approximately from $1–10$. Better estimates of ε have the potential to provide a useful insight into the basinal configurations (and areas of euxinic bottom waters) in which the deposition of ancient black shales occurred.

Wignall (1994) has also shown that ancient black shales developed during the initial stages of transgression occur in topographic hollows. Transgressions are favourable periods for the development of topographically-controlled basins because the initial flooding can occur over incised river valleys (Wignall 1991, 1994) or irregular palaeo-relief surfaces shaped by combinations of glacial, erosive and tectonic processes (e.g. Luning *et al.* 2000). These hollows serve as traps for fine-grained sediment (such as colloidal iron oxides) and organic matter (Huc 1988) and the irregular topography inhibits bottom-water circulation so that the influx of organic matter increases oxygen demand and assists the development of euxinicity (Wignall 1991, 1994). An irregular bottom topography which confines euxinic bottom waters to an area surrounded by oxic shelf sediments might also be characterized by relatively large S/B ratios, compared to those typically found in the enclosed and semi-enclosed basins listed in Table 2. A combination of these factors is likely to be responsible for the notably high levels of reactive iron that are found in many transgressive black shales (Wignall pers. comm.).

Ultimately, the mobilized iron exported from shelf areas will always tend to accumulate by physical trapping within the deeper areas of a basin, as does all fine-grained material, e.g. organic matter (Huc 1988). This will be true whether or not the deep-basin area is euxinic, and enclosed oxic deep basins might therefore show iron enrichments subject to the constraints of sedimentation rate and their S/B ratio. Note that anoxic, non-sulphidic basins may also physically trap reactive iron but the iron will tend to be dissolved and retained within the water column. Thus the sediments of the Orca Basin do not appear to show reactive iron enrichment (Raiswell & Canfield 1998). However, the existence of euxinicity also provides a chemical trap for reactive iron which might otherwise be exported (provided the recycled iron can be brought in contact with the sulphidic waters and sedimented as iron sulphides). Export from the deep areas of an enclosed/semi-enclosed basin is unlikely but chemical trapping may be critical in preventing export from shallow topographic hollows surrounded by oxic shelf sediments. Source generating areas adjacent to euxinic basins may also exist in an oceanographic regime that prevents transport of recycled iron to the basinal sink area. Local oceanographic conditions (wind, currents, bottom topography) will dictate whether, and how effectively, the source and sink areas are connected.

Conclusions

Reactive iron enrichment in the Black Sea results at least partly from the diagenetic mobilization of iron from shelf sediments. A significant fraction of the mobilized iron is exported from the shelf and transported into the deep basin where precipitation of sulphides occurs. A diagenetic model has been used to show that iron can be mobilized from shelf sediments deposited from oxygenated bottom waters at rates of $100–1,000$ µg cm^{-2} a^{-1} under typical diagenetic conditions ($T = 0–20$ °C, pH $= 7–8$, porewater $Fe^{2+} = 10^{-6}$ g cm^{-3}). These rates are essentially independent of pH and temperature provided the thickness of the oxygenated surface layer of the sediments is less than 1.0 cm. The mobilized iron may be either exported or redeposited on the shelf. In the Black Sea the export efficiency (ε) is at least 30% and may be as high as 80–90% but the exported iron is dispersed over a deep basinal area (B) which is larger than the shelf generating area (S). In the Black Sea the ratio $S/B = 0.37$, for a shelf area confined to depths < 200 m, below which euxinic bottom waters occur and diagenetic mobilization of iron is impossible. The S/B ratio of the Black Sea is not unusually large compared to other modern basins which have S/B ratios ranging from 0.25–13 approximately for a shelf generating area at depths of

< 200 m. However, the critical factor that defines the shelf area is the depth at which euxinic bottom waters occur. In general, surface waters are well-mixed by wind and wave action to depths of approximately 200 m (Chester 2000) in the absence of sills or other constraints on circulation. Where such constraints exist, the S/B ratio of any particular basin may be reduced substantially. Estimates of $\varepsilon \times S/B$ for modern euxinic sediments in the Black Sea (0.1–0.3), the Baltic (0.1–1), and the Cariaco Basin (0.5–5) appear possibly similar to those in ancient euxinic sediments (0.3 to 3.0 in the Jurassic Jet Rock), assuming that iron is mobilized from shelf sediments deposited from oxygenated bottom waters. Further studies of these modern euxinic basins are urgently needed to identify the range of variation in ε and the S/B ratio.

The authors are very grateful to Jack Middelburg and Tim Lyons for their valuable reviews. Jim McManus and Will Berelson helpfully supplied Californian Basin data and Will Bradbury assisted in the estimation of S/B ratios.

References

ALLER, R.C. 1977. *The influence of macrobenthos on chemical diagenesis of marine sediments.* PhD thesis, Yale University.

ALLER, R.C., MACKIN, J.E. & COX, R.T. JR 1986. Diagenesis of Fe and S in Amazon inner shelf muds; apparent dominance of Fe reduction and implications for the genesis of ironstones. *Continental Shelf Research*, **6**, 263–289.

ANDERSON, T.F. & RAISWELL, R. 2004. Sources and mechanisms for the enrichment of highly reactive iron in euxinic Black Sea sediments. *American Journal of Science*, **304**, 203–233.

ARTHUR, M.A., DEAN, W.E., NEFF, E.D., HAY, B.J., KING, J. & JONES, G. 1994. Varve calibrated records of carbonate and organic carbon accumulation over the last 2000 years in the Black Sea. *Global Biogeochemical Cycles*, **8**, 195–217.

BELMANS, F., VAN GRIEKEN, R. & BRUGMANN, L. 1993. Geochemical characterization of recent sediments in the Baltic Sea by bulk and electron microprobe analysis. *Marine Chemistry*, **42**, 223–236.

BENSON, B.B. & KRAUSE, D.K. JR 1984. The concentration and isotopic fractionation of oxygen dissolved in freshwater and seawater in equilibrium with the atmosphere. *Limnology and Oceanography*, **29**, 620–632.

BEN-YAAKOV, S. 1973. pH buffering of pore water of Recent anoxic marine sediments. *Limnology and Oceanography*, **18**, 86–94.

BERG, P., RYSGAARD, S. & THAMDRUP, B. 2003. Dynamic modeling of early diagenesis and nutrient cycling: a case study in an Arctic marine sediment. *American Journal of Science*, **303**, 905–955.

BERNER, R.A. 1970. Sedimentary pyrite formation. *American Journal of Science*, **268**, 2–23.

BERNER, R.A. 1980. *Early Diagenesis: A Theoretical Approach*. Princeton University Press, Princeton, N.J.

BERNER, R.A. 1984. Sedimentary pyrite formation: an update. *Geochimica et Cosmochimica Acta*, **48**, 606–616.

BOESEN, C. & POSTMA, D. 1988. Pyrite formation in anoxic environments of the Baltic. *American Journal of Science*, **288**, 575–603.

BOUDREAU, B.P. 1996. *Diagenetic Models and their Interpretation*. Springer-Verlag, Berlin.

BOUDREAU, B.P. & CANFIELD, D.E. 1988. A provisional diagenetic model for pH in anoxic porewaters: Application to the FOAM site. *Journal of Marine Research*, **46**, 429–455.

BOUDREAU, B.P. & CANFIELD, D.E. 1993. A comparison of closed-system and open-system models for porewater pH and calcite saturation state. *Geochimica et Cosmochimica Acta*, **57**, 317–334.

BOUDREAU, B.P. & SCOTT, M.R. 1978. A model for the diffusion-controlled growth of deep-sea manganese nodules. *American Journal of Science*, **278**, 903–929.

BREWER, P.G. & SPENCER, D.W. 1974. Distribution of some trace elements in the Black Sea and their flux between dissolved and particulate phases. *In*: DEGENS, E.T. & ROSS, D.A. (eds) *The Black Sea – Geology, Chemistry, and Biology*. American Association of Petroleum Geologists, Tulsa, Oklahoma, 137–143.

BUESSLER, K.O., LIVINGSTON, H.D. & CASSO, C. 1991. Mixing between oxic and anoxic waters in the Black Sea as traced by Chernobyl cesium isotopes. *Deep Sea Research*, **38**, S725–745.

BURKE, I.T. & KEMP, A.E.S. 2002. Microfabric analysis of Mn-carbonate laminae deposition and Mn-sulfide formation in the Gotland Deep, Baltic Sea. *Geochimica et Cosmochimica Acta*, **66**, 1589–1600.

CALVERT, S.E., THODE, H.G., YOUNG, D. & KARLIN, R.E. 1996. A stable isotope study of pyrite formation in the late Pleistocene and Holocene sediments of the Black Sea. *Geochimica et Cosmochimica Acta*, **60**, 1261–1270.

CANFIELD, D.E. 1989. Reactive iron in marine sediments. *Geochimica et Cosmochimica Acta*, **53**, 619–632.

CANFIELD, D.E. & RAISWELL, R. 1991. Pyrite formation and fossil preservation. *In*: ALLISON, P.A. & BRIGGS, D.E.G. (eds) *Taphonomy: Releasing the Data Locked in the Fossil Record*. Plenum, New York, 337–387.

CANFIELD, D.E., RAISWELL, R. & BOTTRELL S.H. 1992. The reactivity of sedimentary iron minerals towards sulfide. *American Journal of Science*, **292**, 659–683.

CANFIELD, D.E., THAMDRUP, B. & HANSEN, J.W. 1993. The anaerobic degradation of organic matter in Danish coastal sediments: Iron reduction, manganese reduction and sulfate reduction. *Geochimica et Cosmochimica Acta*, **57**, 3867–3993.

CANFIELD, D.E., LYONS, T.W. & RAISWELL, R. 1996. A

model for iron deposition to euxinic Black Sea sediments. *American Journal of Science*, **296**, 818–834.

CHANTON, J.P. 1985. *Sulfur mass balance and isotopic fractionation in an anoxic marine sediment*. PhD thesis, University of North Carolina.

CHESTER, R. 2000. *Marine Chemistry*. 2nd edn. Unwin Hyman, London.

DAVISON, W. & SEED, G. 1983. The kinetics of the oxidation of ferrous iron in synthetic and natural waters. *Geochimica et Cosmochimica Acta*, **47**, 67–79.

DIETRICH, G., KALLE, K., KRAUSS, W. & SIEDLER, G. 1980. *General Oceanography*. 2nd edn. Wiley, New York.

DEFLANDRE B., MUCCI, A., GAGNE, J., GUIGNARD, C. & SUNDBY, B. 2002. Early diagenetic processes in coastal marine sediments disturbed by a catastrophic sedimentation event. *Geochimica et Cosmochimica Acta*, **66**, 2547–2558.

ELDERFIELD, H., MCCAFFREY, J., LUEDTKE, N., BENDER, M. & TRUESDALE, V.W. 1981. Chemical diagenesis in Narragansett Bay sediments. *American Journal of Science*, **281**, 1021–1055.

ELROD, V.A., BERELSON, W.M., COALE, K.H. & JOHNSON, K.S. 2004. The flux of iron from continental shelf sediments: a missing source for global budgets. *Geophysical Research Letters*, **31**, L12307, doi:10.129/2004GL020216.

FERRO, J., VAN NUGTEREN, P., MIDDELBURG, J.J., HERMAN, P.M.J. & HEIP, C.H.R. 2003. Effect of macrofauna, oxygen exchange and particle reworking on iron and manganese sediment biogeochemistry: a laboratory experiment. *Vie Milieu*, **53**, 211–220.

FRIEDL, G., DINKEL, C. & WERHLI, B. 1998. Benthic fluxes of nutrients in the northwestern Black Sea. *Marine Chemistry*, **62**, 77–88.

FRIEDRICH, J., DINKEL, C., ET AL. 2002. Benthic nutrient cycling and diagenetic pathways in the north-western Black Sea. *Estuarine, Coastal and Shelf Science*, **54**, 369–383.

GARDNER, L.R. 1973. Chemical models for sulfate reduction in closed anaerobic marine environments. *Geochimica et Cosmochimica Acta*, **37**, 53–68.

GOLDHABER, M.B. & KAPLAN, I.R. 1974. The sulfur cycle. *In*: GOLDBERG, E.D. (ed.) *The Sea*, **5**, 569–655, Wiley Interscience, New York.

HAESE, R.R., SCHRAMM, J., RUTGERS VAN DER LOEFF, M.M. & SCHULTZ, H.D. 2000. The reactivity of iron. *International Journal of Earth Sciences*, **88**, 619–629.

HONJO, S., HAY B.J., ET AL. 1987. Seasonal cyclicity of lithogenic particle fluxes at a southern Black Sea sediment trap station. *In*: DEGENS, E.T., IZDAR, E. & HONJO, S. (eds) *Particle Flux in the Ocean*. Mitteilungen aus dem Geologisch Palaontologischen Insitut der Universitat Hamburg, **6**, 19–39.

HUC, A.Y. 1988. Aspects of depositional processes of organic matter in sedimentary basins. *Organic Geochemistry*, **13**, 263–272.

JENKYNS, H.C. 1988. The early Toarcian (Jurassic)

anoxic event: stratigraphy, sedimentary and geochemical evidence. *American Journal of Science*, **288**, 101–151.

JORGENSEN, B.B. & BOUDREAU, B.P. 2001. Diagenesis and sediment-water exchange. *In*: BOUDREAU, B.P. & JORGENSEN, B.B. (eds) *The Benthic Boundary Layer*. Oxford University Press, New York, 211–244.

JORGENSEN, B.B., FOSSING, H., WIRSEN, C.O. & JANNSCH, H.W. 1991. Sulfide oxidation in the anoxic Black Sea chemocline. *Deep-Sea Research*, **38**, S1083–S1103.

KEMPE, S., LIEBEZEIT, G., DIERCKS, A.R. & ASPER, V. 1990. Water balance in the Black Sea. *Nature*, **346**, 419.

KRISTENSEN, E. 2000. Organic matter diagenesis at the oxic/anoxic interface in coastal marine sediments, with emphasis on the role of burrowing animals. *Hydrobiologia*, **426**, 1–24.

KROM, M.D., MORTIMER, R.J.G., POULTON, S.W., HAYES, P., DAVIES, I.M., DAVISON, W. & ZHANG, H. 2002. In-situ determination of dissolved iron production in recent marine sediments. *Aquatic Science*, **64**, 282–291.

LANDING, W.M. & BRULAND, K.W. 1987. The contrasting biogeochemistry of iron and manganese in the Pacific Ocean. *Geochimica et Cosmochimica Acta*, **51**, 29–43.

LANDING, W.M. & LEWIS, B.L. 1991. Thermodynamic modeling of trace metal speciation in the Black Sea. *In*: IZDAR, E. & MURRAY, J.W. (eds) *Black Sea Oceanography*. Dordrecht, Kluwer, 125–160.

LEWIS, B.L. & LANDING, W.M. 1991. The biogeochemistry of manganese and iron in the Black Sea. *Deep-Sea Research*, **38**, S773–S803.

LUNING, S., CRAIG, J., LOYDELL, D.K., STORCH, P. & FITCHES, B. 2000. Lowe Silurian 'hot shales' in North Africa and Arabia: regional distribution and depositional model. *Earth Science Reviews*, **49**, 121–200.

LUTHER, G.W. (III). 1991. Pyrite synthesis via polysulfide compounds. *Geochimica et Cosmochimica Acta*, **55**, 2839–2849.

LYONS, T.W. 1991. Upper Holocene sediments of the Black Sea: Summary of Leg 4 box cores (1988 Black Sea Oceanographic Expedition). *In*: IZDAR, E. & MURRAY, J.W. (eds) *Black Sea Oceanography*. NATO ASI Series, Kluwer, Dordrecht, 401–441.

LYONS, T.W. 1997. Sulfur isotopic trends and pathways of iron sulfide formation in upper Holocene sediments of the anoxic Black Sea. *Geochimica et Cosmochimica Acta*, **61**, 3367–3382.

LYONS, T.W. & BERNER, R.A. 1992. Carbon-sulfur-iron systematics of the uppermost Holocene sediments of the anoxic Black Sea. *Chemical Geology*, **99**, 1–27.

LYONS, T.W., BERNER, R.A. & ANDERSON, R.F. 1993. Evidence for large pre-industrial perturbations of the Black Sea chemocline. *Nature*, **265**, 538–540.

LYONS, T.W., WERNE, J.P., HOLLANDER, D.J. & MURRAY, R.W. 2003. Contrasting sulfur geochemistry and Fe/Al and Mo/Al ratios across the last oxic-to-anoxic transition in the Cariaco Basin, Venezuela. *Chemical Geology*, **195**, 131–157.

MCARTHUR, J.M., DONOVAN, D.T., THIRLWALL, M.F., FOUKE, B.W. & MATTEY, D. 2000. Strontium isotope profile of the early Toarcian (Jurassic) oceanic anoxic event, the duration of ammonite biozones, and belemnite paleotemperatures. *Earth and Planetary Science Letters*, **179**, 269–285.

MCMANUS, J., BERELSON, W.M., COALE, K.H. & KILGORE, T.E. 1997. Phosphorus regeneration in continental margin sediments. *Geochimica et Cosmochimica Acta*, **61**, 2891–2907.

MILLERO, F.J. 2001. *The Physical Chemistry of Natural Waters*. Wiley Interscience, New York.

MILLERO, F.J., SOTOLONGO, S. & IZAGUIRRA, M. 1987. The oxidation kinetics of Fe(II) in seawater. *Geochimica et Cosmochimica Acta*, **51**, 793–891.

MULLER, G. & STOFFERS, P. 1974. Mineralogy and petrology of Black Sea sediments. *In*: DEGENS, E.T. & ROSS, D.A. (eds) *The Black Sea – Geology, Chemistry, and Biology*. American Association of Petroleum Geologists, Tulsa, Oklahoma, 200–248.

MURAMOTO, J.A., HONJO, S., FRY, B., HAY, B.J., HOWARTH, R.W. & CSNE, J.L. 1991. Sulfur, iron and organic carbon fluxes in the Black Sea: sulfur isotopic evidence for origin of sulfur fluxes. *Deep-Sea Research*, **38**, S1151–S1187.

MURRAY, J.W., JANNASCH, H.W., HONJO, S., ANDERSON, R.F., REEBURGH, W.S., TOP, Z., FRIEDRICH, G.E., CODISPOTI, L.A. & IZDAR, E. 1989. Unexpected changes in the oxic/anoxic interface in the Black Sea. *Nature*, **338**, 411–413.

MURRAY, J.W., CODISPOTI, L.A. & FRIEDERICH, G.E. 1995. Oxidation–reduction environments: the suboxic zone in the Black Sea. *In*: HUANG, C.P, O'MELIA, C.R. & MORGAN, J.J. (eds) *Aquatic Chemistry: Interfacial and Interspecies Processes*. American Chemical Society Advances in Chemistry Series **224**, 157–176.

NEALSON, K.H. 1997. Sediment bacteria: who's there, what are they doing and what's new? *Annual Reviews of Earth and Planetary Science*, **25**, 403–434.

POULTON, S.W. & RAISWELL, R. 2002. The low temperature geochemical cycle of iron: From continental fluxes to marine sediment deposition. *American Journal of Science*, **302**, 774–805.

POULTON, S.W., KROM, M.D. & RAISWELL, R. 2004. A revised scheme for the reactivity of iron (oxyhydr)oxide minerals towards dissolved sulfide. *Geochimica et Cosmochimica Acta*, **68**, 3703–3715.

RAISWELL, R. & BERNER, R.A. 1985. Pyrite formation in euxinic and semi-euxinic sediments. *American Journal of Science*, **285**, 710–724.

RAISWELL, R. & CANFIELD, D.E. 1996. Rates of reaction between silicate minerals and dissolved sulfide in Peru Margin sediments. *Geochimica et Cosmochimica Acta*, **60**, 2777–2787.

RAISWELL, R. & CANFIELD, D.E. 1998. Sources of iron for pyrite formation. *American Journal of Science*, **298**, 219–245.

RAISWELL, R., CANFIELD, D.E. & BERNER, R.A. 1994.

A comparison of iron extraction methods for the determination of degree of pyritisation and the recognition of iron-limited pyrite formation. *Chemical Geology*, **111**, 101–111.

RAISWELL, R., NEWTON, R. & WIGNALL, P.B. 2001. An indicator of water-column anoxia: resolution of biofacies variations in the Kimmeridge Clay (Upper Jurassic, U.K.). *Journal of Sedimentary Research*, **71**, 286–294.

RICKARD, D.T. 1975. Kinetics and mechanisms of pyrite formation at low temperatures. *American Journal of Science*, **275**, 636–652.

RICKARD, D.T. 1997. Kinetics of pyrite formation by the H_2S oxidation of iron (II) monosulfide in aqueous solutions between 25 °C and 125 °C: the rate equation. *Geochimica et Cosmochimica Acta*, **61**, 115–134.

RICKARD, D.T. & LUTHER, G.W. (III). 1997. Kinetics of pyrite formation by the H_2S oxidation of iron (II) monosulfide in aqueous solutions between 25 °C and 125 °C: the mechanism. *Geochimica et Cosmochimica Acta*, **61**, 135–147.

ROEKENS, E.J. & VAN GRIEKEN, R.E. 1983. Kinetics of iron oxidation in seawater of various pH. *Marine Chemistry*, **13**, 195–202.

SAAGER, P.M., DE BAAR, H.J.W. & BURKILL, P.H. 1989. Manganese and iron in Indian Ocean waters. *Geochimica et Cosmochimica Acta*, **53**, 2259–2267.

SANTSCHI, P., HOHENER, P., BENOIT, G. & BUCHOLTZ-TEN-BRINK, M. 1990. Chemical processes at the sediment-water interface. *Marine Chemistry*, **30**, 269–315.

SCHIEBER, J. 1995. Anomalous iron distribution in shales as a manifestation of 'non-clastic iron' supply to sedimentary basins: relevance for pyritic shales, base-metal mineralisation, and oolitic ironstone deposits. *Mineralium Deposita*, **30**, 294–302.

SCHOONEN, M.A.A. & BARNES, H.L. 1991. Reactions forming pyrite and marcasite from solution. *Geochimica et Cosmochimica Acta*, **55**, 1505–1514.

SHEN, Y., CANFIELD, D.E. & KNOLL, A. 2002. Middle Proterozoic ocean chemistry: evidence from the McArthur Basin, Northern Australia. *American Journal of Science*, **302**, 81–109.

SHEN, Y., KNOLL, A.H. & WALER, M.R. 2003. Evidence for low sulphate and anoxia in a mid-Proerozoic marine basin. *Nature*, **423**, 632–636.

SORENSEN, J. & JORGENSEN, B.B. 1987. Early diagenesis in sediments from Danish coastal waters: Microbial activity and Mn–Fe–S geochemistry. *Geochimica et Cosmochimica Acta*, **51**, 1583–1590.

STUMM, W. & MORGAN, J.J. 1980. Kinetics and products of ferrous iron oxygenation in aqueous solutions. *Environmental Science and Technology*, **14**, 561–568.

THAMDRUP, B. 2000. Bacterial manganese and iron reduction in aquatic sediments. *In*: SCHINK, B. (ed.) *Advances in Microbial Ecology*. Kluwer, New York, 41–84.

THAMDRUP, B. & CANFIELD, D.E. 1996. Pathways of

carbon oxidation in continental margin sediments off central Chile. *Limnology and Oceanography*, **41**, 1629–1650.

THAMDRUP, B., FINSTER, K., FOSSING, H., HANSEN, J.W. & BARKER, B.B. 1994. Thiosulfate and sulfite distributions in porewater of marine sediments related to manganese, iron and sulfur geochemistry. *Geochimica et Cosmochimica Acta*, **58**, 67–73.

THEIS, T.L. & SINGER, P.C. 1974. Complexation of iron (III) by organic matter and its effect on iron (II) oxygenation. *Environmental Science and Technology*, **8**, 569–573.

TREFRY, J.H. & PRESLEY, B.J. 1982. Manganese fluxes from Mississippi Delta sediments. *Geochimica et Cosmochimica Acta*, **46**, 1715–1726.

ULLMAN, W.J. & ALLER, R.C. 1982. Diffusion coefficients in nearshore marine sediments. *Limnology and Oceanography*, **27**, 552–556.

WANG, Y. & VAN CAPPELLEN, P. 1996. A multicomponent reactive transport model of early diagenesis: application to redox cycling in coastal marine sediments. *Geochimica et Cosmochimica Acta*, **60**, 2993–3014.

WERNE, J.P., SAGEMAN, B.B., LYONS, T.W. & HOLLANDER, D.J. 2002. An integrated assessment of a type 'euxinic' deposit: Evidence for multiple controls on black shale deposition in the Middle Devonian Oatka Creek Formation. *American Journal of Science*, **302**, 110–143.

WIGNALL, P.B. 1991. Model for transgressive black shales? *Geology*, **19**, 167–170.

WIGNALL, P.B. 1994. *Black Shales*. Clarendon Press, Oxford.

WIJSMAN, J.W.M. 2001. *Early diagenetic processes in northwestern Black Sea sediments*. PhD thesis, Netherlands Institute of Ecology.

WIJSMAN, J.W.M., MIDDELBURG, J.J. & HEIP, C.H.R. 2001. Reactive iron in Black Sea sediments: implications for iron cycling. *Marine Geology*, **172**, 167–180.

WIJSMAN, J.W.M., HERMAN, P.M.J., MIDDELBURG, J.J. & SOETAERT, K. 2002. A model for early diagenetic processes in sediments of the continental shelf of the Black Sea. *Estuarine and Coastal Shelf Science*, **54**, 403–421.

WILKIN, R.T. & ARTHUR, M.A. 2001. Variations in pyrite texture, sulfur isotope compositions, and iron systematics in the Black Sea: evidence for Late Pleistocene to Holocene excursions of the O_2–H_2S redox transition. *Geochimica et Cosmochimica Acta*, **65**, 1399–1416.

Distinguishing biological from hydrothermal signatures via sulphur and carbon isotopes in Archaean mineralizations at 3.8 and 2.7 Ga

N. V. GRASSINEAU[1], P. W. U. APPEL[2], C. M. R. FOWLER[1] & E. G. NISBET[1]

[1]*Department of Geology, Royal Holloway, University of London, Egham, Surrey TW20 0EX, UK (e-mail: nathalie@gl.rhul.ac.uk)*
[2]*Geological Survey of Denmark and Greenland (GEUS), Øster Voldgade 10, 1350 Copenhagen, Denmark*

Abstract: Carbon and sulphur isotopes have been analysed in mineralization from two Archaean greenstone belts at 3.8 Ga and 2.7 Ga, with the aim of distinguishing between the inorganic and organic processes that occurred. Despite an obvious overprinting by metamorphism (in the early belt), or hydrothermal fluids, there are convincing differences between the values of carbon and sulphur in inorganic formations and those formed by biological processes. An attempt is made to estimate the changes that occurred in the early life activities over this 1 Ga year period. Life in the Isua Greenstone Belt (3.8 Ga) was most likely present in transitory settings, probably under high temperature conditions. This was very different from the life at 2.7 Ga in the Belingwe Greenstone Belt, indicated by ranges of 38‰ for $\delta^{34}S$ and 37‰ for $\delta^{13}C_{red}$. By this time, the biological sulphur and carbon cycles seem to have reached almost full operation, with the presence of well-established photosynthetic microbial mat communities.

Mineralization by chemical reaction can be formed by various processes. Beyond high-temperature magmatic systems, mineral deposits can be produced by the release of fluids from diagenesis or metamorphism, but also at low to moderate temperature conditions by sea- or meteoric-water circulation. They can also be formed by biological activities in sediments. Stable isotopes are powerful tools to determine these processes involved in mineralization, as well as their source. In inorganic systems, it is possible to measure the extent of exchange between fluids and host rocks, by estimating the interaction, fluid circulation, temperatures of crystallization and the general conditions of mineralization. $\delta^{34}S$ and $\delta^{13}C$ values can be widely fractionated from the sources, especially in low to moderate temperature hydrothermal systems, whereas smaller isotopic fractionations will result under high-temperature conditions unless Rayleigh distillation has occurred due to an open system. Mineralization from organic activity can also record important S- and C-isotopic variations. This is the result of metabolic processes producing specific fractionation from distinctive and well-defined reservoirs.

Archaean mineralizations are widespread, particularly in greenstone belts. Carbon and sulphur isotopic analyses enable the determination of the origin and the type of hydrothermal fluids from which they formed. They are also able to identify the bacterial activity that might have been present from the isotopic signatures left behind at the time of the deposition, which are commonly the only traces of Archaean life remaining. Consequently, these sulphur and carbon isotopic fingerprints can be used to interpret early microbial communities, and therefore their mutual interactions within particular ecosystems (Nisbet & Fowler 1996, 1999; Nisbet & Sleep 2001; Grassineau *et al.* 2001*a*, 2002). Looking in the geological record for evidence is necessary, as the rRNA studies on the initial evolution of prokaryotes are controversial (Woese 1987; Pace 1997).

Life in modern hydrothermal systems is abundant, with hyperthermophiles in hot places and mesophiles more distally. In deep locations, extremophiles exploit hot acid environments, resulting in a redox potential of sulphur creating a wide range of $\delta^{34}S$ compositions but with little effect on $\delta^{13}C$. Meanwhile, photosynthesis dominates at the surface. Can similar activities be found in Archaean hydrothermal systems? This study investigates the issue with specific attention to material from two Archaean greenstone belts, separated by 1 Ga of age, Isua (3.7–3.8 Ga) and Belingwe (2.7 Ga). In particular, the discrimination of isotopic signatures that may distinguish early life from those that represent contemporaneous or later mineralization events will be discussed.

From: McDonald, I., Boyce, A. J., Butler, I. B., Herrington, R. J. & Polya, D. A. (eds) 2005. *Mineral Deposits and Earth Evolution.* Geological Society, London, Special Publications, **248**, 195–212. 0305-8719/$15.00

Sulphur and carbon isotopic studies

Sulphur and carbon isotopic variations are due to different processes acting on primary or secondary sources. The two main modern sulphur reservoirs are restricted, with $\delta^{34}S$ seawater sulphate at $+21 \pm 0.2‰$ (Faure 1986) and the mantle at $+0.3 \pm 0.5‰$ (e.g. Sakai *et al.* 1984). The sediment reservoir has a wide range of -60 to $+20‰$ (Hoefs 1997) produced by fractionation processes. Some of these processes are inorganic, resulting from hydrothermal fluid circulation, but the largest ranges are from repeated biological processing (e.g. Schidlowski *et al.* 1983; Canfield & Teske 1996). The sulphur cycle evolution in the Archaean is subject to controversy. The microbial 'standard' model shows indirect evidence for very early S-oxidation and S-reduction processes (Woese 1987; Pace 1997). Ohmoto (1992; Ohmoto *et al.* 1993) showed evidence for microbial seawater sulphate reduction at 3.4–3.2 Ga, but other authors suggest that the cycle evolved only from the Proterozoic (Blank 2004), with a full cycle appearing only by 0.86–1.0 Ga (Canfield & Teske 1996), and high seawater sulphate concentrations occurring only with the rise of atmospheric oxygen. The $\delta^{34}S$ of Archaean seawater is still unknown, and until recently, the range of Archaean sulphide minerals in sediments seemed to be less than 11‰ (Cameron 1982; Habicht & Canfield 1996) or 13‰ in mineral deposits (Ohmoto 1992). Conversely, Grassineau *et al.* (2001*a*, 2002) obtained a range of 37‰ at 2.7 Ga, and Shen *et al.* (2001), a range of 16‰ for 3.47 Ga. This shows the antiquity of at least some of the S-processes, with an almost, if not fully, operating cycle by the late Archaean. On the other hand, sulphate-reduction by biological activity in the middle Archaean has been suggested by Ohmoto & Felder (1987), Ohmoto *et al.* (1993), Kakegawa & Ohmoto (1999), and most recently by Shen & Buick (2004).

Carbon reservoirs are variable, $\delta^{13}C$ in the mantle is from -8 to $-5‰$, but graphites in igneous rocks vary from -33 to $-6‰$ (in Taylor 1986). Marine carbonates are at $+0.56 \pm 1.55‰$ (Faure 1986). Biological fractionations create wide ranges of isotopic values in modern sediments, commonly -30 to $-10‰$ for organic matter. Two distinct values dominate the organic carbon: *c.* -27 to $-22‰$ due to Rubisco I enzyme using preferentially ^{12}C in CO_2 leaving residual inorganic carbon at *c.* 0‰ (Pierson 1994; in Jahnke *et al.* 2001) in aerobic and microaerobic environments, and $-30‰$ or lighter methanogen residue, mainly in anaerobic

conditions (Coleman *et al.* 1981). Early Archaean reservoirs are less known, but the reduced carbon range of -52 to $-13‰$ suggests evidence of early life (in Strauss & Moore 1992). $\delta^{13}C$ studies helped to assess the Archaean biology (e.g. Schidlowski *et al.* 1983; Hayes *et al.* 1983), especially the global carbon management by Rubisco I. Once started, the oxygenic photosynthetic process, mainly used by cyanobacteria, spread fast and widely. The moment of its commencement in the Archaean is controversial, but stable isotopic studies can estimate it. They can also help to identify the anoxygenic photosynthesis that seems to predate the oxygenic process (e.g. Nisbet & Sleep 2001) and the appearance of methanogenesis, estimated to be no older than late Proterozoic by some biochemists (Cavalier-Smith 2002).

Analytical techniques

Sulphur and reduced carbon were analysed with a VG/Fisons/Micromass 'Isochrom-EA' system, an elemental analyser (EA1500 Series 2) on line to an Optima mass spectrometer in He continuous flow mode (Matthews & Hayes 1978 for carbon; Grassineau *et al.* 2001*b* for sulphur). This high-resolution technique can measure isotopic fractionations at the millimetre scale, allowing the determination of biological activities at micro-community scale, that are otherwise homogenized in larger analysed samples.

The hand-picked sulphide minerals, with quantities as small as 0.8 mg for pyrite, have a reproducibility of $\pm 0.1‰$ for $\delta^{34}S$. The mineral standards, including NBS123, NBS127 and IAEA-S3, cover a total range from -32 to $+20‰$. Hand-picked samples for reduced carbon were treated for 12 hours in 20%HCl at 120 °C. Samples of 0.07 mg for pure carbon to 30 mg for whole rock with 0.1 wt%C have a reproducibility better than $\pm 0.1‰$. The standards, including NBS 21 and IAEA-CO9 cover a $\delta^{13}C$ range from -47 to $+3‰$. Samples with less than 200 ppmC are discarded to prevent confusion with possible contamination of the blank coming from the <34 ppmC, measured in the laboratory in the tin capsule containers.

Isua Greenstone Belt, West Greenland, Early Archaean *c.* 3.8–3.7 Ga

The Isua Greenstone Belt (IGB), West Greenland, consists of remains of volcanosedimentary rocks in an arcuate shape surrounded and locally intruded by diverse tonalitic gneisses (Fig. 1). With ages of 3.8–3.65 Ga, the belt is the oldest sedimentary complex known on Earth. The most commonly occurring rocks are thick volcanic sequences (Appel *et al.* 1998; Myers 2001; Polat *et al.* 2003), principally pillow lavas with interlayed sedimentary sequences, comprising mica shists and garnet mica shists ± staurolite and chemical sediments such as cherts and banded iron formations. Sulphide mineralizations are numerous (Appel 1979). Several carbonate alteration episodes occurred from 3.7–2.8 Ga, with the earliest during the extrusion of lava flows and sediment depositions (Myers, pers. comm.), the later events happened mainly in the western part (Moorbath & Whitehouse 1996; Rose *et al.* 1996; Frei & Rosing 2001). Five structural domains, determined by different metamorphic effects, have been defined by various authors and Rollinson (2002) (Fig. 1). Domain I is a low-strain area identified from the study of metapelites that occurred at 3.69 Ga. Domains II and V record two high-grade metamorphic events, both at 3.74 Ga.

Domains III and IV underwent the same events but with higher intensity, and were also affected by a late metamorphic episode at 2.8 Ga. Despite multiple events, Domain IV had preserved some low-strain zones. Therefore, Domains I and IV are areas where early life traces might have been preserved.

Sedimentary sequences have been studied from these two areas, along with igneous rocks, graphite-bearing amphibolites, gold and quartz-fuchsite deposits in order to identify high and low temperature mineralization fractionations. The sampling for this study took place during 1998 and 1999 and was part of the Isua Multidisciplinary Research Project.

Lithologies studied

Sedimentary protoliths. BIFs are common in the IGB; they are mainly actinolite-bearing metachert and iron formation of magnetite-rich layers interbanded with grunerite-rich beds or

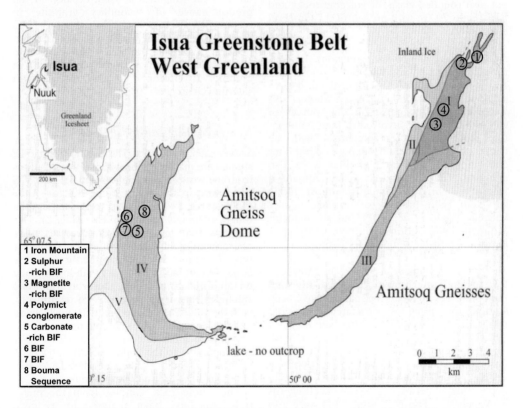

Fig. 1. Map of Isua Greenstone Belt (3.7–3.8 Ga) showing the five structural domains defined by various authors (compiled and modified by Rollinson 2002). Four of the sedimentary formations studied are in the eastern part in Domain I, and four in the west in Domain IV, the two areas that most likely preserved some low-strain zones. The slightly younger Amîtsoq gneisses at 3.7 Ga enclose the IGB.

quartz-bands. The eastern BIFs of Domain I include a sulphide- and magnetite-rich BIF, a magnetite-rich BIF, and a chert-rich BIF, a massive formation named the Iron Mountain, consisting of alternating layers of quartz and magnetite with locally amphibole-rich bands (Frei *et al.* 1999; Myers 2001). One of the western BIFs includes a carbonate facies, with magnetite and siderite bands, and also contains graphite and pyrrhotite.

A 3.78 Ga turbidite from the NW area (Domain IV) was sampled. It has been inter-preted as a Bouma sequence, affected by low-grade metamorphism. It consists of successions of medium-grained quartzites grading into fine-grained metapelites, with abundant small carbon particles in all layers (Rosing 1999). Fedo *et al.* (2001) contest the Bouma interpret-ation, but agreed that the deposition was below wave base. A metaconglomerate from the NE (Domain I) was also sampled (Nutman 1986). It comprises poorly sorted polymict conglomerate, mainly of metacherts pebbles, but also rounded clasts of volcanic rocks and BIF (Fedo 2000), with an age of 3.7 Ga (Frei *et al.* 1999).

Igneous and amphibolitic formations. Igneous rocks, including basalts, ultramafic, pillow lavas and cross-cutting dykes were sampled. $\delta^{34}S$ of sulphide minerals and $\delta^{13}C_{red}$ of whole rocks were measured to constrain the sources of carbon and sulphur involved in high tempera-ture processes, and for comparison with the results of the sedimentary deposits. Graphites and sulphides from amphibolites were also analysed. They are mainly from Domain I, where the original signatures most likely survived. Their primary origins are mainly volcanic (Fedo *et al.* 2001), but some meta-sediment compositions (Nutman 1986) may indicate a possible sedimentary deposition.

Hydrothermal mineral deposits. The effects of low to moderate temperature hydrothermal fluid processes on the $\delta^{34}S$ signatures have been studied for sulphides associated with gold mineralization in the SW of the belt, dated at 3.81 Ga (Frei & Rosing 2001). This zinc–lead-rich deposit, present in intrusive tonalitic sheets, hosts galena, sphalerite, and sparse pyrite in veins (Appel 2000). Four quartz–fuchsite deposits dated between 3.55–3.74 Ga (Richards & Appel 1987; Frei & Rosing 2001) were also sampled. These are widespread in the IGB, and associated with ultramafic rocks (Dymek *et al.* 1983).

Previous stable isotope work at Isua

Monster *et al.* (1979) obtained homogeneous $\delta^{34}S$ around 0 ± 0.5‰ on sulphides from mafic volcanic rocks and BIFs (Fig. 2), and Strauss (2003), values from –3 to +1‰ in BIFs. The conclusions drawn were that all $\delta^{34}S$ are in the range of hydrothermal sulphides of igneous origin with no clear evidence for sulphur-utilizing bacterial activity during deposition.

There have been many carbon isotope studies of the IGB (Perry & Ahmad 1977; Schidlowski *et al.* 1979; Hayes *et al.* 1983; Naraoka *et al.* 1996; Rosing 1999; Ueno *et al.* 2002; van Zuilen *et al.* 2002) (Fig. 3). A wide range for $\delta^{13}C_{red}$ (reduced carbon) has been obtained from –28 to –6‰. Rosing (1999) suggested that a value of –19‰ obtained from graphite grains from the Bouma sequence was close to the primary depositional value and thus indicated a biological origin. On the other hand, Perry & Ahmad (1977), Naraoka *et al.* (1996) and van Zuilen *et al.* (2002) contested a biogenic interpretation for most of the graphites at Isua, because of the predominance of secondary graphite in metacarbonates with $\delta^{13}C$ mostly around –12‰. They defined it as the result of siderite decom-position to graphite during a high temperature metamorphic event, with ^{13}C-enrichment in graphite by isotopic re-equilibration with siderite (Ueno *et al.* 2002).

New isotopic results for the Isua Belt

Sedimentary sequences. The sulphide minerals studied are mainly pyrite, with a few pyrrhotites from the western carbonate facies BIF. All $\delta^{34}S$ values are in the range from –3.8 to +3.4‰ (Table 1; Fig. 2). This 7.2‰ spread is larger than ranges obtained before for Isua (Fig. 2). This suggests that more diverse processes were operational than hitherto thought. However this variation is only found in the three eastern BIFs, present in the Domain I with lower grade metamorphism, including the Iron Mountain sulphides that have positive $\delta^{34}S$, contrasting with the values of the two other BIFs, which are mainly negative. The homogeneity of the polymict conglomerate with $\delta^{34}S$ at +0.4 ± 0.5‰ is comparable to the findings of Monster *et al.* (1979). In Domain IV, the carbonate facies BIF at –0.7 ± 0.1‰ is also homogeneous.

The whole rock samples from the meta-sediments have been analysed for reduced carbon after acid treatment (Table 1; Fig. 3), giving a $\delta^{13}C_{red}$ range from –29.6 to –6.5‰ with 0.02–0.18 wt%C. The type of carbon analysed

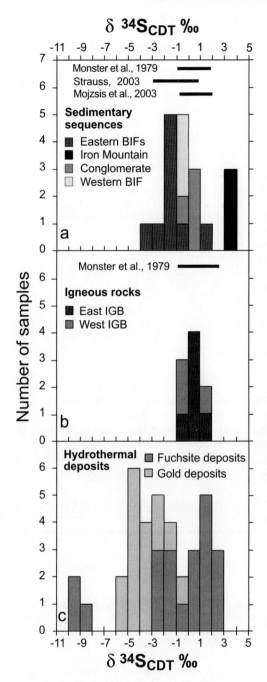

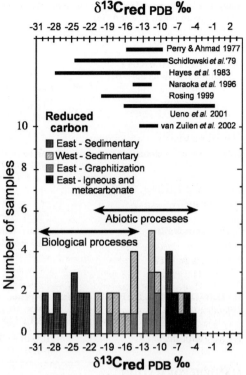

Fig. 3. Distributions of $\delta^{13}C_{red}$ for whole rocks from the IGB. The histogram shows $\delta^{13}C_{red}$ for the sedimentary rocks, compared to the values for igneous and metamorphic rocks in the eastern belt. Previous studies are represented by thick horizontal bars. Data are given in $\delta^{13}C$ notation relative to the Pee Dee Belemnite (PDB) standard.

Fig. 2. $\delta^{34}S$ distributions for sulphide minerals in the IGB. Histogram (**a**) shows the 7.2‰ range in the metasediments, in comparison with the range of sulphides associated with (**b**) igneous and (**c**) fuchsite or gold deposits. Previous studies shown by thick horizontal bars. Data are given in $\delta^{34}S$ notation relative to the Cañon Diablo Troilite (CDT).

has not been fully identified for this study. It will be defined here as reduced carbon instead of graphite. The lightest values of –29.6‰ are from the east (Domain I), within the polymict conglomerate and the sulphide-rich BIF. Except for two values, all $\delta^{13}C_{red}$ are below –18.8‰. In the west (Domain IV), the $\delta^{13}C_{red}$ are from –20.1 to –6.5‰. The carbonate-rich BIF covers the entire range of values found in the west. The data of –18.4 to –14.7‰ from the dark layers in the Bouma sequence are in the same range as reported by Rosing (1999). Finally, the two small BIFs intercalated with ultramafic rocks are very homogeneous with $\delta^{13}C_{red}$ at –11.1 ± 0.8‰.

Igneous and amphibolitic formations. Samples from igneous rocks and a later Tarssartôq dyke (3.47 Ga) show $\delta^{34}S$ values from –0.5 to +1.0‰, which indicates that the sulphur source for the

Table 1. $\delta^{13}C_{red}$ and $\delta^{34}S$ ranges obtained in this study of the Isua Greenstone Belt (3.8 Ga), Manjeri and Cheshire Formations (2.7–2.6 Ga)

Locality – Lithology	$\delta^{34}S$‰	N	$\delta^{13}C_{red}$‰	Carbon Wt%	N
ISUA – EAST DOMAIN I					
Iron Mountain Metachert–BIF	+3.1 to +3.4‰ (+3.2 ± 0.2‰)	3	–	–	–
Sulphide-rich BIF	–3.8 to +1.1‰ (–1.5±1.3‰)	8	–29.6 to –22.6‰ (–25.75±2.7‰)	0.02 to 0.03%	8
Magnetite-rich BIF	–1.2‰	1	–	–	–
Polymict metaconglomerate	–0.4 to +0.9‰ (+0.4 ± 0.5‰)	4	–28.6 to –8.6‰ (–20.2 ± 8.3‰)	0.02 to 0.12%	6
ISUA – WEST DOMAIN IV					
Bouma sequence	–	–	–18.4 to –14.7‰ (–16.4 ± 1.5‰)	0.09 to 0.21%	7
Carbonate-rich BIF	–0.8 to –0.6‰ (–0.7 ± 0.1‰)	3	–20.2 to –6.5‰ (–14.2 ± 5.7‰)	0.05 to 0.11%	3
Other western BIFs	–	–	–12.0 to –10.1‰ (–11.1±0.8‰)	0.02 to 0.18%	3
HYDROTHERMAL DEPOSIT mainly west Isua					
Gold deposits	–5.4 to +0.6‰ (–2.5 ± 1.8‰)	25	–	–	–
Fuchsite deposits	–10.0 to +3.0‰ (–1.1±3.8‰)	21	–	–	–
MANJERI FORMATION					
Jimmy Member S- and C-rich sediments	–21.1 to +16.7‰ (–3.4 ± 6.2‰)	206	–38.4 to –20.7‰ (–32.8 ± 4.2‰)	0.04 to 20.0%*	27
Shavi Member S- and C-rich sediments	–18.4 to +5.4‰ (–3.5 ± 4.5‰)	53	–32.0 to –7.2‰ –25.0 ± 7.62‰	0.13 to 3.18%	16
Spring Valley Member S- and C-rich sediments	+1.2 to +2.9‰ (+2.2 ± 0.7‰)	6	–31.9 to –23.3‰ –26.2 ± 4.0‰	1.36 to 2.66%	3
CHESHIRE FORMATION					
Dark shales	–6.0‰ (–39.5 ± 3.0‰)	1	–43.8 to –32.0‰	0.11 to 3.19%	39[†]

Averages are in parenthesis with 1sd.
* high carbon content is found in samples containing bitumen.
[†] Compilation of Yong (1991) and this study.

high-temperature fluids circulating throughout the belt is at +0.4 ± 0.5‰ (Fig. 2). This is corroborated by the small $\delta^{34}S$ range of +0.5 ± 0.1‰ from the pyrites in the metabasalts.

All the igneous rocks and graphites analysed for $\delta^{13}C_{red}$ have been collected in the eastern part of the belt. Carbon contents are from 212 ppm to 3.5 wt% (Fig. 3). The graphites in the amphibolites give a range of –20.3 to –10.2‰ (–14.8 ± 5.4‰), which is similar to the graphite values obtained in earlier studies. Other $\delta^{13}C_{red}$ measured in this study give a range of –6.3 ± 1.3‰ for the metacarbonates, values of –24.2 and –8.6‰ in two volcano-felsic rocks, and –8.5

and –5.8‰ for two ultramafic samples. Despite a spread of 18‰, it is possible to determine two main sources for these formations. The range from –8.6 to –5.8‰ indicates an origin from the mantle, whereas the $\delta^{13}C_{red}$ from –14.6 to –10.2‰ is the signature of the high-grade metamorphic fluids for the formation of graphite by siderite decomposition (Ueno *et al.* 2002). The –24.2‰ value associated with very low C-content in the volcanic rocks probably represents residual carbon after degassing.

Hydrothermal mineral deposits. Larger sulphur ranges have been found in low to moderate temperature hydrothermal deposits (Fig. 2). These mineralizations were possibly formed by fluid circulation through the upper crustal sequences, especially in the west due to the two major metamorphic events at 3.74 Ga and c. 2.8 Ga (Nutman et al. 1997; Frei et al. 1999). In the gold deposit, $\delta^{34}S$ in sphalerite, galena and pyrite ranges from –5.4 to +0.5‰, and the pyrite in the quartz–fuchsite deposits from –10.0 to +2.6‰, the largest spread yet found in Isua. The hydrothermal fluids carried relatively homogenized sulphur sourced within the belt, mainly of magmatic origin. This sulphur formed a range of sulphide minerals during mineralization with significant fractionations, most notably a 13‰ range in one fuchsite deposit. Such a range was most likely produced because of a high $\Sigma SO_4^{2-}/H_2S$ ratio. One possible way of creating such ratios is in submarine hydrothermal systems where the seawater sulphate is partially reduced by basalts to H_2S (Ohmoto 1992; Ohmoto & Goldhaber 1997).

To summarize, the preservation of primary sedimentary and some of the igneous features in Domain I (Appel et al. 1998) indicates the least metamorphic effects in the belt and no deformation after 3.69 Ga (Frei et al. 1999), an age that might represent the depositional period (Moorbath & Kamber 1998). Some traces of primary $\delta^{34}S$ and $\delta^{13}C_{red}$ may then have been preserved in this area.

Discussion and interpretation of Isua results

By comparing the isotopic ranges from the inorganic formations and sedimentary sequences, it is possible to distinguish those rocks that were most likely to have been overprinted isotopically and those containing mineralization not of organic provenance, from the ones that might host traces of life. Values around 0‰ for $\delta^{34}S$ and higher than –15 to –10‰ for $\delta^{13}C_{red}$ cannot be linked to a biological origin, although they may represent an overprinted bacterial signature.

Different metamorphic events affected some formations more than others, bringing values close to homogenization. This is particularly true in Domain IV, for the western carbonate facies BIF, which was altered by high-temperature metamorphism and metasomatism (Rose et al. 1996; Rollinson 2002). The sulphide present in veins and along fractures is pyrrhotite and the carbonate is mainly siderite. $\delta^{34}S$ of the secondary pyrrhotites was re-homogenized by the high-grade metamorphic overprinting, indicated by the narrow range around 0‰. Two

main sulphur sources are possible: one from the pillow lavas and one from sulphur in the seawater during the BIF deposition on the seafloor. The $\delta^{34}S$ of –0.7‰ is slightly lighter than the magmatic source (+0.5‰), and this difference can be produced by hydrothermal processes on the primary igneous sulphur, but it can be also interpreted as a mixing with another source, most likely the sulphur already present in the BIF.

The large $\delta^{13}C_{red}$ range in the carbonate facies BIF indicates that the re-homogenization was incomplete, unlike for $\delta^{34}S$. Despite significant metamorphic overprinting, the lightest value obtained is –20.1‰. This may be original, indicating that organic traces might be present in this BIF. It is also close to the biological signature suggested by Rosing (1999) in the nearby turbidites. This origin is unlikely for the –6.5‰ value that seems to be the result of the local decomposition of siderite to graphite under high-grade metamorphic conditions. This argument is even more convincing for the two other western BIFs at –11.1 ± 0.8‰, close to the graphite signature of –12‰ obtained by metamorphism in the rest of the belt.

The $\delta^{34}S$ values of +3.2 ± 0.2‰ measured in the Iron Mountain metacherts are the heaviest found in the study. The idiomorphic pyrites present in or close to the veins are of a secondary crystallization generated by the circulating hydrothermal fluids from the early 3.69 Ga metamorphic event (Frei et al. 1999). The fluids responsible for the pyrite crystallization probably had a mixed origin. Sulphur from the IGB volcanic formations (+0.5‰) possibly mixed with sulphur already present in the host formation. This may have been in the form of barite later reduced by the hydrothermal fluids. REE patterns in apatites within Iron Mountain confirm this as they represent pervasive fluids from mixed sources (Lepland et al. 2002).

The $\delta^{34}S$ of pyrites from the clasts and detrital pyrites from the matrix in the polymict conglomerate are homogeneous. The origin of the sulphur is most likely magmatic with $\delta^{34}S$ of +0.4 ± 0.5‰, which is expected as the clasts are from mafic volcanic rocks and metacherts (Fedo 2000). This is also confirmed by some of the $\delta^{13}C_{red}$, at –8.6‰, close to metabasalt values (Fig. 3). Conversely, the $\delta^{13}C_{red}$ of –28.6 to –23.4‰ from the matrix indicate another carbon source that might be organic. Organic matter can be preserved in conglomerate, especially if the metamorphic grade is low enough. Fedo (2000) suggested a deposition in a shallow-water environment. This might have supported life, with a primary biological signature around

−28‰. It is tentatively suggested that this value represents a signature of oxygenic photosynthesis. This could have been possible if the oxidation state on the ocean surface and continental margins was moderately high as Rosing & Frei (2004) imply.

In the Bouma sequence, the $\delta^{13}C_{red}$ of −18.4‰ is close to the proposed signature for plankton (Rosing 1999). The values are homogeneous (−16.4 ± 1.5‰) despite the variation in carbon contents (0.02–0.12 wt%). A combination of two carbon sources with a partial overprinting cannot be ruled out, but such mixing should be small. Therefore, the least transformed $\delta^{13}C_{red}$ at −20.2 and −18.4‰ may indicate organic activity at 3.8 Ga.

The better-preserved eastern metasediments show a larger $\delta^{34}S$ range than the west. This is particularly true for the sulphide-rich BIF (including the nearby magnetite-rich BIF) having mainly lighter $\delta^{34}S$ with a range of 4.9‰ (Fig. 2). The temperature of metamorphism in Domain I was no more than 520 °C (Rollinson 2002), so the $\delta^{34}S$ re-homogenization might have been only partial. On the other hand, the sulphide minerals are in both quartz and magnetite bands (up to 5% locally), and the agglomerated pyrite crystals show no obvious recrystallization, suggesting that the primary minerals and signatures might have been partly preserved. In fact, with a lower metamorphic grade and wider isotopic range, the sulphide-rich BIF is a good candidate for bearing a preserved primary organic signature, such as for sulphate-reducing bacteria, despite a probable reduction of the original range by a partial homogenization. The $\delta^{34}S$ values may have been derived from magmatic sulphur at +0.5‰ by hydrothermal activity, but are most likely a result of mixing of biogenic and hydrothermal sulphur that occurred during the metamorphism.

The $\delta^{13}C_{red}$ in this sulphide-rich BIF, with an average of −25.8 ± 2.7‰, is different from the ranges in the igneous rocks and the graphites in the amphibolites, and includes the lightest $\delta^{13}C_{red}$ at −29.6‰ found in this study of Isua (Fig. 3). They are similar to the metaconglomerate range and the lower metamorphic grade of the area is the most likely reason. A 6‰ spread for samples with 0.02–0.03 wt%C disagrees with a high-temperature fluid source that would have homogenized the $\delta^{13}C_{red}$. Therefore the carbon source might be at least partially biological. $\delta^{13}C_{red}$ then could indicate two origins, a small inorganic CO_2 supply from hydrothermal fluids of the belt and the synsedimentary organic carbon. The organic source might have been

even lighter than −29‰, considering the possible ^{13}C-enrichment by metamorphism (Schidlowski 2001). Consequently, this suggests that anaerobic activity, and possibly methane-utilizing processes, were occurring at that time.

Despite overlapping of the $\delta^{13}C_{red}$ values (Fig. 3), it is possible to differentiate abiotic isotopic signatures, with values mainly between −21 and −4‰, from the biological signatures most likely between −30 and −14‰ (some values from the western BIF and the conglomerate have been overprinted).

In summary, most of the Isua original isotopic signatures were overprinted by metamorphism, but locally, some original $\delta^{13}C_{red}$ and $\delta^{34}S$ from the metasediments might have survived partially as their ranges are wider (and lighter for carbon) than in ultramafic and amphibolitic rocks. Where better preserved, the results suggest that biogenic processes (related to methanogenesis and photosynthesis) were working at 3.8 Ga.

Late Archaean: the Belingwe Greenstone Belt

Belingwe (2.7 Ga), Zimbabwe, is one of the least metamorphosed and deformed of all Archaean greenstone successions. The belt was laid down in a continental basin (Bickle & Nisbet 1993; Hunter *et al.* 1998), unconformably upon a subsiding floor of an eroded 3.5 Ga tonalitic gneiss (Fig. 4). Two well-preserved formations were studied, Manjeri and Cheshire, respectively the lowest and highest parts of Ngezi Group. They comprise carbon- and sulphide-rich sediments and stromatolitic limestones, that have been affected only by low-grade metamorphism (Martin *et al.* 1980; Abell *et al.* 1985; Bickle & Nisbet 1993). These formations were dated at 2706 ± 49 Ma and 2601 ± 49 Ma respectively (Bolhar *et al.* 2002). This is coherent with the Pb–Pb and Sm–Nd ages of 2692 ± 9 Ma for the directly overlying Reliance volcanic rocks (Chauvel *et al.* 1993).

The Manjeri Formation contains four sedimentary units deposited on the *c.* 3.5 Ga gneiss: a basal 40 m of carbon-rich deposits intercalated with thin sulphide layers (Spring Valley and Shavi Members), 56 m of volcaniclastic rocks (Rubweruchena Member), and a top 10 m of basal massive sulphide-iron, with pyrite- and carbon-rich black shales above (Jimmy Member). These are overlain by a thick mafic lava sequence, the Reliance Formation. This study is on the three sulphide- and carbon-rich units, sampled in the Nercmar drillcore: the

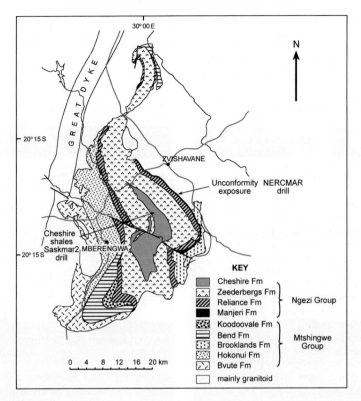

Fig. 4. Map of the Belingwe Greenstone Belt (2.7 Ga) (modified from Grassineau *et al.* 2002). The samples studied are from the Manjeri Formation of the upper greenstones (NERCMAR core), and the Cheshire Formation (Saskmar2 core).

Spring Valley and Shavi Members from shallow-water environments, and the Jimmy Member, showing well-preserved structures, despite being deformed at the contact with Reliance Formation (Hunter *et al.* 1998; Grassineau *et al.* 2002).

The 2 km thick Cheshire formation, deposited at the top of the Ngezi Group, is mainly composed of shallow-water sediments, including conglomerates, ferruginous and dark shales, ironstones, and large limestone reefs, inter-calated with lava and tuff. The 1500 m thick dark shales, showing ripple mark structures, have been sampled from the 50 m Saskmar2 core (Fig. 4). Their deposition was above basal volcanic sequences in a subsiding basin. These well-preserved shales suffered only low-grade metamorphic effects.

Previous stable isotope work at Belingwe

Initially, the few $\delta^{13}C_{red}$ results published for the Manjeri units were from −32.2 to −8.8‰ (in Strauss & Moore 1992), and for the Cheshire shales, from −40.8 to −32.0‰ (Yong 1991; *n* = 15)

(Table 1). More recently, $\delta^{13}C_{red}$ from the three Manjeri carbon- and sulphide-rich sediments have been analysed by Grassineau *et al.* (2001*a*; 2002), with a range of −38 to −17‰ (*n* = 24). They also report a detailed study on sulphides giving a $\delta^{34}S$ range of −17.6 to +16.7‰ (*n* = 135) in these sulphide-rich units.

Results obtained in this study and interpretation

Manjeri Formation. More $\delta^{13}C_{red}$ and $\delta^{34}S$ analyses have been carried out on the shallow water and sub-tidal facies sediments of Shavi and Jimmy Members to complete the profiles of the Nercmar core (Table 1; Grassineau *et al.* 2002). Some $\delta^{13}C_{red}$ and $\delta^{34}S$ in the two adjacent volcanic formations, the Rubweruchena Member and the base of Reliance Formation, have been also analysed to compare with the other three Manjeri units (Fig. 5).

The results expand the large $\delta^{34}S$ range, now from −21.1 to +16.7‰ (Table 1). This spread of

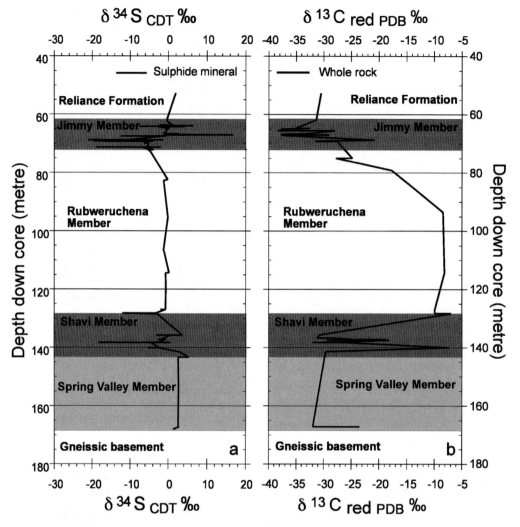

Fig. 5. Vertical stratigraphic profiles of (**a**) $\delta^{34}S$ and (**b**) $\delta^{13}C_{red}$ in the NERCMAR drill core (modified from Grassineau *et al.* 2002), showing the large isotopic variations in the Shavi and Jimmy Members.

almost 38‰ is the widest yet recorded in Archaean time. The total range is found within the Jimmy Member, where there are more common light values below –6‰ and also most of the enriched values, particularly above +6‰. Comparatively, Shavi and Spring Valley Members show smaller variations for $\delta^{34}S$ (Fig. 5). In contrast, the $\delta^{34}S$ values obtained in Rubweruchena Member and Reliance Formation, the adjacent volcanic units, are very homogeneous and close to 0‰ (average of –0.5 ± 0.9‰), with a sulphur source typically of magmatic origin. Similar values were also found in the Jimmy Member, but mostly in the upper part toward the contact with overlying volcanic

unit. Due to the stratigraphic closeness of the volcanism right above the Manjeri Fm. (discussed in Grassineau *et al.* 2002), it is proposed that the hydrothermal processes related to the extrusion of the volcanic rocks or occurring shortly afterwards, locally re-homogenized the original $\delta^{34}S$.

Although the $\delta^{34}S$ range in Shavi Member sulphides is smaller, the spread of *c.* 24‰ is still important with only few values near 0‰. The two main units have wide $\delta^{34}S$ ranges indicating the presence of biological activity, like microbial sulphate-reduction, and possibly restricted areas of sulphide oxidation (Grassineau *et al.* 2002). Locally, $\delta^{34}S$ reach up to +17‰ (TR51, 67.05 m) for pyrites with concentric bleb-like shapes that could be a biogenic feature, and the presence of light $\delta^{13}C_{carb}$ (–12 to –10‰).

Manjeri sediments have a spread of 31‰ for $\delta^{13}C_{red}$ with an average of –29.8‰ (Table 1; Fig. 5), and 0.04 to 20.0 wt%C for the whole rocks. Some samples within Jimmy Member having a high content of reduced carbon (14–20 wt%) resemble a tar-like material. The average of –32.8‰ for Jimmy Member is lighter and with less variation than for the Shavi and Spring Valley members that have an average of –25.2‰. The $\delta^{13}C_{red}$ obtained in the volcaniclastic middle unit are mostly much heavier (around –10‰) with carbon contents between 0.03–1.4 wt%.

Primary bacterial material is suggested by the wide $\delta^{13}C_{red}$ range. The overall average around –29‰ indicates a fractionation by Rubisco I during oxygenic photosynthesis (Pierson 1994). In more detail, the Shavi and Spring Valley Members show mainly a Rubisco I signature, whereas the Jimmy Member with values averaging –33‰ suggests most likely methanogenic and methanotrophic processes in addition to photosynthesis (Grassineau et al. 2001a; 2002).

Cheshire Formation. The only pyrite found in the Cheshire dark shales has a $\delta^{34}S$ of –6‰. The results obtained for $\delta^{13}C_{red}$ are very light, with an average of –39.5‰ (Table 1), and homogeneous if compared with the Manjeri samples. Light $\delta^{13}C_{red}$ strongly indicate a biological origin, most likely with methanogenic activity (Yong 1991; Grassineau et al. 2002), but the smaller range suggests that fewer bacterial processes were operating during the deposition of these sediments than in the Manjeri.

Detailed study of BES50. Microbial consortia have been estimated by studying in detail sample sections across the three units. They are illustrated by significant $\delta^{34}S$ and $\delta^{13}C_{red}$ variations occurring at the millimetre scale. Such is the case of BES50, a core section 8 cm long, in the middle part of the Jimmy Member. The dark shale with different layers of sulphides shows strong evidence of diagenetic compaction (Fig. 6).

All sulphides are pyrite, in cataclased bleb-like shape, formed of agglomerated micro-crystals with micro-intercavities. The $\delta^{34}S$ range is from –18.2 to –4.2‰, with some 14‰ of variation in a single sulphide bleb. Each value in the bleb centre seems to have preserved the primary isotopic signature from the time of the deposition, sealed from possible future exchange. The lighter values and large

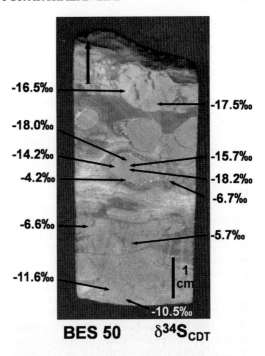

Fig. 6. Detailed sulphur isotopic study of the BES50 sample, a core section from the Jimmy Member. BES50 is a black shale with several layers of rounded bleb-like sulphides, with $\delta^{34}S$ values from –18.2 to –4.2‰.

variations are within the c. 1–2 cm size blebs. This suggests that the rounded shapes are the remains of the bacterial communities whose high activity created further fractionations (e.g. Canfield & Teske 1996). Conversely, the heavy values, –6.7 to –4.2‰, are only found at the edges of the blebs or in the scattered small pyrite crystals. Two explanations are possible, a partial re-homogenization by hydrothermal fluids that was most effective on isolated small pyrites, or more likely, genuine biological signature from secluded activity as values around –5‰ are typical of isolated inter-bedded pyrites. The total spread of BES50 indicates very active bacterial communities at 2.7 Ga. The various micro-ecosystems once present in the sediments interacted with each other generating different fractionations, conditional on the activities of their promiscuous bacterial neighbours and the environment.

No $\delta^{13}C_{red}$ was measured for this section but close by, values of –30‰ and –37‰ were obtained under and above BES50 respectively, showing evidence for biological activity. They suggest that below BES50, the signature was more likely of Rubisco I with oxygenic photosynthetic processes, whereas above the lighter value records more than one activity, such as oxygenic and anoxygenic processes, or even methanogenesis. The isotopic variations at the sample scale give an instant view of bacterial communities existing at 2.7 Ga and their interactions with each other.

Discussion of the Belingwe results

The $\delta^{34}S$ values obtained for the two volcanic sections, Rubweruchena Member and Reliance Formation are close to 0‰ and very different to the range observed in the three carbon- and sulphide-rich sedimentary units. Nevertheless, at the top of the Jimmy Member, where values are around 0‰, $\delta^{34}S$ could have been re-homogenized by a post-depositional hydrothermal event penecontemporaneous to the extrusion of the Reliance sequence. Consequently, this inorganic process would have overprinted the bacterial $\delta^{34}S$ signatures that most likely existed previously, and in such a scenario, this part of the section would not provide any evidence for biological activity. However, as the $\delta^{13}C_{red}$ are preserved in this section, it is unlikely that there was any significant degree of thermal alteration. The other possibility is that a change in the sulphur supply to a hydrothermal source, directly providing elemental sulphur or H_2S, would have resulted in very small fractionations as some bacteria fixed this sulphur as pyrite; however, evidence for a hydrothermal vent has still to be found.

$\delta^{13}C_{red}$ values in the Jimmy and Shavi Members are generally much lighter than the ones recorded in the volcanic Rubweruchena Member, where they are heavier than –10‰ for the lower two-thirds of the unit, but lighter than –20‰ only in the top 5 m. This change in values might mean that the depositional environment was starting to become organism-friendly well before deposition of the black shales, or that the organisms penetrated the underlying volcanic rocks during shale deposition. These isolated values are not typical of the volcanosedimentary Rubweruchena Member, so the very different results obtained for the three units of Manjeri Formation can all be used to provide evidence for biological processes.

With a large $\delta^{34}S$ range of 38‰ in the Manjeri pyrites, the biological sulphur cycle may have been fully or almost fully operational at 2.7 Ga (Grassineau *et al.* 2001*a*), before the rise of the atmospheric oxygen observed at 2.2 Ga (Habicht & Canfield 1996; Blank 2004). The study of the samples at millimetre scale

indicate that different sulphur-dependant organisms existed, especially sulphate-reducing bacteria, but also possibly sulphide oxidizers. The sulphide-oxidizing bacteria evolve in anoxygenic conditions (Cohen *et al.* 1989), and are associated generally with carbonate formation. In the Jimmy Member, the increase of carbonates in cluster-shape with light $\delta^{13}C_{carb}$ down to –12‰ (Grassineau *et al.* 2002), corresponds to a change in the $\delta^{34}S$ values. Such features exist in modern settings, for example near methane seeps, where the anaerobic oxidation of methane produces isotopically light carbonate (Peckmann *et al.* 2004). Sulphur-oxidizing bacteria are also present at the oxic–anoxic boundary where sulphur is released and then possibly reduced back to sulphide. It seems that such conditions could have occurred temporarily in some sections of Jimmy Member. The sharp change in the $\delta^{34}S$ values is most likely explained by the abrupt modification of the environment, rather than the result of a Rayleigh fractionation, which is another explanation for positive $\delta^{34}S$ (Fig. 7). However, the difference in the $\delta^{34}S$ ranges between the Shavi and Jimmy Members could be the result of local changes in SO_4 supply due to modification of the environment. Shavi Member shows a smaller range and most likely indicates a high rate of SO_4 reduction (e.g. Schidlowski *et al.* 1983). In contrast the results obtained from Jimmy Member suggest that locally a temporary decrease in the rate of sulphate supply has slowed the SO_4 reduction rate, favouring organisms that preferably used sulphur or sulphide.

The same observation can be made with $\delta^{13}C_{red}$. A fully-functioning carbon cycle is illustrated by the wide range obtained in the three carbon-rich units. Shavi and Spring Valley Members indicate oxygenic photosynthetic activity controlled by Rubisco I, consistent with their description as shallow-water sediments (Hunter *et al.* 1998). The more complex mixed signatures from Jimmy Member imply multiple processes from composite communities. The finely laminated material observed in the sediments gently settled down below wave base in an anaerobic environment, deeper than Shavi Member. In the upper part of the Jimmy Member, the previously aerobic conditions appeared to change as a sharp depletion of the $\delta^{13}C_{red}$ from –37 to –30‰ occurred with an important increase of the carbon content from *c.* 1 wt%–18 wt% (Fig. 5). This may be in agreement with a rapid input of methane, as discussed in the paragraph above. The Rubisco I signatures obtained in the Jimmy Member may

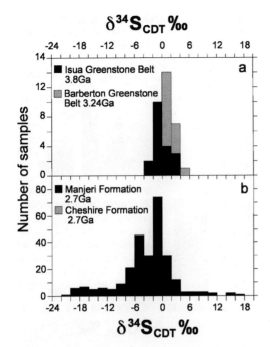

Fig. 7. Comparison between $\delta^{34}S$ of sulphide minerals in sedimentary sequences in (**a**) Early, Middle and (**b**) Late Archaean. The $\delta^{34}S$ range expanded through Early and Middle Archaean, but widened considerably (38‰) during the Late Archaean.

Comparisons between the evidence for early life at 3.8 and 2.7 Ga

The isotopic results for Isua at 3.8 Ga and Belingwe at 2.7 Ga can be used to suggest the rate of evolutionary advancement in the carbon and sulphur cycles over this one billion year period. The results are further constrained by comparison with published and new isotopic data for sulphide-rich carbonaceous shales from the well-preserved Barberton Greenstone Belt in South Africa. The samples are from the Fig Tree Group (Siebert 2003), with an age of 3.26–3.22 Ga (Lowe & Byerly 1999), and therefore constraining processes in the Middle Archaean.

Development of the sulphur cycle

The $\delta^{34}S$ distributions for Isua and Barberton Belts are unimodal with a restricted dispersion (Fig. 7). The peak around 0‰ for Isua sediments indicates that the effects of metamorphism, including in Domain I, may have partly homogenized a larger original range. Nevertheless, a biological source cannot be ruled out despite the lack of direct evidence, especially for the sulphide-rich BIF where the $\delta^{34}S$ spread of 4.9‰ is most likely partly primary. Another possible option for this reduced range could be a low SO_4 supply. In a possible high temperature environment above 85 °C (Nisbet & Fowler 1996; Nisbet & Sleep 2001), the values for the inorganic elemental sulphur formed at seafloor vents are of 0 to +1‰. Appel *et al.* (2001) showed evidence for hydrothermal activity at Isua from least deformed pillow lavas where they found methane and carbonate-rich inclusions in vesicles, that are similar to the ones in modern off-ridge axis vent fields (Touret 2003). Sulphate was probably rare in this setting, as in the global ocean at that time. Even if the existence of sulphate-reducers was still possible, the sulphur-oxidizing could have happened due to deficient sulphate supply (Jannasch 1989). In this case, the biogenic fractionation of sulphur would probably have been smaller.

$\delta^{34}S$ on sulphides from the middle Archaean Fig Tree Group range from –0.4 to +4.0‰ (de Ronde *et al.* 1992; Strauss & Moore 1992), and from –0.9 to +4.4‰ found by Kakegawa & Ohmoto (1999). For this study, conducted at millimetre scale, the $\delta^{34}S$ in pyrites and pyrrhotites are between +1.0‰ and +5.6‰ (Fig. 7). Although, like Isua, the primary $\delta^{34}S$ range may have been partly reduced by a low-grade metamorphism, there is still a significant spread of 6‰, which is strong evidence for some

represent the dead organisms falling from the water surface, although these are mostly in the lower part of Jimmy Member before the change to anoxic conditions.

The change in environments created diversities in the prokaryotic mat ecosystems, from the tidal microbial mats of the Shavi Member to below wave base anaerobic processes in the Jimmy Member, with a complex reduction–oxidation sulphur cycle. The change in the complex metabolic processes is observed directly in the isotopic profiles of the core. Light $\delta^{13}C_{red}$ in Shavi Member corresponds with light $\delta^{34}S$, but in Jimmy Member lighter $\delta^{13}C_{red}$ is related to heavier $\delta^{34}S$, due to an underlying change in conditions.

In the Cheshire Formation, the very light $\delta^{13}C_{red}$, down to –44‰, indicates an anaerobic environment during the dark shale deposition. The complex processes might have included anoxygenic photosynthesis, and most likely methanogenesis and methanotrophy.

form of microbial activity, probably bacterial sulphate reduction in sediments or in the anoxic water column (Kakegawa & Ohmoto 1999), or possibly a sulphur-oxidizing process.

By the late Archaean, the wide $\delta^{34}S$ range for the Manjeri sedimentary units indicates diverse microbial activities (Fig. 7). The $\delta^{34}S$ distribution is bi-modal, with the largest peak at around 0‰ that seems to record hydrothermal activity in the upper part of the Jimmy Member, either as a primary signature or overprinting the signature of bacterial sulphides. Conversely, the second peak with a value averaging around −5‰ is smaller. $\delta^{34}S$ values show a skewed distribution around this peak (once the hydrothermal peak is removed), illustrating interaction of different processes in a well evolved sulphur cycle.

Isotopic evidence therefore suggests that the sulphur cycle developed gradually between Early and Middle Archaean, but expanded greatly by the Late Archaean, particularly during the 2.8–2.6 Ga period.

Development of the carbon cycle

Many of the $\delta^{13}C_{red}$ data obtained for Isua (Fig. 8) may have been modified by metamorphic effects and secondary processes, and such values should be adjusted in consequence to estimate the primary biological signatures (Schidlowski 2001). However the range of 23‰ with values lighter than −22‰ for the metasediments, suggests that some original values may have been preserved locally in low grade, low strain zones. Even if these values of −30‰ have been reset by metamorphism they would originally have been lighter and in both cases are characteristic of biological processes. Therefore, bacterial communities related to plankton-like life (Rosing 1999) including oxygenic photosynthesis (Rosing & Frei 2004), to anaerobic processes such as in the metaconglomerate and sulphide-rich BIF, including methanogenesis (leaving residual carbon with −29.6‰), could have evolved prior to 3.8 Ga.

Much stronger evidence for a well developed biological carbon cycle exists by the time of the deposition of the Fig Tree Group of Barberton at 3.24 Ga. $\delta^{13}C_{red}$ in the middle Archaean has a wider spread toward lighter values than at Isua. New $\delta^{13}C_{red}$ values of −32.4‰ to −5.7‰ obtained here (Fig. 8), on lightly metamorphosed black shales, are consistent with the isotopic studies already carried out on this sequence. These report $\delta^{13}C_{red}$ from −35.4 to −24.3‰ (Strauss & Moore 1992) and −29.5 to −9.5‰ (de Ronde & Ebbesen

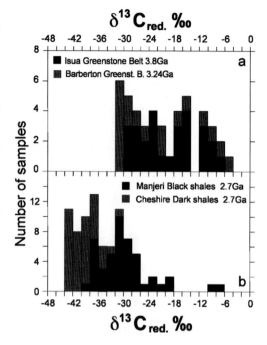

Fig. 8. Comparison between $\delta^{13}C_{red}$ values for sedimentary rocks in (**a**) Early, Middle and (**b**) Late Archaean. There is a probable isotopic shift to the right (^{13}C enrichment) due to metamorphism in the IGB and BGB. The $\delta^{13}C_{red}$ values are much lighter for the Belingwe belt. Values lighter than −18‰ most likely represent biological signatures.

1996). Such values are most likely evidence of organic activity, despite low-grade metamorphic overprinting and accompanying enrichment in ^{13}C. The values from −32 to −26‰ suggest that more than one biological process was operational, including oxygenic photosynthesis. The lighter carbon values possibly represent anoxygenic photosynthesis, or even methanogenesis and methanotrophy occurring at 3.24 Ga.

Furthermore, the $\delta^{13}C_{red}$ measured in the three Manjeri carbon-rich units implies that a highly functional carbon cycle existed by 2.7 Ga. The distribution is slightly bi-modal (Fig. 8). The main peak around −30‰ is characteristic of the Rubisco I signature, with evidence of oxygenic photosynthesis and thus cyanobacterial activity. This peak corresponds mainly to the Shavi and Spring Valley Members. The distribution for the second peak is lighter than −36‰ and these samples are mainly in the upper part of the Jimmy Member, indicating anoxygenic photosynthesis, and methanogenic activity. This is also the case for the dark shales of the Cheshire

Formation, the main $\delta^{13}C_{red}$ values are lighter (c. $-40‰$) also suggesting anaerobic conditions, the presence of anoxygenic photosynthesis, and certainly methanogenic and methanotrophic processes.

Therefore a well developed biological carbon cycle may have been operational by 3.8 Ga, but due to a high grade of metamorphism in most of the Isua belt only glimpses of what might be biogenic carbon can be seen, although this suggests that both oxygenic and methane-utilizing bacteria were already present. By 3.2 Ga the isotopic evidence for these bacterial acitivities is stronger and by 2.7 Ga it is unequivocal.

Conclusions

This study has been carried out on two greenstone belts separated in age by one billion years. A few results obtained for the middle Archaean Barberton Greenstone belt have been added to try to constrain the intermediate evolution. The isotopic signatures of inorganic sources and processes have been identified, including secondary hydrothermal mineralization and overprinting. All data clearly outside the ranges of these signatures have been assessed for their biological potential. $\delta^{34}S$ and $\delta^{13}C_{red}$ results suggest that these activities expanded from 3.8 Ga to 2.7 Ga.

In the Early Archaean, at 3.8 Ga, it is proposed that life activity was mainly around high-temperature hydrothermal vents. Methanogenesis and sulphur-processing organisms seem to have existed and there may also have been photosynthetic activity. It is difficult to find definite evidence for life at 3.8 Ga, but the stable isotope data advocate it.

In the Middle Archaean, at 3.24 Ga, both $\delta^{34}S$ and $\delta^{13}C_{red}$ show changes from Isua. The sulphur range is still restricted, but sulphate-reducing processes have been suggested. Microbial activity was probably mainly anoxygenic, but methane oxidation and reduction reactions were probably occurring as indicated by light $\delta^{13}C_{red}$.

In the Late Archaean, at 2.7 Ga, more diverse and complex metabolic processes were present, indicated by the large $\delta^{13}C_{red}$ and $\delta^{34}S$ ranges, respectively of c. 37‰ and 38‰. These variations, found also at millimetre scale, suggest interactions between the prokaryotic mat communities using different pathways. The oxygenic photosynthetic and sulphate-reduction processes were active in shallow-water reefs. In deeper water, anoxygenic photosynthesis and methanogenesis were operational, as possibly local sulphide-oxidizing processes. In more anaerobic settings, methanogenesis and methanotropy were most likely the main processes.

Although a limited biological sulphur cycle may have been operational at 3.8 Ga and 3.2 Ga, the isotopes suggest a flourishing of diverse bacterial processes by 2.7 Ga. In contrast, much of the biological carbon cycle was operational by 3.2 Ga and probably at 3.8 Ga, but evidence has been limited by the high grade of metamorphism.

NERC and the Leverhulme trust have financed the Belingwe work, the Geological Survey of Denmark and Greenland, the Bureau of Minerals and Petroleum in Greenland, Isua Multidisciplinary Research Project, the Commission of Scientific Research in Greenland, Danish research Council, GEODE and Royal Society for the Isua work. Jan Kramers and Christopher Sierbert (University of Bern) are thanked for supplying the Fig Tree samples. The authors thank Simon Bottrell and Stephen Grimes for their constructive reviews of the manuscript. And a special thank you to Dave Lowry for his comments.

References

ABELL, P.I., McCLORY, J., MARTIN, A. & NISBET. E.G. 1985. Archaean stromatolites from the Ngesi Group, Belingwe Greenstone Belt, Zimbabwe. Preservation and stable isotopes – preliminary results. *Precambrian Research*, **27**, 357–383.

APPEL, P.W.U. 1979. Stratabound copper sulfides in a banded iron-formation and in basaltic tuffs in the Early Precambrian Isua supracrustal belt, West Greenland. *Economic Geology*, **74**, 45–52.

APPEL, P.W.U. 2000. Gahnite in the ~3.75 Ga Isua Greenstone Belt, West Greenland. *Mineralogical Magazine*, **64**, 121–124.

APPEL, P.W.U., FEDO, C.M., MOORBATH, S. & MYERS, J.S. 1998. Recognizable primary volcanic and sedimentary features in a low strain domain of the highly deformed, oldest know (3.7–3.8 Ga) greenstone belt, Isua, West Greenland. *Terra Nova*, **10**, 57–62.

APPEL, P.W.U., ROLLINSON, H.R. & TOURET, J.L.R. 2001. Remnants of an early Archaean (>3.74 Ga) sea-floor, hydrothermal system in the Isua Greenstone Belt. *Precambrian Research*, **112**, 27–49.

BLANK, C.E. 2004. Evolutionary timing of the origins of mesophilic sulphate reduction and oxygenic photosynthesis: a phylogenomic dating approach. *Geobiology*, **2**, 1–20.

BICKLE, M.J. & NISBET, E.G. 1993. *The Geology of the Belingwe Greenstone Belt: A study of the evolution of Archaean continental crust*. Geological Society Zimbabwe, Special Publication **2**, A.A. Balkema, Rotterdam.

BOLHAR, R., HOFMANN, A., WOODHEAD, J.D., HERGT, J.M. & DIRKS, P. 2002. Pb- and Nd-isotope systematics of stromatolitic limestones from the 2.7 Ga Ngezi Group of the Belingwe Greenstone Belt: constraints on timing of deposition and provenance. *Precambrian Research*, **114**, 277–294.

CAMERON, E.M. 1982. Sulphate and sulphate reduction in early Precambrian oceans. *Nature*, **296**, 145–148.

CANFIELD, D.E. & TESKE, A. 1996. Late Proterozoic rise in atmospheric oxygen concentration inferred from phylogenetic and sulphur-isotope studies. *Nature*, **382**, 127–132.

CAVALIER-SMITH, T. 2002. The neomuran origin of archaebacteria, the negibacterial root of the universal tree and bacterial megaclassification. *International Journal of Systematic and Evolutionary Microbiology*, **52**, 7–76.

CHAUVEL, C., DUPRE, B. & ARNDT, N.T. 1993. Pb and Nd isotopic correlation in Belingwe komatiites and basalts. *In*: BICKLE, M.J. & NISBET, E.G. (eds) *The Geology of the Belingwe Greenstone Belt, Zimbabwe*. Geological Society of Zimbabwe, Harare, A.A. Balkema, Rotterdam, Special Publication **2**, 167–174.

COHEN, Y., GORLENKO, V.M. & BONCH-OSMOLOVSKAYA, E.A. 1989. Interaction of Sulphur and Carbon Cycles in Microbial Mats. *In*: BRIMBLECOMBE, P. & LEIN, A. YU. (eds) *Evolution of the Global Biogeochemical Sulphur Cycle*. J. Wiley & Sons, Scope 39, **8**, 191–238.

COLEMAN, D.D., RISATTI, J.B. & SCHOELL, M. 1981. Fractionation of carbon and hydrogen isotopes by methane-oxidizing bacteria. *Geochimica et Cosmochimica Acta*, **45**, 1033–1037.

DE RONDE, C.E.J. & EBBESEN, T.W. 1996. 3.2 b.y. of organic compound formation near sea-floor hot springs. *Geology*, **24**, 791–794.

DE RONDE, C.E.J, SPOONER, E.C.T., DE WIT, M.J. & BRAY, C.J. 1992. Shear zone-related, Au quartz vein deposits in the Barberton greenstone belt, South Africa: Field and petrographic characteristics, fluid properties, and light stable isotope geochemistry. *Economic Geology*, **87**, 366–402.

DYMEK, R.F., BOAK, J.L. & KERR, M.T. 1983. Green micas in the Archaean Isua and Malene supracrustal rocks, southern West Greenland, and the occurrence of a barian-chromian muscovite. *Rapport Grønlands Geologiske Undersøgelse*, **112**, 71–82.

FAURE, G. 1986. *Principles of isotope geology*. 2nd edn. J. Wiley, New York.

FEDO, C.M. 2000. Setting and origin for problematic rocks from the >3.7 Ga Isua Greenstone Belt, southern west Greenland: Earth's oldest coarse clastic sediments. *Precambrian Research*, **101**, 69–78.

FEDO, C.M., MYERS, J.S. & APPEL, P.W.U. 2001. Depositional setting and paleogeographic implications of earth's oldest supracrustal rocks, the > 3.7Ga Isua Greenstone belt, West Greenland. *Sedimentary Geology*, **141**, 61–77.

FREI, R. & ROSING, M.T. 2001. The least radiogenic terrestrial leads; implications for the early Archean crustal evolution and hydrothermal-metasomatic processes in the Isua supracrustal belt (West Greenland). *Chemical Geology*, **181**, 47–66.

FREI, R., BRIDGWATER, D., ROSING, M. & STECHER, O. 1999. Controversial Pb–Pb and Sm–Nd isotope results in the early Archean Isua (West Greenland) oxide iron formation: Preservation of primary signatures versus secondary disturbances. *Geochimica et Cosmochimica Acta*, **63**, 473–488.

GRASSINEAU, N.V., NISBET, E.G., BICKLE, M.J., FOWLER, C.M.R., LOWRY, D., MATTEY, D.P., ABELL, P. & MARTIN, A. 2001a. Antiquity of the biological sulphur cycle: evidence from S and C isotopes in 2.7Ga rocks of the Belingwe Belt, Zimbabwe. *Proceedings of the Royal Society of London*, **B268**, 113–119.

GRASSINEAU, N.V., MATTEY, D.P. & LOWRY, D. 2001b. Rapid sulphur isotopic analyses of sulphide and sulphate minerals by continuous flow-isotope ratio mass spectrometry (CF-IRMS). *Analytical Chemistry*, **73**, 220–225.

GRASSINEAU, N.V., NISBET, E.G., *ET AL.* 2002. Stable isotopes in the Archaean Belingwe belt, Zimbabwe: evidence for a diverse microbial mat ecology. *In*: FOWLER, C.M.R., EBINGER, C.J. & HAWKESWORTH, C.J. (eds) *The Early Earth: Physical, Chemical and Biological Development*. Geological Society, London, Special Publications, **199**, 309–328.

HABICHT, K. & CANFIELD, D.E. 1996. Sulphur isotope fractionation in modern microbial mats and the evolution of the sulphur cycle. *Nature*, **382**, 342–343.

HAYES, J.M., KAPLAN, I.R. & WEDEKING, K.W. 1983. Precambrian organic geochemistry, preservation of the record. *In*: SCHOPF, J.W. (ed.) *Earth's Earliest Biosphere: its Origin and Evolution*. Princeton University Press, Princeton, **5**, 93–134.

HOEFS, J. 1997. *Stable Isotope Geochemistry*. Springer-Verlag, Berlin.

HUNTER, M.A.H., BICKLE, M.J., NISBET, E.G., MARTIN, A. & CHAPMAN, H.J. 1998 A continental extension setting for the Archaean Belingwe Greenstone belt, Zimbabwe. *Geology*, **26**, 883–886.

JAHNKE, L.L., EDER, W., *ET AL.* 2001. Signature lipids and stable carbon isotope analyses of octopus spring hyperthermophilic communities compared with those of Aquificales representatives. *Applied and Environmental Microbiology*, **67**, 5179–5189.

JANNASCH, H.W. 1989. Sulphur Emission and Transformations at Deep Sea Hydrothermal Vents. *In*: BRIMBLECOMBE, P. & LEIN, A.Y. (eds) *Evolution of the Global Biogeochemical Sulphur Cycle*. J. Wiley & Sons, Scope 39, **7**, 181–190.

KAKEGAWA, T. & OHMOTO, H. 1999. Sulfur isotope evidence for the origin of 3.4 to 3.1Ga pyrite at the Princeton gold mine, Barberton Greenstone Belt, South Africa. *Precambrian Research*, **96**, 209–224.

LEPLAND, A., ARRHENIUS, G. & CORNELL, D. 2002. Apatite in Early Archean Isua supracrustal rocks, southern West Greenland: its origin, association with graphite and potential as a biomarker. *Precambrian Research*, **118**, 221–241.

LOWE, D. & BYERLY, G.R. 1999. Stratigraphy of the west-central part of the Barberton Greenstone Belt, South Africa. *In*: LOWE, D. & BYERLY, G.R. (eds) *Geologic Evolution of the Barberton*

Greenstone Belt, South Africa. Geological Society of America, Special Paper, **329**, 1–36.

MARTIN, A., NISBET, E.G. & BICKLE, M.J. 1980. Archaean stromatolites of the Belingwe Greenstone Belt. *Precambrian Research*, **13**, 337–362.

MATTHEWS, D.E. & HAYES, J.M. 1978. Isotope-ratio-monitoring gas chromatography-mass spectrometry. *Analytical Chemistry*, **50**, 1465–1473.

MONSTER, J., APPEL, P.W.U., THODE, H.G., SCHIDLOWSKI, M., CARMICHAEL, C.M. & BRIDGWATER, D. 1979. Sulfur isotope studies in Early Archaean sediments from Isua,West Greenland: Implications for the antiquity of bacterial sulfate reduction. *Geochimica et Cosmochimica Acta*, **43**, 405–413.

MOORBATH, S. & KAMBER, B.S. 1998. Re-appraisal of the age of the oldest water-lain sediments, West Greenland. *In*: CHELA-FLORES, J. & RAULIN, F. (eds) *Exobiology: Matter, Energy and Information in the Origin of Evolution of Life in the Universe.* Kluwer Academic, Dordrecht, 81–86.

MOORBATH, S. & WHITEHOUSE, M.J. 1996. Age of the Isua supracrustal sequence of West Greenland. *In*: CHELA-FLORES, J. & RAULIN, F. (eds) *Chemical Evolution: Physics of the Origin and Evolution of Life.* Kluwer Academic Publ. Dordrecht, 87–95.

MYERS, J.S. 2001. Protoliths of the 3.8–3.7Ga Isua Greenstone Belt, West Greenland. *Precambrian Research*, **105**, 129–141.

NARAOKA, H., OHTAKE, M., MARUYAMA, S. & OHMOTO, H. 1996. Non-biogenic graphite in 3.8Ga metamorphic rocks from the Isua district, Greenland. *Chemical Geology*, **133**, 251–260.

NISBET, E.G. & FOWLER, C.M.R. 1996. Some like it hot. *Nature*, **382**, 404–406.

NISBET, E.G. & FOWLER, C.M.R. 1999. Archaean metabolic evolution of microbial mats, *Proceedings of the Royal Society of London B*, **266**, 2375–2382.

NISBET, E.G. & SLEEP. N.H. 2001. The habitat and nature of early life. *Nature*, **409**, 1083–1091.

NUTMAN, A.P. 1986. The early Archaean to Proterozoic history of the Isukasia area, southern West Greenland. *Grønlands geologiske Undersøgelse*, Bulletin **154**, pp. 80.

NUTMAN, A.P., BENNETT, V.C., FRIEND, C.R.L. & ROSING, M.T. 1997. ~3710 and >3790Ma volcanic sequences in the Isua (Greenland) supracrustal belt; structural and Nd isotope implications. *Chemical Geology*, **141**, pp. 271–287.

OHMOTO, H. 1992. Biochemistry of sulfur and the mechanisms of sulfide-sulfate mineralization in Archean oceans. *In*: SCHIDLOWSKI, M. *ET AL.* (eds) *Early organic evolution: Implications for mineral and energy resources.* Springer-Verlag, Berlin, 378–397.

OHMOTO, H. & FELDER, R.P. 1987. Bacterial activity in the warmer, sulphate-bearing, Archaean oceans. *Nature*, **328**, 244–246.

OHMOTO, H. & GOLDHABER, M. 1997. Sulfur and carbon isotopes. *In*: BARNES, H.L. (ed.) *Geochemistry of Hydrothermal Ore Deposits*, **3**. John Wiley & Sons, New York, 517–612.

OHMOTO, H., KAKEGAWA, T. & LOWE, D.R. 1993. 3.4 billion-year-old biogenic pyrites from Barberton, South Africa: Sulfur isotope evidence. *Science*, **262**, 555–557.

PACE, N.R. 1997. A molecular view of microbial diversity and the biosphere. *Science*, **276**, 734–740.

PECKMANN, J., THIEL, V., REITNER, J., TAVIANI, M., AHARON, P. & MICHALIS, W. 2004. A microbial mat of a large sulfur bacterium preserved in a Miocene methane-Seep limestone. *Geomicrobiology Journal*, **21**, 247–255.

PERRY, E.C. JR & AHMAD, S.N. 1977. Carbon isotope composition of graphite and carbonate minerals from 3.8-AE metamorphosed sediments, Isukasia, Greenland. *Earth and Planetary Science Letters*, **36**, 280–284.

PIERSON, B.K. 1994. The emergence, diversification and role of photosynthetic eubacteria. *In*: BENGTSON, S. (ed.) *Early Life on Earth*. Nobel Symposium **84**. Columbia University Press, New York, 161–180.

POLAT, A., HOFMANN, A.W., MÜNKER, C., REGELOUS, M. & APPEL, P.W.U. 2003. Contrasting geochemical patterns in the 3.7–3.8Ga pillow basalt cores and rims, Isua greenstone belt, Southwest Greenland: Implications for postmagmatic alteration processes. *Geochimica et Cosmochimica Acta*, **67**, 441–457.

RICHARDS, J.R. & APPEL, P.W.U. 1987. Age of the 'least radiogenic' galenas at Isua, West Greenland. *Chemical Geology*, **66**, 181–191.

ROLLINSON, H. 2002. The metamorphic history of the Isua Greenstone Belt, West Greenland. *In*: FOWLER, C.M.R., EBINGER, C.J. & HAWKESWORTH, C.J. (eds) *The Early Earth: Physical, Chemical and Biological Development.* Geological Society, London, Special Publications, **199**, 329–350.

ROSE, N.M., ROSING, M.T. & BRIDGWATER, D. 1996. The origin of metacarbonate rocks in the Archaean Isua supracrustal belt, West Greenland. *American Journal of Science*, **296**, 1004–1044.

ROSING, M.T. 1999. ^{13}C-depleted carbon microparticles in >3700-Ma sea-floor sedimentary rocks from West Greenland. *Science*, **283**, 674–676.

ROSING, M.T. & FREI, R. 2004. U-rich Archaean seafloor sediments from Greenland – indications of >3700 Ma oxygenic photosynthesis. *Earth and Planetary Science Letters*, **217**, 237–244.

SAKAI, H., DES MARAIS, D.J., UEDA, A. & MOORE, J.G. 1984. Concentrations of isotope ratios of carbon, nitrogen and sulfur in ocean-floor basalts. *Geochimica et Cosmochimica Acta*, **48**, 2433–2441.

SCHIDLOWSKI, M. 2001. Carbon isotopes as biogeochemical recorders of life over 3.8 Ga of Earth history: evolution of a concept. *Precambrian Research*, **106**, 117–134.

SCHIDLOWSKI, M, APPEL, P.W.U., EICHMANN, R. & JUNGE, C.E. 1979. Carbon isotope geochemistry of the 3.7x109 yr old Isua sediments, West Greenland: Implications for the Archaean carbon and oxygen cycles. *Geochimica et Cosmochimica Acta*, **43**, 189–199.

SCHIDLOWSKI, M., HAYES, J.M. & KAPLAN, I.R. 1983. Isotopic Interferences of Ancient Biochemistries: Carbon, Sulfur, Hydrogen, and Nitrogen. *In*: SCHOPF, J.W. (ed.) *Earth's Earliest Biosphere: Its Origin and Evolution.* Princeton University Press, Princeton, **7**, 149–186.

SHEN, Y. & BUICK, R. 2004. The antiquity of microbial sulfate reduction. *Earth-Science Reviews*, **64**, 243–272.

SHEN, Y., BUICK, R. & CANFIELD, D.E. 2001. Isotopic evidence for microbial sulphate reduction in the early Archaean era. *Nature*, **410**, 77–81.

SIEBERT, C. 2003. *Molybdenum isotope fractionation and its application to studies of marine environments and surface processes through geological time.* Unpublished PhD Thesis, University of Bern.

STRAUSS, H. 2003. Sulphur isotope and the early Archaean sulphur cycle. *Precambrian Research*, Special Volume, **126**, 349–361.

STRAUSS, H. & MOORE, T.B. 1992. Abundances and isotopic compositions of carbon and sulfur species in whole rock and kerogen samples. *In*: SCHOPF, J.W. & KLEIN, C. (eds) *The Proterozoic Biosphere.* Cambridge University Press, Cambridge, **17**, 711–798.

TAYLOR, B.E. 1986. Magmatic volatiles: isotopic variation of C, H, and S. *In*: VALLEY, J.W., TAYLOR, H.P. & O'NEIL, J.R. (eds) *Reviews in Mineralogy: Stable Isotopes in High Temperature Geological Processes.* Mineralogical Society of America, **16**, 185–226.

TOURET, J.L.R. 2003. Remnants of early Archaean hydrothermal methane and brines in pillow-breccia from the Isua Greenstone Belt, West Greenland. *Precambrian Research*, **126**, 219–233.

UENO, Y., YURIMOTO, H., YOSHIOKA, H., KOMIYA, T. & MARUYAMA, S. 2002. Ion microprobe analysis of graphite from ca. 3.8 Ga metasediments, Isua supracrustal belt, West Greenland: Relationship between metamorphism and carbon isotopic Composition. *Geochimica et Cosmochimica Acta*, **66**, 1257–1268.

VAN ZUILEN, M.A., LEPLAND, A. & ARRHENIUS, G. 2002. Reassessing the evidence for the earliest traces of life. *Nature*, **418**, 627–630.

WOESE, C.R. 1987. Bacterial evolution. *Microbiological reviews*, **51**, 221–271.

YONG, J.N. 1991. *Stable isotope ratios of shales from the Archaean Cheshire Formation, Zimbabwe.* Unpublished PhD thesis, University of Rhode Island.

Diamond mega-placers: southern Africa and the Kaapvaal craton in a global context

B. J. BLUCK[1], J. D. WARD[2] & M. C. J. DE WIT[2]

[1]Division of Earth Sciences, University of Glasgow, Glasgow G12 8QQ
(e-mail: B.Bluck@earthsci.gla.ac.uk)

[2]De Beers, Africa Exploration, Centurion, South Africa

Abstract: Diamond mega-placers, defined as ≥ 50 million carats at ≥ 95% gem quality, are known only from along the coast of southwestern Africa, fringing the Kaapvaal craton, where two are recognized. One is associated with the Orange–Vaal dispersal, the other, to the south, has an uncertain origin. Placers are residual when left on the craton, transient when being eroded into the exit drainage, and terminal. Terminal placers, the final depositories of diamonds, have the greatest probability of being a mega-placer. There are four main groups of controls leading to the development of a mega-placer: the craton, the drainage, the nature of the environment at the terminus and the timing.

Cratons, being buoyant, have a tendency to leak diamonds into surrounding basins; however, being incompressible they may have orogens converge onto them resulting in some lost sediment being returned as foreland basin fills. The craton size, its diamond-fertility and the retention of successive kimberlite intrusions that remain available to the final drainage, are significant to mega-placer development.

Maximum potential recovery is achieved when the drainage delivering diamonds to the mega-placer is efficient, not preceded by older major drainages and focuses the supply to a limited area of the terminal placer. There should be sufficient energy in the terminal placer regime to ensure that sediment accompanying the diamonds is removed to areas away from the placer site. All conditions should be near contemporaneous and most were satisfied in the Orange–Vaal Rivers–Kaapvaal system and mega-placers were consequently generated.

Placer diamonds have been mined in India since 300 BC, more than 2000 years before the discovery of the first primary deposit at Kimberley in 1871. However, in contrast to the volume of research on kimberlites, relatively little has been written about alluvial diamonds, particularly in a global context. The volume of diamonds retained in kimberlite pipes diminishes with depth and following erosion, placer deposits are a significant accumulating residue, potentially carrying a record of some of the history of diamond emplacement on a craton.

The bulk of the natural diamonds are currently produced from primary (kimberlitic, lamproitic) sources. For example, on the Kalahari craton of southern Africa the primary diamond production for the year 2003 stands at c. 40×10^6 carats and the placers on and around this craton yield c. 1.5×10^6 carats or c. 3.8% of the total production. However, placer deposits on or fringing the Kalahari account for >7.5% of the total production value. It follows that placer deposits have to be very large to have any significant impact on total diamond production and they should (and commonly do) have a relatively high value per carat. For this, and other reasons to emerge later, we propose a narrow definition of a diamond mega-placer as being a continuum of linked deposits that are generally the result of a single or continuous process of transportation and deposition and which contain or have contained at least 50×10^6 carats, of which >95% are of gem quality.

The largest single deposit (primary or secondary) in terms of volume is the complex at Mbuji-Mayi in the Democratic Republic of Congo (see Table 1). It is now recognized as a highly weathered primary deposit and, as it is not transported, does not qualify as a true placer. In terms of the definition of a mega-placer the only ones so far identified occur along the coast and immediately offshore of southwestern Africa (Fig. 1). This total region has, over the time of its production, yielded more than 120×10^6 carats.

Types of placers and conditions for their formation

Three categories of placer deposits are recognized (Fig. 2).

From: MCDONALD, I., BOYCE, A. J., BUTLER, I. B., HERRINGTON, R. J. & POLYA, D. A. (eds) 2005. *Mineral Deposits and Earth Evolution.* Geological Society, London, Special Publications, **248**, 213–245. 0305-8719/$15.00

214   B. J. BLUCK *ET AL.*

Table 1. *Statistics for both diamond placer and primary deposits (for source see text)*

Region	Million carats	Value US m$ (production 2001)
Southern Africa	140	633.1
Angola & DRC	120	1124.2
DRC (Mbuji Mayi)	834	–
Central African Republic	20	92.1
W Africa	200	251.9
S America	75	79.2
Others	74	–
TOTAL	1463	
Primary deposits	1500	

Retained: those deposits remaining on the craton and not readily available to the dispersal system;

Transient: those that, during the current cycle of erosion, have been dispersed but have been temporarily stored in or on the margin of the dispersal route; and

Terminal: the final accumulation from the current drainage.

Retained placers

Kimberlites, intruded over a range of time, may disperse their diamonds into deposits remaining on the craton, in some cases for >2.5 Ga, so that they may accumulate there for final release. There are a number of ways in which these secondary deposits may be retained, two of the more significant are in karsts and intra-cratonic basins. (Figs 2–4).

Karsts (including karst surfaces), estimated to occupy 10–20% of the Earth's land area (Palmer 1991), may become areas of substantial diamond retention. Karstification is extensively developed on cratonic cover and adjacent areas where post-Archaean carbonates have been subject to climatic and base level changes since *c.* 2.5 Ga. With diamond-fertile kimberlites being found principally on Archaean cratons there is great potential for long-term diamond

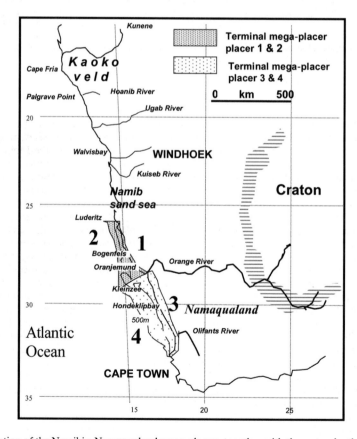

Fig. 1. Distribution of the Namibia–Namaqualand mega-placers, together with the water depths on the shelf.

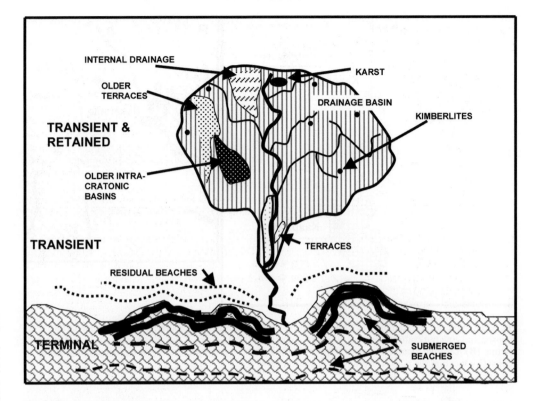

Fig. 2. Types of placers.

retention through subsequent erosion and trapping in and on karsts.

Potential trapping occurs in caves and on the highly irregular karst surfaces, and cavities, hundreds of metres across and tens of metres deep, may run for several kilometres. Cave systems develop both quickly and extensively: extension rates are up to 200–300 m ka^{-1} and flow rates in caves from $c.$ 4 m hr^{-1} to 700 m hr^{-1}. The Carboniferous Madison Limestone of South Dakota illustrates the potential for possible volumes generated, where cave density reaches 120 km of passage-way beneath an area of 1.3 km^2 (Bakalowicz *et al.* 1987).

In the Late Archaean–Palaeoproterozoic Chuniespoort and Ghaap Groups (Transvaal Supergroup), the Kaapvaal craton has one of the oldest known regional carbonates in the world, estimated to cover 500 000 km^2 and up to 1700 m thick (Button 1973; Tankard *et al.* 1982). In the North-West Province (formerly Western Transvaal; Fig. 3), more than 12×10^6 carats have been recovered from karst surfaces retaining diamond-bearing gravels (du Toit 1951; Stettler *et al.* 1995; de Wit 1996). At least four

periods of karstification have affected some of these carbonates and Martini & Kavalieris (1976) suggested that in the Lichtenburg area a palaeo-river system, along with potholes, controlled the distribution of the diamond-bearing gravels (Fig. 3). Some of the gravel lies in potholes and other gravel 'runs' that stand slightly higher than the surrounding land surface. Du Toit (1951) thought that the high standing gravel may originally have been deposited in karst-related depressions but, due to down-wasting of the surrounding carbonates (the 'runs' protecting the carbonate surface from down-wasting), invert to relief. With some of the gravels being Upper Cretaceous in age (Bamford 2000) and with surface reduction rates in karst being recorded elsewhere at 1.5–4.0 m/Ma then this explanation is highly probable.

In places on the craton, and even within the externally draining basin itself, there are areas of internal drainage not yet tapped by the drainage network that are therefore failing to dispense diamonds to the other placer types. Some basins are quite small and temporary, but

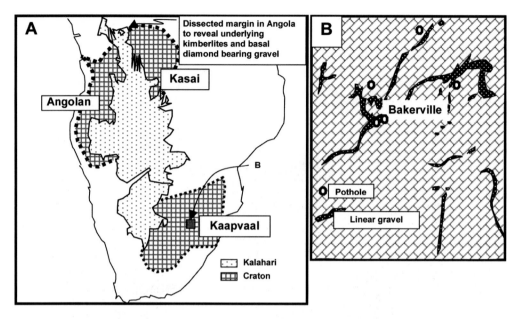

Fig. 3. Examples of residual placers. (**A**) The internal, diamond-bearing, Kalahari basin in relation to the exposed and covered cratons, demonstrating its extent over the Angolan, Kasai, Kaapvaal and Zimbabwe cratons. (**B**) Residual diamond-bearing deposits in the karst topography in the region of Lichtenburg, South Africa (for location see map A) from Stettler *et al.* (1995).

others are large intra-cratonic basins covering thousands of square kilometres. The post-Gondwanan Kalahari basin is still active. It extends widely over southern and central Africa covering a large portion of the Kalahari and adjacent cratons and shields (Fig. 3). This basin has been subject to a number of influences since the break-up of Gondwana in the late Jurassic (Haddon 2000). There are four main depocentres; the sediments, over 300 m thick, begin with coarse basal gravel and terminate in calcrete, aeolian sand and pan sediments, these finer sediments commonly overlapping onto the craton (Haddon 2000).

The potential for diamond retention in this deposit is demonstrated on its northern (Angolan) margin where northerly flowing rivers have dissected a small part of the basin edge in response to the downcutting of the Kasai. Where close to primary sources, the basal gravel of the Calonda Formation and associated younger deposits host high diamond grades (Fig. 4). Continued cut back by large rivers results in dilution of diamond grades unless upgrading is achieved in local scour pools. The life-span of basins of this type is uncertain, but the Kalahari basin may be at least Early Cretaceous in age, thus holding a Mesozoic–Cenozoic record. During this time and during

the initial stages of basin opening it would have derived diamonds from the erosion of kimberlites on the basin floor (when underlain by craton, as occurred in Angola). However, as the basin develops, the peripheral drainage may derive diamonds from the surrounding basement. There is also the possibility of post Early Cretaceous kimberlites having been intruded into the sediment fill.

Other sedimentary basins currently developing on or near to cratons in Africa include the Niger, Congo, Taoudenni and possibly Chad (Burke 1996). The extent to which these basins contain diamonds is uncertain and depends on the fertility of the associated cratons, but they are all depocentres into which drainage crossing Archaean crust is or may have been active in the past. Smaller basins near to or on cratons include the Senegal and Volta basins, although the extent to which Hercynian (Mauritanides) and Pan African Rokelides belts have dominated sediment input to all these basins is yet to be determined.

Extensional episodes were common on most cratonic blocks from Late Archaean–Proterozoic times and intra-cratonic basins thus generated may have been repositories for placers or have provided a host cover for diamond-bearing intrusions. They range from poorly explained

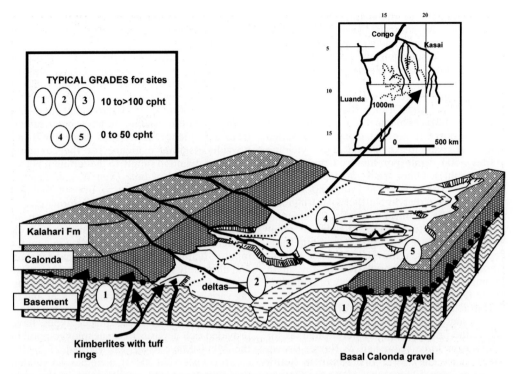

Fig. 4. Residual and transient type placer deposits in the eroded northern edge of the Kalahari basin (and associated kimberlites) in NE Angola (see inset for location). The Calonda Formation, here the base of the Kalahari basin, was deposited by small streams that eroded exposed kimberlites to form a local sheet-like retained placer, 1. 2 is formed by small deltas in the tributary streams to the main channel concentrating diamonds from the Calonda Formation and exposed primaries. Placer 3 occurs in mass-flow deposits, with some fluvial reworking giving exceptional grades; placer 4 occurs in the principal rivers and 5 their terraces. The type 4 deposits often have low grades when being diluted by barren Kalahari sediment.

and long-lived roughly circular depressions such as Hudson Bay and the Williston Basin on the North American shield; Paranaiba on the Central Brazilian shield and Congo, to Triassic to Recent rift-generated basins. The potential of these and related basins to retain diamonds is illustrated on the Kaapvaal craton where placer diamonds have been recovered from Ventersdorp and Witwatersrand Supergroups.

The potential of intra-cratonic and associated basins in preserving kimberlites was recognized by Gold (1984) and clearly applies also to the detritus accumulating from eroded kimberlites (Figs 3, 4 & 12). This potential is revealed when basins have been partially inverted and subsequently truncated, as demonstrated clearly in craton history (for the Kaapvaal see Brandl & de Wit 1997; Cheney & Winter 1995; Cheney 1996) and is seen spectacularly in Angola (Fig. 4).

Transient placers

Transient placers are diamond-bearing sediment piles stored along the dispersal route or within the active drainage basin. The distinction between these and residual placers clearly merge if the drainage density in the drainage basin is low with the result that the removal of sediment is inefficient and diamond bearing sediment remains untouched there for extended periods. In the present Orange River drainage basin, for example, aridification has resulted in low drainage density and run-off and placers have remained in that drainage basin since the Cretaceous.

Transient placers in river terraces have provided some rich placer deposits that depend upon uplift–incision for their formation and stacking. In addition, the attendant rejuvenation is needed to increase the slope and energy in the channel, to effect bedrock incision related trap sites and, in rejuvenated tributaries, to deliver coarse sediment to the main channel which might otherwise comprise fines. Without this coarse sediment load and increase in velocity of flow, hydraulic selection and retention of the diamonds into viable placers may not have occurred.

Fine examples are supplied by the terraces along the Vaal and Orange rivers in southern Africa, and terraces of uncertain age (but almost certainly post-Cretaceous) along the Kwango and other rivers in Angola (in places yielding grades >500 carats per hundred metric tons gravel (cpht)). Other transient deposits occur along the Krishna River in India, the Yuan in China, Smoke Creek in Northern Australia and many deposits in Borneo and South America. All are related to episodes of uplift or other base level changes followed by incision. These deposits, and those removed by erosion, supplement the diamond population of the main channels ensuring a steady supply of diamonds to the terminal placer.

Transient placers, in the form of river terraces, were a source of diamonds probably centuries before the primary deposits were found, as for example in India (Marshall & Baxter-Brown 1995). Since they were first found in South Africa in 1866 at Hopetown, along the Orange River, extensive coarse-grained gravel terraces

along both the Vaal and the Orange river upstream of their confluence have yielded more than 3 million carats (de Wit 1996). These transient placers, developed during the evolving drainage, range in age from Cretaceous to Recent (Partridge & Brink 1967; Bamford 2000; de Wit 2004).

On the Lower Orange River, c. 100 km inland from its mouth, diamonds have been mined from a sequence of terraces since 1967 and provide essential evidence for the changing nature of diamond supply to the terminal placer over an interval of time (Van Wyk & Pienaar 1986; Jacob et al. 1999; Fig. 5). In both Vaal and Orange Rivers, the terraces demonstrate a declining grade and a general increase in diamond size from oldest to youngest (Fig. 6), thought to be the record of a decline in a major input of diamonds fed into the drainage system and ultimately to the terminal placer. This information is an important line of evidence explaining variable grade in the terminal placer and in targetting areas in the offshore environment.

Although all of these placer types have provided rich deposits (a single 'plunge-pool' in the Mbuji-Mayi River has yielded 25×10^6 carats (Oosterveld 2003)), none has yet yielded the abundance and quality seen in the only known substantial terminal placers on the SW

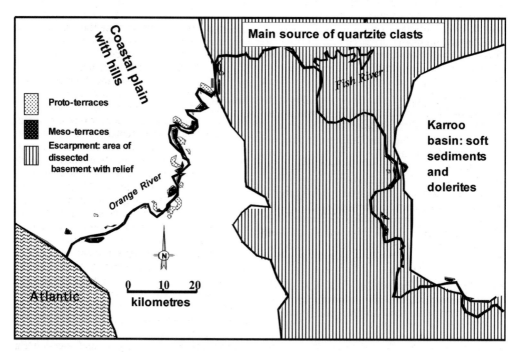

Fig. 5. Distribution of transient placers along the lower Orange River valley. The terraces are particularly common (but not confined to) the river where it has emerged from the area of relief (Jacob et al. 1999).

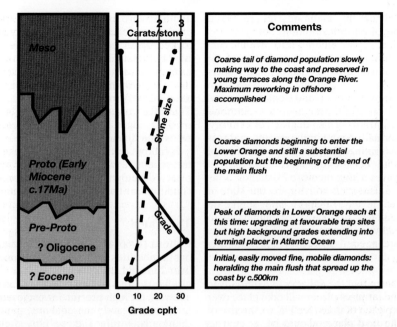

Carats/stone

Stone size

Grade

Meso

Coarse tail of diamond population slowly making way to the coast and preserved in young terraces along the Orange River. Maximum reworking in offshore accomplished

Proto (Early Miocene c.17Ma)

Coarse diamonds beginning to enter the Lower Orange and still a substantial population but the beginning of the end of the main flush

Pre-Proto

? Oligocene

Peak of diamonds in Lower Orange reach at this time: upgrading at favourable trap sites but high background grades extending into terminal placer in Atlantic Ocean

? Eocene

Initial, easily moved fine, mobile diamonds: heralding the main flush that spread up the coast by c.500km

0   10   20   30
Grade cpht

Comments

Fig. 6. Diagrammatic section through the Lower Orange River terraces (Fig. 5) along with the changing grade and diamond size. Cpht, carats per 100 metric tons of gravel; stones refers to diamonds (see Jacob *et al.* 1999).

African coast. The extent to which they and the diamond-bearing kimberlites remain in the drainage basin is a measure of the lack of efficiency of the drainage gathering the diamonds to deliver them finally to the terminal placer. In this respect, the Orange drainage basin, although delivering a mega-placer is, nevertheless, remarkably inefficient as it has left behind some substantial residual placers, and probably the world's greatest reserve of primary diamonds after more than 60 Ma of erosion.

Terminal placers

Terminal placers occur at the extreme end of drainage systems where diamonds and sediments, in a different repository, are subject to further segregation. Processes in the terminal placer need to change successive increments of sediment containing a small population of diamonds into a relatively large, quasi-single population of diamonds hosted in a relatively low volume sediment. The conditions required to achieve this are:

1   The total volume of diamonds transported over a period of time determines the magnitude of the terminal placer. The extent and richness of the craton with respect to both primary and secondary

diamond deposits is an essential prerequisite as is the availability of diamonds to the transport system. The diamonds generated need to be retained on the craton for as long as possible so that yields from successive kimberlite intrusions or from re-erosion of earlier placers, accumulate there and become available for their final exit.

2   In order to focus the maximum number of diamonds into the smallest terminal area, the exit drainage needs to be large-scale and single. The greater the drainage area, given that this correlates with a greater proportion of craton, then the greater the abundance of diamonds being transported to the terminal point. However rivers capable of bringing a substantial quantity of diamonds are also likely to have brought down considerable volumes of diluting sediment. Ideally there should be low volumes of diluting sediment in the fluvial delivery and this sediment should be of a grain size capable of trapping diamonds.

3   The terminal area has to have sufficient energy to separate the diamonds from the other entrained sediment as (with respect to 2, above) the greater the drainage area, the greater the sediment load and therefore the environment in the terminal placer has to be correspondingly more energetic to

concentrate the diamonds. There are a number of critical conditions here:

a. The environment into which the diamond-bearing sediment is deposited needs to be sufficiently energetic to segregate diamonds from other sediment and separate gravel out to form the host for diamond retention. This degree of energy, normally found in the marine realm, is determined by the characteristics of both ocean and shelf. Long-period waves acting on a broad, shallow, low-gradient shelf, will, for any one state of sea level, transport and segregate sediment and farm for diamonds. In addition, wind, wave and tidal energy are needed to remove the diluting sediment from the potential placer site.

b. Given that the delivery of diamonds to the terminal placer will take place over a period of time (see Fig. 6), the size of the final placer should be, as near as possible, the sum of all diamonds brought down by the river since its inception. As many increments of diamonds as possible brought down by the river to the terminus should be available for re-concentration there, i.e. not buried by the accompanying or other sediments so as to be sealed off from further concentration. This demands a number of factors including a wide area of shelf with little or no subsidence over the time of placer development so that sea-level changes can repeatedly rework the sediment. As a corollary to this, unwanted sediment, separated from diamonds, needs to be dispersed to areas away from the site of placer development to nearby regions of accumulation where it can be accommodated and more permanently buried.

4 Conditions 1–3 must be united by time. The processes of concentrating the diamonds from the other sediment delivered are not likely to be maintained for any great length of time. It is therefore important that the delivery of diamonds coincides with the time interval over which the optimum conditions for sediment fractionation and disposal are operating in the environment of the terminal placer.

Timing of the whole dispersal and terminal placer development is also highly significant in

another sense as the diamonds are to be recovered from loose sediment rather than being locked-up in cemented rock. Diamonds are hard but relatively brittle so that, unlike gold for example, recoverable diamond placers are more likely to be found in deposits that have escaped deep burial, orogenesis or severe cementation. Thus Mesozoic or younger sediments on cratons or their edges are favourable sites for easily recovered diamonds.

It follows from the considerations above that the groups of variables concerned with mega-placer development are: the craton, drainage, conditions at the terminus and timing.

The craton

Clifford (1966) showed that most diamondiferous kimberlites are confined to crustal blocks over 2 Ga old, characterized by low geothermal gradients and relatively thick crust, thus allowing for high pressure at comparatively low temperature and, consequently, generation of diamonds within the stability field in the mid–lower portions of cratonic lithosphere. Nevertheless diamond bearing kimberlites and lamproites have been found in Proterozoic belts (e.g. Western Australia (Atkinson 1986), Buffalo Hills, Alberta (Carlson *et al.* 1998)) although these may be cover to older cratons which might have been beneath them at the time of kimberlite emplacement. There are minor occurrences of small diamonds in regional eclogites, mainly in Phanerozoic belts (Nixon 1995).

Janse & Sheahan (1995) estimated that *c.* 500 of the 5000 kimberlites discovered by the mid 1990s were diamond bearing and there is clearly a wide variation in fertility of pipes within the same craton as there is between cratons (Figs 7 & 8). The combination of the extent, fertility and geological history of the craton are part of the fundamental controls on the development and richness of a mega-placer that it might generate.

Extent

It follows from 'Clifford's rule' (above) that the global availability of diamonds is largely related to the global area of Archaean crust, and continents richly endowed with Archaean blocks have commensurately greater potential for diamond yields and likelihood of producing a mega-placer. Goodwin (1996) has calculated that the total area of Archaean craton is 15.5 × 10^6 km^2, 43% of which is exposed. About 24% of the total estimated Archaean crust is in North

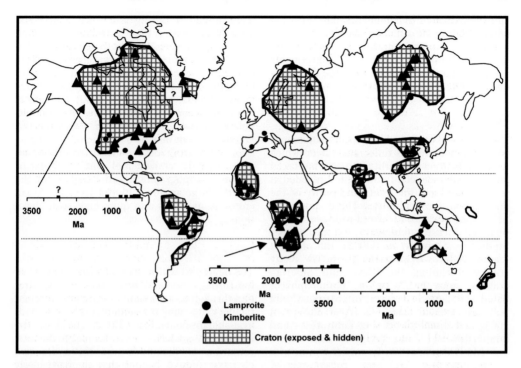

Fig. 7. Distribution of kimberlite/lamproite clusters on the cratons together with their age ranges (largely based on Nixon 1995).

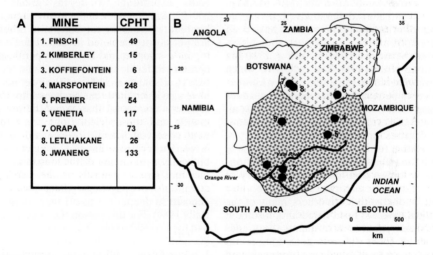

MINE	CPHT
1. FINSCH	49
2. KIMBERLEY	15
3. KOFFIEFONTEIN	6
4. MARSFONTEIN	248
5. PREMIER	54
6. VENETIA	117
7. ORAPA	73
8. LETLHAKANE	26
9. JWANENG	133

Fig. 8. (**A**) Variations in diamond grade between kimberlites mined by De Beers on the Kalahari craton, based on production figures for 1998–2003. (**B**) Location of these De Beers primary deposits on the craton (source: De Beers annual reports).

America and Africa with 32% of the exposed Archaean crust in Africa and 27% in North America. Both these continents therefore have potential to yield high volumes of diamonds (Canada is the country with the largest area of Archaean crust (Janse 1993)). However, it is the interactive craton (i.e. including the craton, now partly buried but which has interacted with younger erosion cycles) that is the key influence on volumes of diamonds generated and the extent of that influence has yet to be determined. In the same way, for any continent, a formerly extensive craton may have been detached during continent break-up, possibly leaving or taking with it some of the diamonds dispersed out of the existing pipes.

Fertility

The fertility of cratons with respect to diamonds is determined by the time range over which diamond-bearing kimberlite intrusion has taken place, the abundance of kimberlites intruded at any one time and the grade and purity of the diamonds in the kimberlites. There appears to be growing evidence for world-wide episodes of diamond formation beneath cratonic blocks. Studies by Shirey *et al.* (2002) indicate that diamonds have been available below many cratons, including the Slave and Siberian cratons, since *c.* 3.2–3.4 Ga. Lamprophyres dated at *c.* 2.67 Ga in the Proterozoic Abitibi belt have yielded diamonds (Ayer & Wyman 2003) and diamond-bearing kimberlites and lamporites >1177 Ma (Argyle) and others *c.* 20 Ma (Ellendale) are found off craton with apparently no Archaean rocks nearby (Atkinson 1986).

In both the North American and South African cratons, kimberlites of *c.* 1100 Ma have been dated and alluvial diamonds (implying the existence of primaries at the surface) have been recovered from the Witswatersrand basin, deposited between 3074 Ma and *c.* 2710 Ma (Armstrong *et al.* 1991). The extent to which other cratonic blocks have these older kimberlites is yet to be revealed, but many have suites extending back to *c.* 500–600 Ma and almost all have suites of Permian, Jurassic and Cretaceous diamondiferous kimberlites (Fig. 7).

The relative fertility of cratons is difficult to establish for a number of reasons. Some cratons are poorly known and have a short history of diamond exploration so that many kimberlites may yet be discovered. On others, many of the kimberlites have not been dated and in some it has not been possible to evaluate the grades within pipes. Even within cratons, kimberlites have a wide range of grades not always related to present erosion levels, so signifying a real variation in the original diamond xenocryst population.

Bearing in mind the caveats listed above, current evidence suggests that the Kaapvaal craton is the most fertile in the world. It has a range of ages of kimberlite intrusions and contains some of the richest pipes ever discovered (Figs 7 & 8). In addition, many kimberlites

were intruded during each time slice and particularly during the Cretaceous. This high degree of fertility is thought to be partly related to the characteristics of the oceanic lithosphere possibly generated by subduction processes during Archaean and later crustal growth (De Wit *et al.* 1992; Carlson *et al.* 2000). Komatiites from the Kaapvaal craton have, for example, volatiles at >4% (Parman *et al.* 1997) and this volatile content may be a key factor in bringing diamond-bearing rocks to the surface. However, it may not be coincidental that the Kalahari craton is, topographically, the highest of the world's large cratonic blocks and the Aldan craton is, in places also quite high, with elevations in excess of 2000 m.

Geological history

The degree to which diamonds are retained or lost through time and the degrees of concentration through down-weathering are all important factors in deciding the quantity of the diamonds available for final release. The unique characteristics of cratons have some relevance to the loss-retention of diamonds and there are two characteristics of cratons that are particularly significant in this respect.

First, cratons typically have seismically fast roots extending to depths greater than 200–250 km. Samples, in the form of xenoliths in kimberlites, from the top 200 km of lithosphere are peridotites depleted in Ca, Al & Fe resulting in less garnet and a residual peridotite less dense than fertile mantle at the same temperature (Carlson *et al.* 2000). O'Reilly *et al.* (2001) have also calculated that Archaean lithospheric mantle is less dense than younger lithospheric mantle and these calculations are consistent with observed seismic properties. Archaean cratons therefore have a long history of buoyancy, a prediction that is consistent with the fact that stratigraphically younger formations commonly overlap onto them from thicker deposits in deeper basins off their margins (e.g. Bally 1989). For this reason they have, and have had for a considerable time, the potential to leak diamonds to surrounding areas.

Secondly, in addition to buoyancy there is the also the effect of the root or tectosphere (Jordan 1988) in controlling the post-formation history of the craton. The tectosphere is considered to be not only a thermal shield (Nyblade 1999) but is also able to resist delamination and fail to compress in the same way as younger crust (O'Reilly *et al.* 2001). However, cratons may break up by extension, as evidenced by the numerous rift valleys crossing them (Sengor &

Natal'in 2001), some eventually to translate into passive margins on craton dispersal.

Margins of cratonic blocks appear to undergo periods of extension when sediment and some of the diamonds eroded from primary and retained placers is dispersed from the comparatively elevated craton into the basins adjacent to it. However extensional events are frequently followed by compression on the craton margin and as the main craton resists this compression fold belts ride onto it to return sediment that was previously lost (Fig. 9).

In North America, orogens, at least as far back as the Neoproterozoic Grenville, have been thrust onto (in that case) the Superior Province and a number of other orogens have also converged onto the North American craton (Fig. 9A). A similar situation is seen in southern Africa where the Kheis, Namaqua-Natal, Damaran and Cape fold belts all converge onto the Kaapvaal craton (Fig. 9B). These converging fold belts are not only instrumental in sediment return but they are also usually accompanied by foreland basins that migrate

toward the interior of the craton as the fold belts advance (Fig. 9C). These basins also have significance for the retention-loss of diamonds.

The foreland basin margin, distal from the fold belt (i.e. interfacing with the craton), receives sediment from the craton and so a potential diamond loss is incurred. In addition, the sediment in the foreland basins offers a host to kimberlite intrusions, as in the western Canada basin and in the Karoo basin of southern Africa. Foreland basin sediment can extend over the larger part of the craton. The Grenville orogen, for example, dispersed sediment 1000s of kilometres to the Slave province in NW Canada (Rainbird et al. 1992; Fig. 9A) as did the Franklinian–Caledonian orogen (Patchett et al. 2004).

Many foreland basins are marine, so placer deposits may develop along the migrating shorelines at the craton–basin margin during both advance and retreat of the basin edge (Fig. 12C). There is potential, for example, for the development of placers on the NE–SW migrating edge of the foreland basin to the

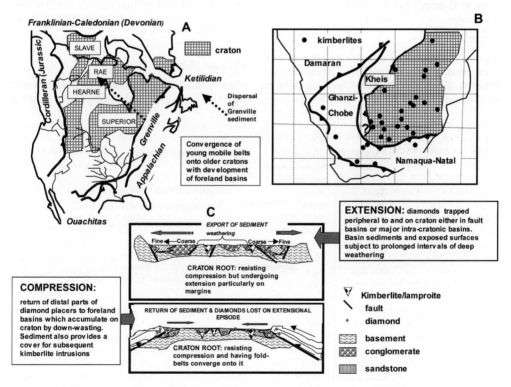

Fig. 9. The convergence of orogens peripheral to the cratons of (**A**) North America and (**B**) South Africa. Most converging orogens are accompanied by foreland basins, which, in the case of the Grenville is known to have dispersed sediment widely over the Canadian shield. (**C**) is a section illustrating some of the processes thought to occur during the extensional–compressional cycle.

Cape fold belt (Behr 1965; Catuneanu *et al.* 1998).

As cratons, away from the zone of convergence, passively receive the sediment rather than actively subside to accommodate it, much of the foreland basin sediment is lost when the converging mountain mass is eroded down and the basin inverts. At this point any newly intruded kimberlite suites are subject to erosion and if the time of inversion is optimum for both large-scale drainage development and conditions at the river terminus then the potential for mega-placer development is enhanced.

In the Kaapvaal instance, late Neoproterozic–Cambrian sediments of the foreland basin to the Damaran–Gariep fold belts are still preserved (Germs 1995) as are the sediments of the foreland basin to the Palaeozoic Cape fold belt (Catuneanu *et al.* 1998). If the sediment from the earlier of these foreland basins had extended well onto the craton then they would have formed a host to kimberlite intrusions. The timing of kimberlite intrusion in relation to foreland basin development assumes some significance in this respect and two examples are given in Figures 10 & 11.

Drainage

Important prerequisites to the development of a mega-placer are: the accessibility to the final drainage network of diamond-bearing deposits on the craton or its borders, whether primary, transient or retained; the area of drainage basin and the proportion of it which is on-craton; the degree to which older drainage networks have robbed the craton of accumulated diamonds; and the sediment loads and efficiency of the drainage network to release diamonds.

The accessibility of diamonds to the drainage network: the nature of diamond retention

In order to synchronize the time of diamond delivery to the terminus with the optimum conditions for their concentration there, diamonds existing on the craton, primary or secondary, should be available to the drainage network in abundance for a specific time interval. Source areas releasing small quantities of diamonds over a long time-span may fail to yield sufficient for mega-placer development. The range of settings in which diamonds may be retained on the craton and their corresponding ease of access to the drainage are illustrated in Figure 12. Three situations are taken to illustrate this point: primary deposits in cover, preassembled placers and down-wasting on planation surfaces.

Primary deposits in cover

Many kimberlites have intruded a cover (often provided by a foreland basin). The rate of its

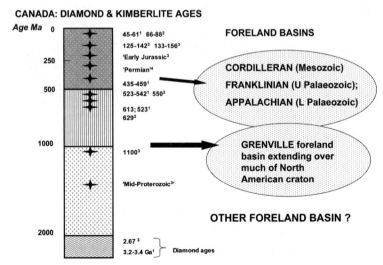

Fig. 10. Distribution of diamond-bearing kimberlites through time for the cratons of the Canadian shield and the presence of foreland basins when there was potential loss of diamond from the craton.

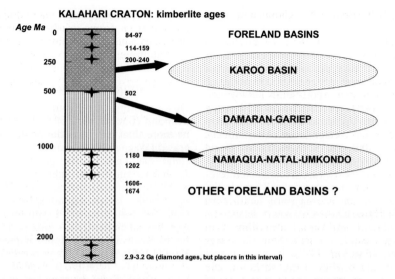

Fig. 11. Distribution of diamond-bearing kimberlites and times of possible leakage, Kalahari Craton.

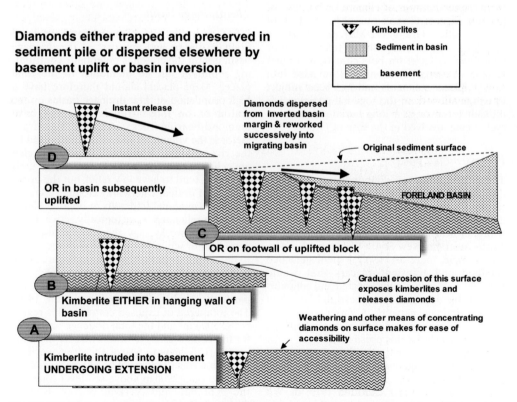

Fig. 12. Means of concentrating and retaining diamonds on cratons. (**A**) Gradual downwasting of the craton surface intruded by kimberlites. (**B**) Kimberlite being intruded into sedimentary cover. (**C**) Diamonds being eroded from uplifted craton for reworking on the margin of a migrating foreland basin. (**D**) Basin sediment uplifted and diamonds dispersed over the craton to find residence elsewhere.

erosion and removal to expose diamond-bearing rocks is important to the delivery time of diamonds to the placers. Exposed kimberlite pipes may continue below the current level of erosion to yield significant quantities of diamonds as the cover continues to be stripped off (Fig. 12). Field & Scott-Smith (1998) estimated kimberlitic breccias and diamond-bearing kimberlitic rocks to extend for more than 2 km below near-surface crater facies. Alternatively, they may be totally buried by sediments. Good examples of both these cases are seen in Angola where a group of northward-flowing tributaries (Luembe, Chiumbe Luachimo, Tchikapa) to the Kasai River are uncovering Early Cretaceous kimberlites buried by the Calonda (Cretaceous) and Kalahari (Cainozoic) Formations even before some had even had their tuff rings eroded away (Figs 4 & 12A,D).

The diamond-bearing kimberlites on the present Kalahari craton exist only because erosion has not yet removed them. In some cases, erosion has been extremely slow and a permanent drainage network has failed to reach them, either because of climate or because of the low rates of erosion caused by the lack of surface relief. Orapa (92 Ma) and Jwaneng (245 Ma) in Botswana probably remained in or near to crater facies for both climatic and relief reasons. Although Jwaneng was intruded into Lower Karoo sediments and has been subject to exhumation, from the regional geology, it is difficult to envisage a long burial–exhumation cycle being involved in the case of Orapa.

Pre-assembled placers

The accessibility of diamonds to the final drainage network of the river generating the terminal mega-placer is greatly enhanced if discrete placer deposits are already gathered within or near to the final drainage network. In North Australia, New South Wales and Borneo, for example, there are several small detached retained and transient placers that would possibly yield larger, concentrated and commercially viable deposits if they were all in reach of a larger drainage network. A variety of sources over a wide area and a dearth of larger drainage networks precludes this possibility.

In marked contrast, abundant commercial diamond placer deposits have existed in the present Orange–Vaal drainage basin since at least Cretaceous times (Stratten 1979; de Wit 1996, 2004; Fig. 17). When this drainage basin was rejuvenated in Tertiary times ample transient and residual placers had already been gathered on the craton, some during a more humid Cretaceous climate, so that whatever the contribution from primary sources, these secondary sources ensured a diamond supply at the right time.

Down-wasting on planation surfaces

Down-wasting significantly increases the diamond concentration of weathered products by more than ten times the grades of the host gravel (Wagner 1914; Marshall 2004). Cratonic relief has been mildly positive since post Archaean times and many unconformities, weathered surfaces and palaeosols have developed on them, some extending back 3.0 Ga (see Gall 1999 for examples). Down-wasting of this type has taken place over long periods of time on the Kaapvaal craton (for example Cheney & Winter 1995; Fig. 13). Many pre-1 Ga kimberlites may have been eroded and the diamonds held in palaeosols, or upgraded and dispersed into intra-cratonic basins or basins peripheral to the craton.

The drainage basin

Large rivers with high discharges have large drainage basins and any part of the drainage basin not on the craton is potentially contributing sediment but no diamonds to the final placer. Mega-placers should therefore have a high proportion of their drainage basins on the craton or on those fringes that may have diamond-bearing rocks. Another significant factor is the volume of sediment transported by the river system, determined largely by relief and climate. Cratonic blocks, whatever their elevation, tend to have low relief, but the upper reaches of drainage networks crossing a craton may be in marginal fold belts or uplifted blocks with high relief and be tectonically active. These drainage networks may therefore transport a good deal of sediment onto and over the craton to the terminal placer. Three examples are taken to illustrate the significance of this point.

The Lena River drains mainly the Aldan and partly the Anabar cratons in Siberia (Fig. 14A). The total area of exposed craton in this region is $482\,500$ km^2 and the total drainage basin area for the Lena is 2.5×10^6 km^2 with c. 25% on the craton. This leaves 75% of the drainage from off-craton rocks ranging in height from sea level to more than 2900 m above sea level (masl) in the Khrebet and Verkonoyansky Ranges. Although these ranges are comparatively old, sediment loads dispersed into the tributary rivers such as the Aldan are likely to be high.

The Congo River illustrates a different

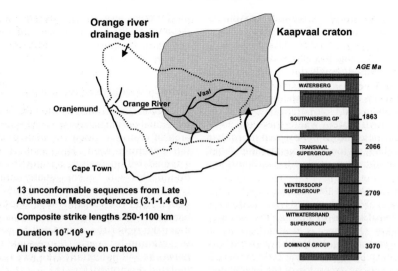

Fig. 13. The early history of the Kaapvaal cratons illustrating the numerous unconformites likely to have been accompanied by intensive weathering on exposed craton surfaces (from Cheney & Winter 1995).

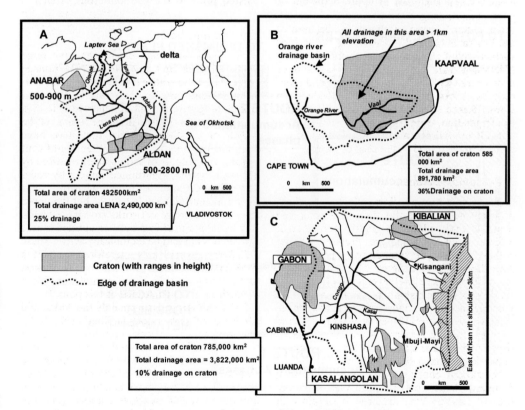

Fig. 14. The relationship between craton area, drainage basin area and the percentage overlap in both for three selected major river systems. (**A**) The Lena River, Siberian craton; fine dotted line marks the subsurface extent of cratons. (**B**) The Orange River, Kaapvaal craton. (**C**) The Congo River, for the Congo craton. The area of the Congo craton is estimated to be 5.7×10^6 km^2 (Goodwin 1996), but is largely covered by the thick sediments of the Congo basin.

228   B. J. BLUCK *ET AL.*

situation. The total area of craton is
c. 785 000 km^2 and the Congo River has a
drainage basin area of 3.7×10^6 km^2 but with
only c. 10% draining the exposed craton. The
Congo River, initiated only during the Cenozoic
(Reyre 1984; Droz *et al.* 1996) crosses the Congo
basin which has low relief, but the headwaters
(via the Lualaba and Uele) drain sections of the
East African rift system where there are some
of the highest elevations in Africa at more than
3000 masl. This region has considerable relief,
continual uplift, substantial sediment input and
is off craton (Fig. 14C).

In marked contrast the Kaapvaal craton has
a total area of 585 000 km^2, and is drained by the
Orange–Vaal Rivers with a drainage basin area
of 891 780 km^2 (Bremner *et al.* 1990), c. 36% of
which is on craton (Fig. 14B). Although the bulk
of the drainage basin lies above 1000 m there is
little relief, with the exception of the headwaters
in the Drakensberg, where the relief is steep,
and elevations reach 3200–3400 m but has little
evidence of a high rate of uplift. Sediment off

the Drakensberg is dispersed westwards to the
Atlantic via the Orange River drainage but also
eastward, in short reach rivers, to the Indian
Ocean.

Time of drainage development

To release the greatest abundance of diamonds,
a regional drainage network covering as wide an
area of craton as possible needs to develop at
the optimum time. The pool of diamonds
retained within the craton and its immediate
boundaries will be continually enhanced by
repeated additions if newer kimberlites are
added to the craton and down-wasting
continues. The younger the drainage development the richer the source will be with respect
to diamonds: early large-scale drainage
networks will potentially take away those accumulated diamonds (Fig. 15).

In the case of North America (Fig. 16),
whatever large drainage networks might have
existed prior to the Carboniferous, Archer &

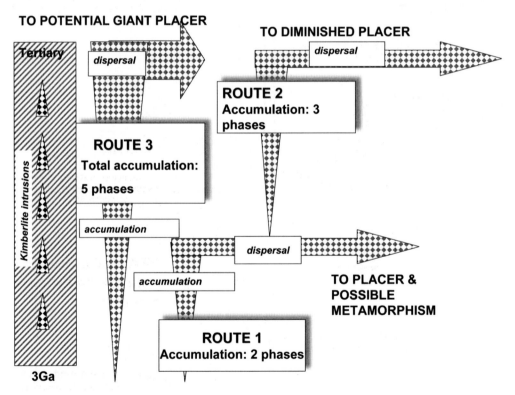

Fig. 15. The importance of retaining diamonds on the craton in order to yield a mega-placer. The left-hand
column has five phases of diamond-bearing intrusions. In Route 1 a large-scaled drainage removes the
accumulated diamonds from two phases of intrusion; then a subsequent drainage removes the other three in
Route 2. Route 3 sees a late-stage drainage having access to the total diamond-bearing phases and is therefore
most likely to yield a mega-placer.

Greb (1995) have demonstrated that Amazon-scale drainages, covering much of the Superior province, were certainly present in the Carboniferous deposits of the mid-West. This drainage probably links with the proto-Mississippi flowing into the Illinois basin-embayment (Potter 1978). Sediment from the Slave craton would almost certainly have dispersed into a series of Palaeozoic basins developed along the route of the McKenzie River.

A Cenozoic river supplying sediment to a basin extending from off Baffin Island to the sea off Labrador had an estimated drainage basin covering parts or a great deal of the Superior, Hearne, Nain and Rae cratons (McMillian 1973). Then, finally, in addition to the earlier Proterozoic glaciations, a major glaciation has occurred in the last *c.* 2 Ma. The extent to which these removed kimberlitic material (discussed in greater detail later) is not known but diamonds have been recovered from glacial outwash sediments in the American mid-continent (Gunn 1968) and kimberlite blocks and tracers in glacial deposits on the Canadian shield (e.g. Baker 1982).

In South America, the Guyanas and Guapore–Sao Fransico cratons (the Central Amazonian Province) are divided by the present Amazon drainage and Amazonian Basin which, together with its extension to the SW, is underlain by a series of basins that have existed since at least Silurian times (Colombo & Macabira 1999; Milani & Zalán 1999). Here, and in the North American craton, where at least four major basins developed on or near to craton edges (the Williston basin, Hudson Bay basin, Michigan basin and Illinois basin; Bally 1989), whatever diamonds were contributed to these basins are now buried.

In the case of the Kaapvaal craton there is no evidence yet for a large-scaled drainage prior to the development of the Orange–Vaal River basin with the possible exception of its eastern margin (Moore & Larkin 2001). Neoproterozoic, Ordovician and Permo-Carboniferous ice ages have all been possible agents of diamond loss. The Permo-Carboniferous glaciation on the Kaapvaal and adjacent cratons may have transported sediment as far west as the Parana basin in Brazil and the Colorado basin, west of

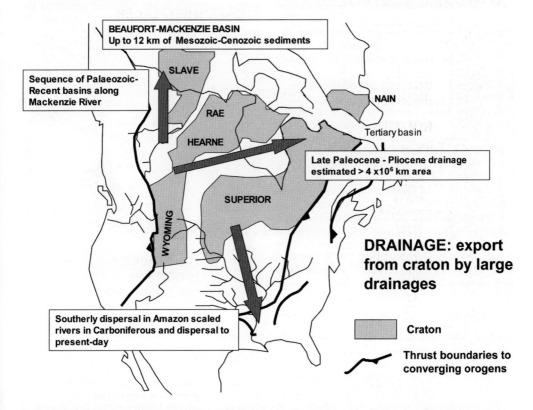

Fig. 16. Large river systems known to have drained parts of the North American cratons and therefore may have removed diamond deposits existing up to the time when the drainage was initiated (see text for sources).

Argentina (Visser 1990; Moore & Moore 2004). In the case of the Kaapvaal, some of that sediment reached the Karoo foreland basin and therefore remained within the subsequently developed Orange River drainage basin.

Sediment loads

Large African rivers have lower sediment loads compared to all others (Milliman & Meade 1983), the result of them draining cratons with relatively low elevation and low relief. Africa, like Australia, also lacks substantial marginal active mountain chains. However the East African rift system, and to a lesser extent the Drakensberg and the Atlas reach more than 3000 masl. In marked contrast to Australia and the Canadian cratons, Africa has very large rivers with equally impressive drainage basins (the Congo being the second largest in the world).

Other cratons with major rivers and very large drainage basins differ from African ones in that the bulk of the discharge and sediment supply comes from adjacent, active uplifted blocks. This unwanted sediment load dilutes the contribution from the craton itself. The Amazon, McKenzie and rivers traversing the cratons in India, for example, already carry a substantial volume of sediment before they cross the cratons.

Although sediment loads are relatively low in African rivers and there are also significant differences between them. The Congo, for example, is sourced in the East African rift flanks; a drainage basin with precipitation of 1000–2000 mm a^{-1} and a sediment discharge of 43×10^6 t a^{-1}, contrasts with the drier, more seasonal Orange River drainage basin. In the latter, the bulk of the drainage basin experiences precipitation of less than 500 mm a^{-1} and a sediment discharge of 17×10^6 t a^{-1}. That state has probably existed for a substantial part of the Cenozoic after the onset of initial aridification began (Ward & Corbett 1990), and at a critical time when the bulk of the diamonds were carried down.

Rejuvenation

Given a constant climate, when base level falls and adjustments are made in the river systems, drainage density increases and more residual placer and kimberlite deposits are brought into the drainage network. In addition, rejuvenation, in providing steeper slopes also increases the energy in the landscape, resulting in both increased transport efficiency and potential for

diamond sorting in the rivers. At the same time, incising rivers cut through any soft cover and may expose primaries and bedrock and both the bedrock topography and the clasts yielded from it provide the trapsites for diamonds.

As discussed previously, the presence of kimberlites on any cratonic block, particularly those in crater facies, is a testament to the inefficiency or lack of available time to scour the craton free of its diamonds. Many of the Angolan Cretaceous pipes for example, were buried by small-scale river systems that laid down the Calonda and its equivalent Formations in central Africa. These streams were either grossly inefficient at erosion (possibly because they were infilling a subsiding basin) or had little time to remove the kimberlitic crater facies, in places tuffs, which spread onto the surface (Fig. 4). On a larger scale, the presence of crater facies kimberlites, e.g. Premier (c. 1250 Ma), Venetia (530 Ma), and Orapa (c. 90 Ma) on the Kalahari craton illustrate a lack of intensive erosion and inefficient removal of diamond-bearing rocks (Fig. 17).

Unusually, coastal uplifted blocks may yield coarse materials to the terminal placer as the river cuts through them before entering the sea.

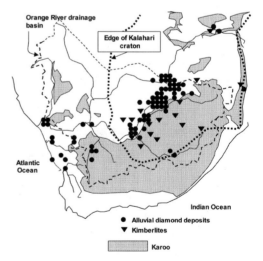

Fig. 17. Distribution of kimberlites and alluvial deposits with respect to the Orange River drainage basin. Kimberlites still yielding diamonds range from *c.* 1200 Ma and workable alluvial (transient) deposits from the Cretaceous. Diamonds have been recovered from Archaean basins, making this drainage near to route three of Figure 15 (based on de Wit 1996).

The terminal placer: the Orange River and related placers

Diamonds in substantial quantities and of exceptional quality are found on the SW African coast and shelf, north and south of the present Orange River mouth. There are two pairs of major diamond accumulates (Fig. 1) both possibly linked to an evolving drainage system related to, or directly the result of, the Orange–Vaal River. The most significant pair consists of gravel beaches and desert deflation deposits along the Namibian coast north of the Orange River, and corresponding submerged equivalents offshore. The second pair comprises gravel beaches and alluvial deposits in onshore Namaqualand and a diminished offshore counterpart. Together these two broad areas have yielded 120×10^6 carats, more than 95% of them being gem quality. The Namibian sector has yielded $c. 80 \times 10^6$ carats and the Namaqualand $c. 40 \times 10^6$ carats to date (Corbett 1996; Oosterveld 2003; Table 1). The additional estimated deposits remaining in Namaqualand bring that total to 50×10^6 carats and the past Namibian production plus estimated resources amounts to more than 100×10^6 carats, making each area a mega-placer deposit.

The total estimated potential for all four placers is 1.5×10^9 carats according to Levinson et al. (1992). This estimate, which depends heavily on the estimated thickness of the Karoo cover (and consequently the former height of the kimberlite pipes in South Africa), is disputed and the total diamond count in the southwestern African placers may be only 30% of that value.

Westward draining river systems, including the Orange, Buffels, Groen and Olifants all have terraces along them containing diamonds and each has made a contribution to the terminal placers. There is a well-documented decline in diamond size northward along the coast from the present Orange mouth (Hallam 1964; Sutherland 1982; Schneider & Miller 1992, Figs 18–20). With no or few other likely contributors, the Orange River has been identified as the point source of the diamonds to the coast in this instance.

There is also a northward decline in the diamond size along the Namaqualand coast, beginning at or near to the mouth of the present Olifants River (Rogers et al. 1990; Fig. 18). However in this case, there are a number of potential in-put points for diamonds, highlighted by Stevenson & McMillan (2004).

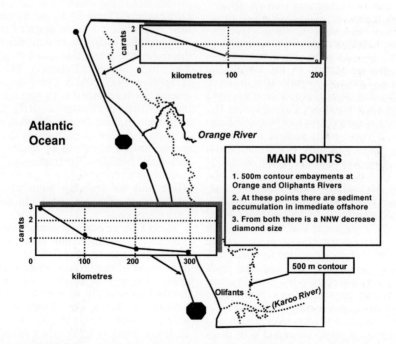

Fig. 18. The variation in diamond size along the SW African coast and some of the evidence for two principal mega-placer systems.

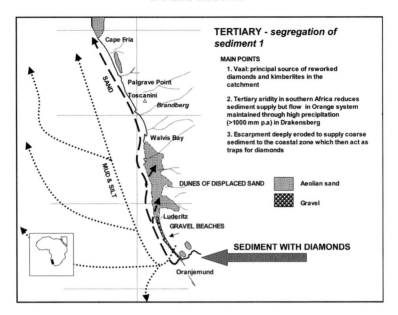

Fig. 19. A reduced sediment input (cf. Fig. 21) enters a neutrally buoyant shelf where there is vigorous coastal energy. The fractionated sediment disperses with fines moving off the immediate shelf, a strong onshore wind returning sand to land along the arid coast and the gravel accreting to the coast immediately up-drift from the river mouth (as it would have done for the various sea-level fluctuations over the broad shallow shelf).

Aspects of the Mesozoic–Recent geological history of the west coast and Orange River drainage basin

The Orange–Vaal drainage is the principal route along which the diamonds have been transported from the interior to the coast (e.g. Hallam 1964). Like many South African rivers, it has all the characteristics of a superimposed meandering river drainage, cutting across the strike of the Neoproterozoic fold belts (Wellington 1958). Offshore of its present mouth and dominating the stratigraphy of the continental shelf lies the Cretaceous Kudu delta of Turonian–Maastrichtian age (*c.* 93–70 Ma; Wickens & McLachlan 1990; Clemson *et al.* 1997; Aizawa *et al.* 2000; Brown *et al.* 1995, Fig. 21 but see Stevenson & McMillan 2004). Ward & Bluck (1997) suggested that the wavelength of the meander loops was consistent with a large river which might have deposited this delta and then in post-Cretaceous time, imposed itself through a Karoo cover onto the local Precambrian basement, retaining an approximate memory for its original meander loop size.

Key exposures at both the river mouth and northwards from it comprise gravel with clasts of, amongst others, agate and chalcedony. These are typical of the Vaal and upper reaches of the Orange drainage with the minerals having a provenance in the Ventersdorp (Archaean) and Drakensberg (Jurassic) lavas. At one locality near Bogenfels, *c.* 150 km north of the present river mouth, a sedimentary sequence containing this distinctive gravel clast assemblage has been dated by its faunas as Eocene (*c.* 43 Ma; Siesser & Salmon 1979; SACS 1980) and this clast assemblage is regarded as the signature of sediments of that age (Kaiser 1926). Agate clasts up to 100 mm in diameter are recorded from these deposits and, close to the current Orange River mouth, are accompanied by quartzite cobbles and small boulders. Their presence implies a considerable change in the slope of the Orange River and its drainage from the one that deposited the fine-grained Kudu delta in the Late Cretaceous.

These Eocene deposits indicate that between *c.* 70 Ma (the estimated final phase of deposition of the fine-grained Kudu delta) and *c.* 43 Ma most or all of the Orange River basin, and by implication most of southern Africa, had been uplifted sufficiently to expose the basement beneath its Karoo cover and access coarse sediment from the eastern margins of its drainage basin. This resulted in the channel becoming steeper and able to transport this substantially coarser sediment. The precise

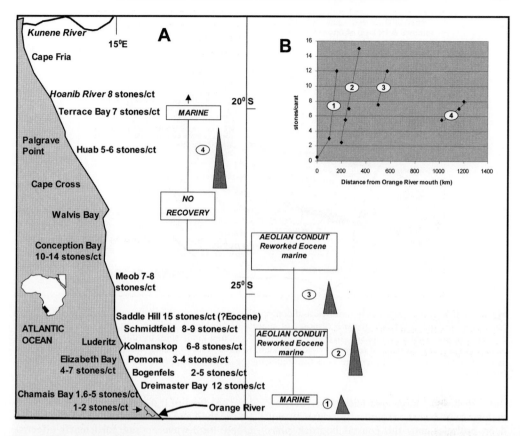

Fig. 20. The changing diamond sizes along the Namibian coast. In **B** the *y*-axis of the graph records the number of diamonds/carat. 1–4 refer to the dispersal units each with a differing rate of diamond-size decline and may reflect both the redistribution processes (such as sheetwash and aeolian influences) as well as differing rates of diamond-size decline for differing periods going back to the Eocene (data from Hallam 1964).

nature and timing of this uplift event has yet to be determined; however, it is partly recorded in a sub-Tertiary unconformity offshore. Aizawa *et al.* (2000) estimated an uplift of *c.* 1000 m on this western margin of SW Africa. The occurrence of a major epeirogenic uplift in this region of southern Africa is supported by the estimated elevation of both undeformed Permo-Carboniferous Dwyka Group shallow marine sediments now at 1200 masl in central Namibia (Martin & Wilczewski 1970) and Permian Ecca beach deposits at similar elevations near Bothaville in South Africa (Behr 1965). This estimate also approximates to the height of the base of the Neoproterozoic–Cambrian Nama Group that crops out *c.* 100 km inland at *c.* 1000 masl and comparatively young uplifts may be partly a reason for the preservation of these exposures.

This uplift event was of great significance in the development of the placer. It rejuvenated an existing river system making it able to cut back into the craton and through any cover to access the diamond-bearing rocks that were widely available there. Concomitantly, in increasing the slope of trunk and contributory streams, coarse sediment (including diamonds) was transported out of the interior and a more vigorous route to the Atlantic was established, resulting in an upgrading in the quality of the transported diamonds.

Through this major uplift, and critically for placer development, the Orange–Vaal becomes a rare example of a river with a drainage area of almost 1 million km^2, with an abundant supply of coarse gravel at its mouth. In cutting through the coastal Neoproterozoic fold belts, the river and its tributaries yielded clasts, up to 5 m in diameter, near to the terminal placer thereby delivering materials for the trapping of the diamonds. It also provided the tools, in the

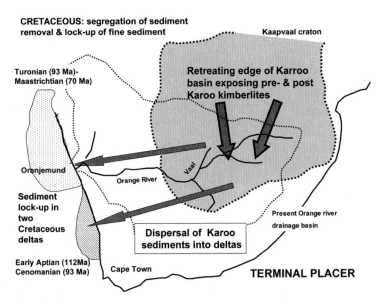

Fig. 21. Soft Karoo cover is easily removed and deposited in two deltas in the offshore. Thus sediment with the potential to dilute the Tertiary placer is disposed of in areas of subsidence and is no longer a contributor to the processes at the terminal placer site. In addition, the latest batch of kimberlites intruded into the craton are exposed to erosion. The rate of sediment yield is thought to have reduced greatly when the river system began its incision into the various basement rocks thus reducing the sediment supply to the terminus. Deltas adapted from Brown *et al.* 1995.

form of durable, large quartzite clasts, derived mainly from Nama Group with which to cut deep potholes into the coastal bedrock platforms (Jacob *et al.* in press) as well as acting as long-lasting hosts for diamond accumulation in the Namibian littoral environments. This Early Tertiary uplift event, in initiating the flush of coarse gravel and accompanying diamonds from the interior craton to the coast, is considered to be the onset of the formation of the Orange River–Namibian mega-placer (Fig. 6).

A more humid climate in Late Cretaceous times allowed the palaeo-Orange drainage to easily erode into the soft Karoo cover and dispose of it in a relatively short time. This resulted in a low gradient river removing unwanted Karoo cover to a delta (where some, if not most, was sealed from further interaction) and also exposing kimberlites, intruded into that cover, to a more vigorous, rejuvenated Tertiary drainage (Fig. 21).

The Atlantic coast of southern Africa is amongst the most energetic in the world. The coast receives almost continuous southwesterly swell from the South Atlantic storm belt and has, in addition, a dominantly southerly quadrant, coastal wind system created by the south Atlantic anticyclone superimposed on a cold ocean fringing a warm landmass. The swell

has the effect of transporting sediment at depth on the shelf and periods of strong onshore winds create local wave energy, particularly effective along the coastal strip. Both swell and locally generated wave orthogonals are oblique to the shore so this energy is converted into a substantial northbound long-shore drift. The arid coastline has limited vegetation to hinder the movement of sand off the coast and onto the land, where it has accumulated in desert sand seas.

In the present coastal regime, 90% of the waves have a height falling within the range 0.75–3.25 m with an average 1.75 m for winter and 1.5 m for summer (Rossouw 1981; de Decker 1988). The coast is subject to long period swell with 53% of the waves having a period over 12 seconds and a height of more than 5 m. Energy from this swell is capable of moving cobble sized gravel, at depths of 15 m (de Decker 1988); submersible dives have also detected wave-generated water movements capable of transporting sediment at 120 m depth.

In addition there are significant ocean currents moving the water mass on the shelf, the most well known and probably most significant of which is the Benguela. The Benguela current has a northward velocity of 288 m/hour; the deeper South Atlantic central water and Atlantic intermediate water currents move south at *c.* 208 m/hour (Bremner & Wills 1993).

Seismic sections reveal that since the post-Cretaceous unconformity, large areas of the shelf, particularly those immediately north and south of the Orange River, have accumulated little sediment and therefore have undergone minor, if any subsidence in that time (Aizawa *et al.* 2000). The present shelf is both wide and shallow (Figs 1, 22) and a substantial proportion of it is under the influence of the vigorous wave and current regimes, factors critical to the ultimate generation of the mega-placer.

The combined effect of the wind–wave system, longshore drift, shallow, neutrally buoyant shelf with little slope and arid coastal climate has been to segregate the sediments by grain size and disperse them into discrete areas of accumulation (Fig. 19). In the Orange River–Namibian coast placer system, gravel is returned to land in the immediate vicinity of the river mouth, where, under the influence of the

strong longshore drift it migrates northward, mainly in the intertidal and nearshore subtidal zone, for a distance over 200 km. The sand also moves mainly along the coast in a narrow, usually less than 3 km wide, subtidal zone (Corbett 1996) and northwards, sand beaches replace gravel, possibly continuing northward for *c.* 1000 km. Sand, transported on-shore from the offshore conveyor, is trapped in J-shaped bays, where wave refraction, in transporting sand around headlands, deposits large volumes onto relatively quiet beaches. Given the long period waves, sand build-up on the beaches would be potentially high but from there it is rapidly transferred onto the land by strong onshore winds and carried in fast moving barchans (up to 100 m a^{-1}; Corbett 1996) to join the main Namib Sand Sea (Rogers 1977; Lancaster & Ollier 1983; Lancaster 2000).

In this highly energetic regime, the mud fraction is dispersed into the marine water column and moves offshore. Although it may settle on the floor temporarily (Bremner &

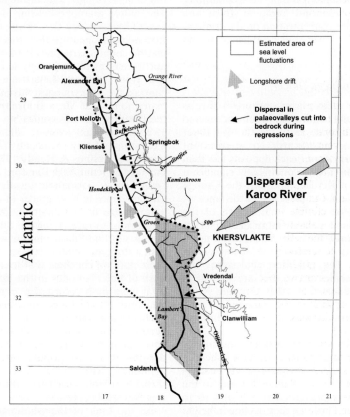

Fig. 22. The area covered by possible sea-level fluctuations and the areal distribution of diamonds over the shelf and bordering land for the Namaqualand mega-placer.

Willis 1993) the mud generally finds a more permanent residence on the continental slope (Aizawa *et al.* 2000), and in subsiding parts of the shelf to the north (Holtar & Frosberg 2000) or to a lesser extent to the south of the Orange River mouth. This fractionation of the sediment is highly significant for placer development as the coarser diamonds move with the gravel fraction and are preserved, relatively free of other diluting sediment, in gravel beaches and their associated wave-cut platforms.

The exposures at the river mouth and to the north, near Buntfeldschuh show that the incision of the Orange drainage basin, the delivery of coarse clasts, the wind system, long-shore drift and little subsidence on the shelf have all been operating to some degree since at least the Middle Eocene (*c.* 43 Ma). Sections dated as Eocene expose shoreface and possible sand–gravel beaches, and a section with strong Eocene clast type signature at Buntfeldschuh also has dunes with northeasterly dipping cross-bedded strata indicative of an onshore southerly wind system at that time (Corbett 1993). The critical point is that these conditions, effecting the diamond removal from the interior and their concentration on the contemporary shorelines, were all established within the same time frame.

The Namibian mega-placer

The Namibian mega-placer onshore is a relatively thin strip of continuous beach and related deposits, *c.* 3 km wide, near the Orange River mouth tapering northwards to intermittent pocket gravel or raised beach deposits less than 300 m wide. The most economically significant sector of this placer lies between the Orange River mouth and Chameis Bay, a distance of *c.* 110 km. However, the total length of the dispersal system is possibly more than 1000 km as small diamonds have been recovered north from the Orange River mouth at Hoanib on the Skeleton Coast (Fig. 20). Here chalcedony and small agate pebbles occur, that are known to have a provenance in the Orange River drainage basin and to be associated with diamonds both there and nearer the mouth of the Orange River.

The placer is at its most complete and continuous near the Orange River mouth, where it comprises a sequence of coarse gravel beaches stacked from old in the east and young to the west. The oldest preserved beach at +10 to +30 masl is Late Pliocene; they decline in height through the Pleistocene to a Mid-Holocene beach at *c.* +2 masl (see Pether *et al.* 2000).

Farther north, preservation of the older (higher) beaches is fragmentary.

The southern beaches, exposed in mine workings and sample trenches, also reveal a sequence of beach types, with spits and barrier beaches in the south, replaced by barrier beaches, linear and finally pocket beaches in the north. Each beach type, and the subenvironments that make them, have a variable potential for retaining diamonds (Spaggiari *et al.* in press). These beaches and the often competent, potholed, Neoproterozoic footwall on which they rest are the essential trap-sites holding the littoral diamond population (Hallam 1964; Apollus 1995; Jacob *et al.* in press). The beach gravel ranges up to boulder size and is dominantly Nama quartzite brought down by the Orange River in the south, augmented by local bedrock in the north.

The transient terrace placers along the lower Orange River, for some 100 km inland, contain diamond grades (measured in carats/hundred tonnes, cpht) and stone sizes (recorded in carats/stone, cts/stone) varying spatially along the channel length and through time (Fig. 6). The oldest diamond-bearing deposits in the lower Orange valley, close to the mouth, have an Eocene clast signature that, together with their marine counterparts preserved onshore further north, demonstrate a low grade (<0.5 cpht) and low stone size (<0.4 cts/stone) deposit. By contrast the next youngest fluvial deposits (estimated to be *c.* 25 Ma, and locally referred to as the Pre-Proto Orange suite) have grades up to 35 cpht and an average diamond size of 2 cts/stone. Early Miocene (*c.* 17 Ma) age terraces (Corvinus & Hendy 1978; SACS 1980; Pickford & Senut 2003; Bamford 2003) of Arries Drift Gravel Formation, locally called Proto-Orange deposits, have grades of *c.* 1–5 cpht and a stone size of 1–2 cts/stone. The topographically lowest terraces, known as the Meso Orange deposits (? Plio-Pleistocene) have <0.5 cpht but with a stone size up to 3 cts/stone (Fig. 6). From these data we conclude that there was an early flush of fine diamonds in the Eocene followed by a main flush recorded in pre-Proto terrace remnants (?Late Oligocene) and thereafter a declining population with the largest diamonds accumulating in the youngest deposits. This conclusion is supported by the increasing density of a clast population through time, beginning with siliceous clasts (Eocene), through epidote rich (pre-Proto) and banded iron formation (Meso).

As the bulk of the Namibian mega-placer comprises littoral gravel deposits of Late Pliocene–Holocene age, it was not supplied directly with diamonds from the Orange River. The immediate source for these deposits was older beaches that had accumulated off the Orange River mouth since the Eocene and which were reworked into the Late Cenozoic–Holocene beaches. Some intraclasts of marine conglomerate recovered from the Late Cenozoic–Holocene beaches contain a Miocene fauna and support this conclusion. In order to be accessible to reworking onto these Late Cenozoic–Holocene beaches, sufficient of the former deposits must have been exposed or not buried too deeply. At the same time, the older regimes needed to have been vigorous enough to remove unwanted sediment and concentrate both diamonds and the host boulder and cobble-bearing gravel, as occurs off the present Orange River mouth (Murray et al. 1970; Corbett & Burrell 2001).

Diamond-bearing beaches, much like the present ones, are likely to have existed since the Eocene: the time of significant diamond release from the craton. Subsequently there have been numerous changes in sea-level with a highstand at c. 170 masl (Eocene) and a low stand at c. 120 m below sea-level in the Pleistocene (ignoring the potentially lower, controversial Oligocene low-stand). To this we can add wave base, calculated by de Decker (1988) to be c. 30 m, but thought to be more than 60 m from submersible dives. Thus the whole width of the shelf possibly down to 200 m may have been within wave base at some time during the last 60 Ma. As the shelf and coastal strip have not undergone any substantial subsidence since the Early Tertiary, this fairly extensive shallow region, c. 150 km wide to 200 m below sea-level and tapering north to c. 30 km wide at Walvis Bay, would have been raked by marine processes. Consequently the diamond-bearing sediment brought down by the Orange River since the Eocene has been available for concentration and reworking at innumerable intervals over this extensive shelf.

Today, divers working at depths c. 20 m record northward moving gravel ('travel gravel') sometimes highly enriched in diamonds. This moving gravel probably derived its diamonds from pre-existing deposits, possibly of beach-type, and may be an illustration of one of the mechanisms by which the diamond population migrated along the coast over the past c. 45 Ma. In addition to contributions to the current coastal placer from the offshore there was almost certainly a diamond contribution from onshore deposits laid down during earlier high stands. Sheetwash action and small ephemeral streams are known to redistribute diamonds from older high stand beaches in the northern areas of the Sperrgebiet (Kaiser 1926).

Any diamonds transported by the Cretaceous rivers, in spite of their fine grain size, are also likely to have been concentrated and maintained on this shelf. The post-Late Cretaceous unconformity truncated the deposits of the proximal Kudu delta so subjecting that sediment to the rigorous reworking on the neutrally buoyant Tertiary shelf.

In Namibia, the coastal placer deposit is a complex one with many indirect sources for its diamonds. However, each increment of diamond-bearing gravel deposited during earlier sea-levels was probably influenced by a similar longshore drift system and each probably retained a northerly clast and diamond-size decline. All subsequent deposits inherited and modified this decline to some degree. The potential of seaward and landward sources for the diamonds are abundantly clear and this is considered in more detail with respect to the Namaqualand mega-placer. This complexity may be an underlying reason for the irregular nature of the diamond-size decline along the coast (Fig. 20).

Since the pioneering work of Sammy Collins in 1960s (reported in Williams, 1996) it has been recognized that the broad continental shelf off the West coast of Namibia contains diamond bearing gravel (Murray et al. 1970). Submersible dives carried out recently, together with sophisticated side-scan sonar surveys have successfully identified patches of gravel beach in this offshore region (Corbett & Burrell 2001). This area, located c. 100 km north and south of the Orange River mouth has now yielded more than 3 million carats of gem quality diamonds and is regarded as an extension of the onshore placer.

The richness of the Tertiary terraces on the lower Orange River implies the widespread availability of diamonds in the source. There are numerous diamond-bearing kimberlites presently exposed, with a range of ages, within the Orange River drainage basin (Figs 8 & 17) and they stand as an obvious source either making a direct contribution to the drainage or indirect contribution (via terraces, etc.). The contribution of the Late Cretaceous Kimberley primary source cluster is evident in the +10 carat octahedral diamonds of Cape Yellow colour in the lower Orange transient and Namibian terminal placers. However, Van Wyk & Pienaar (1986), Maree (1987); Moore & Moore (2004) are amongst many who see earlier sedimentary formations, particularly the Dwyka Group as a probable source or even the main source for the diamonds to the terminal coastal placers.

The Dwyka includes a period of glaciation extensively covering the Kalahari craton and those kimberlites older than it would have been liable to erosion at this time. Episodes of Dwyka glaciation extend from Early Carboniferous (Streel & Theron 1999) through *c.* 300–302 Ma (Bangert *et al.* 2000) into Permian. Although it is impossible to estimate the abundance of diamond-bearing kimberlites intruded in pre-Dwyka times, Premier (*c.* 1202 Ma), Colossus (*c.* 502 Ma), the Kuruman Group (*c.* 1606 Ma) and Venetia (535–505 Ma) are examples of what may have been more widespread kimberlitic intrusions. These are clearly potential diamond sources to the Dwyka diamictites (Moore & Moore 2004).

On the Kaapvaal craton, many of these older kimberlites still retain either crater or near crater facies, some of which are richly diamond bearing (Fig. 8). This clearly indicates that the Dwyka glaciation was not entirely effective in removing diamond-bearing kimberlite. A substantial volume of diamond-bearing kimberlite was left in post-Dwyka times with the potential to add to the later Karoo and abundant Cretaceous kimberlites, many of which were available to the Orange River drainage from post-Gondwana times onwards.

Assuming there was cover to these kimberlites at the time of glaciation, their preservation may be the result of their being relatively soft and often found in hollows. Studies of glacial erosion have demonstrated that glaciers moving over regionally flat terrane, as large tracts of the Kaapvaal craton would have been in Permo-Carboniferous times, are unlikely to have cut deeply into bedrock (unlike glaciers in valleys). Hence, in NW Canada, which has undergone a recent glaciation, many kimberlites are preserved high in crater facies with graded lapilli-tuffs (Leckie *et al.* 1997; Carlson *et al.* 1998; Field & Scott-Smith 1998) although some may have had a pre-glacial cover of variable thickness.

The distribution of placers along the edge of the Karoo (Dwyka) outcrop (Fig. 17) suggests that they may be related to the southeasterly retreat of this outcrop and the consequent down-wasting and concentration of the contained diamonds. In this case not all the diamonds need to have come from the Dwyka: as already discussed there is the potential for all marine Karoo deposits to have shorelines and thus related placers (e.g. Behr 1965).

As the erosion of these Karoo sediments is likely to have commenced in post-Gondwana times, it is possible that an extensive drainage developed on the craton during this wetter period (which built substantial deltas on the Atlantic coast). This may therefore have been the beginning of the period of transient placer build-up both on the craton and adjacent areas prior to the uplift which resulted in the rejuvenation and incision of the Orange–Vaal Rivers and their tributaries. Diamond-bearing gravels dating back to at least to the Cretaceous and located well into the drainage basin (de Wit *et al.* 2000; Bamford 2000), may be remnants of this earlier phase of concentration with sources in pre- and post-Karoo kimberlites as well as the Dwyka Group sediments.

The major concentration of residual and transient placers along the Vaal River and in other drainage areas to the immediate north (Fig. 17) have yielded more than 16 million carats with average stone sizes often greater than 0.5 carats/stone and with maximum reaching 511 carats (de Wit 1996). The spatial and temporal range of these deposits and the heterogeneous diamond populations highlights the range of potential sources for these diamonds, the Dwyka being a possible one.

With respect to the source of Orange River diamonds this leaves two points to consider. First, many diamond-bearing kimberlites were intruded post-Dwyka and a number of these are in the Orange River catchment. Second, pre-Dwyka kimberlites still carry diamonds, thus remaining a potential source for the whole of post-Dwyka times.

The Namaqualand mega-placer

The post-Gondwana geological history of the Namaqualand mega-placer is less well understood than its Namibian counterpart, resulting in some diversity of opinion over its origin. Dingle & Hendy (1984), along with earlier writers, proposed that the present Olifants-Sout drainage marks the Atlantic mouth of an earlier drainage network that included the Orange–Vaal systems. De Wit (1999), extending the case for this Cretaceous outfall, naming this earlier drainage the Karoo River.

The evidence for the existence of a major river mouth in this southern Namaqualand is based on the following:

1   The presence of a deltaic sequence near to the present Olifants mouth (Dingle *et al.* 1983; Brown *et al.* 1995) of Early Aptian–Cenomanian age (112–93 Ma; Fig. 21).

2   A considerable deflection in the coastal escarpment at the Olifants-Sout exit (Fig. 18).

3   Lines of pans that may trace old drainage

routes (Dingle & Hendy 1984; de Wit 1999) and

4   A substantial volume of sand in both the coastal strip and immediately inland, with zircon ages 2.7 Ga–130 Ma (Rosendaal *et al.* 2002), and which could have a provenance partly in Upper Karoo sediments eroded after the break-up of West Gondwana.

The current total drainage area of the streams originating in the escarpment and delivering sediment to the Namaqualand offshore is 30 000 km^2 and for the Karoo River drainage basins is estimated to be 68 400 km^2 (Dingle & Hendy 1984). This compares with 953 200 km^2 for the Orange River, with some evidence that the drainage basin of the latter has changed little since Late Cretaceous (Ward & Bluck 1997).

Seismic sections (Brown *et al.* 1995) are interpreted as a record of a wave-dominated delta with strike-aligned barriers deposited on highstands, and prograding wedges with incised valleys on lowstands. This interpretation is consistent with the absence of characteristic signatures of a major prograding delta (as seen, for example at the Kudu–Orange River delta). In addition, in the younger Kudu–Orange delta (Turonian–Maastrichtian 93–70 Ma), there is a record of dunes building to the NE (Wickens & McLachlan 1990). A similar northeasterly airflow is recorded in the Jurassic Etjo sandstone (Mountney & Howell 2000); the Lower Cretaceous sandstones of the Huab basin (Jerram *et al.* 2000) and the supposed Eocene deposits at Buntfeldschuh (Corbett 1993). It follows that there is every possibility that the sedimentary complex offshore from the Olifants River was deposited under a wave regime and northerly longshore drift similar to, but perhaps less vigorous than, the present one, thus skewing the sedimentary pile to the NNW (Fig. 21). On this basis the Karoo River is seen as a possible source for diamonds in the Namaqualand deposits.

The channels crossing the contemporary Cretaceous alluvial plain were interpreted as lowstand features re-activating possible existing channels in the escarpment by Brown *et al.* (1995). However, Stevenson & McMillan (2004) suggested that these channels that originated in the escarpment, were the principal routes for sediment building the shelf and therefore assigned the principal sediment source to relatively small-headed rivers now seen crossing the coastal plain. If they are the principal sediment source then it is possible that they also delivered diamonds to the coast. These rivers, themselves

not tapping the craton, are required to have derived diamonds from secondary deposits such as the Dwyka Group which covered some of this region.

The total recovered diamonds from the Namaqualand placer up to 1996 was *c.* 42 × 10^6 carats (Oosterveld 2003) but since then, and with unexploited reserves, is >50 × 10^6 and, as with the Namibian placer the diamond population is more than 95% gem quality. Initially discovered in 1926 (Carstens 1962), the placer deposit consists of a compound of beaches, small fluvial channels and weathered gravel of uncertain origin. Diamond grades are highly variable, with some of the highest grades (>200 cpht) in small river channel deposits found in shallow, incised, bedrock valleys and remnant marine basal lag-gravels of Miocene affinity (Pether *et al.* 2000). The deposits have a similar age range to those of the Orange River-derived deposits of Namibia, the oldest dated being Oligocene and the youngest being Mid-Holocene. Remnants of Cretaceous channels onshore (Rogers *et al.* 1990) have yielded no diamonds but diamondiferous gravels are associated with silcretes of possible Late Cretaceous–Early Tertiary age.

As with the Namibian mega-placer, there is a decline in the size of diamonds when traced north from a point near the present Olifants River mouth (Rogers *et al.* 1990; Fig. 18) and, as with the Namibian placer, grades are quite variable along this path of size decline depending upon the local conditions at the time of placer development. Far less numerous than in offshore Namibia, diamonds have also been recovered from the offshore where they are trapped in gravel pockets around ridges of partly lithified, tilted Cretaceous sediment (Kuhns 1995).

There are two possible views on the origin of this placer: that the diamonds are directly derived from rocks formerly covering the coastal plain up to the escarpment and, presumably, were eroded during an assumed scarp retreat. The conduits in this instance are the short reach rivers and the Groen, Swartlintjies and Buffels (Fig. 22) may be the present expressions of them (Stevenson & McMillan 2004). The other sees the interior as the source and the Karoo River as the principal conduit delivering diamonds to the coast for redistribution by reworking.

These two interpretations lead to significantly different view of the source of the diamonds, the origin of the Namaqualand placer and also of the distribution of diamonds in this part of Africa before they were brought to the coast. If the bulk of the sediment came through the Karoo River at the present Olifants mouth, then there is potential for deriving them from a substantial area of the interior, part of which would have contained all primary sources older than and including Group 2 type kimberlites (110–150 Ma). Because of the switch in offshore depocentres at *c.* 95 Ma Kimberley and related Group 1 kimberlites would be excluded.

If they are derived from the small rivers draining the area west of the escarpment then the total potential drainage is small. Given that Namaqualand has yielded *c.* 42 $\times$ 10^6 (up to 1996) carats from a drainage area of *c.* 30 000 km^2, then the source would have yielded 1200 carats km^2 (whatever its thickness). The Orange basin with 953 200 km^2 has yielded >80 $\times$ 10^6 carats (up to 1996) at *c.* 84.4 carats km^2. The Orange basin contains many diamond-bearing kimberlites (including the Kimberly and related clusters), whereas the Namaqualand has no known diamondiferous primaries and in both cases the volumes of diamonds in the offshore are as yet unknown.

In Namaqualand, if the diamonds are derived from the region of the coastal plain to the escarpment, then their source was in a pre-existing Karoo (and particularly the Dwyka) cover or, alternatively, an older sedimentary package such as the Neoproterozic Nama or Gariep rocks now preserved in parts of the escarpment. In order to yield the volumes of diamonds so far recovered, both these deposits would need to have been very rich, indeed far richer in diamonds than previously considered and as such would open up new areas of potential placers which exploration, to date, has failed to find.

The northward diamond size decline would be accounted for if they were transported to the coast by the Cretaceous Karoo River and dispersed northwards under a southerly wind system. Emplacement of the diamonds into structurally controlled, small-scale bedrock channels in the Mesoproterozoic Namaqua gneisses, aligned subparallel and sometimes normal to the vector of size decline, would have occurred during the regional Early–Mid Tertiary uplift. There, small channel systems are interpreted to have redistributed older (Cretaceous and possibly Eocene) diamond-bearing units emplaced formerly in higher coastlines closer to the escarpment. Such a process, although not as pronounced, has been identified in the Orange placer, particularly north of Chameis Bay.

If, on the other hand, the diamonds are derived from the escarpment and related deposits then their northward size decline would either be a reflection of the parent deposit or a response to the longshore drift. In this latter case the size decline would be, with additions of a range of diamond sizes along its length, poorly defined and difficult to achieve if the diamonds were retained within valleys. Moreover, the distal part of the placer would have a diamond distribution with a large spread of sizes.

Conclusions

Taking the mega-placers in Nambia and the Namaqualand as examples, several conditions are required for their formation.

1   The drained craton must be of considerable size, fertile with respect to diamonds and both primary and secondary deposits should be available to the drainage in order to maximize the supply of diamonds. This availability is enhanced if transient placers have already been assembled along the final exit routes.

2   The drainage basin must cover a large proportion of the craton not only to recover the maximum number of diamonds but also to reduce the supply of potentially diluting sediment. Drainage from an uplift adjacent to the craton either by the presence of a destructive plate margin or an incipient rift–passive margin could swamp the potential placer with diluting sediment.

3   There needs to be limited or at best one point of exit for diamonds so that a substantial outfall focuses the delivery to a restricted area.

4   Cratons should have accumulated diamonds over a considerable length of time by deep weathering and recycling within the craton and its immediate boundaries and there should have been few or no previous large-scale drainage systems or glaciations to disperse the diamonds into smaller placers in older deposits.

5   In order to remove the accumulated diamonds efficiently and completely off the craton, the drainage should have sufficient slope and density and, through incision, be able to reach buried residual and primary deposits so as to release them into the drainage network. This is best achieved by regional uplift and preferably through the rejuvenation of a pre-existing river system (to ensure a single or few outlets to the terminal placer). The volume of residual

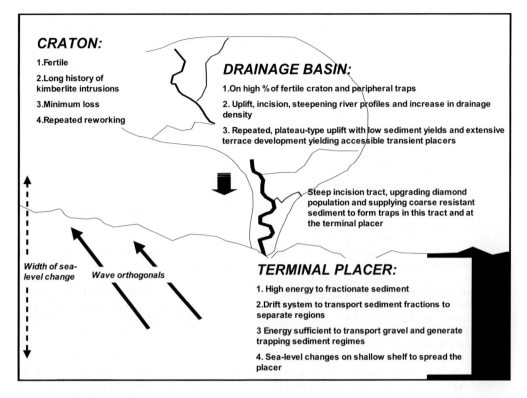

Fig. 23. Summary of the requirements needed to form a diamond mega-placer.

placers and the abundance of diamond-bearing pipes left on the craton is a measure of the inefficiency of the drainage.

6   The drainage should have operated for sufficient time for the drainage network to expand over the surface and cut deeply into it in order to deplete the craton of its diamonds.

7   With respect to this African terminal placer certain critically timed conditions are necessary:
  •   The removal of the Karoo cover in Cretaceous times and its deposition into subsiding basins where it was no longer a potential dilutor to the placer;
  •   The incision of a pre-existing drainage (essentially Cretaceous) to access the newly exposed primary and the already gathered secondary diamond deposits in the Tertiary;
  •   The development of a high-energy coastal regime with a unidirectional wind over a neutrally buoyant, shallow shelf off-the-mouth of the delivery system, during the same time interval and;

  •   that time interval has been of sufficient length to allow a mega-placer to build up.

8   The final dispersal of the diamonds should be relatively young and in largely unlithified deposits so that they can be mined with relative ease.

9   Terminal mega-placer deposits are most likely to have developed along marine coastlines where sufficient energy is need to segregate the grain sizes and concentrate the diamonds.

We have been fortunate to have the unstinting support of two chief geologists, M. Lain and B. Burrell at Oranjemund, as well as generous assistance from De Beers and Namdeb diamond companies respectively. A group of young and dedicated geologists, in particular R. Spaggiari, J. & J. Jacob, and L. Apollus have collected and shared their data and opinions, taken part in numerous discussions and provided great stimulus by their own sense of purpose and focus. Dick Barker shared his extensive knowledge of placer diamonds and set us on the path to a more critical approach to finding them and J. Cartwright made us aware of the offshore record. We are indebted to C. Skinner, G. Alves, P. Hundt, A. Machin & M. Weir

for making the visit of JDW & BJB to Angola in 2000
such a great and enriching experience despite the difficult circumstances. H. Wilson is thanked for drafting
and reviewers K. Leahy & K. S. Viljoen are thanked
for their constructive appraisals.

References

AIZAWA, M., BLUCK, B.J., CARTWRIGHT, J., MILNER,
    S., SWART, R, & WARD, J.D. 2000. *Constraints on the
    geomorphological evolution of Namibia from the
    offshore stratigraphic record*. Geological Survey of
    Namibia (Heno Martin volume) **12**, 337–346.
APOLLUS, L. 1995. *The distribution of diamonds on a
    Late Cainozoic gravel beach, Southwestern
    Namibia*. MSc Thesis. Department of Geology
    and Applied Geology, University of Glasgow.
ARCHER, A.W. & GREB, S.F. 1995. An Amazon scale
    drainage system in the Early Pennsylvanian of
    central North America. *Journal of Geology*, **103**,
    611–628.
ARMSTRONG, R.A., COMPSTON, W., RETIEF, E.A.,
    WILLIAMS, I.S. & WELKE, H.J. 1991. Zircon ion
    microprobe studies bearing on the age and evolu-
    tion of the Witwatersrand triad. *Precambrian
    Research*, **53**, 143–266.
ATKINSON, W.J. 1986. *Diamond exploration phil-
    osophy, practice and promises: a review. Kimber-
    lites and related rocks, 2. Their mantle/crust setting,
    diamonds and diamond exploration*. Geological
    Society of Australia Special Publication **14**,
    1075–1107.
AYER, J.A. & WYMAN, D.A. 2003. *Origin of diamon-
    diferous lamprophyres in the evolution of the
    Michipicoten and Abitibi greenstone belts*.
    Programmes with Abstracts, 8th International
    Kimberlite Conference, Canada, 137.
BAKALOWICZ, M.J., FORD, D.C., MILLER, T.E., PALMER,
    A.N. & PALMER, M.V. 1987. Thermal genesis of
    dissolution caves in the Black Hills, South
    Dakota. *Geological Society of America Bulletin*,
    99, 729–738.
BAKER, C.L. 1982. *Report of the sedimentology and
    provenance of sediments in eskers in the Kirkland
    lake area and the finding of kimberlite float in the
    Gauthier township*. Ontario Department of
    mines, Miscellaneous Paper, **106**, 125–127.
BALLY, A.W. 1989. Phanerozoic basins of North
    America. *In*: BALLY, A.W. & PALMER, A.R. (eds)
    *The Geology of North America volume A. An
    overview*. Geological Society of America,
    Boulder, Colorado, 397–446.
BAMFORD, M.K. 2000. Cenozoic macroplants. *In*:
    PARTRIDGE, T.C. & MAUD, R. (eds) *The Cenozoic
    of Southern Africa*. Oxford Monographs on
    Geology and Geophysics, **40**, 351–356.
BAMFORD, M.K. 2003. Fossil wood from Auchas and
    their Palaeoenvironment. *In*: PICKFORD, M. &
    SENUT, B. (eds) *Geology and Palaeobiology of the
    central and southern Namib, Volume 2. Palaeon-
    tology of the Orange River valley, Namibia*.
    Ministry of Mines and energy. Geological Survey
    of Namibia Memoir, **19**, 23–34.

BANGERT, B., STOLLHOFEN, H., GEIGER, M. & LORENZ,
    V. 2000. Fossil record and high-resolution
    tephrostratigraphy of Carboniferous glacio-
    marine mudstones, Dwyka Group, southern
    Namibia. *Communications of the Geological
    Survey of Namibia*, **12**, 235–245.
BEHR, S.H. 1965. *Heavy mineral beach deposits in the
    Karoo System*. Memoir of the Geological Survey
    of South Africa, **56**, 110 pp.
BRANDL, G. & DE WIT, M.J. 1997. *In*: DE WIT, M.J. &
    ASHWAL, L.D. (eds) *Greenstone Belts*. Oxford
    Monographs in Geology and Geophysics, **35**,
    581–607.
BREMNER, J.M. & WILLS, J.P. 1993. Mineralogy and
    geochemisrtry of the clay fraction of sediments
    from the Namibian continental margin and the
    adjacent hinterland. *Marine Geology*, **115**,
    85–116.
BREMNER, J.M., ROGERS, J. & WILLIS, J.P. 1990. Sedi-
    mentological aspects of the 1988 Orange River
    floods. *Transactions of the Royal Society of South
    Africa*, **47**, 247–294.
BROWN, L.F., BENSON, J.M., BRINK, G.J. DOHERTY, S.,
    JOLLANDS, A., JUNGSLAGER, E.H.A., KEENAN,
    J.H.G., MUNTINGH, A. & VAN WYKE, N.J.S. 1995.
    *Sequence stratigraphy in offshore South African
    divergent basins. An Atlas for exploration of
    Cretaceous low stand traps by Soekor (pty) Ltd*.
    American Association of Petroleum Geologists
    studies in Geology, **41**, 184pp.
BURKE, K. 1996. The African Plate. *South African
    Journal of Geology*, **99**, 339–409.
BUTTON, A. 1973. Early history of the Malmani
    Dolomite in the eastern and northeastern Trans-
    vaal. *Geological Society of South Africa Transac-
    tions*, **76**, 230–247.
CARLSON, J.A., KIRKELY, M.B., THOMAS, E.M. &
    HILLIER, W.D. 1998. Recent Canadian kimberlite
    discoveries. *In*: GURNEY, J.J., GURNEY, J.L.,
    PASCOE, M.D & RICHARDSON, S.H. (eds)
    *Proceedings of the 7th International Kimberlite
    Conference*. Red Roof Designs Ltd, 81–89.
CARLSON, R.W., BOYD, F.R., *ET AL.* 2000. Continental
    growth, preservation and modification in
    Southern Africa. *GSA Today*, **10**, 1–7.
CARSTENS, J. 1962. *A fortune through my fingers*.
    Howard Timmins, Cape Town.
CATUNEANU, O., HANCOX, P.J. & RUBIDGE, B.S. 1998.
    Reciprocal flexural behaviour and contrasting
    stratigraphies: a new basin development model,
    for the Karoo retro arc foreland system, South
    Africa. *Basin Research*, **10**, 417–439.
CHENEY, E.S. 1996. Sequence stratigraphy and plate
    tectonic significance of the Transvaal succession
    of southern African and its equivalent in Western
    Australia. *Precambrian Research*, **79**, 3–24.
CHENEY, E.S. & WINTER, H. DE LA R. 1995. The late
    Archean to Mesoproterozoic major unconfor-
    mity-bounded units of the Kaapvaal province of
    southern Africa. *Precambrian Research*, **74**,
    203–223.
CLEMSON, J. CARTWRIGHT, J. & BOOTH, J. 1997. Struc-
    tural segmentation and influence of basement
    structure on the Namibian passive margin.

Journal of the Geological Society, London, **154**, 477–482.

CLIFFORD, T.N. 1966. Tectono-metallogenic units and metallogenic provinces of Africa. *Earth and Planetary Science Letters,* **1**, 421–434.

COLOMBO, C.G.T. & MACAMBIRA, M.J.B. 1999. Geochronological Provinces of the Amazonian Craton. *Episodes,* **22**, 174–182.

CORBETT, I.B. 1993. The modern and ancient pattern of sand flow through the southern Namib deflation basin. *In*: PYE, K. & LANCASTER, N. (eds) *Aeolian sediments Ancient and Modern.* Special Publications International Association of Sedimentology **16**, 45–60. Blackwell, Oxford.

CORBETT, I.B. 1996. A review of diamondiferous marine deposits of western southern Africa. *African Science Review,* **3**, 157–174.

CORBETT, I.B. & BURRELL, B. 2001. The earliest Pleistocene (?) Orange River fan-delta: an example of successful exploration delivery aided by Quaternary research in diamond placer sedimentology and palaeontology. *Quaternary International,* **82**, 63–73.

CORVINUS, G. & HENDY, Q.B. 1978. A new Miocene vertebrate locality at Arrisdrif in South West Africa. *Neues Jarbuch für Geologie und Paläontologie Monatshefte,* **4**, 193–205.

DE DECKER, R.H. 1988. The wave regime on the inner shelf south of the Orange River and its implications for sediment transport. *South African Journal of Geology,* **91**, 358–71.

DE WIT, M.C.J. 1996. The distribution and stratigraphy of inland alluvial diamond deposits in South Africa. *Africa Geoscience Review* (Special Edition), 19–33.

DE WIT, M.C.J. 1999. Post-Gondwana drainage and the development of diamond placers in western South Africa. *Economic Geology,* **94**, 721–740.

DE WIT, M.C.J. 2004. The diamondiferous sediments on the farm Nooitgedacht (66), Kimberley, South Africa. *South African Journal of Geology,* **107**, 477–488.

DE WIT, M.C.J., MARSHALL, T.R. & PARTRIDGE, T.C. 2000. Fluvial deposits and drainage evolution. *In*: PARTRIDGE, T.C. & MAUD, R.R. (eds) *The Cenozoic of Southern Africa.* Oxford Monographs on Geology and Geophysics, **40**, 55–72.

DE WIT, M.J., ROERING, C., HART, R.J., ET AL. 1992. Formation of an Archaean continent. *Nature,* **357**, 553–562.

DINGLE, R.V. & HENDY, Q.B. 1984. Late Mesozoic and Tertiary sediment supply to the Eastern Cape basins (SE Atlantic) and the palaeo-drainage systems in southwestern Africa. *Marine Geology,* **56**, 13–26.

DINGLE, R.V., SIESSER, W.G. & NEWTON, A.R. 1983. *Mesozoic and Tertiary Geology of Southern Africa.* A.A. Balkema, Rotterdam.

DROZ, L., REGAUT, F., COCHONAT, P. & TOFANI, R. 1996. Morphology and Recent evolution of the Zaire turbidite system (Gulf of Guinea). *Bulletin of the Geological Society of America,* **108**, 253–269.

DU TOIT, A.L. 1951. *The diamondiferous gravels of*

Lichtenberg. Geological Survey of South Africa, Memoir **44**.

FIELD, M. & SCOTT-SMITH, B.H. 1998. Contrasting geology and near surface of kimberlite pipes in southern Africa and Canada. *In*: GURNEY, J.J., GURNEY, J.L.,PASCOE, M.D. & RICHARDSON, S.H. (eds) Proceedings of the 7th International Kimberlite Conference, 214–237.

GALL, O. 1999. Precambrian palaeosols: a view from the Canadian shield. *In*: THIRY, M. & SIMON-COINCON, R. (eds) *Palaeoweathering, Palaeosurfaces and related continental deposits.* International Association of Sedimentologists, Special Publication, **27**, 207–221.

GERMS, G.J.B. 1995. The Neoproterozoic of southwestern Africa with emphasis on platform stratigraphy and palaeontology. *Precambrian Research,* **73**, 137–151.

GOLD, D.P. 1984. A diamond exploration philosophy for the 1980s. *Pennsylvania State University Earth Minerals Science,* **53**, 34–42.

GOODWIN, A.M. 1996. *Principles of Precambrian Geology.* Academic Press, London.

GUNN, C.B. 1968. A descriptive catalogue of the drift diamonds of the Great lakes region, North America. *Gems and Gemology,* **12**, 287–303.

HADDON, I.G. 2000. Kalahari Group sediments. *In*: PARTRIDGE, T.C. & MAUD, R. (eds) *The Cenozoic of Southern Africa.* Oxford Monographs on Geology and Geophysics, **40**, 173–181.

HALLAM, C.D. 1964. The geology of the coastal diamond deposits of southern Africa (1959). *In*: HAUGHTON, S.H. (ed.) *The Geology of some Ore Deposits in Southern Africa.* Geological Society of South Africa, **2**, 671–728.

HOLTAR, E. & FROSBERG, A.W. 2000. Postrift development of the Walvis Basin, Namibia: Results from the exploration campaign in Quadrant 1911. *In*: MELLO, M.R. & KATZ, B.J. (eds) *Petroleum systems of the South Atlantic margins.* American Association of Petroleum Geologists Memoir, **73**, 429–446.

JACOB, R.J., BLUCK, B.J. & WARD, J.D. 1999. Tertiary-age diamondiferous fluvial deposits of the Lower Orange River Valley, Southwestern Africa. *Economic Geology,* **94**, 749–758.

JACOB, J., WARD, J.D. & BLUCK, B.J., SCHOLZ, R.A. & FRIMMEL, H.E. (in press). Some observations on diamondiferous bedrock gully trapsites on Late Cainozoic, marine-cut platforms of the Sperrgebiet, Namibia. *Ore Geology Reviews.*

JANSE, A.J.A. 1993. The aims and economic parameters of diamond exploration. *In*: SHEAHAN, P. & CHATER, A. (eds) *Diamonds: Exploration, Sampling and Exploration.* Proceedings of the Short Course on diamond exploration. Prospectors and Developers Association of Canada, Toronto, 173–184.

JANSE, A.J.A. & SHEAHAN, P.A. 1995. Catalogue of worldwide diamond and kimberlite occurrences. A selective and annotative approach. *Journal of Geochemical Exploration,* **53**, 73–111.

JERRAM, D.A., MOUNTNEY, N., HOWELL, J. & STOLLHOFEN, H. 2000. The fossilized desert:

recent developments in our understanding of the Lower Cretaceous deposits in the Huab basin, NW Namibia. *Communications of the Geological Society of Namibia*, **12**, 269–178.

JORDAN, T.H. 1988. Structure and formation of the continental tectosphere. *Journal of Petrology*, **29**, 11–37.

KAISER, E. 1926. *Die Diamantenwueste Suedwestafrikas*. Dietrich Reimer (Ernst Vohsen), Berlin. Volumes 1 & 2.

KUHNS, R. 1995. *Sedimentological and geomorphological environments of the South African continental shelf and its control on the distribution of alluvial fluvial and marine diamonds*. Society for Mining, Metallurgy and Exploration inc. Proceedings of Annual Meeting, Denver Co.

LANCASTER, N. 2000. Eolian deposits. *In*: PARTRIDGE, T.C. & MAUD, R.R. (eds) *The Cenozoic of Southern Africa*. Oxford Monograph on Geology and Geophysics, **40**, 73–87.

LANCASTER, N. & OLLIER, C.D. 1983. Sources of sand for the Namib sand sea. *Zeitschrift für Geomorphologie N.F.*, **45**, 71–83.

LECKIE, D.A., KJARSGAARD, B.A., BLOCH, J., McINTYRE, D., McNEIL, D., STASUIK, L. & HEAMAN, L. 1997. Emplacement and reworking of Cretaceous, diamond-bearing, crater facies kimberlite of central Saskatchewan, Canada. *Bulletin of the Geological Society of America*, **109**, 1000–1020.

LEVINSON, A.A., GURNEY, J.J. & KIRKLEY, M.B. 1992. Diamond sources and production, past present and future. *Gem and Gemology*, **28**, 234–252.

McMILLIAN, N.J. 1973. Shelves of Labrador Sea and Baffin Bay, Canada. *In*: McCROSSAN, R.G. (ed.) *The future petroleum provinces of Canada – their geology and potential*. Canadian Society of Petroleum Geologists, Memoir **1**, 473–517.

MILANI, E.J. & ZALÁN, P.V. 1999. An ouline of the geology and petroleum systems of the Paleozoic interior basins of South America. *Episodes*, **22**, 199–205.

MAREE, B.D. 1987. Die afsetting en verspreiding van spoeldiamante in Suid-Afrika. *South African Journal of Geology*, **90**, 428–447.

MARSHALL, T.R. 2004. Rooikoppie gravels. *Rough Diamond Review*, **6**, 21–26.

MARSHALL, T.R. & BAXTER-BROWN, R. 1995. Basic principles of alluvial diamond exploration. *Journal of Geochemical Exploration*, **53**, 277–292.

MARTIN, H. & WILCZEWSKI, N. 1970. Paleoecology, conditions of deposition and palaeogeography of the marine Dwyka beds of South West Africa. Second Gondwana Symposium, South Africa. *Geological Society of South Africa*, 225–232.

MARTINI, J. & KAVALIERIS, I. 1976. The karst of the Transvaal (SA). *International Journal of Speliotherms*, **8**, 229–251.

MILLIMAN, J.D. & MEADE, R.H. 1983. World-wide delivery of river sediment to the Oceans. *Journal of Geology*, **91**, 1–21.

MOORE, A.E. & LARKIN, P.A. 2001. Drainage evolution in south-central Africa since the breakup of Gondwana. *South African Journal of Geology*, **104**, 47–68.

MOORE, J.M. & MOORE, A.E. 2004. The roles of primary kimberlitic and secondary Dwyka glacial sources in the development of alluvial and marine diamond deposits in southern Africa. *Journal of African Earth Sciences*, **38**, 115–134.

MOUNTNEY, N. & HOWELL, J. 2000. Aeolian architercture, bedforms climbing and preservation space in the Cretaceous Etjo Formation, NW Namibia. *Sedimentology*, **47**, 825–849.

MURRAY, L.G., JOYNT, R.H., O'SHEA, D.O'C., FOSTER, A.W. & KLEINJAN, L. 1970. The geological environment of some diamond deposits off the coast of South West Africa. *In*: DELANY, M. (ed.) *The geology of the East Atlantic Continental Margin*. ICSU/SCOR Working Party 31 Symposium, Cambridge 1970. Institute of Geological Science Report, 70/3, 110–141.

NIXON, H. 1995. The morphology and nature of primary diamondiferous occurrances. *Journal of Geochemical Exploration*, **53**, 41–71.

NYBLADE, A.A. 1999. *In*: VAN DER HILST, R.D. & McDONOUGH, W.F. (eds) Composition, deep structure and the evolution of continents. *Lithos*, **48**, 81–91.

OOSTERVELD, M.M. 2003. Evaluation of alluvial diamond deposits. *Alluvial diamonds in South Africa workshop*. Geological Society of South Africa. 1–7.

O'REILLY, S.Y., GRIFFIN, W.L., DJOMANI, Y.H.P. & MORGAN, P. 2001. Are lithospheres for ever? Tracking changes in subcontinental lithospheric mantle through time. *Geological Society of America Today*, **11**, 4–10.

PALMER, A.N. 1991. Origin and morphology of limestone caves. *Geological Society of America Bulletin*, **103**, 1–21.

PARMAN, S.W., DANN, J.C., GROVE, T.L., & DE WIT, M.J. 1997. Emplacement conditions for komatiite magmas from the 3.49 Ga Komati Formations, Barberton greenstone belt, South Africa. *Earth and Planetary Science Letters*, **150**, 303–323.

PARTRIDGE, T.C. & BRINK, A.B.A. 1967. Gravels and terraces of the Lower Vaal River basin. *South African Geographical Journal*, **49**, 21–34.

PATCHETT, P.A., EMBRY, A.F., *ET AL*. 2004. Sedimentary cover of the Candadian shield through Mesozoic time reflected by Nd Isotopic and Geochemical results for the Sverdrup Basin, Arctic Canada. *Journal of Geology*, **112**, 39–57.

PETHER, J., ROBERTS, D.L. & WARD, J.D. 2000. Deposits of the West Coast. *In*: PARTRIDGE, T.C. & MAUD, R.R. (eds) *The Cenozoic of Southern Africa*. Oxford Monographs on Geology and Geophysics, **40**, 33–49.

PICKFORD, M. & SENUT, B. 2003. Miocene palaeobiology of the Orange River Valley, Namibia. *In*: PICKFORD, M. & SENUT, B. (eds) *Geology and Palaeobiology of the central and southern Namib. Volume 2 Palaeontology of the Orange River valley, Namibia*. Ministry of Mines and energy, Geological Survey of Namibia, Memoir **19**, 1–22.

POTTER, P.E. 1978. Significance and origin of big rivers. *Journal of Geology*, **86**, 13–33.

RAINBIRD, R.H., HEAMAN, L.M. & YOUNG, G.M. 1992. Sampling Laurentia: Detrital zircon chronology offers evidence of an extensive Neoproterozoic river system originating from the Grenville Orogen. *Geology*, **20**, 351–354.

REYRE, D. 1984. Petroleum characteristics and geological evolution of a passive margin. Example of the Lower Congo-Gabon basin. *Bulletin of the Center for Research, Exploration, Production, Elf Aquitaine*, **8**, 303–332.

ROGERS, J. 1977. Sedimentation on the continental margin off the Orange River and the Namib Desert. *Bulletin Geological Survey/University of Cape Town Marine Geoscience Unit*, **7**, 1–212.

ROGERS, J., PETHER, J., MOLYNEUX, R., HILL, R.S., KILHAM, J.L.C., COOPER, G. & CORBETT, I.B. 1990. *Cenozoic geology and mineral deposits along the west coast of South Africa and the Sperrgebiet*. Guidebook, Geocongress '90. Geological Society of South Africa, 1–111.

ROSENDAAL, A., PHILANDER, C. & ARMSTRONG, R.A. 2002. *Characteristics and age of zircons from diamondiferous and heavy mineral placer deposits along the west coast of South Africa: indicators of sediment provenance*. International Sedimentological Congress Abstracts, Johannesburg, South Africa, 315.

ROSSOUW, J. 1981. *Wave conditions at Oranjemund: Summary of wave rider data*. March 1976–April 1980. Rept.C.S.I.R. Stellenbosch, T/SEA 8106: 1–4.

SCHNEIDER, G.I.C. & MILLER, R. McG. 1992. *Diamonds*. The Mineral Resources of Namibia, 1st edition. Ministry of Mines and Energy, 5.1-1–5.1-32.

SENGOR, A.M.C. & NATAL'IN, B.A. 2001. Rifts of the world. *In*: ERNST, R.E. & BUCHAN, K.L. (eds) *Mantle plumes: Their identification through time*. Geological Society of America Special Paper, **352**, 389–482.

SHIREY, S.B., HARRIS, J.W., *ET AL*. 2002. Diamond genesis, Seismic structure, and evolution of the Kaapvaal-Zimbabwe Craton. *Science*, **297**, 1683–1686.

SIESSER, W.G. & SALMON, D. 1979. Eocene marine sediments in the Sperrgebiet, South West Africa. *South African Museum Annals*, **79**, 9–34.

SOUTH AFRICAN COMMITTEE FOR STRATIGRAPHY (SACS). 1980. *Stratigraphy of South Africa*. Kent, L.E. (comp.); *Part 1. Lithostratigraphy of the Republic of South Africa, South West Africa/Namibia, and the Republics of Bophuthatswana, Transkei and Venda*. Handbook of the Geological Survey of South Africa, **8**, 690 pp.

SPAGGIARI, R.I., BLUCK, B.J. & WARD, J.D. (in press). Characteristics of diamondiferous Plio-Pleistocene littoral deposits within the palaeo-Orange River mouth, Namibia. *Ore Geology Reviews*.

STETTLER, E.H., KLEYWEGT, R.J. & DE WIT, M.C.J.

1995. Geophysical prospecting for diamonds in the Lichtenberg district, Western Transvaal. *Southern African, Geophysical Review*, **1**, 55–69.

STEVENSON, I.B. & MCMILLAN, I.K. 2004. Incised valley fill stratigraphy of the Upper Cretaceous succession, proximal Orange Basin, Atlantic margin of southern Africa. *Journal of Geological Society, London*, **161**, 185–208.

STRATTEN, T. 1979. *The origin of the diamondiferous gravels in the southwestern Transvaal*. Geokongress 77 Geological Society of South Africa Special Publications, **6**, 218–228.

STREEL, M. & THERON, J.N. 1999. The Devonian–Carboniferous boundary in South Africa and the age of the earliest episode of the Dwyka glaciation: new palynological result. *Episodes*, **22**, 41–44.

SUTHERLAND, D.G. 1982. The transport and sorting of diamonds by fluvial and marine processes. *Economic Geology*, **77**, 1613–1620.

TANKARD, A.J., JACKSON, M.P.A., ERIKSSON, K.A., HOBDAY, D.K., HUNTER, D.R. & MINTER, W.E.L. 1982. *Crustal Evolution of Southern Africa 3.8 Billion Years of Earth History*. Springer-Verlag, New York.

VAN WYK, J.P. & PIENAAR, L.F. 1986. Diamondiferous gravels of the Lower Orange River, Namaqualand. *In*: ANHAUSSER, C.R. & MASKE, S. (eds) *Mineral Deposits of Southern Africa*. Geological Society of South Africa, **2**, 2309–2321.

VISSER, J.N.J. 1990. A review of the Permo-Carboniferous glaciations in Africa. *In*: MARTINI, I.P. (ed.) *Global changes in postglacial times: Quaternary and Permo-Carboniferous*. Oxford University Press, Oxford.

WAGNER, P.A. 1914 (reprinted 1971). *The diamond fields of southern Africa*. C. Struik (Pty) Ltd.

WARD, J.D. & BLUCK, B.J. 1997. *The Orange River – 100 million years of fluvial evolution in southern Africa* (abs). International Association of Sedimentologists, International Fluvial Conference, Cape Town, South Africa Abstract, 92.

WARD, J.D. & CORBETT, I. 1990. Towards an age for the Namib. *In*: SEELEY, M.K. (ed.) *Namib Ecology: 25 years of Namib research*. Transvaal Museum Monograph, **7**, 17–26.

WELLINGTON, J.H. 1958. *The evolution of the Orange River Basin: some outstanding problems*. The South African Geographical Society, 3–30.

WICKENS, H., DEV. & MCLACHLAN, I.R. 1990. The stratigraphy and sedimentology of the reservoir interval of the Kudu 9A-2 and 9A-3 boreholes. *Communications of the Geological Society of Namibia*, **6**, 9–22.

WILLIAMS, R. 1996. *King of Sea Diamonds. The saga of Sam Collins*. W.J. Flesch & Partners Cape Town. 176pp.

The formation of economic porphyry copper (–gold) deposits: constraints from microanalysis of fluid and melt inclusions

C. A. HEINRICH, W. HALTER, M. R. LANDTWING & T. PETTKE

Isotope Geochemistry and Mineral Resources, Department of Earth Sciences, Swiss Federal Institute of Technology, ETH Zentrum NO, CH-8092 Zürich, Switzerland
(e-mail: heinrich@erdw.ethz.ch)

Abstract: This paper summarizes our current understanding of the formation of porphyry-style Cu ± Au ± Mo deposits, in the light of data obtained by direct analysis of the ore metals in individual fluid and melt inclusions using laser-ablation ICP mass spectrometry. An integrated study of the evolution of the calcalkaline Farallón Negro Volcanic Complex hosting the Bajo de la Alumbrera porphyry Cu–Au deposit (Argentina), and supplementary fluid-chemical data from Bingham (Utah) and other examples, permit a quantitative re-assessment of the fundamental processes controlling the key economic parameters of porphyry-style ore deposits. Deposit size (total metal content) is optimized by exsolution of a relatively dense (>0.3 g cm^{-3}) single-phase fluid or a two-phase brine + vapour mixture from a moderately large hydrous pluton, possibly with an intermediate step involving the scavenging of the ore-forming elements in a magmatic sulphide melt. Emplacement mechanism, magma-chamber dynamics and possibly an additional source of sulphur are probably more decisive for the formation of a large deposit than sheer pluton volume and elevated Cu contents in the melts. Primary bulk ore grade is determined by temperature-controlled precipitation of ore minerals, which is optimized where a large magmatic fluid flux is cooled through 420–320 °C over a restricted vertical flow distance. Bulk metal ratios of the deposits, exemplified by the economically important Au/Cu ratio in the ore, are primarily controlled by the magmatic source defining the composition of the fluids before they reach the deposit site, although selective precipitation may contribute to metal zoning within orebodies.

Porphyry-style ore deposits constitute our main source of copper and molybdenum and they contribute significant resources of gold and other rare metals such as rhenium. Economically, they are the most important type of hydrothermal deposits that are related directly to hydrous magmas, through a process that is simple in principle but complex in detail – thus giving rise to many variants in composition, age, tectonic setting, structural style, and dominant host rocks among different porphyry ore deposits (Sillitoe 1973, 1997; Burnham & Ohmoto 1980; Titley & Beane 1981; Hedenquist & Lowenstern 1994). In broad terms, all porphyry-style deposits owe their origin to hydrous mantle-derived or lower-crustal melts, usually forming near convergent ocean-to-continent plate boundaries, followed by their ascent and emplacement as intermediate-composition magmas in the upper crust (Richards 2003). Here, magma crystallization and decompression drive the exsolution of excess volatiles, in the form of magmatic-hydrothermal fluids composed dominantly of water and chloride salts. This fluid exsolution is a first main step in the selective enrichment of ore metals and sulphur, compared to the parent magma and its melting source region (Dilles 1987; Candela 1989; Cline & Bodnar 1991). For the second metal-enrichment step in the formation of an economic orebody, the magmatic-hydrothermal fluids must be focused from a large volume of magma into a confined fracture network, which often develops in or above an apophysis or subvolcanic stock in the roof of an upper-crustal intrusion (Titley & Beane 1981; Proffett & Dilles 1984; Dilles & Einaudi 1992). Metals are concentrated by precipitation of ore minerals in a relatively small volume of intensely fractured and chemically altered rock, by a combination of physical and chemical processes that may include fluid decompression and throttling at the lithostatic to hydrostatic transition, phase separation and adiabatic cooling, heat exchange and mixing of magmatic with surface-derived fluids, and chemical reaction of the fluids with the wall-rocks (Burnham & Ohmoto 1980; Fournier 1999).

Successive steps of chemical redistribution of ore metals – first from silicate magma to ore fluids, and finally from ore fluids to metallic minerals – are the key to economic ore formation

From: McDonald, I., Boyce, A. J., Butler, I. B., Herrington, R. J. & Polya, D. A. (eds) 2005. *Mineral Deposits and Earth Evolution.* Geological Society, London, Special Publications, **248**, 247–263. 0305-8719/$15.00
© The Geological Society of London 2005.

(before supergene upgrading which is important for some deposits but not discussed in this paper). Abundant analytical data about ore-metal concentrations in magmatic source rocks and hydrothermal ore samples exist, but information about the concentration and speciation of metals in the mobile phases (silicate melts, locally occurring igneous sulphide melts, and various types of hydrothermal fluids) are more limited. A considerable level of understanding of the chemical behaviour of ore metals has been achieved by high-temperature experimental studies of chloride, sulphur and metal partitioning between silicate melts and saline fluids (e.g. Burnham 1979; Bodnar *et al.* 1985; Candela 1989; Carroll & Webster 1994; Williams *et al.* 1995), and by measurements of the volatility and solubility of ore minerals at subsolidus temperatures (Zotov *et al.* 1995; Williams-Jones *et al.* 2002; Stefánsson & Seward 2004; Liu *et al.* 2001).

Only recently, extensive and systematic data about the actual concentrations of ore metals in natural magmatic melts and evolving hydrothermal fluids have become available thanks to technical progress in the development of microanalytical techniques for analysing fluid and melt inclusions. Texturally-controlled analysis of individual inclusions is now possible using micro-PIXE (Ryan *et al.* 1991), synchrotron-XRF (Vanko & Mavrogenes 1998), and by laser-ablation ICP mass-spectrometry (LA-ICPMS; Günther *et al.* 1998; Heinrich *et al.* 2003), which is currently the most efficient and sensitive technique for studying metal concentrations in single fluid and melt inclusions at parts-per-million or even lower levels (Fig. 1). Following initial application of this technique to a granite-related Sn–W system (Audétat *et al.* 1998, 2000), LA-ICPMS has been used for studying several porphyry-type ore deposits, permitting analysis not only of Cu, Mo and other major and trace-elements, but also of Au in single fluid inclusions (Ulrich *et al.* 1999, 2001; Landtwing *et al.* 2005; Rusk *et al.* 2004). Thanks to the ability of representative sampling and effective homogenization of the bulk element content even of heterogeneous inclusions, LA-ICPMS is suited for quantitative analysis of fluid inclusions with daughter crystals and glassy as well as crystallized melt inclusions (Halter *et al.* 2002*a*; Heinrich *et al.* 2003; Fig. 1).

This paper presents a summary of observations and conclusions obtained from quantitative studies of fluids in porphyry deposits, as well as silicate and sulphide melts in associated magmatic rocks. The main focus is on processes in the Farallón Negro Volcanic Complex in NW Argentina (Fig. 2), where detailed geological data about long-term magma evolution from melt inclusions (Halter *et al.* 2002*b*, 2004*a*, *b*; Harris *et al.* 2004) can be combined with mine-scale observations of the geology and compositional evolution of ore fluids in the Bajo de la Alumbrera porphyry Cu–Au–(Mo) deposit (Sasso & Clark 1998; Ulrich *et al.* 2001; Ulrich & Heinrich 2001; Harris *et al.* 2003; Proffett 2003; Fig. 3). These observations are supplemented by cathodoluminescence petrography and fluid inclusion analyses from Bingham (Utah; Landtwing *et al.* 2005; Redmond *et al.* 2004) and reconnaissance studies of other porphyry-style deposits of variable metal proportion (Ulrich *et al.* 1999; Kehayov *et al.* 2003; Rusk *et al.* 2004). Conclusions regarding the basic economic parameters of porphyry-style ore deposits will be discussed in turn: deposit size or total metal content; bulk ore grade or degree of metal enrichment; and proportion of economic metals such the Au/Cu ratio.

Magma-chamber evolution and deposit size

The size of a porphyry-style ore deposit, i.e. the total mass of metal enriched in a mineable orebody volume, varies from typical values of about 1 Mt copper, 0.1 Mt molybdenum or 10 t gold in presently operating small to medium-sized mines, to 100 Mt Cu (El Teniente; Camus & Dilles 2001), 1.6 Mt Mo (Climax; Carten *et al.* 1993), or 1000 t Au + 20 Mt Cu (Bingham; Krahulec 1997) in some of the worlds biggest deposits of their kind. Explaining these variations and the formation of several of the very largest deposits within certain provinces and during short periods of the Earth's history has remained one of the most elusive questions in global metallogeny. For example, how could 9 of the 16 biggest porphyry Cu deposits of the circumpacific belt be formed in Chile within two short time periods between 40–30 and between 9–5 Ma ago (Camus & Dilles 2001)? On a global to province-scale, the answer probably lies in the tectonic architecture of interacting lithospheric plates and underlying asthenosphere, enabling the emplacement and evolution of hydrous magma reservoirs in the middle to upper crust (Richards 2003). The size of a porphyry deposit depends on the dimensions of this magmatic source for fluids and ore-forming components, but also on the efficiency by which ore-metals are extracted into a fluid with the right physical and chemical properties for final

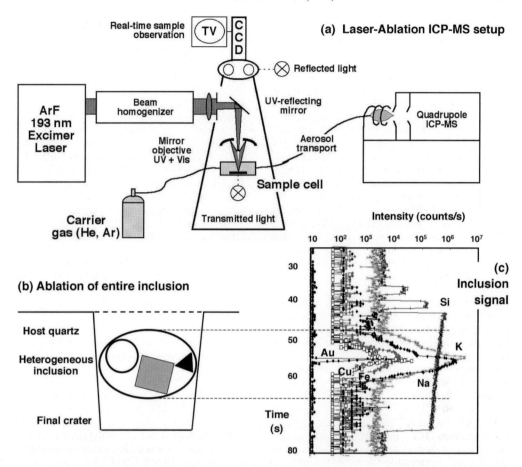

Fig. 1. The principle of laser-ablation ICP mass spectrometric analysis of microscopic fluid and melt inclusions. (**a**) Experimental set-up, using a UV laser whose high-energy photon beam (193 nm wavelength) is imaged onto the sample surface, using a mirror-optical microscope that allows concurrent observation of the sample in the visible wavelength range. (**b**) The sample is a small mineral plate containing fluid or melt inclusions a few tens of microns under its polished surface. The circular ablation spot can be adjusted in size to that of the inclusion, typically 8–40 µm in diameter. Successive ablation from a circular, flat-bottomed crater generates an aerosol in the small sample chamber, which is continually flushed by a carrier gas stream (pure He and/ or Ar) to the ICP torch of a commercial quadrupole ICP mass spectrometer, where the multi-element composition of the aerosol is recorded. Successive ablation of host mineral and inclusion from the deepening crater thus generates a time-resolved record of element intensities (**c**), starting with the pure host mineral (quartz in this case) followed by ablation of the inclusion contents. The inclusion represented by (**c**) contained a vapour bubble, a saline liquid phase and at least two daughter crystals that were precipitated internally during cooling from magmatic-hydrothermal temperatures (like the brine inclusions in Fig. 4c, below). Although the halite (Na peak) and the Au-bearing copper sulphide daughter crystals (overlapping Au and Cu spikes) are clearly resolved from the broader Na and Cu signals originating from the elements dissolved in the aqueous liquid, time integration of element intensities and calibration using an external standard yields consistent element ratios in an assemblage of several coeval inclusions. Absolute element concentrations are calculated by combination of these ratios with the NaCl (wt% equivalent) salinity determined by microthermometry prior to ablation of the inclusion (modified from Heinrich *et al.* 2003).

concentration in an economic orebody. In the Farallón Negro Volcanic Complex of NW Argentina, we have followed the ore-forming components through the compositional evolu-

tion of melt, mineral and fluid phases, to learn more about the processes in the source of porphyry-mineralizing fluids and to constrain the metal mass-balance in the formation of the

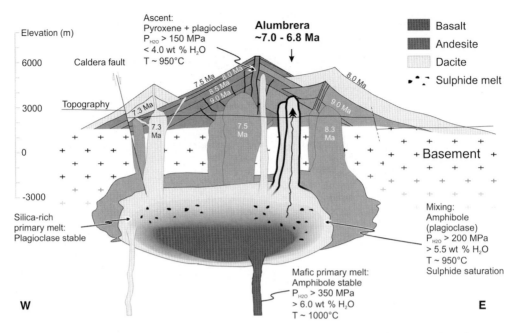

Fig. 2. Reconstructed cross-section of the Farallón Negro Volcanic Complex at the time of porphyry Cu–Au mineralization at Bajo de la Alumbrera, interpreted from stratigraphic mapping relations and geochronological constraints among extrusive and subvolcanic intrusive stocks. Bulk-rock and melt-inclusion compositional evidence indicates generation of andesitic rocks (grey) during magma ascent through evolving, compositionally structured reservoirs in which dacitic (white) and hydrous basaltic magmas (black) mixed in variable proportions. A complex stratovolcano, whose main feeder had been west of Alumbrera during preceding volcanic activity, generated an estimated lithostatic fluid pressure of c. 1 kb for pre-sulphide quartz–magnetite veining and potassic alteration, and variable lower temperatures during subsequent Cu–Fe sulphide precipitation at hydrostatic or even vapourstatic conditions (Ulrich et al. 2001). The approximate size and location of the subvolcanic magma chamber at the time of mineralization is inferred from petrological and mass-balance constraints, and from field indications for a partial caldera collapse which terminated extrusive activity some 0.4 Ma before porphyry Cu–Au formation (after Halter et al. 2004a, modified according to new drilling evidence for crystalline basement around 500 m below Bajo de la Alumbrera; S. Brown, pers. comm. 2004).

associated Cu–Au deposit of Bajo de la Alumbrera.

Magma evolution and ore fluid generation in the Farallón Negro Volcanic Complex

The Farallón Negro Volcanic Complex is a Late Miocene stratovolcano that has been eroded almost down to its base above Palaeozoic basement in the interior of the Andes of NW Argentina. Geological mapping, isotope geochemistry and geochronology (Llambías 1972; Sillitoe 1973; Sasso & Clark 1998; Proffett 2003; Halter et al. 2004a; Harris et al. 2004) document a history of dominantly basaltic–andesitic to later dacitic–rhyolitic, extrusive and subvolcanic intrusive activity (Fig. 2). Silicate melt inclusions reveal that the volumetrically

predominant high-K andesites are hybrid magmas formed by mixing of a hydrous basaltic to lamprophyric melt (selectively recorded by melt inclusions in amphibole phenocrysts) and a dacitic to rhyolitic melt (recorded by melt inclusions in plagioclase and quartz phenocrysts) occurring together in rocks of intermediate andesitic bulk composition (Fig. 4a; Halter et al. 2004b). The dacitic to rhyolitic melt component becomes more prominent in porphyries emplaced during the later stages of c. 2.1 Ma of extrusive andesite volcanism. This compositional evolution is interpreted to reflect the gradual build-up of a subvolcanic magma chamber, which is not exposed but was probably internally heterogeneous (Fig. 2). The more mafic melts contain about 200 ppm Cu, and the copper content of all melt inclusions is about five times higher than the bulk copper content

Fig. 3. View of the Bajo de la Alumbrera deposit in 1994, before mining started. The centre of intense quartz–magnetite veining, potassic alteration and high-grade Cu–Au mineralization stands out as reddish-brown hills, surrounded by a wide halo of later feldspar-destructive alteration (phyllic-argillic; cream white). Bedded andesites, intruded and propylitically altered by the polyphase porphyry stock, crop out in the ring of dark hills around the Alumbrera depression. In the background, stratigraphically higher volcanics dip away towards the northern flank of the former stratovolcano.

of extrusive or intrusive rocks of similar silica content (Fig. 5; Halter *et al.* 2002*b*). All types of phenocrysts, constituting together about 30 vol% of most volcanic rocks, were analysed by LA-ICPMS and found to contain negligible concentrations of copper. These observations and the consistent difference between Cu in melts and in bulk rocks require that 60–80% of the copper initially present in the matrix melt of the andesitic rocks was lost during volcanic eruptions, probably to a low-density volatile phase.

The matrix of all fresh rocks is sulphide free, in contrast to most of the barren subvolcanic intrusions which preserve minute inclusions of iron-rich sulphide melt in amphibole phenocrysts. These sulphide melt inclusions are enriched in copper and gold (*c.* 0.8% Cu and 1 ppm Au; Halter *et al.* 2002*b*, 2005; see Fig. 8, below), and coexist with almost totally Cu-depleted silicate melt inclusions in the same phenocrysts (Fig. 5, black triangles). This

indicates that Cu and Au were scavenged, at least transiently in some parts of the magma chamber, by sulphide melt droplets exsolving from an initially more Cu- and probably Au-rich mafic to intermediate silicate melt. Unless protected inside phenocrysts, the sulphide melt droplets in the silicate melt were decomposed again during extrusive eruption and subvolcanic emplacement, as a result of desulphidation attending fluid saturation (Keith *et al.* 1997).

The Bajo de la Alumbrera deposit was formed by several pulses of porphyry emplacement and focused magmatic fluid flow towards the end of this protracted igneous evolution. Ore formation required large-scale fluid saturation of the subvolcanic magma chamber and exsolution of a major quantity of a relatively dense magmatic fluid. This fluid is recorded by brines of about 50% NaCl (eq.) containing around 0.3 wt% Cu, which were trapped in high-temperature inclusions found as widespread and compositionally consistent inclusion

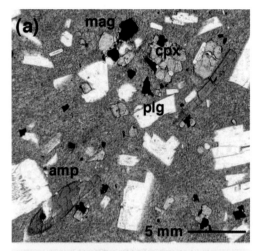

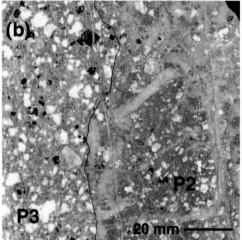

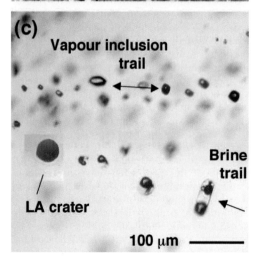

assemblages in quartz ± magnetite stockwork veins (Ulrich *et al.* 2001). Such veins successively cut three distinct porphyry phases, exhibiting decreasing intensities of veining and chalcopyrite ± bornite mineralization (Fig 4b; Ulrich *et al.* 2001, based on deposit-scale geology by Proffett 2003). These 50% salinity fluids were probably trapped initially in the single-phase fluid stability field at >600 °C and confining pressures of >1 kbar, whereas later brines were entrapped together with vapour inclusions on 'boiling trails', at lower pressures indicated by even higher brine salinities up to 70 wt% NaCl equivalent (similar to Fig. 4c). Assuming lithostatic conditions for the earliest brines, a minimum overburden of approximately 3 km above the deposit is required, irrespective of whether these brines exsolved directly from the underlying magma or were derived from an even deeper fluid precursor of

Fig. 4. Typical rock, mineral and fluid inclusion relationships observed in porphyry copper systems: (**a**) thin section of a barren subvolcanic andesite porphyry from the Farallón Negro Volcanic Complex, containing plagioclase (plg) and amphibole (amp) phenocrysts that are effective 'xenocrysts', because they grew from different melts prior to magma mixing: amphibole contains basaltic–andesite melt inclusions, plagioclase contains dacitic melt inclusions; only clinopyroxene (cpx) contains intermediate-composition andesite melts similar to the bulk rock composition because it crystallized after magma mixing (Halter *et al.* 2004*b*). (**b**) Two generations of successively intersecting magmatic intrusions and stockwork veins from the Bajo de la Alumbrera deposit; quartz–magnetite ± bornite ± chalcopyrite veins mineralizing the earlier P_2 porphyry are truncated by a later P_3 porphyry intrusion (left side of photo, emphasized by thin black line) affected by less intense alteration and veinlets with chalcopyrite + pyrite + minor quartz that cuts the igneous contact; in both vein generations, Cu–Fe sulphides precipitated late compared with most of the quartz. (**c**) Typical association of high-temperature fluid inclusions in porphyry stockwork veins (Rosia Poieni, Romania): brine inclusions (containing liquid, a gas bubble, a halite cube, a red hematite crystal and a smaller Cu-rich opaque daughter crystal), and vapour inclusions (large dark bubble and a small invisible fraction of liquid phase wetting the inclusion wall, reflecting the lower density of the high-temperature vapour phase). One of the brine inclusions has been 'drilled out' for quantitative laser-ablation ICP-MS microanalysis (dark crater with a diameter of 40 μm; inset photographed with image plane on the sample surface).

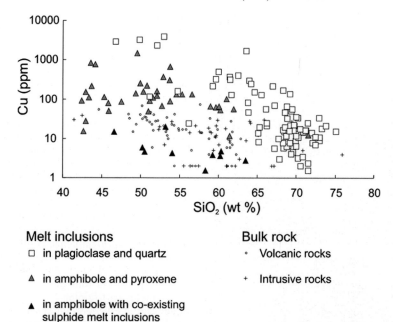

Melt inclusions

□ in plagioclase and quartz

▲ in amphibole and pyroxene

▲ in amphibole with co-existing
sulphide melt inclusions

Bulk rock

∘ Volcanic rocks

+ Intrusive rocks

Fig. 5. Cu concentration (on a logarithmic scale) against silica content, comparing bulk rocks and silicate melt inclusions from the Farallón Negro Volcanic Complex. For a given SiO_2 content, bulk magmatic rocks (small symbols for volcanics and subvolcanic intrusives) typically contain 3–10× lower Cu concentrations than silicate melt inclusions in amphibole and pyroxene (grey triangles) and quartz or plagioclase (squares). By contrast, silicate melts in amphiboles from barren intrusions, associated with igneous sulphide inclusions in the same phenocrysts (black triangles), are almost totally depleted in Cu, due to ore-metal extraction into a sulphide melt prior to fluid saturation (Halter *et al.* 2002*b*, 2004*b*).

intermediate salinity (cf. Bingham; Redmond *et al.* 2004). This pressure estimate is consistent with a projected stratigraphic thickness of 2.5–2.8 km at a position located about 3 km off the main centre of the preceding volcanic activity (Llambías 1972; Proffett 2003). Geochronological data by U–Pb on zircons (Harris *et al.* 2004) and stepwise heating Ar/Ar methods (Sasso & Clark 1998; Halter *et al.* 2004*a*) are consistent with porphyry emplacement and hydrothermal copper introduction some 0.4 Ma after cessation of extrusive volcanism, which had been terminated by dacitic volcanism and an inferred partial caldera collapse (Llambías 1972; Halter *et al.* 2004*a*). This intervening period of volcanic quiescence probably allowed the magma chamber to evolve towards the final event of large-scale fluid saturation, metal extraction, and Cu–Au mineralization. Harris *et al.* (2004) suggested that the first, incompletely preserved but most richly mineralized porphyry event may have occurred around 1 Ma earlier, but the compositional and petro-

graphic similarity of all mineralized porphyries (Proffett 2003) and the possibility of zircon growth significantly before magma emplacement (Lowenstern *et al.* 2000; Oberli *et al.* 2004) are consistent with the interpretation that all three mineralized porphyry phases have been emplaced in relatively short succession, after completion of extrusive magmatism (Halter *et al.* 2004*a*).

Fluid source processes controlling the size of porphyry deposits

The bulk Cu content of the Farallón Negro igneous rocks (Fig. 5) is typical for or even lower than that of many calcalkaline basaltic to dacitic rocks world-wide (e.g. Kesler 1997; Kamenetsky *et al.* 2001), implying that bulk ore-metal contents even of fresh lavas are not a good indicator for the ore-forming potential of a magmatic complex. The observation that copper contents of the melts are significantly higher than those of the equivalent bulk rocks, without significant incorporation into phenocryst phases, shows that melt inclusions are a much better measure of ore metal availability. Our observations also imply that copper dispersion by low-density fumarolic vapour is probably the norm in calcalkaline magmatism, consistent with significant copper transport in violently erupted volcanic gases (Taran *et al.* 1995) and local Cu enrichment in low-density vesicle fillings in lavas and melt inclusions (Lowenstern *et al.* 1991; Lowenstern 1993). By contrast, the generation of a sufficiently dense ore fluid (*c.* 0.3–1 g cm^{-3}, depending on salinity) suitable for ore formation requires specific conditions of wholesale volatile saturation in a confined magma source volume at elevated pressures. The analysed copper contents in the silicate melt inclusions imply that quite moderate melt volumes are sufficient to source the copper contained in a large porphyry deposit. For example *c.* 15 km^3 of mixed magma with a density of 2.4 g cm^{-3}, containing 70% melt that loses 120 ppm Cu to an ore fluid upon crystallization, would be adequate to source the 3 Mt Cu contained in the Alumbrera orebody. This minimum estimate is lower than the actual magma volume involved, because the geological Cu resource is at least twice this value and sourcing the gold may well require a larger magma source (Ulrich *et al.* 1999). Nevertheless, the required magma volume is only a small fraction of the total quantity of magma produced during the lifetime of the Farallón Negro Volcanic Complex, which is estimated to be in the order of 300 km^3 (Halter *et al.* 2004a). From the total volume of volcanics, about 80 Mt Cu (comparable to the metal content of the giant El Teniente deposit) must have been dispersed in the atmosphere or the local groundwater, prior to the event that formed the Alumbrera deposit.

The complete extraction of Cu from a silicate melt into a sulphide melt containing about 0.8% Cu requires an amount of sulphur that must be around forty times higher than the initial Cu content of the silicate melts. The estimated concentration of about 5000 ppm S probably exceeds the equilibrium solubility of sulphur in any upper-crustal silicate melt, be it as sulphide in the mafic melt or as sulphate in the felsic mixing component (Halter *et al.* 2005). The presence of the Fe–S-rich sulphide melt inclusions associated with almost totally Cu-depleted silicate melts thus requires addition of sulphur to at least some parts of the complex magma chamber, and this unknown process could well be decisive for the formation of a large porphyry copper deposit. Sulphur addition may be effected by an SO_2 and/or H_2S-rich volatile phase ascending from depth (e.g. from a large volume of underlying mafic magma; cf. Wallace & Gerlach 1994), or by addition of solid anhydrite from an unusually sulphate-rich felsic magma and its reduction to sulphide + magnetite attending magma mixing with the FeO-rich mafic magma (Halter *et al.* 2005; see Fig. 2).

Combined observations from the Farallón Negro Volcanic Complex show that the physical and chemical processes of metal extraction from silicate melt into an effective ore fluid, and perhaps the availability of an additional sulphur source, are probably more decisive for the formation of a large porphyry deposit than sheer magma volume or total ore-metal availability in an exceptionally Cu ± Au-rich magma. The potential chemical influence of sulphide melts as an intermediate metal scavenger (Spooner 1993; Keith *et al.* 1997; Halter *et al.* 2002b) challenges the interpretation that optimal conditions for direct metal partitioning from the melt to the fluid occur in magmas that are so oxidized to be sulphide undersaturated (Dilles 1987; Cline & Bodnar 1991; Core & Kesler 2001; Audétat *et al.* 2004). Possibly more important than the presence or absence of igneous sulphide are the salinity, density and phase state of the hydrothermal fluid during volatile exsolution (Audétat & Pettke 2003). Efficient metal partitioning of chloride-complexed metals including Cu, together with volatile sulphur species required for later Cu–Fe-sulphide precipitation, are favoured by initial fluid saturation of a magma with high Cl/OH at elevated pressure (probably >1 kbar in general; cf. Cline & Bodnar 1991), yet not too deep to prevent efficient focusing into brittle structures by hydraulic fracturing in the upper crust (Burnham & Ohmoto 1980; Fournier 1999). Wholesale or convection-driven (Shinohara *et al.* 1995) fluid-saturation from a significant magma reservoir, at elevated pressures and without continual loss of magmas and low-density volatiles to the surface, seems to be a key physical prerequisite for generating a large volume of relatively dense (>0.3 g cm^{-3}) fluid able to produce a large porphyry Cu-(Mo–Au) deposit. Magma ponding, the exsolution of dense magmatic fluids, and transient accumulation of fluids in the top of the magma chamber, are all favoured by a neutral to compressive upper-crustal stress regime (Tosdal & Richards 2001). As a result, many large porphyry deposits correlate with subduction reversals or tectonic basin inversions, and with the cessation of extrusive activity prior to hydrothermal ore formation (Solomon 1990; Sillitoe 1997; Camus & Dilles 2001; Rohrlach 2003; Halter *et al.* 2004a).

Metal precipitation efficiency and ore grade

The bulk ore grade of a deposit is defined as the fraction, in wt% Cu for example, of total metal contained in a certain mass of economically mineable ore. Primary ore grades in porphyry copper deposits are affected by the typical complexity of intrusions containing variable metal concentrations introduced in multiple magma and fluid pulses (Fig. 4b). Commonly, the earliest mineralized porphyry contains the highest metal concentrations, and subsequent less-intensely mineralized porphyries contribute to overall ore dilution rather than to upgrading of copper concentrations, despite additional metal introduction (e.g. Gustafson & Hunt 1975; Proffett 2003; Redmond et al. 2004). Thus, Bingham partly owes its exceptionally high grade to a single dominant pulse of hydrothermal activity that mineralized the main quartz monzonite porphyry, followed by volumetrically minor less-mineralized dykes (Redmond et al. 2004). The lower bulk grade of Alumbrera is partly due to the dilution of the most strongly mineralized 'P2' porphyry by volumetrically dominant but somewhat less intensely mineralized later porphyries. Sharp grade differences at intrusive contacts indicate that each mineralizing magmatic-hydrothermal pulse was essentially completed before the next one started (Proffett 2003). For one pulse of hydrothermal copper introduction into a fractured porphyry intrusion and its surrounding and overlying wall rocks, the resulting ore grade is a function of the efficiency of metal deposition within a certain ore volume (Fig. 6). For an approximately cylindrical orebody of any horizontal cross-section, upward focusing of magmatic fluid generates an integrated fluid flux corresponding to the total mass of available fluid (m_{fluid}) divided by the cross-section (A) of the mineralized ore zone, as indicated by Equation (2) in Figure 6. Ore grade is related to the amount of initially dissolved metal focused into this orebody volume ($m_{fluid} \times C_{Cu}^{fluid}$), and to the efficiency by which the advected metal is trapped within this volume by precipitation of hydrothermal ore minerals (precipitation efficiency, X_{Cu}). Equation 2 in Figure 6 further illustrates that ore grade is maximized by high precipitation efficiency over a preferably short flow distance (d), for a given cross-section of the orebody.

Fluid evolution and Cu–Fe–sulphide precipitation at Alumbrera and Bingham

LA-ICPMS microanalysis of fluid inclusions has provided extensive data on the initial concentration of Cu (and sometimes Au and Mo) as well as many other element concentrations in porphyry-mineralizing fluids of highly variable density and total salinity. Single-phase fluids of

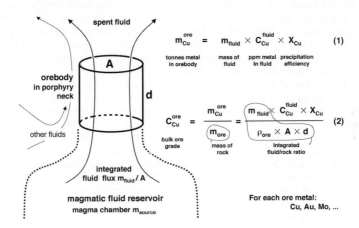

Fig. 6. Schematic cartoon of a cylindrical porphyry copper orebody forming in a zone of intensely hydrofractured rocks above a larger pluton, illustrating the basic proportionalities relating the size (metal tonnage m_{Cu}; Eqn 1) and the metal grade of an ore deposit ($C_{Cu,ore}$; Eqn 2) to the key parameters of ore formation: ore-fluid composition (e.g. copper concentration in the incoming fluid, C_{Cu}^{fluid}), metal extraction efficiency by ore-mineral precipitation (X_{Cu}) and the dimension of the orebody in terms of horizontal cross-section (A) and vertical extent along the flow distance (d) of the fluids transporting the metals into the orebody. Equations relate to mass-balance only and fundamentally apply irrespective of orebody shape, which commonly is not a circular cylinder and has lateral boundaries that may or may not coincide with intrusive contacts of a porphyry stock.

near-critical density and a total salinity of 3–4 wt% NaCl (eq.), trapped in early stockwork quartz veins at Butte (Montana) contain about 1 wt% Cu as one of four dominant cations (Na, K, Fe and Cu; Rusk *et al.* 2004). Comparable compositions with regard to salinity and major cations including Cu are observed in vapour inclusion assemblages from Grasberg (Ulrich *et al.* 1999: *c.* 1.2% Cu), Bajo de la Alumbrera (Ulrich *et al.* 2001: up to 3% Cu), Bingham (Landtwing *et al.* 2005: Cu/Na *c.* 0.8, with absolute concentrations less certain due to the presence of CO_2), and Elatsite (Bulgaria; Kehayov *et al.* 2003; *c.* 2% Cu). In all orebodies except Butte, pre-sulphide quartz also contains high-salinity brine inclusions, indicating fluid phase separation well before the onset of ore-mineral precipitation, or introduction of hyper-saline brine as an original single-phase fluid (Ulrich *et al.* 2001). These brines contain higher concentrations of Na, K and Fe in the form of chloride salts, but comparable or slightly lower Cu concentrations compared with coexisting vapour inclusions (average Alumbrera *c.* 0.3, Grasberg *c.* 0.3, Bingham *c.* 0.8, Elatsite *c.* 3 wt% Cu, with overlapping ranges; Heinrich *et al.* 1999). Locally, even higher Cu concentrations up to 10 wt% were measured (see also Harris *et al.* 2003), but such extreme values probably reflect local re-dissolution of existing copper sulphides by subsequent fluid pulses (Landtwing *et al.* 2005). Extensive LA-ICPMS data indicate that the input concentration of Cu in porphyry-copper forming ore fluids, of either phase state, is between 0.3 and 3 wt%. The copper content in the fluids is therefore of the same order of magnitude as typical ore grades in porphyry Cu deposits (both in weight percent), implying that the minimum integrated fluid to rock ratio attending porphyry mineral-ization can be as small as 1 (Ulrich *et al.* 2001). (The fluid:rock ratio is the mass of fluid required for ore formation, divided by the mass of rock in which all the advected and precipitated metal is trapped ($m_{fluid}/m_{ore} \sim m_{fluid}/\rho_{ore} \times A \times d$).) Similarly modest integrated fluid:rock ratios were estimated for porphyry-related alteration reactions (Ulrich & Heinrich 2001), which contrasts with the much higher minimum inte-grated fluid:rock ratios envisaged for metamor-phic gold or sediment-hosted base metal deposits where metal concentrations in the fluid are assumed to be lower and therefore quanti-ties of fluid required for ore formation are much higher. The relatively small amount of extremely metal-rich magmatic fluids involved in the formation of porphyry copper (–Au–Mo) deposits further emphasizes the importance of

processes at the magmatic source of such compositionally specialized fluids, but it also emphasizes that ore formation requires an effi-cient and possibly quite selective metal-precipi-tation process.

In conjunction with petrographic and microthermometric data, quantitative LA-ICPMS microanalysis of fluids has helped us to clarify the mechanism and conditions of ore mineral precipitation at Bajo de la Alumbrera (Ulrich *et al.* 2001) and at Bingham (Landtwing *et al.* 2005). In both deposits, the precipitation of chalcopyrite and/or bornite with gold occurred after most of the quartz deposition in stockwork veins, as shown most clearly at Bingham thanks to extensive cathodoluminescence petrography (Redmond *et al.* 2004). Gold and Cu–Fe sulphide co-precipitation was caused by cooling of a two-phase magmatic fluid (coexisting brine + vapour) within a narrow temperature interval. Sulphide precipitation is recorded by a decrease of two orders of magnitude in the copper concentration, in fluids containing essentially constant concentrations of all other, non-precip-itating elements including chloride and Na, K, Rb, Cs, Zn, Pb and others. The variation in Cu content in the fluids trapped at one location is therefore interpreted to reflect the larger-scale evolution of a fluid of essentially constant input composition flowing through the region of ore deposition. Although textural relations of the analysed fluids at Alumbrera were somewhat ambiguous due to the complex intrusion history (Ulrich & Heinrich 2001), more clearcut relations between fluids and minerals are documented at Bingham (Redmond *et al.* 2004; Landtwing *et al.* 2005), as summarized in Figure 7.

Cathodoluminescence petrography of main stockwork veins at Bingham (Fig. 7a) reveals a consistent separation of an early bright lumi-nescing quartz generation that predates the deposition of bornite, chalcopyrite and textu-rally associated gold, from a later generation of dull-luminescing quartz that largely postdates metal precipitation (Fig. 7a), that is similar to observations in quartz–molybdenite veins at Butte made by Rusk & Reed (2002). A steep decrease in Cu concentration in otherwise identical fluids with falling homogenization temperatures (Fig. 7b) is recorded by numerous assemblages of secondary brine inclusions in the pre-sulphide quartz. A decrease of similar magnitude in copper is also observed among coeval vapour inclusions hosted by the early quartz, and the vapour phase probably repre-sented the dominant ore fluid in terms of absolute   mass   and   metal   contribution

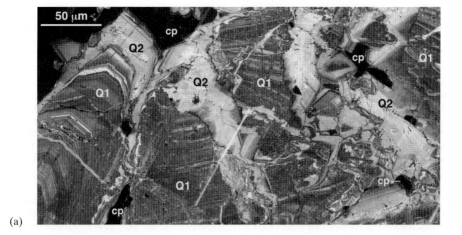

(a)

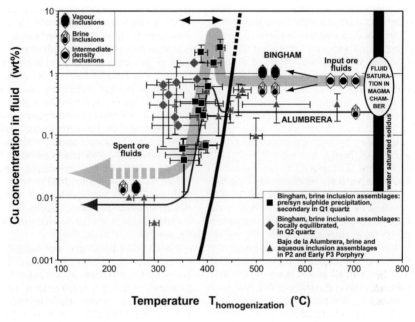

(b)

Fig. 7. (**a**) Cathodoluminescence (SEM-CL) image of quartz crystals in a main-stage stockwork vein in the Bingham Cu–Au orebody (Utah), differentiating an otherwise invisible early generation of vein quartz (Q1) from a later generation of quartz Q2 overgrowing the Cu–Fe sulphide grains, which precipitated during an intervening time of vein fracturing and secondary permeability creation by quartz dissolution. Cu precipitation and quartz dissolution can occur simultaneously by fluid cooling through the 410–350 °C temperature interval at near-hydrostatic pressures below *c.* 500 bar, as recorded by secondary inclusions trapped in the earlier Q1 quartz. (**b**) Concentrations of Cu in brine inclusions in main-stage stockwork vein from the high-grade centre of the Bingham orebody, as a function of homogenization temperature *T* (closely corresponding to trapping temperature as indicated by locally coexisting vapour inclusions). Symbols subdivided according to host quartz generation Q1 and Q2 mapped using CL petrography (see a), with error bars denoting standard deviation of several inclusions in one assemblage (Landtwing *et al.* 2005). Similar data from Alumbrera (without CL-petrographic control; Ulrich *et al.* 2001) are shown by small triangles and the black trendline with arrow. The heavy steep line parallel to Q1-hosted inclusion trend marks the strong temperature dependence of calculated chalcopyrite + bornite solubility (absolute position depending on assumptions about Cl, S and O_2 activities; Hezarkhani *et al.* 1999).

(Landtwing *et al.* 2005). The estimated pressure (140–210 bar) and temperature interval in which Cu is extracted from the fluid (425–350 °C) corresponds to the *P–T* region where quartz solubility is retrograde (i.e. quartz becomes more soluble with cooling; Fournier & Thompson 1993; Manning 1994) while Cu–Fe sulphide solubility decreases steeply with falling temperature (Hezarkhani *et al.* 1999; Liu *et al.* 2001). Ore deposition was facilitated by textural reactivation of the stockwork veins and the generation of secondary permeability, into which the ore metals were precipitated without the obstruction of co-precipitating gangue minerals. After fluid throughput had waned, post-ore dull-luminescing quartz finally rece-mented the veins by local re-equilibration of pore fluids at slightly lower temperature, with local grain boundary adjustment between quartz crystals and the typically anhedral Cu–Fe sulphides. Some of these fluids still have variably high copper contents, interpreted to reflect final cooling during occlusion of the remaining pore space, by increasingly acid fluids in contact with Cu sulphides but isolated from pH buffers in the wall rock (Rusk & Reed 2002; Landtwing *et al.* 2005; Redmond *et al.* 2004).

Fluid cooling as the first-order factor controlling ore grade

The recent results from Bingham confirm the original conclusion from Bajo de la Alumbrera (Ulrich *et al.* 2001) and reconnaissance data from Elatsite (Kehayov *et al.* 2003), indicating that fluid cooling through a small interval at relatively low temperatures is the main driving force for copper ± gold deposition in porphyry deposits. This is in agreement with experimental evidence, which is reliable at least with regard to the steep temperature dependence of Cu–Fe sulphide solubility (Crerar & Barnes 1976; Hemley *et al.* 1992). Experimental solubility data would have to be wrong by many orders of magnitude to permit Cu–Fe sulphide saturation at temperatures above 700 °C, as suggested by Harris *et al.* (2004) on the basis of a few high-temperature inclusions with highly variable copper contents. At Butte, single-phase fluids of intermediate density precipitated chalcopyrite + pyrite at somewhat higher pressures, prior to phase separation into brine and vapour. Ore deposition was associated with sericite alteration of plagioclase-rich rocks, which is indicated by an antithetic correlation between Na and Cu in the ore-depositing fluids (Rusk *et al.*

2004). Temperatures are less well-defined in the single-phase field and may be a little higher than 420 °C, but fluid cooling is again likely to drive acid-consuming alteration reactions, which in turn are likely to drive Cu-precipitation by H^+ neutralization (Rusk & Reed 2002 cf. Hemley *et al.* 1992). None of the deposits studied so far show any evidence for fluid mixing by incursion of meteoric water into the forming orebody, before or during Cu–Fe sulphide and gold precipitation.

Evidence for temperature-controlled sulphide precipitation and the principles defined in Figure 6 lead to the conclusion that ore grade is maximized, to a first order, by steep temperature gradients ($\Delta T/\Delta d$) along the pathway of the focused magmatic fluids. This conclusion is based on observed fluid-compositional constraints and locally recorded temperature variations alone, and therefore does not say anything about the physical mechanism by which fluid cooling is effected at the scale of the ore-forming fluid system. Fluid cooling in a region of focused magmatic fluid flow is severely limited by the slow rate of heat transfer to surrounding rocks of low thermal conductivity (Barton & Toulmin 1961; Hayba & Ingebritsen 1997). Efficient metal precipitation thus requires a sensitive balance between solution chemistry and physical processes. Mechanisms contributing to fluid cooling include adiabatic expansion in the two-phase fluid stability field (which is particularly effective at low pressures where vapour predominates) and heat exchange with externally convecting, but probably not admixing meteoric water. Our fluid-chemical findings imply more generally that the hydrology and thermal structure of the magmatic fluid plume flowing through the porphyry-style vein network has a primary influence on ore grade. This will be influenced strongly by the actual geometry of the system (e.g. cylindrical or dyke-shaped; vein permeability within, surrounding and above the magmatic feeder intrusion) and by the hydraulic characteristics of the surrounding country rocks. Geological situations permitting highly efficient cooling of a large but chemically isolated flux of magmatic fluid over a short flow distance along the vein network will maximize ore grade. Fluid and rock dynamics modelling including two-phase hydrosaline fluid convection is in progress (Geiger *et al.* 2005; Heinrich *et al.* 2004*b*), to investigate the key physical parameters including rate of fluid flow from the pluton, initial fluid salinity, depth of mineralization, and the pressure evolution from lithostatic to hydrostatic conditions.

Bulk metal ratios in porphyry deposits

We have investigated the causes leading to variable metal ratios, in particular the economically important and geologically intriguing Au:Cu ratio of porphyry-style ore deposits, by studies of mine- to grain-scale ore metal distribution, and by comparing metal ratios in the ore with microanalyses of Au:Cu in fluid and melt inclusions. Extensive fluid-chemical data from Alumbrera (Ulrich *et al.* 2001) and reconnaissance data from Grasberg (Ulrich *et al.* 1999), Bingham (Landtwing, unpublished data) and Elatsite (Kehayov *et al.* 2003), consistently indicate that the bulk Au:Cu ratio of each ore deposit is similar to the average Au:Cu ratio of the incoming ore fluids, trapped at high temperature prior to ore mineral saturation. Figure 8 shows the close correspondence between the bulk Au:Cu ratio in the Bajo de la Alumbrera orebody (Au:Cu = 1.2×10^{-4}; Proffett 2003) together with the highly correlated element ratios in individual assay samples from one drill section (Ulrich & Heinrich 2001) and the Au:Cu concentrations in the interpreted pre-precipitation ore brines analysed by LA-ICPMS (Ulrich *et al.* 2001). In conjunction with the fluid-chemical evidence for copper sulphide precipitation at relatively low temperatures

discussed above (Fig. 7b), this correlation between metal ratios in fluid and ore indicates that the magmatic source of the ore fluid is the first-order factor determining the metal ratios of the deposit. Fluid inclusion analyses indicating relatively low precipitation temperatures for copper (and by inference also of the texturally associated gold; Fig. 8) in the Alumbrera and Bingham porphyry Cu–Au deposits do not support the suggestion of Kesler *et al.* (2002), that systematically higher ore-deposition temperatures are the main cause for the formation of gold-rich porphyry deposits, by incorporation of gold into chalcopyrite and particularly into bornite solid solution at higher temperatures. Selective precipitation can lead to zoning of mineralogy and Au:Cu ratios within porphyry orebodies (e.g. Redmond *et al.* 2004), but the overall gold-rich or gold-poor nature of porphyry deposits is determined by processes in the magmatic source of the fluids.

This conclusion from fluid inclusion analyses at the deposit scale, and published evidence for a first-order source control on metal ratios in magmatic–hydrothermal ore deposits at the province scale (e.g. Kesler 1997), was one of the motivations for our study of the evolution of the Farallón Negro Volcanic Complex at the scale of an entire magmatic–hydrothermal system. The

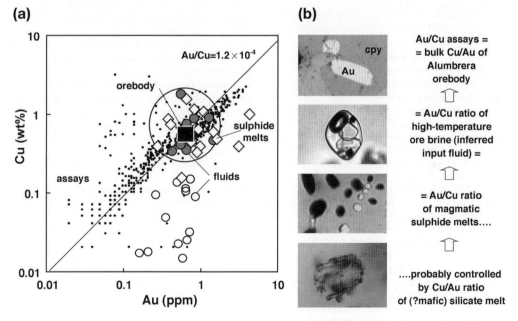

Fig. 8. (**a**) Log–log plot comparing the Cu and Au concentrations in successively generated melts, fluids and ore in the Farallón Negro Volcanic Complex, shown by (**b**) small polished thin section micrographs summarizing the main steps in magmatic control leading to the bulk Cu–Au ratio of the Bajo del la Alumbrera deposit (data from Ulrich *et al.* 1999, 2001; Halter *et al.* 2002*a*),

discovery by Halter *et al.* (2002*b*), that sulphide melt inclusions in barren porphyries pre- and postdating hydrothermal mineralization in this magmatic complex also contain Au and Cu in the same proportions as the ore fluids and as the bulk ore at Bajo de la Alumbrera (Fig. 7), clearly confirms that magmatic source processes, operating at the scale of the subvolcanic magma chamber, are the first-order control on the Au:Cu ratio of porphyry-style ore deposits.

Irrespective of whether igneous sulphide saturation contributes an essential intermediate metal enrichment step for economic ore formation or not, LA-ICPMS analysis of magmatic sulphide inclusions in phenocrysts of barren porphyries and dykes can be applied as a practical exploration indicator, for predicting the potential Au:Cu ratio of undiscovered porphyry deposits in a magmatic complex (Halter *et al.* 2005).

Summary and conclusions

Laser ablation ICPMS microanalysis has proven to be not only sensitive but also efficient enough for the systematic study of ore metal concentrations in assemblages of fluid and melt inclusions preserved in porphyry-mineralizing magmatic–hydrothermal systems. Sample selection and interpretation of microanalytical results in petrographic context has been improved further with the recent use of SEM cathodoluminescence mapping of inclusion generations. Results have allowed the first documentations of spatial variability and temporal evolution of interacting fluids and melts in selected porphyry-mineralizing hydrothermal systems, at scales from individual crystals to entire magmatic complexes. Much work remains to be done before we understand the processes of mixing and unmixing of up to four mobile phases in subvolcanic magma chambers – silicate melt, sulphide melt, brine and vapour – as well as their interaction with non-magmatic fluids.

Physical processes in an evolving magma chamber, generating a large transient flux of relatively dense single- or two-phase magmatic fluid, without gradual loss of metals to low-density volatiles attending volcanic eruption, is the key to generating economic deposits of large metal content; dispersed loss of volatiles and metals is probably the norm in calcalkaline magmatism. The mechanism of metal extraction into a relatively dense volatile phase is probably more important for generating large deposits than the sheer volume of magma, which can be smaller than previously thought in light of the high Cu content of undevolatilized, mafic to intermediate hydrous melts.

The key to accumulating high ore grade in a restricted volume of rock is the high metal content of magmatic ore fluids (in the order of 1% Cu and 1 ppm Au), and the precipitation of these metals by efficient cooling of a focused fluid flux over a relatively small vertical flow distance. Ore deposition ideally happens at relatively low temperature (*c.* 400 °C) and low pressure where Cu–Fe sulphide solubility decreases exponentially while quartz solubility increases with cooling, leading to vein reactivation and the generation of secondary permeability in previously formed quartz veins. The physical mechanism of efficiently cooling a focused fluid flux is a key to high ore grade, but remains poorly understood until hydrodynamic modelling of two-phase fluid flow can be applied to realistic geological scenarios.

Microanalytical results show that Au:Cu ratios in porphyry ores are determined by the metal ratio in the ore fluids entering the forming orebody, and therefore by the magmatic fluid source, but brine/vapour phase separation and selective mineral precipitation may lead to spatial zoning of metal ratios within porphyry orebodies or between porphyry-style and associated epithermal deposits (Heinrich *et al.* 2004*a*). A similar magmatic source control is indicated for Sn/W ratios in granite-related vein deposits (Audétat *et al.* 2000) but other element ratios including Mo/Cu remain to be investigated. Metal ratios in upper-crustal magma reservoirs are in turn controlled by the sources of contributing melts and fluids in the mantle and crust, and the new microanalytical techniques should ultimately lead to an understanding of transport processes and ore-metal variations at the scale of global metallogenic provinces.

We would like to thank Iain McDonald and David Polya for organizing the 2003 Fermor Flagship Meeting and for inviting us to write this review. The results and ideas summarized here have evolved from collaborations with Andreas Audétat, Detlef Günther and Thomas Ulrich, and from discussions with many other colleagues. John Dilles and Steven Kesler helped with detailed and thoughtful reviews of the manuscript. Research supported by Swiss National Science Foundation, and earlier by MIM Exploration (Australia), Minera Alumbrera, and Yacimientos Mineros Agua de Dionisio (Argentina).

References

AUDÉTAT, A. & PETTKE, T. 2003. The magmatic-hydrothermal evolution of two barren granites: A melt and fluid inclusion study of the Rito del Medio and Canada Pinabete plutons in northern New Mexico (USA). *Geochimica et Cosmochimica Acta*, **67**, 97–121.

AUDÉTAT, A., GÜNTHER, D. & HEINRICH, C.A. 1998. Formation of a magmatic-hydrothermal ore deposit: Insights with LA-ICP-MS analysis of fluid inclusions. *Science*, **279**, 2091–2094.

AUDÉTAT, A., GÜNTHER, D. & HEINRICH, C.A. 2000. Causes for large-scale metal zonation around mineralized plutons: Fluid inclusion LA-ICP-MS evidence from the Mole Granite, Australia. *Economic Geology*, **95**, 1563–1581.

AUDÉTAT, A., PETTKE, T. & DOLEJS, D. 2004. Magmatic anhydrite and calcite in the ore-forming quartz-monzonite magma at Santa Rita, New Mexico (USA): genetic constraints on porphyry-Cu mineralization. *Lithos*, **7**, 147–161.

BARTON, P.B. & TOULMIN, P. 1961. *Some mechanisms for cooling hydrothermal fluids*. U.S. Geological Survey Professional Paper, **424-D**, 348–352.

BODNAR, R.J., BURNHAM, C.W. & STERNER, S.M. 1985. Synthetic fluid inclusions in natural quartz. III. Determination of phase equilibrium properties in the system H_2O–NaCl to 1000 °C and 1500 bars. *Geochimica et Cosmochimica Acta*, **49**, 1861–1873.

BURNHAM, C.W. 1979. Magmas and hydrothermal fluids. *In*: BARNES, H.L. (ed.) *Geochemistry of hydrothermal ore deposits*, 2nd edn. New York, Wiley, pp. 71–136.

BURNHAM, C.W. & OHMOTO, H. 1980. *Late-stage processes of felsic magmatism*. Mining Geology Special Issue, **8**, 1–11.

CAMUS, F. & DILLES, J.H. 2001. A special issue devoted to porphyry copper deposits of northern Chile – Preface. *Economic Geology*, **96**, 233–237.

CANDELA, P.A. 1989. Felsic magmas, volatiles, and metallogenesis. *In*: WHITNEY, J.A. & NALDRETT, A.J. (eds) *Ore deposition associated with magmas. Volume 4.* Reviews in Economic Geology, Socorro, NM, United States, Society of Economic Geologists, 223–233.

CARROLL, M.R. & WEBSTER, J.D. 1994. Solubilities of sulfur, noble-gases, nitrogen, chlorine, and fluorine in magmas. *MSA Reviews in Mineralogy*, **30**, 231–279.

CARTEN, R.B., WHITE, W.H. & STEIN, H.J. 1993. *High-grade granite-related molydenum systems: classification and origin*. Geological Association of Canada Special Paper **40**, 521–554.

CLINE, J.S. & BODNAR, R.J. 1991. Can economic porphyry copper mineralization be generated by a typical calc-alkaline melt? *Journal of Geophysical Research*, **96**, 8113–8126.

CORE, D. & KESLER, S.E. 2001. Oxygen fugacity and sulfur speciation in felsic intrusive rocks from the Wasatch and Oquirrh Range, Utah. *In*: 11th Annual Goldschmidt Conference. Hot Springs, Virginia, Abstract no. 3455.

CRERAR, D.A. & BARNES, H.L. 1976. Ore solution chemistry 5. Solubilities of chalcopyrite and chalcocite assemblages in hydrothermal solutions at 200°C to 350°C. *Economic Geology*, **71**, 772–794.

DILLES, J.H. 1987. The petrology of the Yerington Batholith, Nevada: evidence for the evolution of porphyry copper ore fluids. *Economic Geology*, **82**, 1750–1789.

DILLES, J.H. & EINAUDI, M.T. 1992. Wall-rock alteration and hydrothermal flow paths about the Ann-Mason porphyry copper deposit, Nevada – a 6 km vertical reconstruction. *Economic Geology*, **87**, 1963–2001.

FOURNIER, R.O. 1999. Hydrothermal processes related to movement of fluid from plastic into brittle rock in the magmatic–epithermal environment. *Economic Geology*, **94**, 1193–1211.

FOURNIER, R.O. & THOMPSON, J.M. 1993. Composition of steam in the system NaCl–KCl–H_2O–quartz at 600 °C. *Geochimica et Cosmochimica Acta*, **57**, 4365–4375.

GEIGER, S., DRIESNER T., MATTHÄI, S.K. & HEINRICH, C.A. 2005. Multiphase thermohaline convection in the Earth's crust: I. a novel finite element – finite volume solution technique combined with a new equation of state for NaCl–H_2O. II. Benchmarking and application. *Transport in Porous Media* (in press).

GÜNTHER, D., AUDÉTAT, A., FRISCHKNECHT, R. & HEINRICH, C.A. 1998. Quantitative analysis of major, minor and trace elements in fluid inclusions using laser ablation inductively coupled plasma mass spectrometry. *Journal of Analytical Atomic Spectrometry*, **13**, 263–270.

GUSTAFSON, L.B. & HUNT, J.P. 1975. Porphyry copper deposit at El-Salvador, Chile. *Economic Geology*, **70**, 857–912.

HALTER, W.E., PETTKE, T., HEINRICH, C.A. & ROTHEN-RUTISHAUSER, B. 2002a. Major to trace element analysis of melt inclusions by laser-ablation ICP-MS: methods of quantification. *Chemical Geology*, **183**, 63–86.

HALTER, W.E., PETTKE, T. & HEINRICH, C.A. 2002b. The origin of Cu/Au ratios in porphyry-type ore deposits. *Science*, **296**, 1844–1846.

HALTER, W.E., BAIN, N., BECKER, K., ET AL. 2004a. From andesitic volcanism to the formation of a porphyry Cu–Au mineralizing magma chamber: The Farallón Negro Volcanic Complex, north-western Argentina. *Journal of Volcanology and Geothermal Research*, **136**, 1–30.

HALTER, W.E., PETTKE, T. & HEINRICH, C.A. 2004b. Laser-ablation ICP-MS analysis of silicate and sulfide melt inclusions in an andesitic complex II: evidence for magma mixing and magma chamber evolution. *Contributions to Mineralogy and Petrology*, **147**, 397–412.

HALTER, W.E., PETTKE, T. & HEINRICH, C.A. 2005. Magma evolution and the formation of porphyry Cu–Au fluids: evidence from silicate and sulphide melt inclusions. *Mineralium Deposita*, **39**, 845–863.

HARRIS, A.C., KAMENETSKY, V.S., WHITE, N.C., VAN ACHTERBERGH, E. & RYAN, C.G. 2003. Melt inclusions in veins: Linking magmas and porphyry Cu deposits. *Science*, **302**, 2109–2111.

HARRIS, A.C., ALLEN, C.M., BRYAN, S.E., CAMPBELL, I.H., HOLCOMBE, R.J. & PALIN, J.M. 2004. ELA-ICP-MS U–Pb zircon geochronology of regional volcanism hosting the Bajo de la Alumbrera Cu-Au deposit: implications for porphyry-related mineralization. *Mineralium Deposita*, **39**, 46–67.

HAYBA, D.O. & INGEBRITSEN, S.E. 1997. Multiphase groundwater flow near cooling plutons. *Journal of Geophysical Research*, **102**, 12 235–12 252.

HEDENQUIST, J.W. & LOWENSTERN, J.B. 1994. The role of magmas in the formation of hydrothermal ore-deposits. *Nature*, **370**, 519–527.

HEINRICH, C.A., GÜNTHER, D., AUDÉTAT, A., ULRICH, T. & FRISCHKNECHT, R. 1999. Metal fractionation between magmatic brine and vapor, determined by microanalysis of fluid inclusions. *Geology*, **27**, 755–758.

HEINRICH, C.A., PETTKE, T., HALTER, W.E., *ET AL.* 2003. Quantitative multi-element analysis of minerals, fluid and melt inclusions by laser-ablation inductively-coupled-plasma mass-spectrometry. *Geochimica et Cosmochimica Acta*, **67**, 3473–3497.

HEINRICH, C., DRIESNER, T., STEFÁNSSON, A. & SEWARD, T.M. 2004*a*. Magmatic vapor contraction and the transport of gold from porphyry to epithermal ore deposits. *Geology*, **32**, 761–764.

HEINRICH, C.A., DRIESNER, T., GEIGER, S. & MATTHÄI, S.K. 2004*b*. *Modelling magmatic-hydrothermal ore systems: fluid data and process simulation*. Extended Abstract, SEG-SGA Conference 'Predictive Mineral Discovery under Cover', Perth September 2004, 57–61.

HEMLEY, J.J., CYGAN, G.L., FEIN, J.B., ROBINSON, G.R. & D'ANGELO, W.M. 1992. Hydrothermal ore-forming processes in the light of studies in rock-buffered systems. 1. Iron–copper–zinc–lead sulfide solubility relations. *Economic Geology*, **87**, 1–22.

HEZARKHANI, A., WILLIAMS-JONES, A.E. & GAMMONS, C.H. 1999. Factors controlling copper solubility and chalcopyrite deposition in the Sungun porphyry copper deposit, Iran. *Mineralium Deposita*, **34**, 770–783.

KAMENETSKY, V.S., BINNS, R.A., GEMMELL, J.B., CRAWFORD, A.J., MERNAGH, T.P., MAAS, R. & STEELE, D. 2001. Parental basaltic melts and fluids in eastern Manus backarc basin: implications for hydrothermal mineralisation. *Earth and Planetary Science Letters*, **18**, 685–702.

KEHAYOV, R., BOGDANOV, K., FANGER, L., VON QUADT, A., PETTKE, T. & HEINRICH, C.A. 2003. The fluid chemical evolution of the Elatiste porphyry Cu–Au–PGE deposit, Bulgaria. *In*: ELIOPOULOS, D.G. (ed.) *Mineral Exploration and Sustainable Development*. Rotterdam, Millpress, 1173–1176.

KEITH, J.D., WHITNEY, J.A., HATTORI, K., *ET AL.* 1997. The role of magmatic sulfides and mafic alkaline magmas in the Bingham and Tintic mining districts, Utah. *Journal of Petrology*, **38**, 1679–1690.

KESLER, S.E. 1997. Arc evolution and selected ore deposit models. *Ore Geology Reviews*, **12**, 62–78.

KESLER, S.E., CHRYSSOULIS, S.L. & SIMON, G. 2002. Gold in porphyry copper deposits: its abundance and fate. *Ore Geology Reviews*, **21**, 103–124.

KRAHULEC, K.A. 1997. History and production of the West Mountain (Bingham) mining district, Utah. *In*: JOHN, D.A. & BALLANTYNE, G.H. (eds) *Geology and Ore Deposits of the Oquirrh and Wasatch Mountains, Utah*. Guidebook Series of the Society of Economic Geologists, **29**, 189–217.

LANDTWING, M.R., PETTKE, T., HALTER, W.E., HEINRICH, C.A., REDMOND, P.B. & EINAUDI, M.T. 2005. Copper deposition during quartz dissolution by cooling magmatic-hydrothermal fluids: The Bingham porphyry. *Earth and Planetary Science Letters*, in press.

LIU, W.H., MCPHAIL, D.C. & BRUGGER, J. 2001. An experimental study of copper(I)-chloride and copper(I)-acetate complexing in hydrothermal solutions between 50 °C and 250 °C and vapor-saturated pressure. *Geochimica et Cosmochimica Acta*, **65**, 2937–2948.

LLAMBÍAS, E.J. 1972. Estructura del grupo volcanico Farallon Negro, Catamarca, Republica Argentina. *Revista de la Asociacion Geologica Argentina*, **27**, 161–169.

LOWENSTERN, J.B. 1993. Evidence for a copper-bearing fluid in magma erupted at the Valley of Ten-Thousand-Smokes, Alaska. *Contributions to Mineralogy and Petrology*, **114**, 409–421.

LOWENSTERN, J.B., MAHOOD, G.A., RIVERS, M.L. & SUTTON, S.R. 1991. Evidence for extreme partitioning of copper into a magmatic vapor-phase. *Science*, **252**, 1405–1409.

LOWENSTERN, J.B., PERSING, H.M., WOODEN, J.L., LANPHERE, M., DONNELLY-NOLAN, J. & GROVE, T.L. 2000. U–Th dating of single zircons from young granitoid xenoliths: new tools for understanding volcanic processes. *Earth and Planetary Science Letters*, **183**, 291–302.

MANNING, C.E. 1994. The solubility of quartz in H_2O in the lower crust and upper mantle. *Geochimica et Cosmochimica Acta*, **58**, 4831–4839.

OBERLI, F., MEIER, M., BERGER, A., ROSENBERG, C.L. & GIERE, R.. 2004. U–Th–Pb and Th-230/U-238 disequilibrium isotope systematics: Precise accessory mineral chronology and melt evolution tracing in the Alpine Bergell intrusion. *Geochimica et Cosmochimica Acta*, **68**, 2543–2560.

PROFFETT, J.M. 2003. Geology of the Bajo de la Alumbrera porphyry copper-gold deposit, Argentina. *Economic Geology*, **98**, 1535–1574.

PROFFETT, J.M. & DILLES, J.H. 1984. *Geological map of the Yerington district, Nevada*. Nevada Bureau of Mines and Geology Map 77.

REDMOND, P.B., EINAUDI, M.T., INAN, E.E., LANDTWING, M.R. & HEINRICH, C.A. 2004. Copper deposition by fluid cooling in intrusion-centered systems: New insights from the Bingham porphyry ore deposit, Utah. *Geology*, **32**, 217–220.

RICHARDS, J.P. 2003. Tectono-magmatic precursors for porphyry Cu–(Mo–Au) deposit formation. *Economic Geology*, **98**, 1515–1533.

ROHRLACH, B.D. 2003. Tectonic evolution, petrochemistry, geochronology and palaeohydrology of the Tampakan porphyry and high-sulphidation epithermal Cu-Au deposit, Mindanao, Philippines.

Unpubl. PhD Thesis, Australian National University.

RUSK, B. & REED, M. 2002. Scanning electron microscope-cathodoluminescence analysis of quartz reveals complex growth histories in veins from the Butte porphyry copper deposit, Montana. *Geology*, **30**, 727–730.

RUSK, B., REED, M.H., DILLES, J.H. & KLEMM, L. 2004. Compositions of magmatic-hydrothermal fluids determined by LA-ICPMS of fluid inclusions from the porphyry copper-molybdenum deposit at Butte, Montana. *Chemical Geology*, **210**, 173–199.

RYAN, C.G., COUSENS, D.R., HEINRICH, C.A., GRIFFIN, W.L., SIE, S.H. & MERNAGH, T.P. 1991. Quantitative PIXE microanalysis of fluid inclusions based on a layered yield model. *Nuclear Instruments & Methods in Physics Research, Section B – Beam Interactions with Materials and Atoms*, **54**, 292–297.

SASSO, A.M. & CLARK, A.H. 1998. The Farallón Negro Group, northwest Argentina: Magmatic, hydrothermal and tectonic evolution and implications for Cu–Au metallogeny in the Andean back-arc. *Society of Economic Geologists Newsletter*, **34**, 8–17.

SHINOHARA, H., KAZAHAYA, K. & LOWENSTERN, J.B. 1995. Volatile transport in a convecting magma column – Implications for porphyry Mo mineralization. *Geology*, **23**, 1091–1094.

SILLITOE, R.H. 1973. The tops and bottoms of porphyry copper deposits. *Economic Geology*, **68**, 799–815.

SILLITOE, R.H. 1997. Characteristics and controls of the largest porphyry copper–gold and epithermal gold deposits in the Circum-Pacific region. *Australian Journal of Earth Sciences*, **44**, 373–388.

SOLOMON, M. 1990. Subduction, arc reversal, and the origin of porphyry copper–gold deposits in island arcs. *Geology*, **18**, 630–633.

SPOONER, E.T.C. 1993. Magmatic sulphide/volatile interaction as a mechanism for producing chalcophile element enriched, Archean Au–quartz, epithermal Au–Ag and Au skarn hydrothermal ore fluids. *Ore Geology Reviews*, **7**, 359–379.

STEFÁNSSON, A. & SEWARD, T.M. 2004. Gold(I) complexing in aqueous sulphide solutions to 500 °C at 500 bar. *Geochimica et Cosmochimica Acta*, **68**, 4121–4143.

TARAN, Y.A., HEDENQUIST, J.W., KORZHINSKY, M.A., TKACHENKO, S.I. & SHMULOVICH, K.I. 1995. Geochemistry of magmatic gases from Kudryavy Volcano, Iturup, Kuril Islands. *Geochimica et Cosmochimica Acta*, **59**, 1749–1761.

TITLEY, S.R. & BEANE, R.E. 1981. Porphyry copper deposits. *In*: SKINNER, B.J. (ed.) *Economic

Geology. Seventy-Fifth anniversary volume. Society of Economic Geologists, 214–234.

TOSDAL, R.M. & RICHARDS, J.P. 2001. Magmatic and structural controls on the development of porphyry Cu ± Mo ± Au deposits. *In*: RICHARDS, J.P. & TOSDAL, R.M. (eds) *Structural controls on ore genesis*. Reviews in Economic Geology, **14**, 157–181.

ULRICH, T. & HEINRICH, C.A. 2001. Geology and alteration geochemistry of the porphyry Cu–Au deposit at Bajo de la Alumbrera, Argentina (v. 96, pp. 1719–1741, 2001). *Economic Geology*, **96**, 1719 (correctly reprinted in 2002, **97**, 1863–1888).

ULRICH, T., GÜNTHER, D. & HEINRICH, C.A. 1999. Gold concentrations of magmatic brines and the metal budget of porphyry copper deposits. *Nature*, **399**, 676–679.

ULRICH, T., GÜNTHER, D. & HEINRICH, C.A. 2001. The evolution of a porphyry Cu–Au deposit, based on LA-ICP-MS analysis of fluid inclusions: Bajo de la Alumbrera, Argentina. *Economic Geology*, **96**, 1743–1774 (correctly reprinted in 2002, **97**, 1888–1920).

VANKO, D.A. & MAVROGENES, J.A. 1998. Synchrotron-source x-ray fluorescence microprobe: analysis of fluid inclusions (X00ppm DL). *In*: McKIBBEN, M.A., SHANKS, W.C., III & RIDLEY, W.I. (eds) *Applications of microanalytical techniques to understanding mineralizing processes*. Reviews in Economic Geology, Society of Economic Geologists, **7**, 251–263.

WALLACE, P.J. & GERLACH, T.M. 1994. Magmatic vapor source for sulfur dioxide released during volcanic eruptions: evidence from Mount Pinatubo. *Science*, **265**, 497–499.

WILLIAMS, T.J., CANDELA, P.A. & PICCOLI, P.M. 1995. The partitioning of copper between silicate melts and 2-phase aqueous fluids – an experimental investigation at 1 kbar, 800 °C and 0.5 kbar, 850 °C. *Contributions to Mineralogy and Petrology*, **121**, 388–399.

WILLIAMS-JONES, A.E., MIGDISOV, A.A., ARCHIBALD, S.M. & XIAO, Z.F. 2002. Vapor-transport of ore metals. *In*: HELLMANN, R. & WOOD, S.A. (eds) *Water–rock interaction: a tribute to David A. Crerar*. The Geochemical Society, Special Publication, **7**, 279–305.

ZOTOV, A.V., KUDRIN, A.V., LEVIN, K.A., SHILINA, N.D. & VAR'YASH, L.N. 1995. Experimental studies of the solubility and complexing of selected ore elements (Au, Ag, Cu, Mo, As, Sb, Hg) in aqueous solution. *In*: SHMULOVICH, K.I., YARDLEY, B.W.D. & GONCHAR, G.G. (eds) *Fluids in the crust*. London, Chapman & Hall, 95–137.

Index

Page numbers in *italic* denote figures, page numbers in **bold** denote tables.